Spring Cloud 标准服务架构

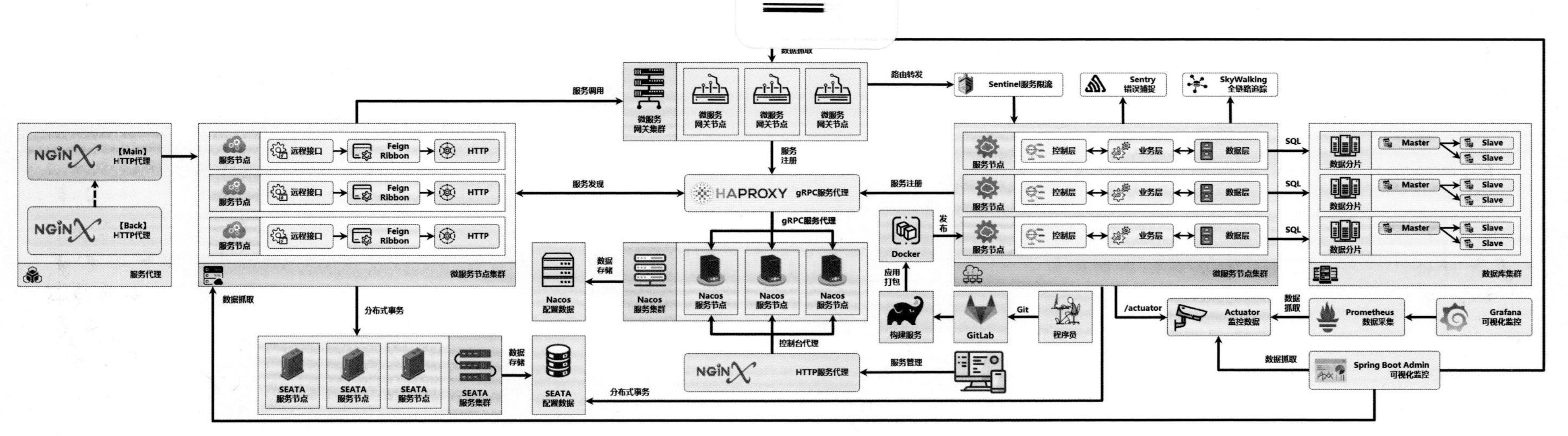

微服务集群架构人事管理系统重构实战项目架构

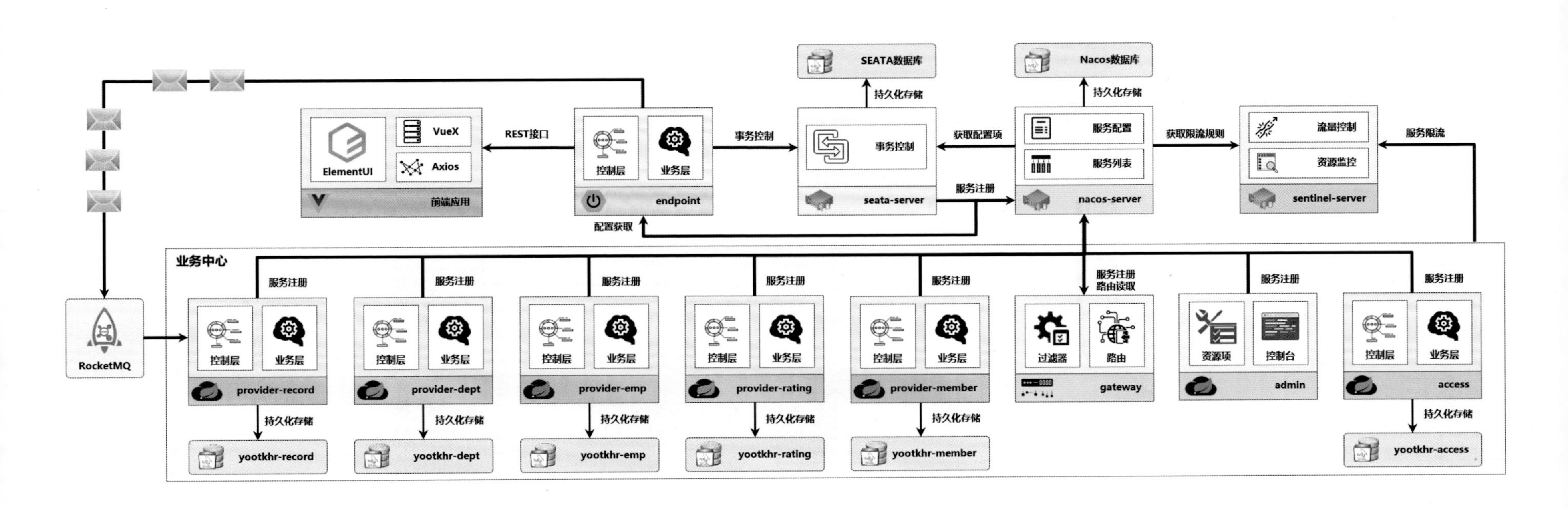

软件开发人才培养系列丛书

Spring Cloud 开发实战

视频讲解版

李兴华 马云涛 / 编著

人民邮电出版社

北京

图书在版编目（CIP）数据

Spring Cloud开发实战 : 视频讲解版 / 李兴华, 马云涛编著. -- 北京 : 人民邮电出版社, 2022.7（2023.10重印）
（软件开发人才培养系列丛书）
ISBN 978-7-115-58866-1

Ⅰ. ①S… Ⅱ. ①李… ②马… Ⅲ. ①互联网络－网络服务器 Ⅳ. ①TP368.5

中国版本图书馆CIP数据核字(2022)第043560号

内 容 提 要

Spring Cloud 是当今 Java 开发行业最为流行的分布式开发架构之一，可用于搭建高可用、高性能、分布式的系统服务架构，本书基于 Spring Cloud Alibaba 的套件进行架构实现的完整讲解。

本书共 10 章内容，基于 IDEA 开发工具讲解，并通过 Linux 系统实现服务的部署。全书的主要内容为 Spring Cloud 简介、Spring Cloud 编程起步、Nacos 注册中心、Spring Cloud 集群服务、Sentinel、Spring Cloud Gateway、微服务安全与监控管理、RocketMQ、微服务辅助技术和服务跟踪，除此之外还包含与面试有关的组件以及核心源代码的讲解。

本书附有配套视频、源代码、习题、教学课件等资源。为了帮助读者更好地学习，作者还提供在线答疑服务。本书适合作为高等教育本、专科院校计算机相关专业的教材，也可供广大计算机编程爱好者自学使用。

◆ 编　　著　李兴华　马云涛
　责任编辑　刘　博
　责任印制　王　郁　陈　犇
◆ 人民邮电出版社出版发行　　北京市丰台区成寿寺路 11 号
　邮编　100164　　电子邮件　315@ptpress.com.cn
　网址　https://www.ptpress.com.cn
　三河市中晟雅豪印务有限公司印刷

◆ 开本：787×1092　1/16　　彩插：1
　印张：25　　2022 年 7 月第 1 版
　字数：699 千字　　2023 年10月河北第 2 次印刷

定价：99.80 元

自　序

从最早接触计算机编程到现在，已经过去 24 年了，其中有 17 年的时间，我在一线讲解编程开发。我一直在思考一个问题：如何让学生在有限的时间里学到更多、更全面的知识？最初我并不知道答案，于是只能大量挤占每天的非教学时间，甚至连节假日都给学生补课。因为当时的我想法很简单：通过多花时间去追赶技术发展的脚步，争取教给学生更多的技术，让学生在找工作时游刃有余。但是这对于我和学生来讲都实在过于痛苦了，毕竟我们都只是普通人，当我讲到精疲力尽，当学生学到头昏脑涨，我知道自己需要改变了。

技术正在发生不可逆转的变革，在软件行业中，最先改变的一定是就业环境。很多优秀的软件公司或互联网企业已经由简单的需求招聘变为能力招聘，要求从业者不再是培训班“量产”的学生。此时的从业者如果想顺利地进入软件行业，获取自己心中的理想职位，就需要有良好的技术学习方法。换言之，学生不能只是被动地学习，而是要主动地努力钻研技术，这样才可以具有更扎实的技术功底，才能够应对各种可能出现的技术挑战。

于是，怎样让学生们以尽可能短的时间学到最有用的知识，就成了我思考的核心问题。对于我来说，教育两个字是神圣的，既然是神圣的，就要与商业的运作有所区分。教育提倡的是付出与奉献，而商业运作讲究的是盈利，盈利和教育本身是有矛盾的。所以我拿出几年的时间，安心写作，把我近 20 年的教学经验融入这套编程学习丛书，也将多年积累的学生学习问题如实地反映在这套丛书之中，丛书架构如图 0-1 所示。希望这样一套方向明确的编程学习丛书，能让读者学习 Java 不再迷茫。

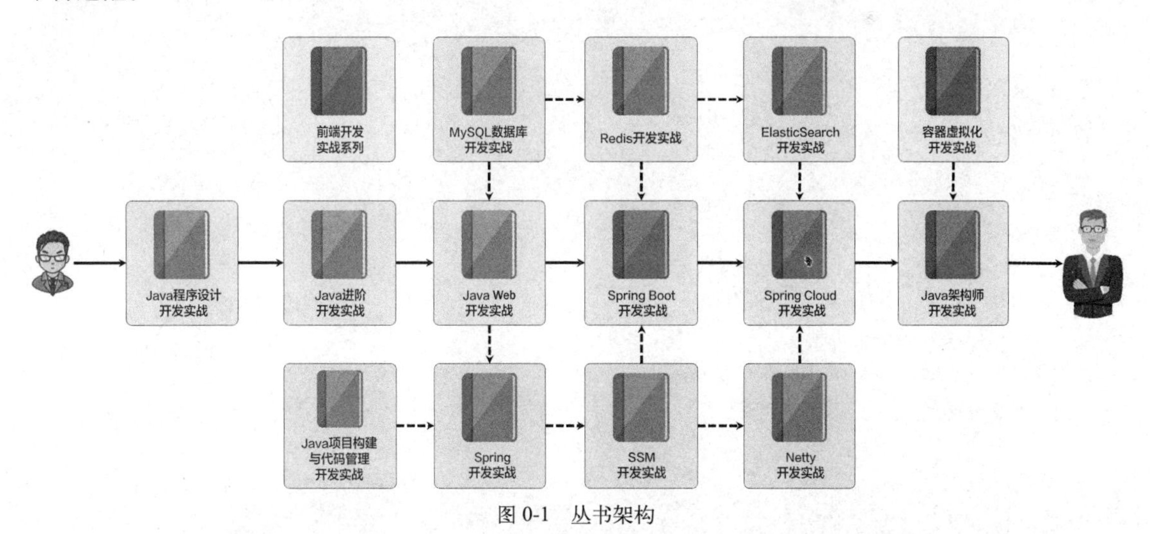

图 0-1　丛书架构

我的体会是，编写一本讲解透彻的图书真的很不容易。在写作过程中我翻阅了大量图书，有些书查看之下发现内容竟然是和其他图书重复的，网上的资料也有大量的重复，这让我认识到“原创”的重要性。但是原创的路途上满是荆棘，这也是我编写一本书需要很长时间的原因。

仅仅做到原创就可以让学生学会吗？很难。计算机编程图书之中有大量晦涩难懂的专业性词汇，不能默认所有的初学者都清楚地掌握了这些词汇的概念，如果那样，可以说就已经学会了编程。为了帮助读者扫除学习障碍，我在书中绘制了大量图形来进行概念的解释，此外还提供了与章节内容相符的视频资料，所有的视频讲解中出现的代码全部为现场编写。我希望用这一次又一次的重复劳动，帮助大家理解代码，学会编程。本套丛书所提供的配套资料非常丰富，可以说抵得上花几万元学费参加的培训班的课程。本套丛书的配套视频累计上万分钟，对比培训班的实际讲课时间，相信读者能体会到我们所付出的心血。我们希望通过这样的努力给大家带来一套有助于学懂、学会的图书，帮助大家解决学习和就业难题。

前　言

本书是《Spring Boot 开发实战（视频讲解版）》一书的延续，而两本图书之中也有很多概念上的交集，两本书的架构如图 0-2 所示，所以在学习本书之前读者一定要对 Spring Boot 有完整且正确的理解。

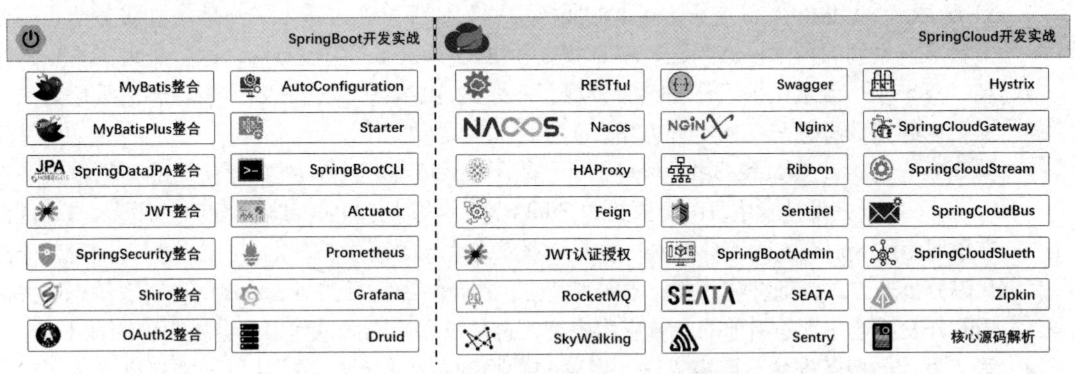

图 0-2　两本书的架构

本书旨在帮助读者进入 Java 架构学习领域，Spring Cloud 也是与实际工作最贴近的架构开发技术之一。在编写此书时，我不仅在编写一个简单易学的 Spring Cloud 知识体系，而且在想办法通过更多的图示和更多的范例帮助大家深刻地理解当前应用中可能存在的各种形式，同时分析每一项技术到底会为整个微服务设计带来何种帮助。

为了便于理解，本书降低了微服务的业务逻辑复杂度，基于最简单的 CRUD 数据操作的形式讲解 RESTful 设计原型的搭建，并在此原型上不断地进行实现设计的缺陷分析，以及进行更多应用组件的引入。本书采用“案例教学”的方式编写，目的也是帮助读者更好地理解微服务这门技术。读者按照本书介绍的步骤就可以实现服务的开发，收获一个完整的微服务技术应用体系。

2017 年我曾经写书介绍 Spring Cloud Netflix 技术架构，而随着时间的推移，该架构不再被广泛采用。本书介绍的是 Spring Cloud Alibaba 的全部套件，把技术选型全部设定在“阿里系”的范畴之中，尽可能地为读者阐述每项技术的使用、核心原理，并对核心源代码进行解释。除此之外，考虑到读者的学习与面试需要，本书还对 Spring Cloud Netflix 的部分内容进行分析并实现了改造。

在开始编写本书时我就已经预料到会有 3 个难关，分别是新版的 Nacos、Sentinel 防护、RocketMQ。编写本书时正好赶上新版 Nacos 的推出，编写相关内容时我在简化概念和排查 bug 上花费了比较多的时间；Sentinel 操作相关源代码的分析难度较高（需要绘制大量的图），章节顺序的安排较复杂；而 RocketMQ 最麻烦的就是架构图的绘制。这 3 个难关被我一一攻克后，我也就知道这本图书即将完成了。

本书的全部案例设计和视频讲解均为沐言科技的原创内容。本书采用递进的方式编写，共 10 章，主要内容如下。

- **第 1 章　Spring Cloud 简介**　Spring Cloud 本质上是微服务的集群管理工具，而微服务背后实现的是业务拆分的逻辑。本章重点为读者分析业务中心的作用以及开发技术的演进过程，同时分析 Spring Cloud 两种实现架构的区别。
- **第 2 章　Spring Cloud 编程起步**　Spring Cloud 是 Spring Boot 的概念升级，所以要想理解 Spring Cloud 就必须搭建 RESTful 应用架构。本章介绍通过 MyBatis/MyBatisPlus 和 Druid

实现部门微服务的搭建。

- **第 3 章　Nacos 注册中心**　Spring Cloud Alibaba 体系排除已经不再更新的 Eureka 注册中心，转而提供性能更强的 Nacos 注册中心。本章介绍基于 Nacos 2.x 实现服务搭建、NacosClient 原生开发以及微服务注册管理。
- **第 4 章　Spring Cloud 集群服务**　为了提供可靠的服务，需要进行服务集群的搭建，本章将模拟微服务集群（RESTful 集群、Nacos 2.x 集群）的创建，同时分析 Nacos 2.x 集群设计的改进与具体实现，而后在这一集群架构的基础之上进行 Ribbon 负载均衡、Feign 转换技术以及 Hystrix 技术的讲解。通过本章的学习，读者可以清楚地掌握集群设计原理与具体实现方法，为后续的学习提供核心理论支持。
- **第 5 章　Sentinel**　高并发访问下的服务安全如何保障？难道要一味地增加服务节点吗？带着这一系列的技术设计问题，本章将全面讲解 Sentinel 流控组件，这也是大家常见的淘宝“双十一”流量保护组件。读者读懂了本章内容之后，就会开始逐步理解亿级用户的开发架构的设计思想与具体实施方案。
- **第 6 章　Spring Cloud Gateway**　Netty 是在 Java 架构设计中被广泛使用的开发框架，在 Spring Cloud 开发技术中，由于原始的 Zuul 出现了难以突破的性能瓶颈以及闭源风波，因此有了原生的 Spring Cloud Gateway 技术。本章为读者详细分析网关的主要作用，同时基于源代码分析的方式详细讲解网关技术实现特点，并介绍如何将其与 Nacos 整合实现动态网关应用的开发，这些都是当前的主流应用模型，也是求职者面试过程中常会被问到的技术项。
- **第 7 章　微服务安全与监控管理**　完善的微服务必然要有合理的认证与授权机制，所以本章继续引用《Spring Boot 开发实战（视频讲解版）》一书所包含的 JWT 技术实现微服务操作管理的内容，同时为了实现微服务的有效监控，又引入 Spring Boot Admin 的组件，介绍利用 Actuator 实现服务监控管理。
- **第 8 章　RocketMQ**　高吞吐量的微服务技术必然要使用消息组件，在“阿里系”中最为重要的就是 RocketMQ 消息组件。本章为读者详细分析当前主流的消息组件的技术特点与执行性能，同时详细地讲解 RocketMQ 消息服务的部署、实现架构、服务集群以及应用开发的操作。
- **第 9 章　微服务辅助技术**　Spring Cloud 在内部提供了更高一级的消息组件整合服务——Spring Cloud Stream，而本章将针对 RocketMQ 消息组件讲解如何实现该服务，详细讲解 Spring Cloud Config 原生技术实现方案以及基于 Nacos 2.x 配置更新的实现方案。
- **第 10 章　服务跟踪**　服务跟踪是微服务技术的完善环节，同时也是微服务应用中必不可少的技术环节。本章介绍如何在已有的微服务基础上增加运维监控的技术管理，以实例的形式讲解 Zipkin、SkyWalking 以及 Sentry 的应用整合。

内容特色

由于技术类的图书所涉及的内容很多，同时考虑到读者对于一些知识的理解盲点与认知偏差，作者在编写图书时设计了一些特色栏目和表示方式，现说明如下。

（1）提示：对一些知识核心内容的强调以及与之相关知识点的说明。这样做的目的是帮助读者扩大知识面。

（2）注意：点明对相关知识进行运用时有可能出现的种种“深坑”。这样做的目的是帮助读者节约理解技术的时间。

（3）问答：对核心概念理解的补充，以及可能存在的一些理解偏差的解读。

（4）分步讲解：清楚地标注每一个开发步骤。技术开发需要严格的实现步骤，我们不仅要教读者知识，更要给大家提供完整的学习指导。由于在实际项目中会利用 Gradle 或 Maven 这样的工具来进行模块拆分，因此我们在每一个开发步骤前会使用“【项目或子模块名称】”这样的标注方式，这样读者在实际开发演练时就会更加清楚当前代码的编写位置，提高代码的编写效率。

本书主要特点如下。

- 项目驱动型的讲解模式，让读者学完每一节课程后都能有所收获，都可以将其应用到技术开发之中。
- 为避免晦涩的技术概念所带来的学习困难，本书绘制了 300 余张原创结构图进行概念解释，进一步降低学习难度。
- 全书配套 300 多个样例代码，帮助读者轻松理解每一个技术知识点。

配套资源

本书提供配套资源如下。

- 全书配套 100 多节讲解视频，时间总数超过 1000 分钟。
- 配套完整 PPT、源代码、教学大纲、工具软件等，轻松满足高校教师的教学需要。

读者如果需要获取本课程的相关资源，可以登录人邮教育社区（www.ryjiaoyu.com）下载，也可以登录沐言优拓的官方网站通过资源导航获取下载链接，如图 0-3 所示。

图 0-3　获取图书资源

答疑交流

本书中难免存在不妥之处，希望读者可以将意见和建议以邮件的形式发送给我们，我们将在后续的版本中进行更正。邮箱地址为 784420216@qq.com。万分感谢大家的支持！

同时也欢迎各位读者加入图书交流群（QQ 群号码为 570104062，群满时请根据提示加入新的交流群）进行沟通互动。

为了更好地帮助读者学习，以及为读者进行技术答疑，我们会提供一系列的公益技术直播课，有兴趣的读者可以访问我们的抖音（ID：muyan_lixinghua）或“B 站”（ID：YOOTK 沐言优拓）直播间。对于每次直播的课程内容以及技术话题，我也会在我个人的微博（ID：yootk 李兴华）之中进行发布。同时，我们欢迎广大读者将我们的视频上传到各个平台，把我们的教学理念传播给更多有需要的人。

最后我想说的是，因为写书与各类公益技术直播，我错过了许多与家人欢聚的时光，内心感到非常愧疚。我希望不久的将来能为我的孩子编写一套属于他自己的编程类图书，这也将帮助所有有需要的孩子进步。我喜欢研究编程技术，也勇于自我突破，如果你也是这样的一位软件工程师，也希望你加入我们这个公益技术直播的行列。让我们抛开所有的商业模式的束缚，一起将自己学到的技术传播给更多的爱好者，以我们微薄之力推动整个行业的技术发展，就如同我说过的，教育的本质是分享，而不是赚钱的工具。

沐言科技 —— 李兴华

2022 年 3 月

目　录

视频目录

第 1 章　Spring Cloud 简介

第 2 章　Spring Cloud 编程起步

第 3 章　Nacos 注册中心

第 4 章　Spring Cloud 集群服务

第1章 Spring Cloud 简介

本章学习目标

1. 理解业务中心设计意义及其在项目中的主要作用；
2. 理解 EJB 技术与业务中心搭建，并清楚地理解 EJB 技术中的各个组成元素；
3. 理解 RPC 开发技术以及 RPC 技术的发展过程；
4. 理解 Spring Cloud 标准设计架构以及各个组件的作用；
5. 了解 Spring Cloud Alibaba 的实现架构及其服务组成。

Spring Boot 是一种微服务的设计框架，可以直接基于 RESTful 设计架构实现分布式的项目开发。然而现实中有各类烦琐的项目业务需求，一个简单的微服务是很难满足的。本章将为读者详细地阐述业务中心的基本作用以及相关技术的发展历史，同时为读者展示 Spring Cloud 中的相关技术架构。

1.1 业务中心

业务中心

视频名称 0101_【理解】业务中心

视频简介 业务中心是项目设计开发的"灵魂"，从最初的 MVC 设计模式到现在的微服务设计架构，全部都是围绕着业务中心展开的。本视频为读者详细地阐述业务中心的基本作用。

软件项目存在的前提条件就是有其需要解决的问题，例如，为解决公民信息的管理问题而建立的公民信息管理系统，为解决企业客户信息留存管理而创建的客户关系管理系统。而每一个系统除了精美的界面展示功能和丰富的设备支持功能之外，最为重要的就是相应的业务中心的设计与开发，如图 1-1 所示。

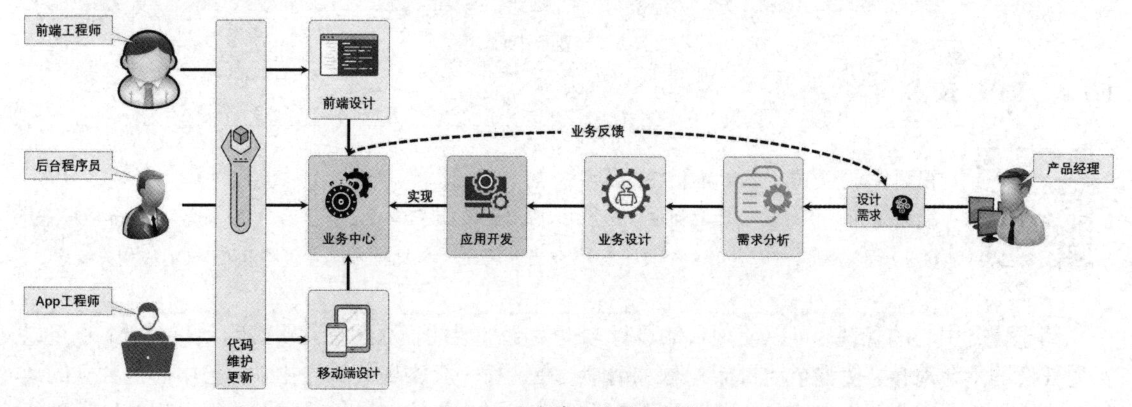

图 1-1 业务中心设计与开发

要进行系统开发，首先要由产品经理提出设计需求，随后由开发人员依据设计需求进行代码实现，并将若干个不同的业务集中在一起，形成一个完整的业务中心，而后根据客户体验和要求不断地进行业务中心的完善。只有业务中心设计完善了，才可以进行前端及移动端设计，从而形成一个广大用户可以使用的软件项目产品。

在现代项目开发中，很多业务处理都离不开数据的支持。在传统的单 Web 实例系统的开发环境中，为了可以清楚地描述出业务中心的概念，往往会为其设计专属的业务层与数据层，这样在控制层进行请求处理时，就可以依据业务层的方法并结合实例的注入管理进行业务层的功能调用，如图 1-2 所示。

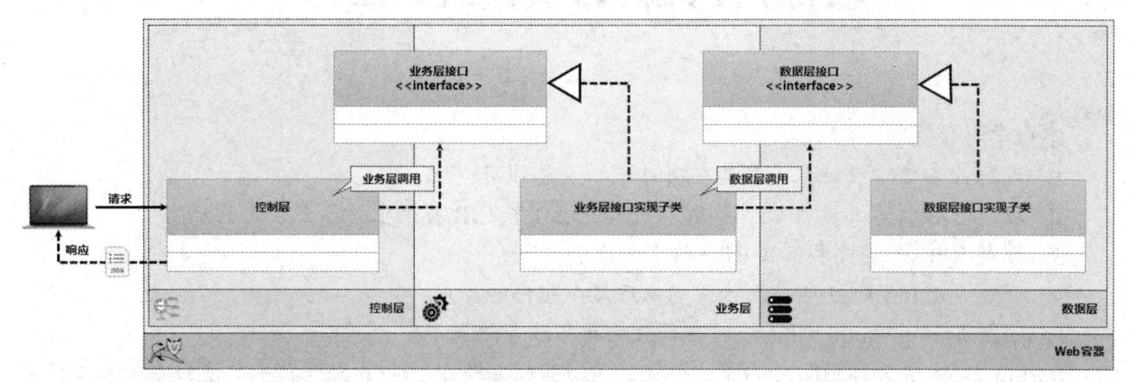

图 1-2　传统项目开发

随着项目业务的不断完善以及并发量的不断提升，如果使用单一的 Web 容器，那么程序的可维护性必然会降低，同时也不便于外部平台的对接。在这样的背景下，可将单 Web 实例中的业务逻辑抽取出来，使之形成一套独立的业务中心，如图 1-3 所示。这样就可以由不同的团队进行各自代码的维护，使得项目的分工更加明确。

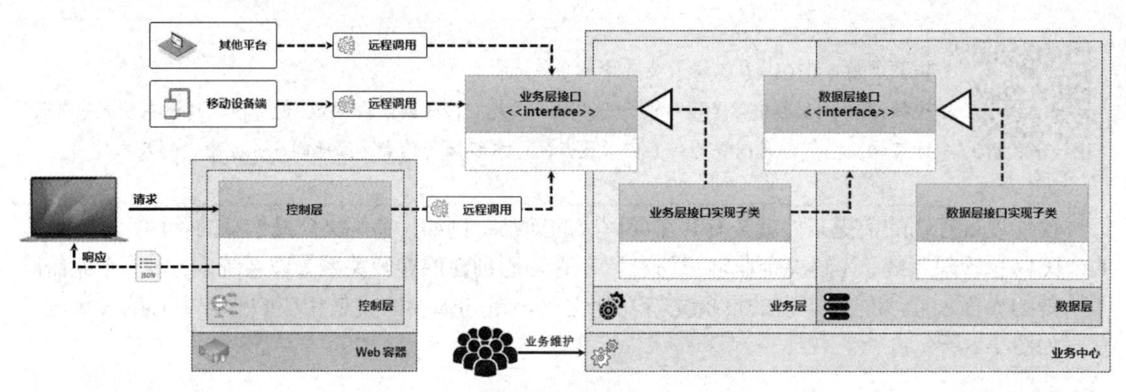

图 1-3　业务中心

1.1.1　RPC 技术

RPC 技术

视频名称　0102_【理解】RPC 技术

视频简介　良好的项目架构中，业务中心是独立的运行单元，符合客户-服务器（Client-Server，C-S）网络架构，需要有一个技术标准来实现调用。本视频为读者讲解 RPC 技术的相关概念。

基于业务中心的管理，可以使项目的设计与开发更加灵活，不仅有着明确的项目分工，也可以实现任意的平台对接，使整个项目架构更加清晰。在这样一套实现架构之中，业务中心是独立的运行单元，业务中心的调用者（客户端）需要通过网络调用业务接口提供的方法，而调用的过程就像调用本地接口一样，如图 1-4 所示。这就是 RPC（Remote Procedure Call，远程过程调用）技术的核心处理形式。

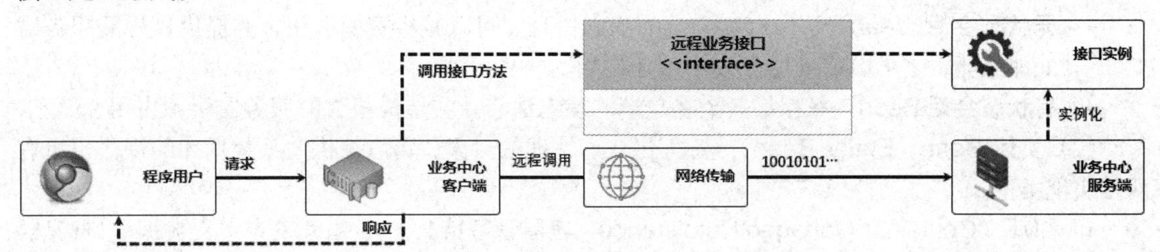

图 1-4 调用远程业务中心

RPC 的调用过程中一般有服务器端（简称服务端）节点以及客户端节点两个组成部分。接口的具体实现子类保存在服务端，这样在服务端就需要维护一个方法映射表；而客户端在进行远程方法调用时，也是依据此映射表中的 ID 来发送调用请求的。由于服务端和客户端属于两个不同的进程，这样就需要将所有请求的参数以及方法的相应结果以二进制数据流的形式进行传输，因此还需要提供对象序列化与反序列化的机制才可以实现最终的远程调用，如图 1-5 所示。

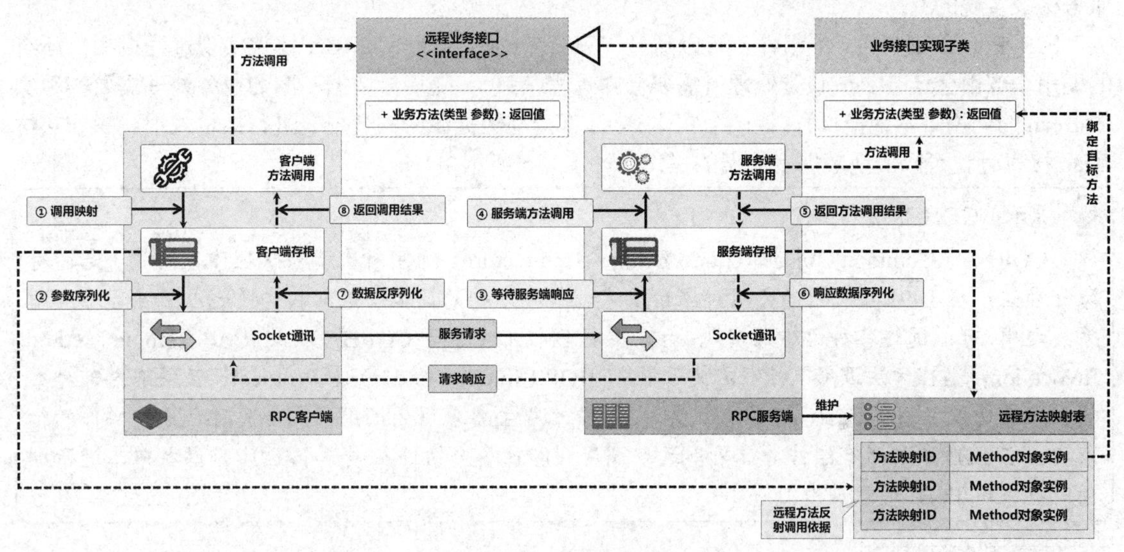

图 1-5 RPC 调用过程

> **提示：Java 的 RPC 技术可以依靠 Netty 实现。**
>
> 在使用 Java 进行 RPC 框架开发时，由于其内部需要进行稳定可靠的网络传输处理，所以往往会利用 Netty 开发框架来实现。本套丛书中有专门讲解 Netty 开发框架的书，该书详细地讲解了如何实现自定义的 RPC 开发框架，有兴趣的读者可以继续深入学习。

1.1.2 EJB 技术

EJB 技术

视频名称 0103_【理解】EJB 技术

视频简介 EJB 是由 SUN 公司推出的适合于 Java 搭建业务中心的核心技术，同时也是现代 Java 框架设计与开发理论的奠基者。本视频为读者分析 EJB 技术的主要组成部分，并分析 RMI 与 EJB 技术的关联，以及 EJB 技术实现的缺陷。

EJB（Enterprise JavaBean，企业级 JavaBean）是早期 Java EE 设计与开发之中非常常用的一个设计组件，可以将 JavaBean 技术扩展到独立的服务端领域，适合于开发和部署多层结构的、分布式的业务中心。在 EJB 程序开发完成后，可以方便地将其部署到任何支持 EJB 技术标准的容器之中（如 WebSphere、WebLogic、JBoss 等）。一个完整的 EJB 架构包含以下 3 个核心组件。

（1）**会话 Bean（Session Bean）**：EJB 的业务中心，根据应用环境不同分为如下两种类型。

- **无状态会话 Bean**：不保存客户端的状态信息，可以减少资源占用，并提供良好的服务端处理性能，是 EJB 在实际开发中最为常用的一种组件。
- **有状态会话 Bean**：保存每一个客户端的状态信息，会造成极大的服务器资源占用。

（2）**实体 Bean（Entity Bean）**：持久化状态处理的对象，可以提供关系数据库的操作，也有两种实现类型。

- **CMP（Container Managed Persistence，容器管理持久化）**：由开发者定义数据接口标准，随后由容器生成相应的数据库操作代码，由于不需要直接编写代码，所以有较强的数据库可移植性。Hibernate 开发框架就是模拟 CMP 技术理论实现的。
- **BMP（Bean Managed Persistence，Bean 管理持久化）**：开发者除了要定义数据接口标准之外，还需要通过 JDBC 技术标准编写数据接口的实现子类，程序的可移植性不强。

（3）**消息驱动 Bean（Message Driven Bean）**：提供异构数据平台的整合处理，采用消息的异步处理机制提供数据交互，基于 JMS 服务标准构建（Apache 提供的 ActiveMQ 组件也是基于 JMS 服务标准实现的）。

EJB 是一个完全独立的组件，可以通过 Web 容器或者通过 CORBA 协议实现远程调用。所有的调用全部由会话 Bean 负责处理（需要暴露远程接口），随后会依据不同的业务处理需要实现实体 Bean 的调用（本地接口）。每一个实体 Bean 都会与数据源中的数据表进行结构映射，操作实体 Bean 接口时就可以实现数据表数据的操作，如图 1-6 所示。

提示：CORBA。

CORBA（Common Object Request Broker Architecture，通用对象请求代理体系结构）是由对象管理组织在 1991 年提出的公用对象请求代理程序结构的技术规范，被广泛地应用在各个大型系统之中，可以通过各种语言来实现。后来在 EJB 技术中基于 CORBA 以及 RMI（Remote Method Invocation，远程方法调用）标准定义了 RMI-IIOP（Internet Inter-ORB Protocol，互联网内部对象请求代理协议）通信标准，而这也成为 EJB 技术中主要采用的调用形式，同时考虑到操作的标准化，在 EJB 中的所有组件全部采用远程调用（即使多个组件在同一个 EJB 容器之中，进行调用时也采用 JNDI 远程调用）。

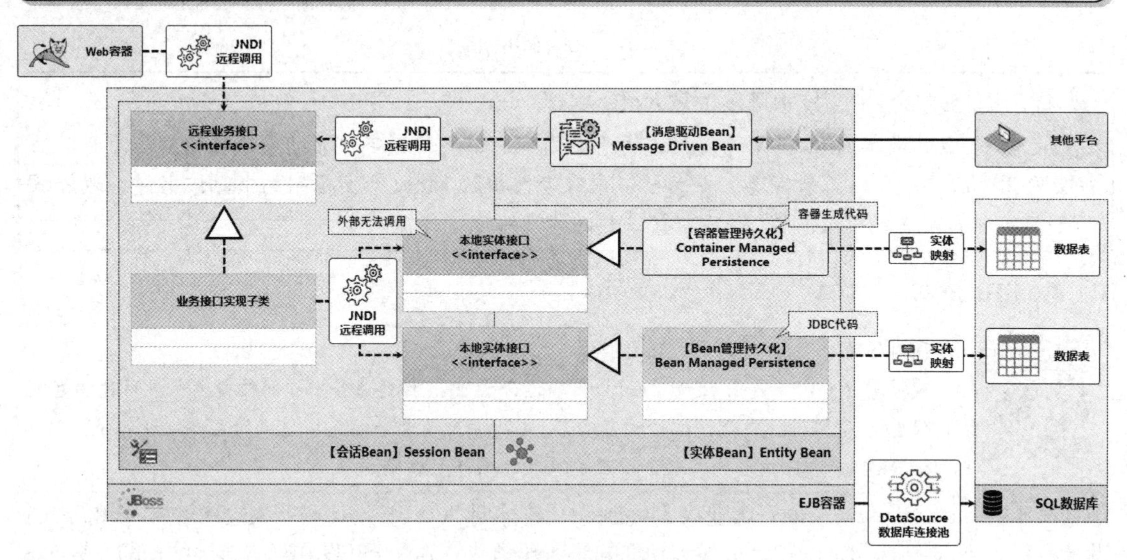

图 1-6　标准的 EJB 实现架构

虽然 EJB 技术设计理念良好，也有众多的大型厂商对 EJB 技术提供了良好的支持，但是由于设计架构复杂以及性能较差，EJB 技术并没有得到广泛的应用。后续推出的 EJB 3.0 开发标准解决了原始设计中的各个性能问题，但是由于其复杂性，最终并没有成为开发技术的主流。

1.1.3 Web Service

视频名称 0104_【理解】Web Service

视频简介 不同平台的整合需要公共的实现标准，XML 的出现提供了统一的通信标准，Web Service 技术随之被提出。本视频为读者讲解 Web Service 技术的主要特点以及缺陷。

在技术发展多元化的今天，项目平台的构建手段也越来越多。而各个公司随着业务的不断扩张，除了需要不断维护自身的技术平台之外，也需要不断对接其他技术平台（如短信平台、存储平台、其他信息平台等），这样才可能满足当前的业务需要，如图 1-7 所示。

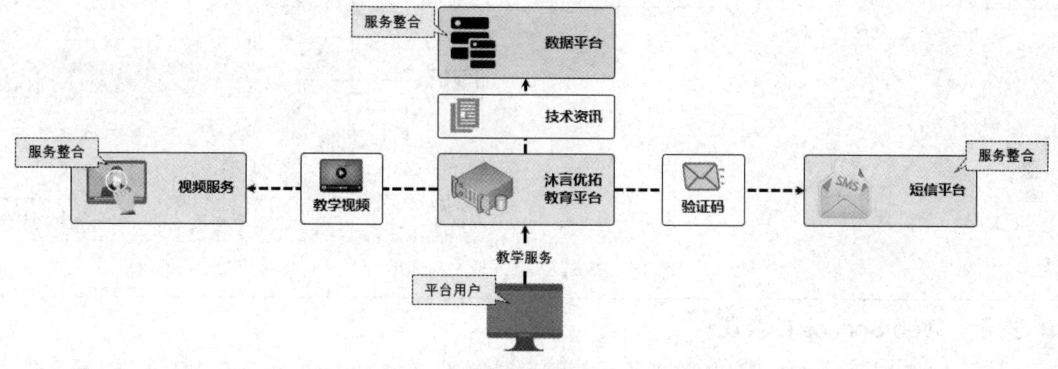

图 1-7 平台整合

此时如果要对接的平台与当前平台采用的是同一种开发技术，那么整合处理会非常容易。然而，实际上不同的平台可能会由不同的开发语言编写，如 PHP、.NET、Python、Go、Java、Ruby 等，这样一来就需要提供一个平台独立的通信标准，于是就有了 Web Service 技术，如图 1-8 所示。

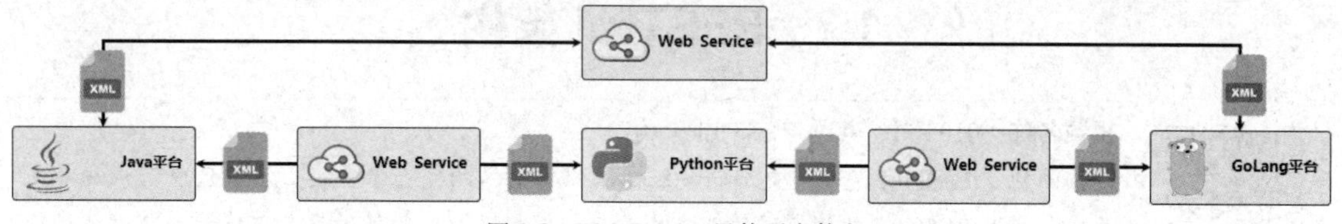

图 1-8 Web Service 异构平台整合

提示：Web Service 产生背景。

早期的 Web Service 标准是在 Java EE 与.NET 两大开发阵营的竞争中产生的。当时构建技术平台可选的两个方案是 SUN 的 Java EE 和微软的.NET，于是很多开发者不得不面对大量且烦琐的服务整合。在这样的背景下，Web Service 技术得到了良好的发展空间，而后基于 Web Service 技术的形式产生了 SOA（Service-Oriented Architecture，面向服务的体系结构）技术架构。

Web Service 是一种利用网络进行应用集成的解决方案，主要是对外发布应用服务接口，所有的服务数据统一被 UDDI（Universal Description Discovery and Integration，通用描述、发现与集成服务）管理。在客户端访问时，可以通过 UDDI 实现服务接口的调用处理。

UDDI 只提供接口服务的管理功能，但是客户端在调用接口之前需要明确该接口的传递参数以及返回的数据类型，这样就需要为接口提供 WSDL（Web Services Description Language，Web 服务描述语言）文件。该文件基于 XML 语法标准，实现接口的完整描述，最终客户端就可以基于 SOAP（Simple Object Access Protocol，简单对象访问协议）采用 XML 数据格式实现接口数据传输，如图 1-9 所示。

使用 Web Service 可实现良好的跨平台的技术架构，利用 Web Service 技术搭建的业务中心可以便于所有的技术平台进行调用。由于 Web Service 基于 XML 实现数据传输以及处理流程上的问题，其性能不高，所以在一些互联网项目平台中使用较少，而在传统的项目开发中使用较为广泛。

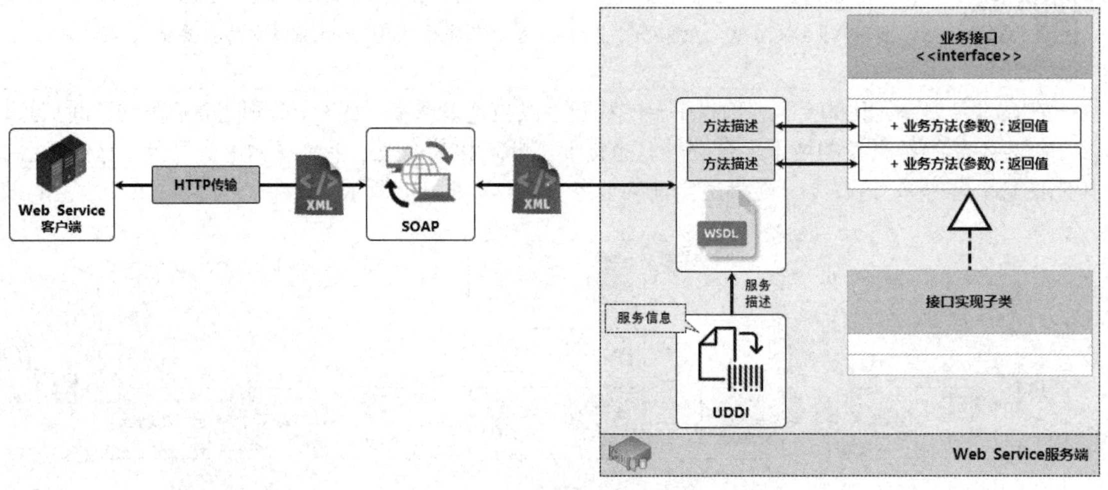

图 1-9　Web Service 发布与调用

提示：Web Service 的实现。

本套丛书中关于 Spring Boot 的书已经为读者讲解了 Web Service 服务的具体实现，有需要的读者可以自行翻阅。需要说明的是，Web Service 技术依然是一种实用的开发技术，也有许多项目还在大量使用 Web Service 技术进行开发，但是如果新开发的是基于互联网环境的项目，一般不建议使用 Web Service 技术。

1.2　RESTful 架构

RESTful 架构

视频名称　0105_【掌握】RESTful 架构

视频简介　RESTful 是一种更加轻量级的数据传输结构，可以直接基于 HTTP 完成分布式开发。利用 RESTful 架构实现的业务中心可以提供高效、简洁的运行环境。本视频为读者分析 RESTful 架构的产生背景以及实现说明。

RESTful，或者简称为 REST（Representational State Transfer，描述性状态传递）是目前非常流行的一种互联网软件架构。它结构清晰，符合标准，易于理解，扩展方便，所以正得到越来越多网站的采用。

提示：REST 提出者——Roy Thomas Fielding。

REST 软件架构是由 Roy Thomas Fielding（菲尔丁）博士（见图 1-10）在 2000 年首次提出的。REST 软件架构是一个抽象的概念，是实现互联网的超媒体分布式系统的行动指南。

他在论文中提到："我这篇文章的写作目的，就是在符合架构原理的前提下，理解和评估以网络为基础的应用软件的架构设计，得到一个功能强、性能好、适宜通信的架构。REST 指的是一组架构约束条件和原则。"如果一个架构符合 REST 的约束条件和原则，则可以将其称为 REST

架构。

图 1-10 Roy Thomas Fielding

Roy Thomas Fielding 是 HTTP（1.0 版和 1.1 版）的主要设计者、Apache 软件的作者之一、Apache 基金会的第一任主席。

REST 中所描述的表现层主要指的是资源（Resource）的表现层，所谓资源就是指网络上的实体数据信息，这种信息可能是一段文本、一段音频或者一种服务，而每一种服务都会绑定唯一的 URI（Uniform Resource Identifier，统一资源标识符）地址，要获取特定的资源只需要访问其对应的 URI 地址，如图 1-11 所示。

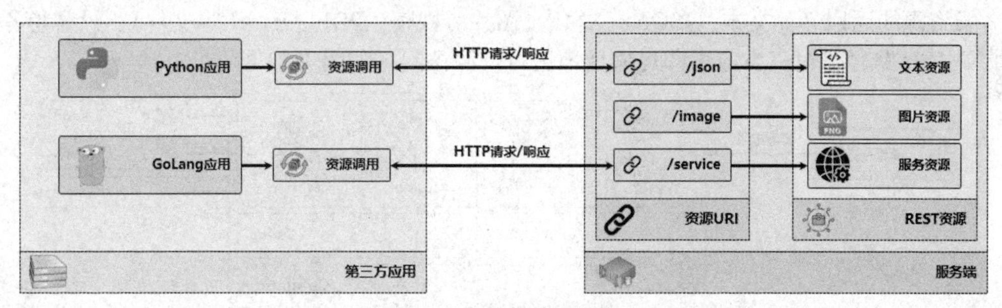

图 1-11 REST 资源访问

提示：REST 与 HTTP。

REST 架构是与 Web 技术捆绑在一起的。从理论上讲，REST 架构并不一定要与 HTTP 绑定在一起，只不过现阶段 HTTP 是 REST 架构的主要展现形式，所以本系列图书中所有的 REST 架构都以 HTTP 为基础进行讲解。

在实际的 REST 开发中较为常见的资源类型是文本资源，这样在进行数据返回时就可以依据不同的环境返回普通文本、XML 文本或 JSON 文本。考虑到数据传输的性能以及数据处理的简洁性，一般都建议采用 JSON 结构实现数据交互，如图 1-12 所示。

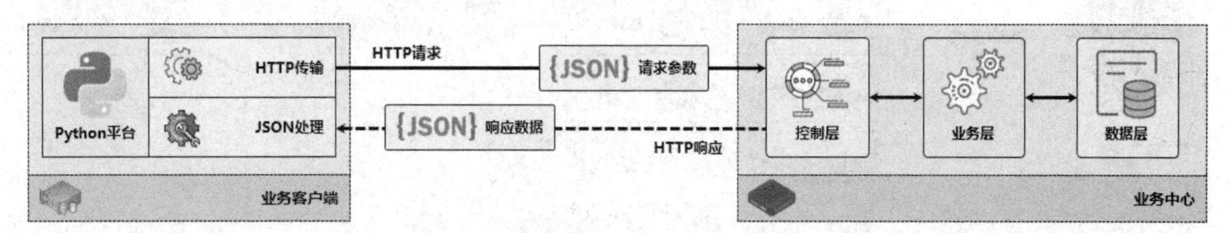

图 1-12 JSON 交互

从严格的意义上来讲，REST 本身并不属于一项新的开发技术，而是对技术的一种规范化处理，其设计的核心理念就是使用现有的 Web 标准中的一些准则和约束搭建系统服务。在使用 RESTful 架构进行项目设计时，应该遵循统一接口的原则，即可以通过一个接口并结合 HTTP 请求模式实现对不同资源的处理操作。例如，现在有一个“/message”的接口，当通过 GET 请求访问时其表示查询，而当通过 POST 请求访问时其表示数据增加，即不同的请求模式对于同一个访问路径有不同的资源处理支持。表 1-1 为读者列出了 HTTP 请求模式与 RESTful 资源操作之间的对应关系。

> **提示：接口的幂等性与安全性。**
>
> **幂等性：** 对同一个 REST 资源的多次访问，得到的资源状态都是相同的。本套丛书中关于 Redis 的书为读者分析了幂等性的控制，读者可自行参阅。
>
> **安全性：** 对 REST 接口的访问不会影响服务端资源的状态。

表 1-1　HTTP 请求模式与 RESTful 资源操作

序号	请求模式	请求路径	资源操作	幂等性	安全性
01	GET	/message?id=yootk	SELECT	是	是
02	POST	/message?id=yootk&title=沐言科技&content=yootk.com	INSERT	否	否
03	PUT	/message?id=yootk&title=李兴华高薪就业编程训练营	UPDATE	是	否
04	DELETE	/message?id=yootk	DELETE	是	否

REST 开发架构可以通过任何语言来实现，如果使用 Java 进行 REST 架构的程序开发，较为常见的就是直接基于 Spring Boot 开发框架来实现。利用 Spring Boot 框架所提供的功能，可以方便地实现各类资源的返回（如文本、JSON、XML、图片、视频、PDF、Excel 等文件），只要为各个资源绑定好 URI 路径，第三方应用即可方便地进行资源访问，如图 1-13 所示。

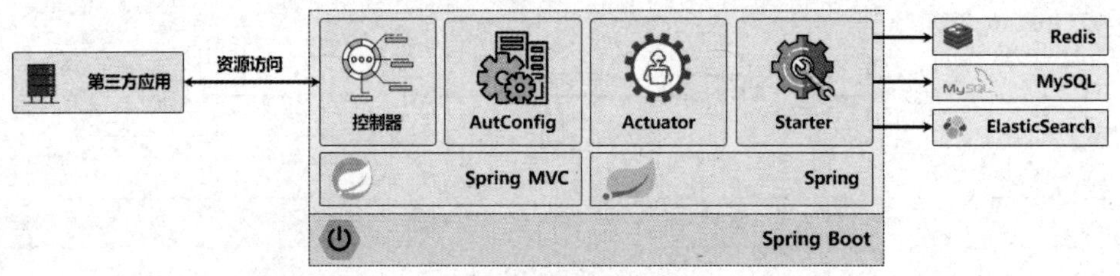

图 1-13　Spring Boot 框架

虽然 Spring Boot 可以方便地实现 REST 开发架构，但是随着项目规模的不断扩大，单一的业务中心很难满足性能以及维护上的要求。这样就需要将一个综合的业务中心拆分为微服务集群，不同的微服务处理不同的业务，如图 1-14 所示。然而这样一来，就需要有一种合理的机制实现微服务的管理，否则会为第三方应用的整合带来极大的麻烦。正是在这样的技术背景下，Spring 推出了 Spring Cloud 技术来解决微服务的设计与管理问题。

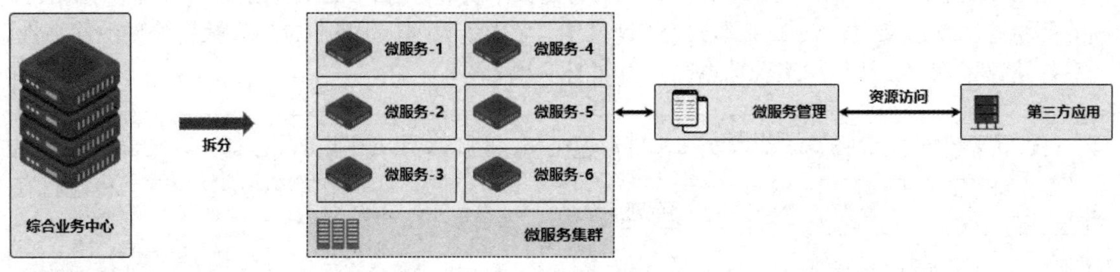

图 1-14　微服务拆分

1.2.1　Spring Cloud Netflix 架构

Spring Cloud
Netflix 技术架构

视频名称　0106_【掌握】Spring Cloud Netflix 技术架构

视频简介　Spring Cloud 在早期主要基于 Netflix 组件实现 REST 微服务的管理。本视频为读者讲解 Spring Cloud Netflix 架构，并分析各个主要组件的作用。

Spring Cloud 是在 Spring Boot 基础之上构建的，是一套完整微服务集群架构管理组件的集合。每一个组件都是独立的，并且不断地提供版本的更新、迭代，其中在 Spring Cloud 早期的技术整合中主要使用了 Netflix 提供的开发套件，这些套件整合形式如图 1-15 所示。

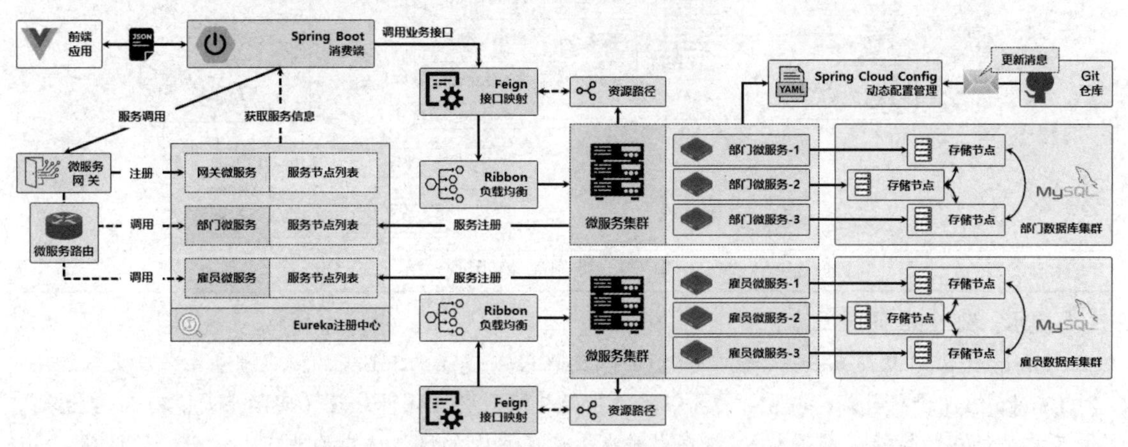

图 1-15　Spring Cloud Netflix 核心架构

提示：HA 机制。

HA（High Availability，高可用性）集群是保证业务连续性的有效解决方案，一般有两个或两个以上的节点，且分为活动节点及备用节点，当一个节点出现故障时可以自动切换到其他节点来提供服务，如图 1-16 所示，这样可以保证服务的稳定性。

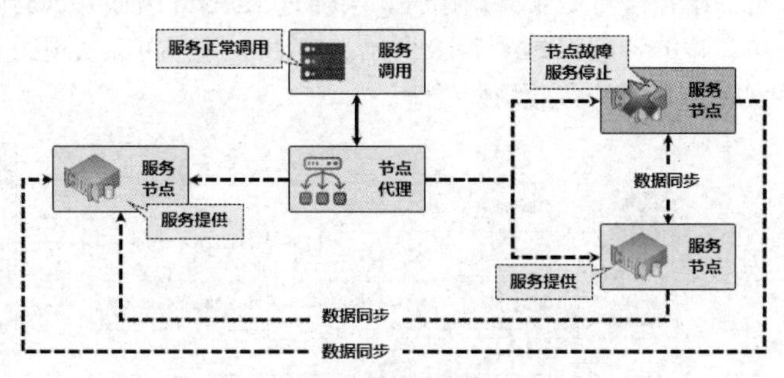

图 1-16　HA 机制

当项目中需要引入 HA 机制时，某一项服务可能会有若干个节点。如果发现某一个服务节点不可用，则自动切换到其他可用节点，并且及时地进行问题反馈。伴随着 HA 机制，往往也会提供负载均衡（Load Balance）机制，利用负载均衡机制可以提高服务的处理性能。

由于需要考虑到微服务的并发处理性能以及 HA 机制，所以微服务一般都会以集群的形式出现，即一个微服务功能由若干个节点提供，同时每一个节点都有可能使用各自的数据库节点进行存储；而为了实现数据的同步，往往会搭建相应的数据库集群。这样如果直接通过微服务的节点地址进行访问，那么失败会造成维护的困难以及设计的缺陷，所以为了便于微服务的访问管理，Spring Cloud 提供了如下核心组件。

1. Eureka 注册中心

为了便于微服务的管理，所有的微服务必须统一在注册中心进行注册，不同的微服务以名称区分（一个微服务的名称会对应若干个节点），在调用时依据微服务的名称进行访问，如图 1-17 所示。利用注册中心可以有效地帮助用户监控微服务节点状态，并且可以更加方便地实现 HA 机制。由于注册中心是整个微服务的核心所在，因此需要为其提供良好的处理性能，并配置 HA 机制以保证稳定提供服务。

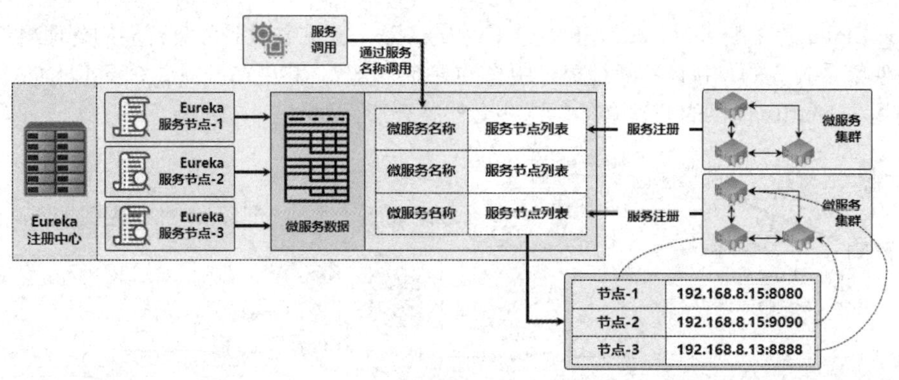

图 1-17　Eureka 注册中心

提示：Eureka 闭源风波。

Spring Cloud 从诞生开始，使用的主要是 Netflix 所提供的 Eureka 1.x 注册中心。后来 Netflix 对 Eureka 组件进行升级，提供了 2.x 版本，但是随后进行了闭源处理（实际上是没有开发出来，因为两个版本的性能差别不大）。这在当时也引起了巨大的风波，成为业内人人讨论的话题。实际上即使 Eureka 不再能被使用了，开发者也可以用 Consul 来代替 Eureka。

2. Ribbon 负载均衡

单个节点微服务的处理性能是有限的，所以为了提升某一个微服务的性能，必须引入若干个相同的微服务。所有的微服务都需要在注册中心进行注册，而为了便于管理，需要为同样功能的微服务设置相同的名称。客户端在进行微服务调用时，会通过 Ribbon 组件并依据微服务的名称找到对应的节点列表，以获取相应的资源，如图 1-18 所示。Ribbon 除了可以实现微服务的调用之外，也可以采用特定的算法实现客户端的负载均衡处理。

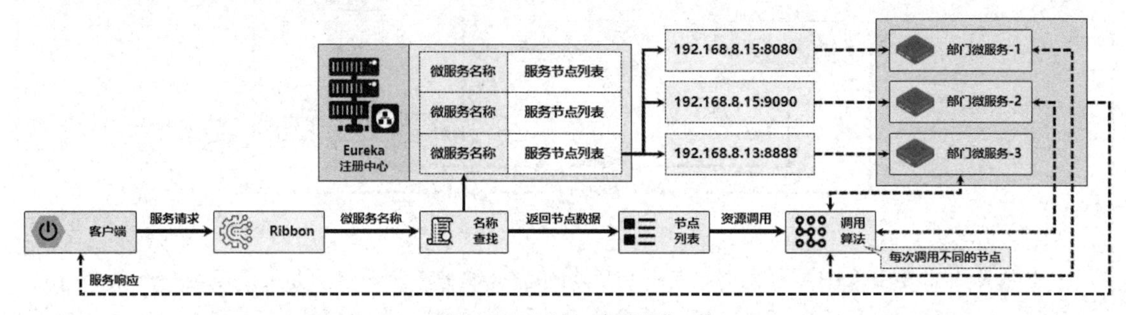

图 1-18　Ribbon 负载均衡

3. Feign 接口映射

每一个微服务都会提供若干访问接口，而为了便于客户端调用，可以将全部的资源接口映射为一个 Java 接口。这样客户端通过 Java 接口即可实现服务端资源调用，如图 1-19 所示。此时就可以依靠 Feign 技术实现接口的映射转换处理。

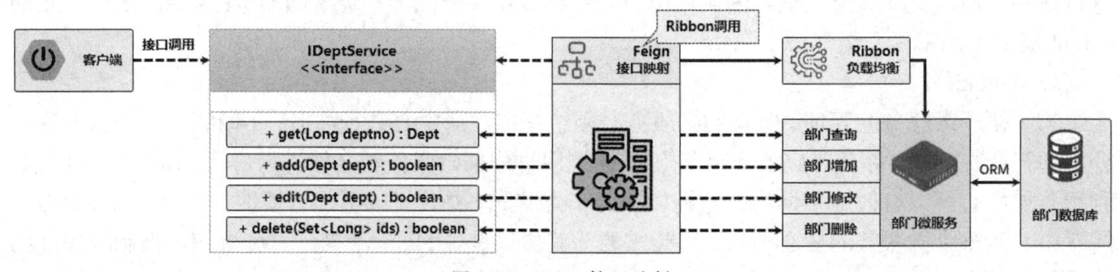

图 1-19　Feign 接口映射

4. 微服务网关

网关可以实现不同类别微服务的统一管理，所有的客户端通过网关路由进行微服务的访问，这样就解决了微服务集群架构中节点过多所造成的服务管理问题，如图 1-20 所示。早期的 Spring Cloud 技术使用 Netflix 中的 Zuul 组件实现了网关微服务的开发，而后由于 Zuul 的性能较差，又基于 Netty 开发框架开发了 Spring Cloud Gateway 网关技术。

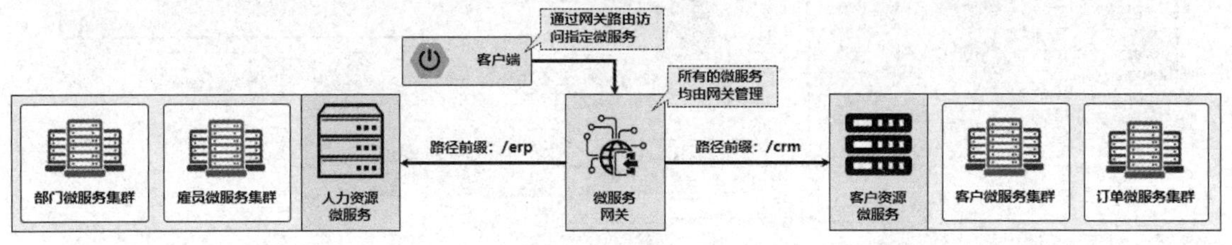

图 1-20　微服务网关

5. Spring Cloud Config

微服务的出现使得业务中心的维护进一步复杂化，同时也增加了服务的维护难度，一个完整的微服务系统可能会包含成百上千个微服务节点，同时也会随着业务的不断发展出现更多的微服务。为了便于微服务的配置管理，Spring Cloud 提供了 Spring Cloud Config 技术，所有的微服务都可以直接通过 Spring Cloud Config 动态实现对服务配置文件的抓取，同时维护者也可以直接将各个微服务所需要的内容保存在 Git 仓库或 SVN 服务器之中，如图 1-21 所示，这样就可以轻松地实现微服务的配置维护。

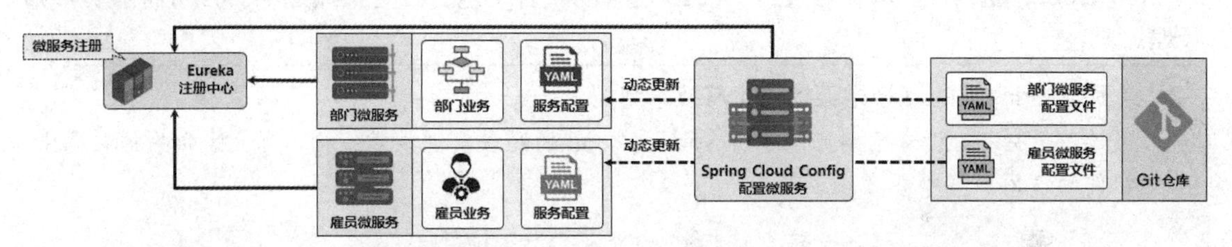

图 1-21　Spring Cloud Config

以上几项是 Spring Cloud 构建微服务的核心实现技术。为了微服务的访问安全，还需要进行 Spring Security 或 JWT 整合；为了便于实现基于消息的微服务设计，也可使用 Spring Cloud Stream 组件、Spring Cloud Sleuth 微服务跟踪执行组件等。当然也可以整合一些第三方常用组件进行开发，本书后文会有所涉及。

1.2.2　Spring Cloud Alibaba 架构

Spring Cloud Alibaba 技术架构

视频名称　0107_【掌握】Spring Cloud Alibaba 技术架构

视频简介　Spring Cloud Alibaba 是现在国内较为流行的 Spring Cloud 解决方案之一。本视频为读者宏观地讲解 Spring Cloud Alibaba 的技术架构，并分析其与 Netflix 技术的区别。

Spring Cloud Alibaba 是阿里巴巴公司结合自身的微服务实践开源的一套微服务管理套件，由若干个不同的应用组件组成，但是其核心依然是围绕着 RESTful 微服务应用展开的，其架构如图 1-22 所示。

通过图 1-22 可以发现，Spring Cloud Alibaba 架构还是以 Ribbon 负载均衡、Feign 接口映射以及网关路由访问为核心，而围绕着这一设计核心又扩展了如下技术。

（1）Nacos 注册中心：提供了微服务的注册与发现服务，基于可视化的方式实现了微服务的管理，与 Erueka 功能类似，但是其处理性能更高。

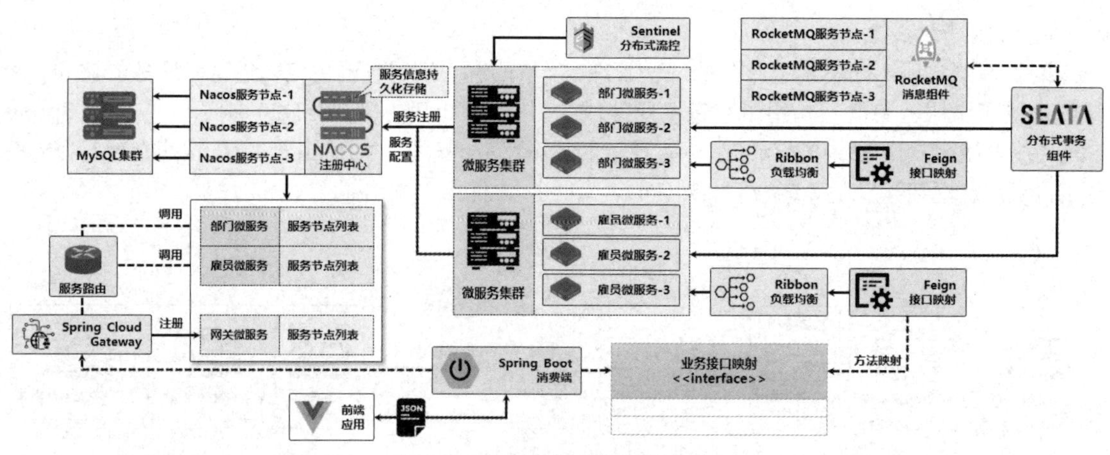

图 1-22　Spring Cloud Alibaba 架构

（2）**Sentinel 分布式流控**：面向分布式微服务架构的轻量级高可用的流控组件，以流量作为切入点，从流量控制，熔断降级，系统负载保护等维度帮助用户保证服务的稳定性。常用于实现流控、熔断降级等策略。

（3）**RocketMQ 消息组件**：基于 Java 的高性能、高吞吐量的消息队列，在 Spring Cloud Alibaba 生态用于实现消息驱动的业务开发。

（4）**Seata 分布式事务组件**：基于 RocketMQ 机制实现了多个微服务的分布式事务管理。

（5）**Dubbo 组件**：可以直接进行 Dubbo 框架的整合，也可以直接通过 Nacos 进行服务注册管理。

提示：Spring Cloud 技术架构对比与技术选型。

通过以上分析，读者已经可以了解 Spring Cloud 的两种实现技术架构，为了区分两种架构中各个组件的作用，可以参考表 1-2。

表 1-2　Spring Cloud 技术架构对比

序号	描述	Spring Cloud Netflix	Spring Cloud Alibaba
01	服务发现	Eureka（停止更新）	Nacos（持续更新）
02	配置中心	Spring Cloud Config	Nacos
03	断路保护	Hystrix 断路保护	Sentinel
04	链路追踪	Sleuth 调用链监控	-
05	负载均衡	Ribbon	-
06	远程调用	Open Feign	Dubbo
07	分布式事务	-	Seata
08	服务网关	Spring Cloud Gateway	-

通过表 1-2 可以发现，Spring Cloud Alibaba 依然需要使用 Spring Cloud Netflix 所提供的很多组件，两者之间是互补的关系，最终可以确定要使用的技术组件为 Nacos + Sentinel + Seata + Ribbon + Feign + Spring Cloud Gateway + Sleuth。

1.3　本章概览

1．大型的分布式系统中最为重要的是业务中心，搭建完善的业务中心应该采用中立的开发技术，并且应可以与任何技术平台实现无缝对接。

2．RPC 技术是分布式技术开发中主要采用的一项技术，利用 RPC 可以实现远程业务接口的调用。

3．EJB 是早期的 SUN 公司（后被 Oracle 公司收购）官方推出的一项业务中心搭建技术，其设计理念非常先进，但是最终在落地环节出现了性能问题。早期的 EJB 技术中仍可以使用的有无状态会话 Bean，同时 EJB 需要有特定的 EJB 容器提供支持，否则无法运行。

4．Web Service 是一个基于 XML 技术实现的分布式开发技术标准，在行业内有较长时间的应用，但是由于其设计及应用时间较早，因此不能满足现代的项目实现要求。

5．RESTful 是一项简单的互联网通信技术标准，利用 RESTful 架构实现的业务中心可以方便地获取资源。Spring Boot 开发框架就是针对 RESTful 实现而推出的。

6．Spring Cloud 是一个基于 RESTful 架构实现的分布式架构集合，提供 Spring Cloud Netflix 及 Spring Cloud Alibaba 两种实现方案。随着时间的推移及技术的进步，有更多的应用厂商提供与之相关的技术架构，但是考虑到当前实际的开发环境，本书将以 Spring Cloud Alibaba 技术方案为主进行讲解。

第2章

Spring Cloud 编程起步

本章学习目标

1. 掌握基于 Spring Boot 框架的 RESTful 服务搭建和数据传输处理技术；
2. 掌握 Spring Boot 和 Spring Cloud 在项目开发中的关联以及区别；
3. 掌握 RestTemplate 处理类的使用方法，并可以依据此类实现微服务接口调用；
4. 掌握 Swagger 工具的使用方法，并可以使用该组件进行 REST 接口的详细描述；
5. 掌握 Spring Security 在 Swagger 组件中的应用方法，并可以提供安全账号保护接口文档。

构建 Spring Cloud Alibaba 项目必须以 Spring Boot 为基础，同时需要搭配正确的开发版本，以提供稳定的服务。本章将从项目的搭建开始介绍微服务提供端和消费端的完整构建。

2.1 RESTful 应用开发

RESTful 实现简介

视频名称 0201_【掌握】RESTful 实现简介

视频简介 RESTful 是微服务实现的理论基础。本视频将为读者介绍本项目要实现的服务端的功能，以及完成本项目需要使用的实现技术。

RESTful 项目设计中最重要的部分就是资源的获取，而在大部分的项目设计之中，对资源一般都需要进行有效的统一管理（如文件系统、关系数据库、NoSQL 数据库等）。本项目将基于 MySQL 数据库实现资源的存储，通过一张部门信息表保存资源数据。

范例：数据库创建脚本

```
DROP DATABASE IF EXISTS yootk8001;
CREATE DATABASE yootk8001 CHARACTER SET UTF8;
USE yootk8001;
CREATE TABLE dept (
   deptno              BIGINT          AUTO_INCREMENT,
   dname               VARCHAR(50),
   loc                 VARCHAR(50),
   CONSTRAINT pk_deptno PRIMARY KEY(deptno)
) ENGINE=InnoDB DEFAULT CHARSET=utf8;
INSERT INTO dept(dname,loc) VALUES ('开发部', database());
INSERT INTO dept(dname,loc) VALUES ('财务部', database());
INSERT INTO dept(dname,loc) VALUES ('市场部', database());
INSERT INTO dept(dname,loc) VALUES ('后勤部', database());
INSERT INTO dept(dname,loc) VALUES ('公关部', database());
COMMIT;
```

本次创建了一个 yootk8001 数据库，随后在该数据库中定义了项目所需要的部门信息表 dept。该表中有 3 个字段，其中 loc 字段的内容将通过 database()函数进行设置，该函数的主要功能是获取当前数据库的名称。

提示：微服务集群环境。

在 Spring Cloud 使用过程中是不能忽视微服务集群环境的，即同样的微服务会部署在不同的节点之中，以提高微服务的处理性能。在实际生产环境下的微服务集群可能会对同一数据库中的数据进行访问（为了提高性能也需要对数据库存储进行集群搭建），如图 2-1 所示。

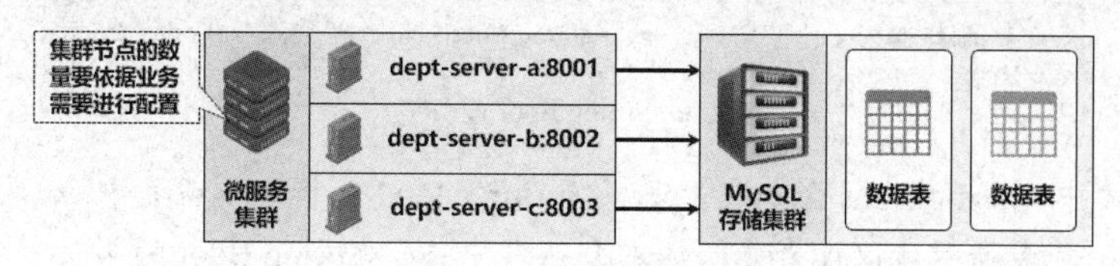

图 2-1 微服务集群环境

但是在学习过程中如果让全部的微服务访问同一个数据库，则不便于读者观察，所以在本书中将定义若干个结构相同的数据库，如 yootk8001、yootk8002、yootk8003（8001～8003 为后续微服务部署的端口编号），同时这些数据库中都存在相同结果的 dept 数据表。为了便于区分，书中会根据当前使用的数据库的不同而进行 loc 字段的填充，这样就可以明确当前是否已经正确开启了集群环境。

为了便于操作数据库中的数据，此处将通过 MyBatis/MyBatisPlus 开发框架实现数据的 CRUD 处理。由于 Spring Cloud 是以 Spring Boot 为基础构建的，因此只需要在构建工具中引入相关的依赖即可实现代码开发。此处的代码需要实现 Spring Cloud 服务端和 Spring Boot 客户端的开发，结构如图 2-2 所示。

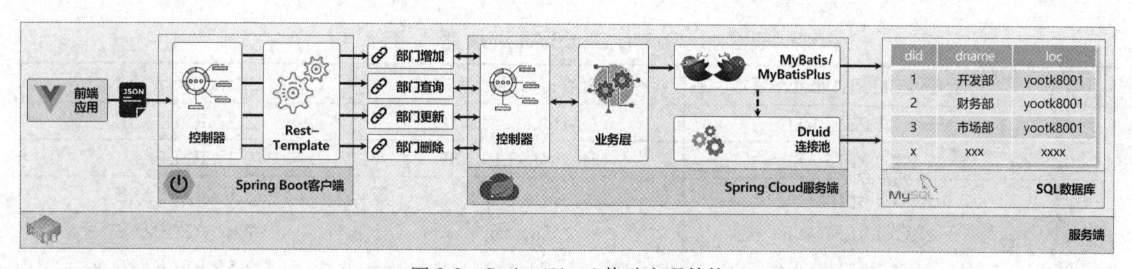

图 2-2 Spring Cloud 基础实现结构

Spring Cloud 所实现的服务端依靠控制器进行 RESTful 接口发布，每一个接口可以使用不同的名称，也可以使用相同的名称（请求模式不同），而此时客户端（微服务客户端，但是本质上依然属于服务端）可以通过 Spring Boot 所提供的 RestTemplate 类进行远程 RESTful 接口调用以获取相关的资源，这样就实现了基础微服务环境。

提问：是否需要 Spring Boot?

在本次所实现的程序架构中，Spring Cloud 开发的服务端可以直接通过控制器进行服务的发布，控制器可以直接以 RESTful 的形式返回数据。既然已经可以进行服务的正常提供，为什么还需要通过 Spring Boot 的客户端进行访问呢?

回答：便于服务整合处理。

微服务的设计目的是对一个完整的服务进行拆分，所以使用 Spring Cloud 开发出来的每一个微服务实际上都只进行简单的资源操作，而在实际的项目中会出现一个业务调用多个微服务的情况，那么此时就需要在服务端通过 Spring Boot 将多个微服务整合后返回，如图 2-3 所示。

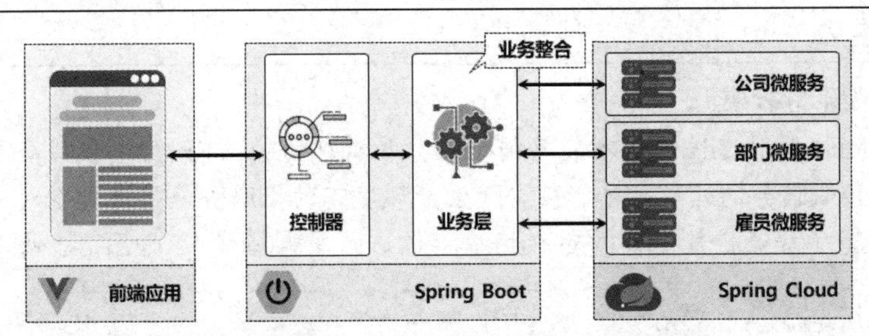

图 2-3　Spring Boot 业务整合

在本书后文中，Spring Boot 会一直出现在 Spring Cloud 的整合处理之中，如果读者对 Spring Boot 还不熟悉，笔者强烈建议读者先阅读本套丛书中关于 Spring Boot 的书。

2.1.1　搭建 Spring Cloud Alibaba 项目

搭建 Spring Cloud Alibaba 项目

视频名称　0202_【掌握】搭建 Spring Cloud Alibaba 项目

视频简介　Spring Cloud 的项目搭建将通过 Gradle 工具完成。本视频将以实际的操作演示如何基于 IDEA 开发工具在 Spring Boot 基础之上搭建 Spring Cloud 项目。

Spring Cloud Alibaba 项目是基于已有的 Spring Boot 和 Spring Cloud 进行构建的，所以在进行项目创建时就必须为其配置正确的组件版本，否则将会出现很多未知的错误而导致开发失败。表 2-1 给出了官方推荐的版本组合。

表 2-1　Spring Cloud Alibaba 版本组合

序号	Spring Cloud Alibaba 版本	Spring Cloud 版本	Spring Boot 版本
01	2.2.1.RELEASE	Hoxton.SR3	2.2.5.RELEASE
02	2.2.0.RELEASE	Hoxton.RELEASE	2.2.x.RELEASE
03	2.1.2.RELEASE	Greenwich	2.1.x.RELEASE
04	2.0.2.RELEASE	Finchley	2.0.x.RELEASE
05	1.5.1.RELEASE	Edgware	1.5.x.RELEASE

截至本书编写时，Spring Cloud Alibaba 最新的套件版本为 2.2.1.RELEASE，根据表 2-1，可以与之组合使用的 Spring Cloud 版本为 Hoxton.SR3、Spring Boot 版本为 2.2.5.RELEASE，如图 2-4 所示。

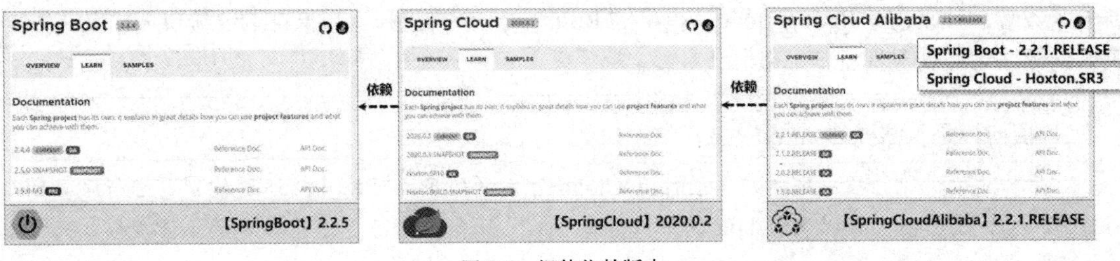

图 2-4　组件依赖版本

提示：不要纠结于版本。

Spring“全家桶”所提供的开发组件随时都在更新，但是并非更新后的组件就一定好用。建议在进行项目版本选型时选用稳定的版本。只要开发版本没有太大的版本号变更，概念与核心功能都是类似的，新版本仅仅做出了一些功能的微调，这些调整都可以在相关项目的 GitHub 或说明文档中查找到。

另外，读者查看 Spring Cloud 版本的时候可能会发现某些 Spring Cloud 版本的编号采用的并非数字标记，而是一些单词标记，这些单词本为伦敦的地名；但是后来的版本又使用了数字标记（名称为“aka llford”），如图 2-5 所示。这也是很有意思的一件事情，正如前面分析的，Netflix 对很多组件进行了闭源处理，所以短期内 Spring Cloud Netflix 是很难有太大的版本变化的。

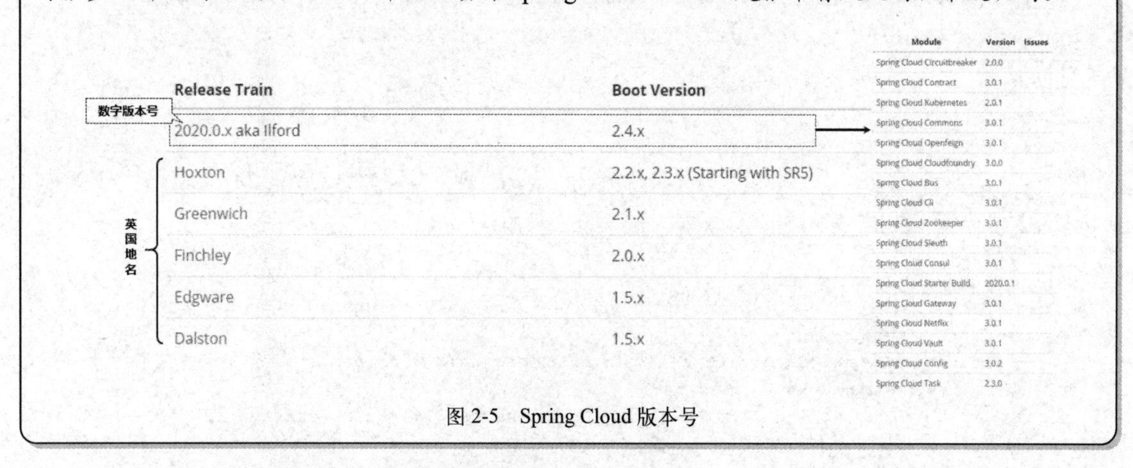

图 2-5　Spring Cloud 版本号

Spring Cloud 程序的构建一般建议使用 Maven 或 Gradle，考虑到未来的技术可延续性，本次将基于 Gradle 进行项目的构建。本章所需要的 Gradle 项目结构如图 2-6 所示。

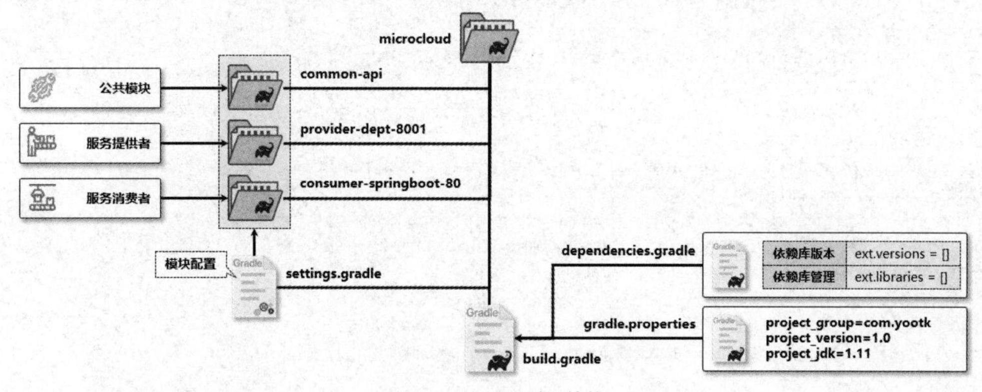

图 2-6　Gradle 项目结构

本节考虑到内容讲解需要进行大量的远程调用模拟，所以将在 IDEA 工具中采用多配置文件和多模块管理的方式进行 Gradle 项目创建，具体的实现步骤如下。

（1）【IDEA 工具】创建一个新的 Gradle 项目，名称为“microcloud”，如图 2-7 所示。

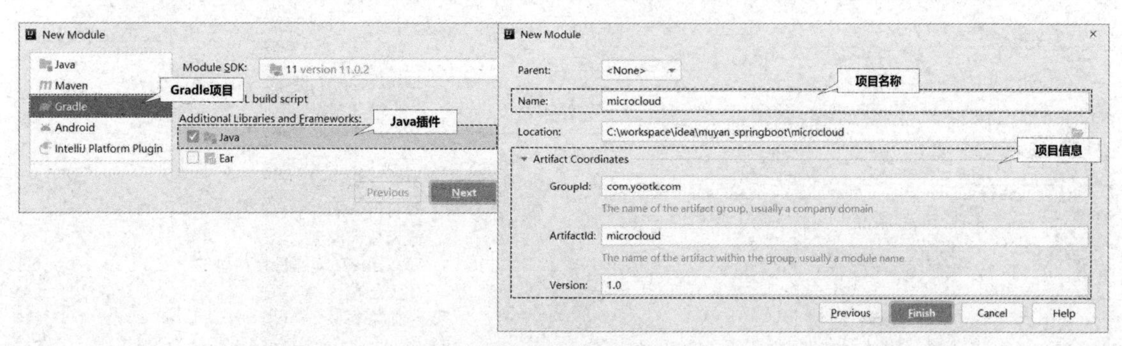

图 2-7　创建 Gradle 项目

（2）【microcloud 项目】创建 gradle.properties 资源文件以保存 Gradle 的核心配置项。

project_group=com.yootk	项目组织名称
project_version=1.0	项目版本信息
project_jdk=11	所使用的 JDK 版本

(3)【microcloud 项目】创建 dependencies.gradle 配置文件以保存所有的项目依赖库。

```
ext.versions = [                      // 定义全部的依赖库版本
    springboot                        : '2.2.5.RELEASE',    // Spring Boot版本
    springcloud                       : 'Hoxton.SR3',       // Spring Cloud版本
    alibabacloud                      : '2.2.1.RELEASE',    // Spring Cloud Alibaba版本
    lombok                            : '1.18.20',          // Lombok版本
    junit                             : '5.6.3',            // 配置JUnit测试工具的版本
    junitPlatformLauncher             : '1.6.3',            // JUnit测试工具运行平台版本
]
ext.libraries = [                                           // 依赖库引入配置
    'spring-boot-gradle-plugin':
        "org.springframework.boot:spring-boot-gradle-plugin:${versions.springboot}",
    'spring-cloud-dependencies':
        "org.springframework.cloud:spring-cloud-dependencies:${versions.springcloud}",
    'spring-cloud-alibaba-dependencies':
        "com.alibaba.cloud:spring-cloud-alibaba-dependencies:${versions.alibabacloud}",
    // 以下配置为与项目用例测试有关的依赖
    'junit-jupiter-api':
        "org.junit.jupiter:junit-jupiter-api:${versions.junit}",
    'junit-vintage-engine':
        "org.junit.vintage:junit-vintage-engine:${versions.junit}",
    'junit-jupiter-engine':
        "org.junit.jupiter:junit-jupiter-engine:${versions.junit}",
    'junit-platform-launcher':
        "org.junit.platform:junit-platform-launcher:${versions.junitPlatformLauncher}",
    'junit-platform-engine':
        "org.junit.platform:junit-platform-engine:${versions.junitPlatformLauncher}",
    'junit-jupiter-params':
        "org.junit.jupiter:junit-jupiter-params:${versions.junit}",
    'junit-bom': "org.junit:junit-bom:${versions.junit}",
    'junit-platform-commons':
        "org.junit.platform:junit-platform-commons:${versions.junitPlatformLauncher}",
    // 以下配置为Lombok组件有关的依赖
    'lombok': "org.projectlombok:lombok:${versions.lombok}",
]
```

(4)【microboot 项目】修改 build.gradle 配置文件，引入 Spring Boot、Spring Cloud、Spring Cloud Alibaba、JUnit 5、Lombok 等核心组件依赖，并进行子模块的结构配置（建议通过代码复制）。

```
buildscript {                                               // 定义脚本使用资源
    apply from: 'dependencies.gradle'                       // 引入所需要的依赖库文件
    repositories {                                          // 脚本资源仓库
        maven { url 'https://maven.aliyun.com/repository/public' }
    }
    dependencies {                                          // 依赖库
        classpath libraries.'spring-boot-gradle-plugin'     // Spring Boot插件
    }
}
group project_group                                         // 组织名称
version project_version                                     // 项目版本
apply from: 'dependencies.gradle'                           // 导入依赖配置
def env = System.getProperty("env") ?: 'dev'                // 获取env环境属性
subprojects {                                               // 配置子项目
    apply plugin: 'java'                                    // 子模块插件
    apply plugin: 'org.springframework.boot'                // 引入Spring Boot插件
    apply plugin: 'io.spring.dependency-management'         // 版本管理
    sourceCompatibility = project_jdk                       // 源代码版本
    targetCompatibility = project_jdk                       // 生成类版本
```

```
repositories {                                                      // 配置Gradle仓库
    mavenLocal()
    maven { url 'http://maven.aliyun.com/nexus/content/groups/public/' }
    mavenCentral()
    jcenter()
}
dependencyManagement {                                              // 版本控制插件
    imports {
        mavenBom libraries.'spring-cloud-dependencies'              // Spring Cloud依赖管理
                cmavenBom libraries.
                        'spring-cloud-alibaba-dependencies'         // Spring Cloud Alibaba依赖管理
    }
}
dependencies {                                                      // 公共依赖库管理
    implementation('org.springframework.boot:spring-boot-devtools') // 项目热部署
    // 以下为测试环境的相关依赖配置
    testImplementation('org.springframework.boot:spring-boot-starter-test') {
        exclude group: 'junit', module: 'junit'                     // 删除JUnit 4
    }
    testImplementation(enforcedPlatform(libraries.'junit-bom')) // 绑定为JUnit 5运行
    testImplementation(libraries.'junit-platform-commons')     // JUnit 5测试组件
    testImplementation(libraries.'junit-platform-engine')      // JUnit 5测试组件
    testImplementation(libraries.'junit-jupiter-api')          // JUnit 5测试组件
    testImplementation(libraries.'junit-vintage-engine')       // JUnit 5测试组件
    testImplementation(libraries.'junit-jupiter-engine')       // JUnit 5测试组件
    testImplementation(libraries.'junit-platform-launcher')    // JUnit 5测试组件
    // 以下为Lombok插件的相关依赖配置
    compileOnly(libraries.'lombok')                                 // 编译时生效
    annotationProcessor(libraries.'lombok')                         // 注解时生效
}
sourceSets {                                                        // 源代码目录配置
    main {                                                          // main及子目录配置
        java { srcDirs = ['src/main/java'] }
        resources { srcDirs = ['src/main/resources', "src/main/profiles/$env"] }
    }
    test {                                                          // test及子目录配置
        java { srcDirs = ['src/test/java'] }
        resources { srcDirs = ['src/test/resources'] }
    }
}
test {                                                              // 配置测试任务
    useJUnitPlatform()                                              // 使用JUnit测试平台
}
task sourceJar(type: Jar, dependsOn: classes) {                     // 源代码打包任务
    archiveClassifier = 'sources'                                   // 设置文件后缀
    from sourceSets.main.allSource                                  // 源代码读取路径
}
task javadocTask(type: Javadoc) {                                   // javadoc打包任务
    options.encoding = 'UTF-8'                                      // 文件编码
    source = sourceSets.main.allJava                                // 所有Java源代码
}
task javadocJar(type: Jar, dependsOn: javadocTask) {                // 先生成javadoc
    archiveClassifier = 'javadoc'                                   // 文件标记类型
    from javadocTask.destinationDir                                 // 查找目标路径
}
tasks.withType(Javadoc) {                                           // 文档编码配置
    options.encoding = 'UTF-8'                                      // 定义编码
}
tasks.withType(JavaCompile) {                                       // 编译编码配置
    options.encoding = 'UTF-8'                                      // 定义编码
}
artifacts {                                                         // 打包操作任务
```

```
    archives sourceJar                                    // 源代码打包
    archives javadocJar                                   // javadoc打包
  }
  gradle.taskGraph.whenReady {                            // 操作准备好后触发
    tasks.each { task ->                                  // 找出所有任务
      if (task.name.contains('test')) {                   // 有test任务
        task.enabled = true                               // 任务启用
      }
    }
  }
  [compileJava, compileTestJava, javadoc]*.options*.encoding = 'UTF-8'// 编码配置
}
```

(5)【本地系统】由于本次需要进行远程调用的模拟操作，建议修改 hosts 主机文件，添加本次的两台主机名称。

服务提供者主机名称	127.0.0.1 provider-dept-8001
服务消费者主机名称	127.0.0.1 consumer-springboot-80

通过以上步骤，就可以创建一个基本的 Spring Cloud Alibaba 项目，而后的开发只需要在此基础上创建相关的子模块，并配置有效的依赖。

2.1.2 REST 公共模块

REST 公共模块

视频名称 0203_【掌握】REST 公共模块

视频简介 Spring Cloud 项目属于 RPC 应用，这样就需要定义服务提供者与消费者之间所可能使用的公共组件。本视频为读者讲解数据传输类的作用，并介绍如何基于 Spring 提供的 BeanUtils 类进行集合数据复制的功能扩展。

Spring Cloud 项目中一般都会有业务生产端和业务消费端，为了便于若干个不同应用的公共代码管理，一般还会创建一个公共模块，可以将数据传输类、业务接口或一些处理工具定义在此模块之中，如图 2-8 所示。

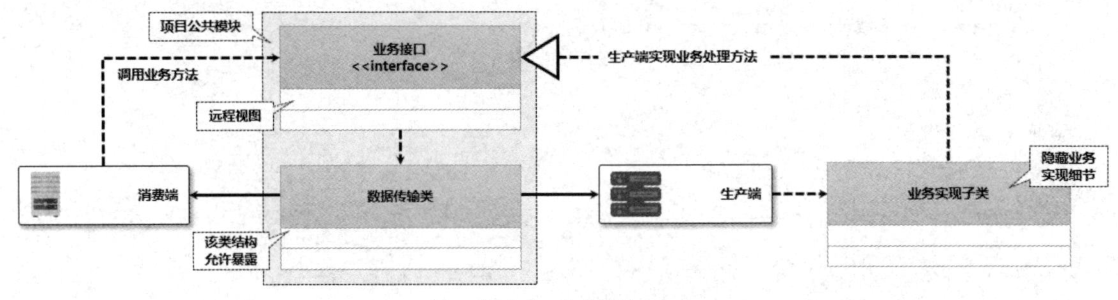

图 2-8 公共模块

利用公共模块可以实现公共组件的有效管理，也便于项目代码的维护。下面通过具体的操作步骤介绍本项目所需要的公共模块的创建。

(1)【microcloud 项目】在 microcloud 项目中，创建一个 common-api 公共模块，如图 2-9 所示。

(2)【microcloud 项目】在公共模块中需要创建一些工具类，所以修改 build.gradle 配置文件，为其追加相关依赖。

```
project('common-api') {                                                  // 子模块
  dependencies {                                                         // 配置子模块依赖
    compile('org.springframework.boot:spring-boot-starter-web')          // Spring Boot依赖
  }
}
```

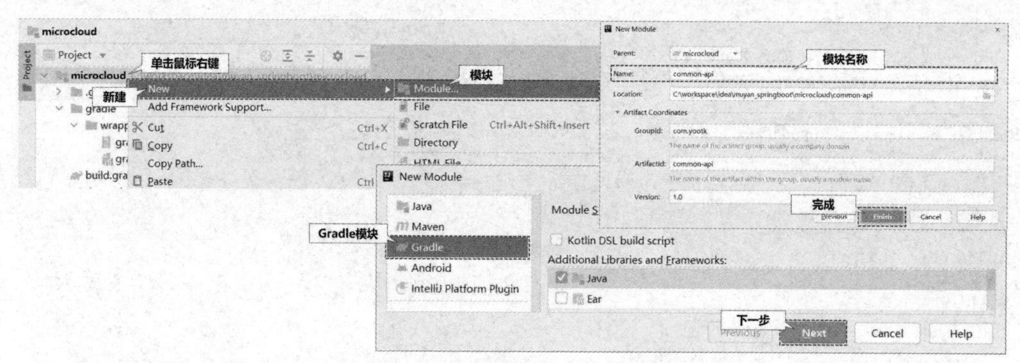

图 2-9 创建 common-api 公共模块

（3）【common-api 公共模块】创建 DeptDTO 数据传输类并实现 Serializable 序列化接口。

```
package com.yootk.common.dto;
@Data                                                         // Lombok结构生成注解
public class DeptDTO implements Serializable {                // 部门数据传输类
    private Long deptno;                                      // 部门编号
    private String dname;                                     // 部门名称
    private String loc;                                       // 部门位置
}
```

此类的结构与部门微服务所使用的部门表结构相同，但是这属于公开的、标准的简单 Java 类，不与任何组件有直接的定义关联，当进行请求或响应需要对象包装时都应该通过此类完成处理。

（4）【common-api】Spring 为了便于开发者使用提供了 Bean 对象的复制操作类，但是这个操作类仅能实现对单一对象的复制处理，而在本次操作中由于涉及集合数据，因此要创建一个工具类。

```
package com.yootk.common.util;
public class DeepBeanUtils extends BeanUtils {                // 继承BeanUtils工具类
    private DeepBeanUtils() {}                                // 构造方法私有化
    /**
     * 实现List集合的复制处理
     * @param sources 复制的原List集合
     * @param target 目标类型
     * @param <S> 源对象类型
     * @param <T> 目标对象类型
     * @return 复制后的List集合
     */
    public static <S, T> List<T> copyListProperties(
                List<S> sources, Supplier<T> target) {
        List<T> list = new ArrayList<>(sources.size());      // 创建List集合
        for (S source : sources) {                            // 源集合迭代
            T obj = target.get();                             // 获取数据
            copyProperties(source, obj);                      // 属性复制
            list.add(obj);                                    // 添加数据
        }
        return list;                                          // 集合返回
    }
}
```

（5）【common-api 子模块】创建 IDeptService 业务接口，该业务接口不与任何应用实现产生关联，所以操作的实体类型为 DeptDTO 传输类。

```
package com.yootk.service;
public interface IDeptService {                               // 部门业务接口
    /**
     * 根据部门ID查询部门详细信息
     * @param id 要查询的部门ID
     * @return 部门传输类实例，如果ID不存在则返回null
     */
    public DeptDTO get(long id);
    /**
```

```
     * 增加部门数据
     * @param dept 部门传输类实例
     * @return 部门增加成功返回true，否则返回false
     */
    public boolean add(DeptDTO dept);
    /**
     * 查询全部部门数据
     * @return 全部数据集合，如果部门数据为空则返回空集合（size() = 0）
     */
    public List<DeptDTO> list();
    /**
     * 部门数据分页查询
     * @param currentPage 当前所在页
     * @param lineSize 每页显示数据行数
     * @param column 模糊查询列
     * @param keyword 查询关键字
     * @return 部门集合数据以及数据统计结果，返回的集合项包括
     * 1. key = allDepts、value = List集合
     * 2. key = allRecorders、value = 总记录数
     * 3. key = allPages、value = 总页数
     */
    public Map<String, Object> split(int currentPage, int lineSize,
                String column, String keyword);
}
```

（6）【common-api 子模块】由于该模块只提供了公共程序类库，因此在进行打包处理时不能够打包为 Spring Boot 程序包。修改 build.gradle 配置文件，关闭 Spring Boot 相关任务。

```
jar { enabled = true }                          // 打包为JAR文件
bootJar { enabled = false }                     // 关闭bootJar任务
javadocTask { enabled = false }                 // 关闭javadocTask
```

（7）【common-api 子模块】子模块创建完成后需要被其他模块所使用，所以需要进行编译处理。

```
gradle build;
```

2.1.3　微服务提供者

微服务提供者

视频名称　0204_【掌握】微服务提供者

视频简介　微服务的提供者需要实现公共的业务接口，同时也需要对外隐藏接口的实现细节。本项目由于是围绕数据库展开业务操作的，因此基于 MyBatisPlus 组件实现数据库的 CRUD 处理开发操作。

在 common-api 子模块中已经定义了本项目所需要使用的业务接口，所以在微服务的提供者项目中，核心任务就是业务接口的实现。为了简化数据层的开发操作，在该应用中会采用 MyBatisPlus 组件实现数据层数据操作。考虑到项目中数据库的操作性能，本次将通过 Druid 数据库连接池实现 JDBC 连接管理，这样就可以得到图 2-10 所示的业务层实现结构。

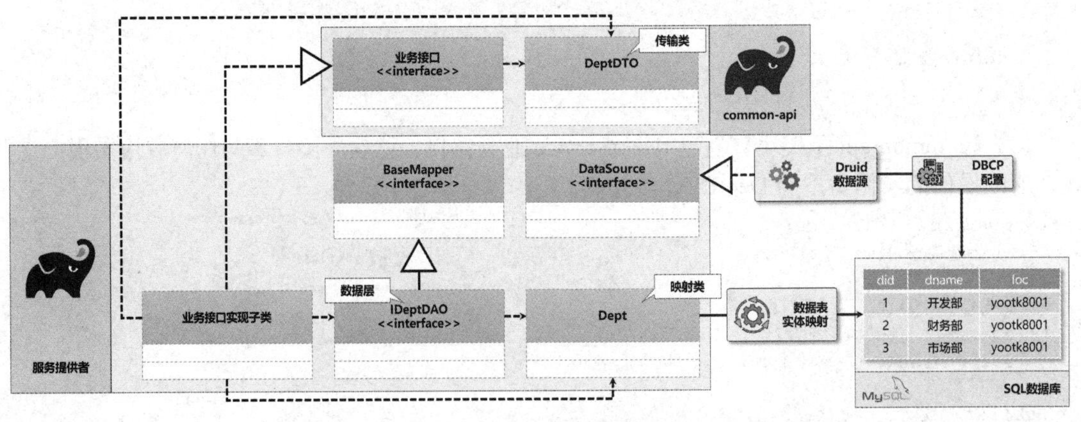

图 2-10　业务层实现结构

Spring Cloud 的分布式架构是以 RESTful 架构为核心进行设计的，对于服务端来讲需要有资源访问接口才可以对外提供服务，而访问接口的实现可以直接通过控制层来完成，这样就可以依据图 2-11 所示的流程进行接口发布，而后具体的实现步骤如下所示。

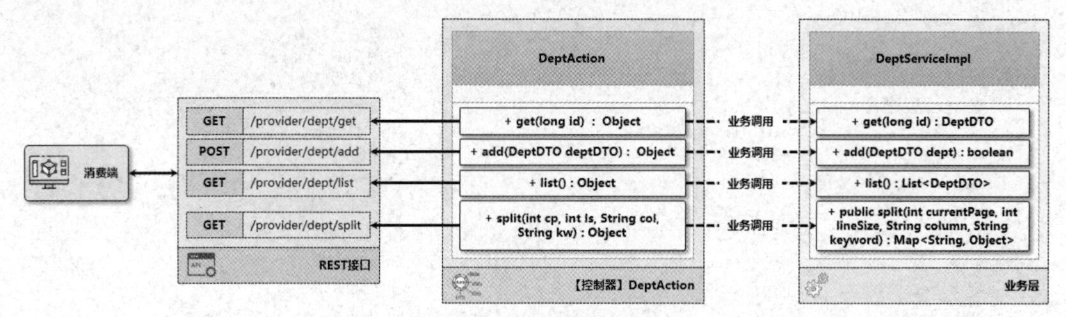

图 2-11　发布 REST 接口流程

（1）【microcloud 项目】创建 provider-dept-8001 服务提供者模块，随后修改 build.gradle 配置文件，进行相关依赖配置。

```
project('provider-dept-8001') {                                        // 子模块
    dependencies {                                                     // 配置子模块依赖
        implementation(project(':common-api'))                         // 引入子模块
        implementation('com.baomidou:mybatis-plus-boot-starter:3.4.2') // MyBatisPlus依赖
        implementation('mysql:mysql-connector-java:8.0.23')            // MySQL依赖
        implementation('com.alibaba:druid:1.2.5')                      // Druid依赖
    }
}
```

（2）【provider-dept-8001 子模块】在 src/main/resources 资源目录中创建 application.yml 配置文件并进行如下配置。

```
server:                                                          # 服务端配置
  port: 8001                                                     # 8001端口监听
mybatis-plus:                                                    # MyBatisPlus配置
  type-aliases-package: com.yootk.provider.vo                    # 别名配置
spring:
  datasource:                                                    # 数据源配置
    type: com.alibaba.druid.pool.DruidDataSource                 # 数据源类型
    driver-class-name: com.mysql.cj.jdbc.Driver                  # 驱动程序类
    url: jdbc:mysql://localhost:3306/yootk8001                   # 连接地址
    username: root                                               # 用户名
    password: mysqladmin                                         # 连接密码
    druid:                                                       # Druid相关配置
      initial-size: 5                                            # 初始化连接池大小
      min-idle: 10                                               # 最小维持连接池大小
      max-active: 50                                             # 最大支持连接池大小
      max-wait: 60000                                            # 最大等待时间
      time-between-eviction-runs-millis: 60000                   # 关闭空闲连接间隔
      min-evictable-idle-time-millis: 30000                      # 连接最小生存时间
      validation-query: SELECT 1 FROM dual                       # 状态检测
      test-while-idle: true                                      # 空闲时检测连接是否有效
      test-on-borrow: false                                      # 申请时检测连接是否有效
      test-on-return: false                                      # 归还时检测连接是否有效
      pool-prepared-statements: false                            # PSCache缓存
      max-pool-prepared-statement-per-connection-size: 20        # 配置PSCache缓存
      filters: stat, wall, slf4j                                 # 开启过滤
      stat-view-servlet:                                         # 监控界面配置
        enabled: true                                            # 启用Druid监控界面
        allow: 127.0.0.1                                         # 访问白名单
        login-username: muyan                                    # 用户名
        login-password: yootk                                    # 密码
        reset-enable: true                                       # 允许重置
        url-pattern: /druid/*                                    # 访问路径
```

```
    web-stat-filter:
      enabled: true                                            # 启动URI监控
      url-pattern: /*                                          # 跟踪全部服务
      exclusions: "*.js,*.gif,*.jpg,*.bmp,*.png,*.css,*.ico,/druid/*" # 跟踪排除
    filter:
      slf4j:                                                   # 日志
        enabled: true                                          # 启用SLF4j监控
        data-source-log-enabled: true                          # 启用数据库日志
        statement-executable-sql-log-enable: true              # 执行日志
        result-set-log-enabled: true                           # ResultSet日志启用
      stat:                                                    # SQL监控
        merge-sql: true                                        # 合并统计
        log-slow-sql: true                                     # 慢执行记录
        slow-sql-millis: 1                                     # 慢SQL执行时间
      wall:                                                    # SQL防火墙
        enabled: true                                          # SQL防火墙
        config:                                                # 防火墙规则
          multi-statement-allow: true                          # 允许执行批量SQL
          delete-allow: false                                  # 禁止执行删除语句
    aop-patterns: "com.yootk.provider.action.*,com.yootk.provider.service.*,
               com.yootk.provider.dao.*"                       # Spring监控
```

（3）【provider-dept-8001 子模块】在 src/main/resources 资源目录中创建 logback-spring.xml 配置文件，随后配置 MyBatis 数据层日志输出包名称。

```
<logger name="com.yootk.provider.dao" level="DEBUG"/>
```

（4）【provider-dept-8001 子模块】创建 MyBatisPlus 配置类。

```
package com.yootk.provider.config;
@Configuration
public class MyBatisPlusConfig {                                        // MyBatisPlus配置类
    @Bean
    public MybatisPlusInterceptor getMybatisPlusInterceptor() {
        MybatisPlusInterceptor interceptor = new MybatisPlusInterceptor();  // 拦截器
        interceptor.addInnerInterceptor(
                new PaginationInnerInterceptor(DbType.MYSQL));           // 分页处理
        return interceptor;
    }
}
```

（5）【provider-dept-8001 子模块】创建 Dept 数据表映射类。

```
package com.yootk.provider.vo;
@TableName("dept")                                            // 映射数据表
@Data                                                         // Lombok结构生成注解
public class Dept {
    @TableId(type = IdType.AUTO)                              // 设置ID映射
    private Long deptno;                                      // 映射deptno字段
    private String dname;                                     // 映射dname字段
    private String loc;                                       // 映射loc字段
}
```

（6）【provider-dept-8001 子模块】创建 IDeptDAO 数据层接口。

```
package com.yootk.provider.dao;
@Mapper                                                       // MyBatis映射
public interface IDeptDAO extends BaseMapper<Dept> {}         // 数据层接口
```

（7）【provider-dept-8001 子模块】创建 DeptServiceImpl 业务实现子类，在该实现类中所有的数据层操作通过 Dept 映射类来完成。而在业务方法进行数据接收和数据返回时考虑到程序安全性要以 DeptDTO 类的实例为主，所以就需要通过 DeepBeanUtils 工具类进行实例属性复制处理。

```
package com.yootk.provider.service.impl;
@Service
public class DeptServiceImpl implements IDeptService {    // 业务实现类
    @Autowired
    private IDeptDAO deptDAO;                              // 注入IDeptDAO接口实例
    @Override
```

```
    public DeptDTO get(long id) {
        DeptDTO deptDTO = new DeptDTO();                                    // 获取DTO对象
        BeanUtils.copyProperties(this.deptDAO.selectById(id),deptDTO);      // 对象复制
        return deptDTO;                                                     // 返回DTO实例
    }
    @Override
    public boolean add(DeptDTO dept) {
        Dept deptVO = new Dept();                                           // VO对象
        BeanUtils.copyProperties(dept, deptVO);                             // 对象复制
        return this.deptDAO.insert(deptVO) > 0;                             // 数据增加
    }
    @Override
    public List<DeptDTO> list() {
        QueryWrapper<Dept> wrapper = new QueryWrapper<>();                  // 查询包装器
        List<DeptDTO> allDepts = DeepBeanUtils.copyListProperties(
          this.deptDAO.selectList(wrapper), DeptDTO::new);                  // 集合数据复制
        return allDepts;
    }
    @Override
    public Map<String, Object> split(int currentPage, int lineSize,
                String column, String keyword) {
        QueryWrapper<Dept> wrapper = new QueryWrapper<>();                  // 查询包装
        wrapper.like(column, keyword);                                      // 模糊查询
        int count = this.deptDAO.selectCount(wrapper);                      // 统计数据行数
        IPage<Dept> page = this.deptDAO.selectPage(new Page<>(currentPage, lineSize,
                count), wrapper);                                           // 分页配置
        Map<String, Object> map = new HashMap<>();                          // 定义Map集合
        map.put("allDepts", DeepBeanUtils.copyListProperties(
                page.getRecords(), DeptDTO::new));                          // 数据复制
        map.put("allRecorders", page.getTotal());                           // 获取总记录数
        map.put("allPages", page.getPages());                               // 获取总页数
        return map;
    }
}
```

（8）【provider-dept-8001 子模块】创建程序启动类。

```
package com.yootk.provider;
@SpringBootApplication
public class StartProviderDept8001Application {
    public static void main(String[] args) {
        SpringApplication.run(StartProviderDept8001Application.class, args); // 服务启动
    }
}
```

（9）【provider-dept-8001 子模块】业务层编写完成后，需要先进行业务功能测试再发布。本次将通过 spring-test 并结合 JUnit 5 编写测试用例。

```
package com.yootk.test;
@ExtendWith(SpringExtension.class)                                  // 使用JUnit 5测试工具
@WebAppConfiguration                                                // 启动Web运行环境
@SpringBootTest(classes = StartProviderDept8001Application.class)   // 配置程序启动类
public class TestDeptService {                                      // 业务测试类
    @Autowired
    private IDeptService deptService;                               // 注入业务层实例
    @Test
    public void testGet() {                                         // 查询测试
        DeptDTO deptDTO = this.deptService.get(1);                  // 调用业务方法
        Assertions.assertNotNull(deptDTO);                          // 结果测试
    }
    @Test
    public void testAdd() {                                         // 增加测试
        DeptDTO deptDTO = new DeptDTO();                            // 实例化DTO对象
        deptDTO.setDname("数据部");                                  // 属性设置
        deptDTO.setLoc("北京");                                      // 属性设置
        Assertions.assertTrue(this.deptService.add(deptDTO));       // 调用业务方法
```

```
    }
    @Test
    public void testList() {                                    // 查询测试
        List<DeptDTO> deptList = this.deptService.list();  // 调用业务方法
        System.out.println(deptList);
        Assertions.assertTrue(deptList.size() > 0);        // 结果测试
    }
    @Test
    public void testSplit() {                                   // 查询测试
        Map<String, Object> result = this.deptService.split(
                1, 2, "dname", "部");                           // 调用业务方法
        System.out.println(result);
    }
}
```

（10）【provider-dept-8001 子模块】创建控制器发布 RESTful 业务接口。

```
package com.yootk.provider.action;
@RestController                                                 // REST控制器
@RequestMapping("/provider/dept/*")                             // 父路径
public class DeptAction {
    @Autowired
    private IDeptService deptService;                           // 注入业务接口实例
    @GetMapping("get/{id}")                                     // 子路径
    public Object get(@PathVariable("id") long id) {
        return this.deptService.get(id);                        // 部门信息加载
    }
    @PostMapping("add")                                         // 子路径
    public Object add(DeptDTO deptDTO) {                        // 对象转换
        return this.deptService.add(deptDTO);                   // 部门信息添加
    }
    @GetMapping("list")                                         // 子路径
    public Object list() {
        return this.deptService.list();                         // 部门信息列表
    }
    @GetMapping("split")                                        // 子路径
    public Object split(int cp, int ls, String col, String kw) {
        return this.deptService.split(cp, ls, col, kw);   // 部门信息分页
    }
}
```

Action 类定义完成并启动 Spring Boot 应用后，当前的部门微服务会自动在 8001 端口提供服务，此微服务所提供的 RESTful 接口路径以及描述如表 2-2 所示。

表 2-2　部门微服务接口

序号	请求模式	RESTful 接口	描述
01	GET	http://provider-dept-8001:8001/provider/dept/list	部门列表
02	GET	http://provider-dept-8001:8001/provider/dept/get/{部门 ID}	部门查询
03	GET	http://provider-dept-8001:8001/provider/dept/split? cp={当前页}&ls={每页显示记录行数}&col={查询列名称}&kw={查询关键字}	部门分页
04	POST	http://provider-dept-8001:8001/provider/dept/add?dname={部门名称}&loc={部门位置}	部门增加

2.1.4　Postman 接口测试

Postman 接口测试

视频名称　0205_【掌握】Postman 接口测试

视频简介　微服务创建完成后，为了保证其对应的 RESTful 接口可以提供正确的服务，往往会进行接口的访问测试以及压力测试，而这些可以通过 Postman 工具来实现。本视频为读者讲解 Postman 的下载、安装以及基本使用方法。

部门微服务开发完成后，在正式交付使用前必须要进行严格的接口访问测试以及压力测试。在实际的开发过程中，较为常用的工具是 Postman。

Postman 是一款强大的页面调试工具，可以直接实现 Web API 与 HTTP 请求调试功能，同时也可以模拟任意 HTTP 请求模式，实现请求参数以及头信息的发送。Postman 为开源工具，如果需要使用该工具，可在 Postman 官网进行下载，如图 2-12 所示。

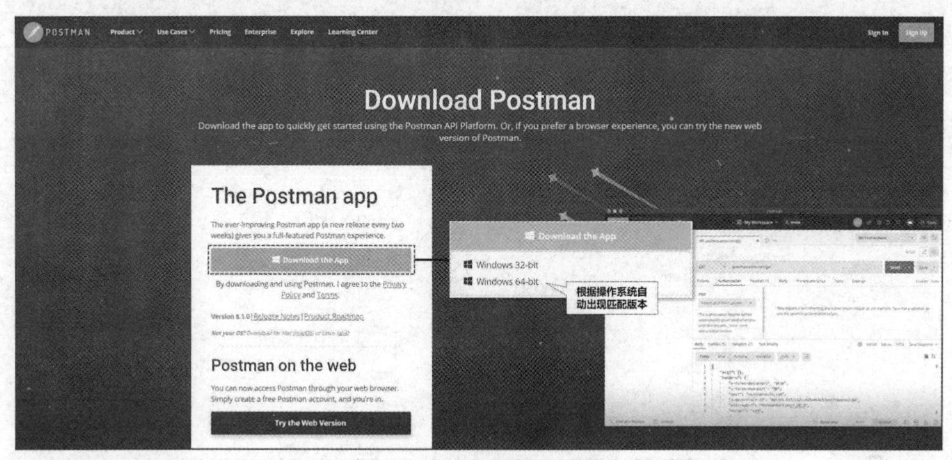

图 2-12 下载 Postman 工具

Postman 官网会自动识别当前用户所使用的操作系统并给出匹配的下载链接，由于当前笔者所使用的操作系统为 Windows，因此会自动出现 Windows 的相关链接。下载后可以自动进行简化安装，并且在安装完成后会自动启动 Postman 工具软件，一般会提示用户需要进行“注册与登录”，如图 2-13 所示。

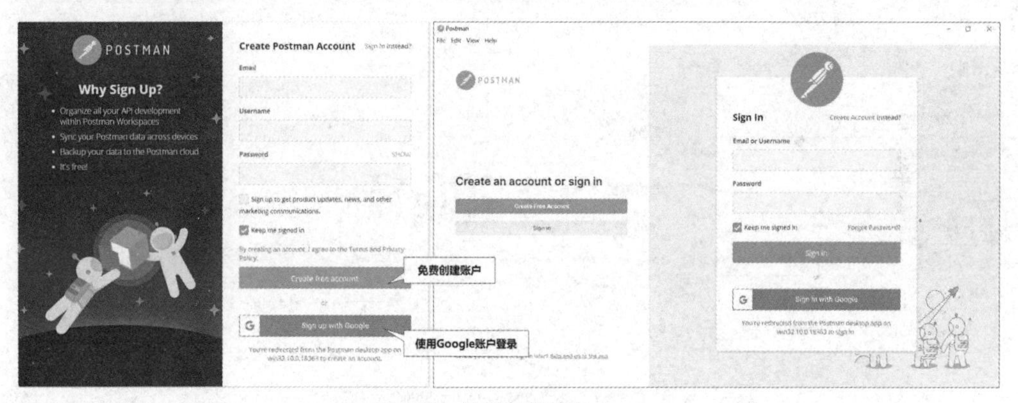

图 2-13 注册与登录 Postman

提示：创建属于自己的 Postman 账户。

Postman 的所有测试配置都是基于云方式进行管理的，即同一个账户下的所有测试环境都会自动在云端存储，在不同的客户端上只要登录账户就会自动进行配置同步，如图 2-14 所示。

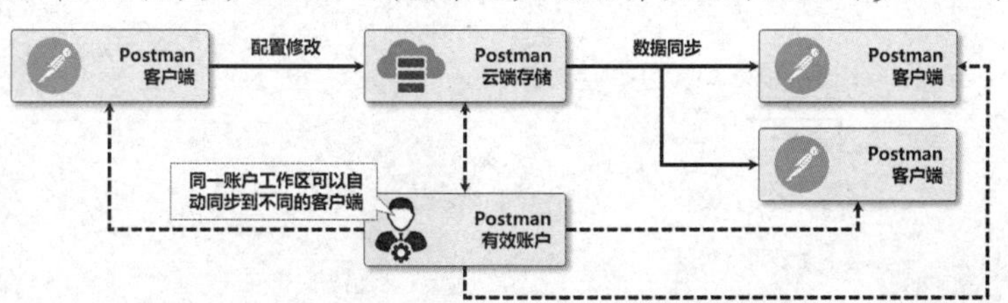

图 2-14 Postman 账户管理

Postman 中的所有配置数据都是基于 JSON 结构管理的，所以非常便于云端存储与数据同步。

Postman 启动之后的主界面如图 2-15 所示，为了便于不同的测试环境管理，使用者可以根据需要创建若干个工作区（Workspace）。此处将创建一个"Spring Cloud 就业编程实战"工作区，实现本书中所有相关测试环境的管理。

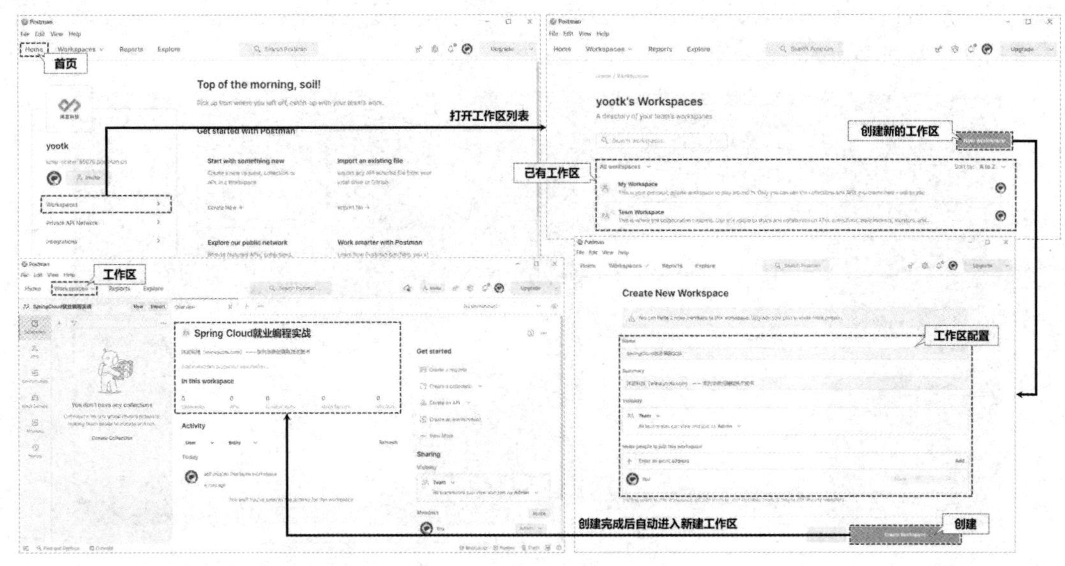

图 2-15　创建 Postman 工作区

在不同的项目之中，会存在不同的接口分类，这些分类在 Postman 中可以称为集合（Collection）。例如，本次要进行测试的是部门微服务，所以创建一个新的集合，名称为"【服务提供者】部门微服务"，如图 2-16 所示。

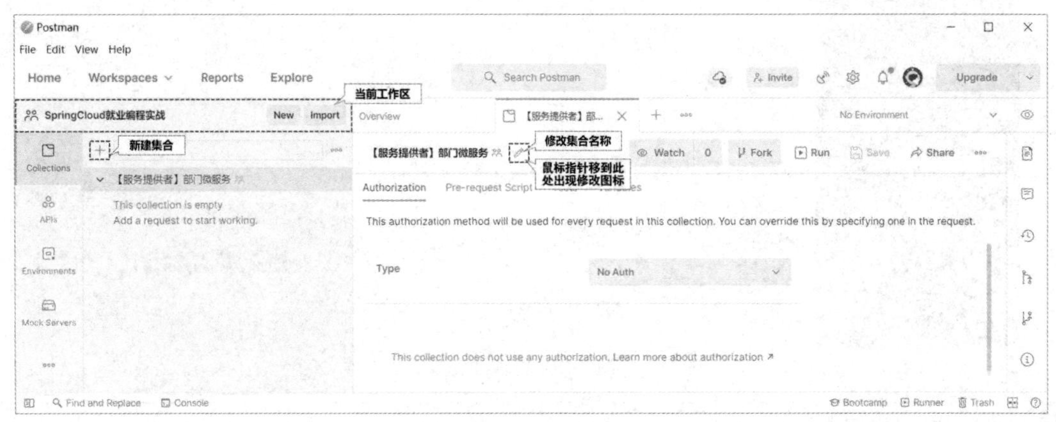

图 2-16　创建集合

在一个集合之中可以保存若干个要发送的请求，单击"New"就可以启动新建界面，建立新的用户请求，如图 2-17 所示。

图 2-17　新建请求

在进行请求创建时，需要输入请求的名称，同时将其保存在相关的集合之中。在请求创建完成之后才可以配置请求路径、请求模式、请求参数等，如图 2-18 所示。配置完成后单击“Send”即可发送请求，请求的响应结果会在相应的位置显示。

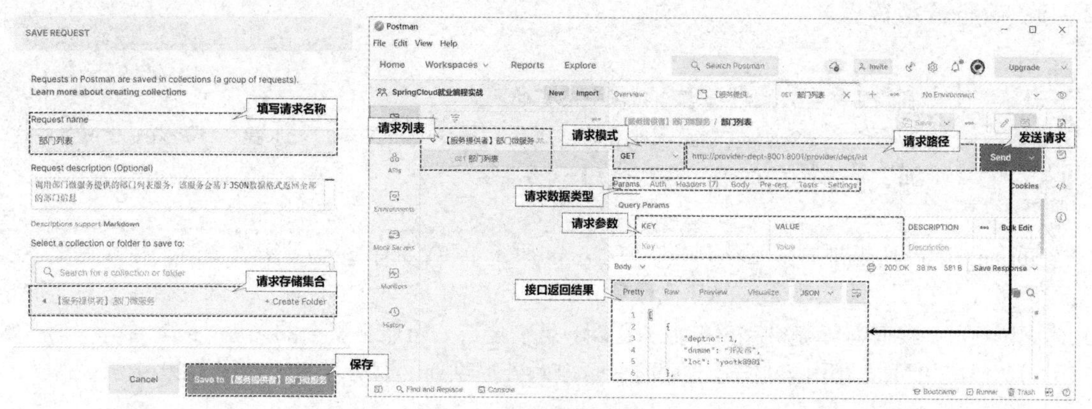

图 2-18 请求设置与发送

随后按照同样的方式分别创建其余部门接口的测试请求，最后得到图 2-19 所示的界面。

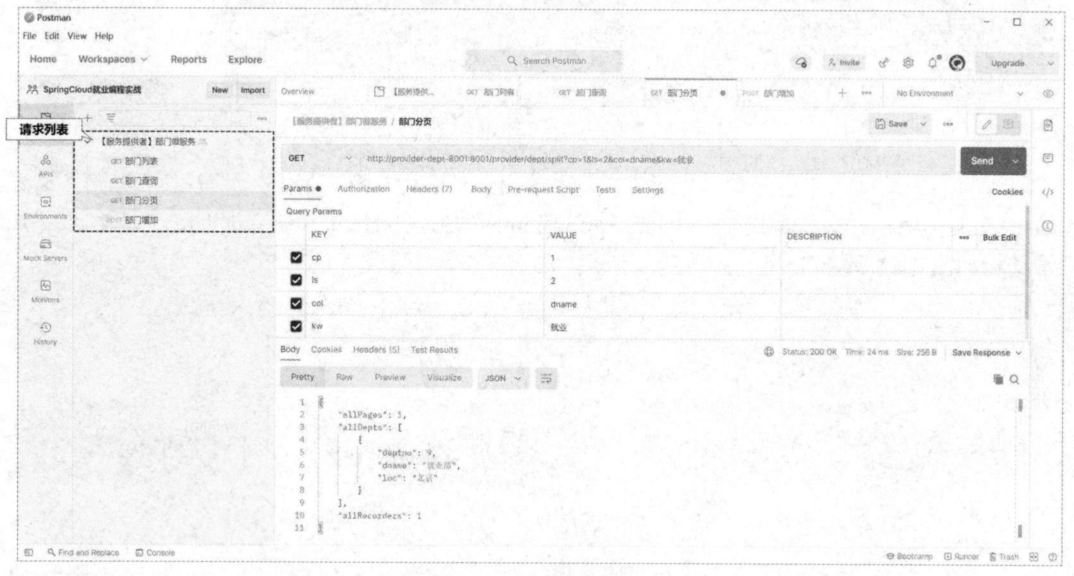

图 2-19 请求列表

提示：curl 命令测试。

curl 命令是一个利用 URL 语法在命令行下工作的文件传输工具，可以方便地实现文件的上传和下载，当然也可以直接在 Shell 脚本程序中使用 curl 命令进行接口服务调用。对于此处的部门微服务，开发者可以采用如下命令进行调用测试。

范例：测试部门微服务

部门列表	`curl http://provider-dept-8001:8001/provider/dept/list`
部门查询	`curl http://provider-dept-8001:8001/provider/dept/get/1`
部门分页	`curl -X GET "http://provider-dept-8001:8001/provider/dept/split?cp=1&ls=2&col=dname&kw=muyan"`
部门增加	`curl -X POST -d "dname=muyan&loc=yootk" "http://provider-dept-8001:8001/provider/dept/add"`

虽然 curl 命令可以方便地使用，但是仅适合于小型的简单测试，如果进行大规模的接口测试，则测试环境的管理会较为麻烦。这也是本书使用 Postman 工具进行测试的主要原因。

2.1.5　微服务消费者

视频名称　0206_【掌握】微服务消费者

视频简介　Spring Cloud 微服务需要通过 Spring Boot 来调用。本视频将介绍创建消费端应用，并通过 RestTemplate 类所提供的方法实现远程服务调用。

Spring Cloud 所提供的微服务为一个个独立的原子单位，而在进行业务处理时，往往需要通过若干个微服务的调用来实现完整的处理业务。同时，考虑到安全性的问题，大部分微服务只允许定义为内网访问。要想对外提供完整的业务逻辑，就需要有一个服务整合的应用（或者称为“消费端应用”），该应用可以调用内网服务，同时又可以对外提供新的服务接口，如图 2-20 所示。

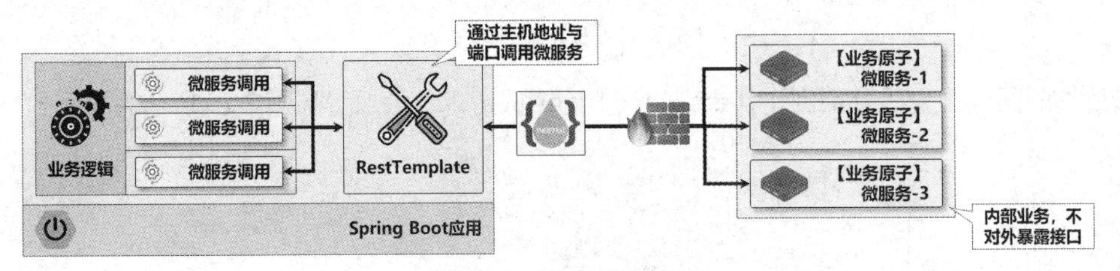

图 2-20　REST 消费端

服务的消费端可以使用任意编程语言实现。在 Java 开发中为了简化，消费端可以通过 Spring Boot 框架来实现，该框架提供了一个 org.springframework.web.client.RestTemplate 工具类，该类的继承结构如图 2-21 所示。

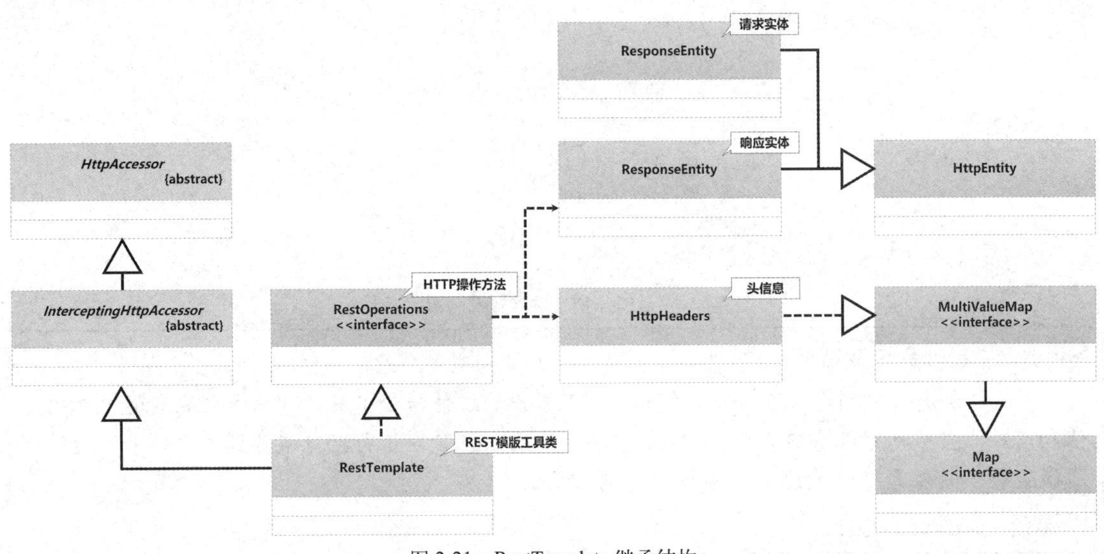

图 2-21　RestTemplate 继承结构

> **提示：HttpEntity 类的作用。**
>
> Java 是面向对象的语言，所以在进行远程接口调用时，最佳的做法就是利用对象包装接口所需要的数据，而返回的数据也可以自动转为所需要的对象，这样才便于程序的开发处理。
>
> 为了便于实现这样的对象转换处理操作，RestTemplate 类内部使用了 HttpEntity 类来表示

HTTP 请求和响应实体，该类有两个子类：RequestEntity（请求实体）、ResponseEntity（响应实体）。在请求时通过 RequestEntity 将对象转为服务端接口所需要的参数结构；而在响应时，则通过 ResponseEntity 将返回的 JSON 数据转为对象实例，如图 2-22 所示。

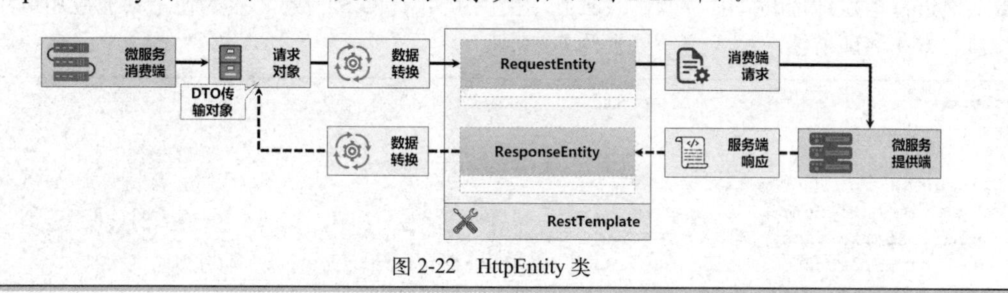

图 2-22　HttpEntity 类

在使用 RestTemplate 类的实例调用服务时，可以发送不同的 HTTP 请求（GET、POST、HEAD 等），这些请求在发送时需要传递参数，也需要将返回的结果转换为对象实例，常用方法如表 2-3 所示。

表 2-3　RestTemplate 类的常用方法

序号	方法	类型	描述
01	public <T> T getForObject(String url, Class<T> responseType, Map<String, ?> uriVariables) throws RestClientException;	普通	发送 GET 请求，设置请求参数以及响应数据类型
02	public <T> T getForObject(URI url, Class<T> responseType) throws RestClientException;	普通	发送 GET 请求，设置响应数据类型
03	public HttpHeaders headForHeaders(String url, Map<String, ?> uriVariables) throws RestClientException;	普通	发送 HEAD 请求，设置头数据
04	public <T> T postForObject(String url, @Nullable Object request, Class<T> responseType, Map<String, ?> uriVariables) throws RestClientException	普通	发送 POST 请求，设置请求参数以及响应数据类型
05	public <T> ResponseEntity<T> postForEntity(String url, @Nullable Object request, Class<T> responseType, Map<String, ?> uriVariables) throws RestClientException;	普通	发送 POST 请求，设置请求参数以及响应数据类型

表 2-3 列出的方法主要定义在 RestOperations 父接口之中，该接口内部针对不同的请求模式会提供相应的方法重载。为了便于读者理解这些方法，下面将通过具体的操作步骤介绍如何进行实现。

（1）【microcloud 项目】创建 consumer-springboot-80 消费端子模块，随后修改 build.gradle 配置文件添加相关依赖。

```
project('consumer-springboot-80') {                    // 子模块
    dependencies {                                      // 配置子模块依赖
        implementation(project(':common-api'))          // 引入子模块
    }
}
```

（2）【consumer-springboot-80 子模块】创建 src/main/resources/application.yml 配置文件，定义服务端口。

```
server:                                                 # 服务端配置
  port: 80                                              # 微服务端口
```

（3）【consumer -springboot-80 子模块】如果要想访问此时服务端所提供的 REST 接口，则必须获取 RestTemplate 类的对象实例，这样可以创建一个配置类并进行所需 Bean 的注册。

```
package com.yootk.consumer.config;
@Configuration                                          // 配置类
public class RestTemplateConfig {
    @Bean                                               // Bean注册
    public RestTemplate getRestTemplate() {             // 获取RestTemplate类的实例
        return new RestTemplate() ;
```

```
    }
}
```

(4)【consumer-springboot-80 子模块】创建 Action 控制器类，随后通过注入 RestTemplate 类的对象实例进行指定地址的微服务调用（服务端提供了 4 个 REST 接口）。为便于其他应用调用消费端接口，本次将所有请求模式统一设置为 GET。

```
package com.yootk.consumer.action;
@RestController
@RequestMapping("/consumer/dept/*")                                      // 消费端父路径
public class DeptConsumerAction {
    // 将需要调用的服务端地址进行统一定义
    public static final String DEPT_ADD_URL =
        "http://provider-dept-8001:8001/provider/dept/add";         // 部门增加REST地址
    public static final String DEPT_GET_URL =
        "http://provider-dept-8001:8001/provider/dept/get/";        // 部门查询REST地址
    public static final String DEPT_LIST_URL =
        "http://provider-dept-8001:8001/provider/dept/list";        // 部门列表REST地址
    public static final String DEPT_SPLIT_URL =
        "http://provider-dept-8001:8001/provider/dept/split";       // 部门分页REST地址
    @Autowired
    private RestTemplate restTemplate;                                  // 注入对象实例
    @GetMapping("add")
    public Object addDept(DeptDTO dept) {
        // 部门增加，发送POST请求，新增的部门数据以DTO对象实例进行传输，返回数据类型为Boolean
        return this.restTemplate.postForObject(DEPT_ADD_URL, dept, Boolean.class);
    }
    @GetMapping("get")
    public Object getDept(Long deptno) {
        // 根据编号查询部门数据，发送GET请求，利用地址拼凑的形式传递部门编号，返回类型为DTO对象实例
        return this.restTemplate.getForObject(DEPT_GET_URL + deptno, DeptDTO.class);
    }
    @GetMapping("list")
    public Object listDept() {
        // 部门数据列表，发送GET请求，返回类型为List集合
        return this.restTemplate.getForObject(DEPT_LIST_URL, List.class);
    }
    @GetMapping("split")
    public Object splitDept(int cp, int ls, String col, String kw) {
        // 部门分页查询，发送GET请求，返回类型为Map集合
        return this.restTemplate.getForObject(DEPT_SPLIT_URL + "?cp=" + cp + "&ls=" +
                ls + "&col=" + col + "&kw=" + kw, Map.class);
    }
}
```

部门列表地址	consumer-springboot-80/consumer/dept/list
部门增加地址	consumer-springboot-80/consumer/dept/add?dname={新部门名称}&loc={新部门位置}
部门查询地址	consumer-springboot-80/consumer/dept/get?deptno={查询部门编号}
部门分页地址	consumer-springboot-80/consumer/dept/split?cp={当前页}&ls={行数}&col={列名称}&kw={关键字}

(5)【consumer-springboot-80 子模块】创建消费端程序启动类。

```
package com.yootk.consumer;
@SpringBootApplication
public class StartConsumerApplication {
    public static void main(String[] args) {
        SpringApplication.run(StringConsumerApplication.class, args);     // 服务启动
    }
}
```

(6)【Postman】启动微服务，随后通过 Postman 工具创建消费端的请求测试，如图 2-23 所示。

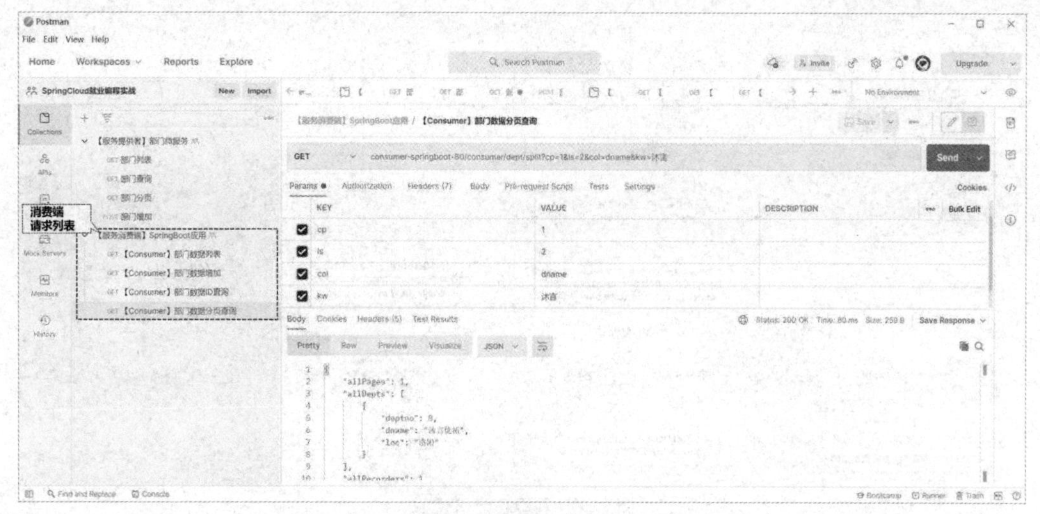

图 2-23 消费端请求列表

注意：服务端对象接收。

此处的消费端在进行部门增加时，需要传递 DTO 对象实例，而为了保证服务端可以正确地接收到此数据，需要修改相应服务端接口定义。

范例：修改部门服务提供端接口参数

```
package com.yootk.provider.action;
@RestController                                          // REST控制器
@RequestMapping("/provider/dept/*")                      // 父路径
public class DeptAction {
    @PostMapping("add")
    public Object add(@RequestBody DeptDTO deptDTO) {
        return this.deptService.add(deptDTO);            // 部门信息添加
    }
}
```

此处在 DeptDTO 参数前使用了@RequestBody，这样就可以接收 RestTemplate 类所传递的对象实例。如果没有添加此注解，接收的属性内容为 null。

2.1.6 HTTP 请求拦截

HTTP 请求拦截

视频名称 0207_【掌握】HTTP 请求拦截

视频简介 在消费端使用 Ribbon 进行服务调用时，可以对服务的请求进行拦截处理。本视频为读者讲解 ClientHttpRequestInterceptor 拦截器接口的使用方法，并通过实例分析 Ribbon 拦截器的具体应用。

REST 是基于 HTTP 实现的服务应用，在客户端通过 RestTemplate 类实现微服务调用处理时，可以基于 HTTP 拦截器的形式对请求进行拦截，这样就可以根据需要设置请求的头信息或日志记录等，如图 2-24 所示。

Spring 开发框架提供了 ClientHttpRequestInterceptor 拦截器接口，开发者可基于此接口实现拦截器的配置，而后将此拦截器注入 RestTemplate 即可。下面通过具体的步骤介绍如何实现。

（1）【consumer-springboot-80 子模块】创建 HTTP 拦截器。

```
package com.yootk.consumer.interceptor;
@Slf4j
@Component
public class MicroServiceHTTPInterceptor
        implements ClientHttpRequestInterceptor {          // 客户端拦截器
```

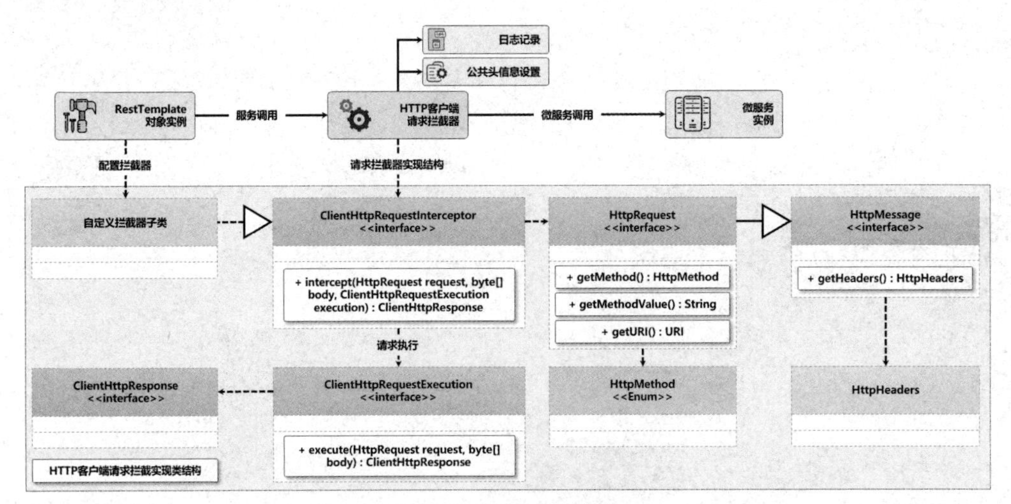

图 2-24 HTTP 请求拦截

```
    @Override
    public ClientHttpResponse intercept(HttpRequest request, byte[] body,
            ClientHttpRequestExecution execution) throws IOException { // 拦截处理
        log.info("【HTTP请求拦截】服务主机：{}, REST路径：{}", request.getURI().getHost(),
                request.getURI().getPath());                          // 日志输出
        // 可以利用此拦截器向服务器端传递一些认证头信息，如JWT数据
        request.getHeaders().set("token", "www.yootk.com");           // 设置头信息
        return execution.execute(request, body);                      // 请求发送
    }
}
```

（2）【consumer-springboot-80 子模块】修改 RestTemplateConfig 配置类，为 RestTemplate 实例添加拦截器。

```
package com.yootk.consumer.config;
@Configuration                                                  // 配置类
public class RestTemplateConfig {
    @Autowired                                                  // 注入拦截器
    private MicroServiceHTTPInterceptor interceptor;
    @Bean                                                       // Bean注册
    public RestTemplate getRestTemplate() {                     // 获取RestTemplate实例
        RestTemplate restTemplate = new RestTemplate();         // 实例化RestTemplate对象
        restTemplate.setInterceptors(Collections.singletonList(
                this.interceptor));                             // 添加拦截器
        return restTemplate;
    }
}
```

后台日志输出	【HTTP请求拦截】服务主机：provider-dept-8001, REST路径：/provider/dept/list

此时已在 RestTemplate 实例中配置了自定义拦截器，这样就可以在每次请求时获取相关的日志数据，同时也可以为用户的请求添加额外的头信息。

2.2 Swagger 接口文档工具

Swagger 接口描述

视频名称 0208_【理解】Swagger 接口描述

视频简介 为了便于使用者进行微服务接口的调用，必然需要有详细且完善的接口文档。为了便于文档的开发与维护，Spring 引入了 Swagger 项目。本视频为读者讲解接口文档的作用，以及 Swagger 与 Spring Boot 的快速整合应用。

项目业务的不断完善一定会带来更多的微服务节点，而为了便于微服务的调用，就需要对微服务提供详细且完善的使用说明。为了解决这类问题，在项目开发中可以通过 Swagger 项目来实现 REST 接口文档生成与维护操作，如图 2-25 所示。

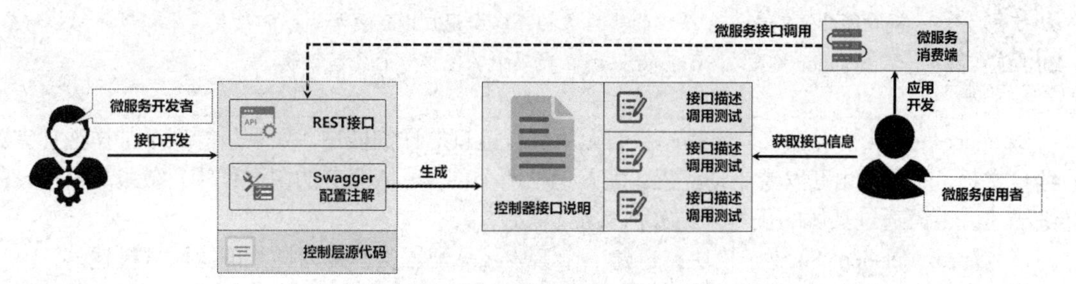

图 2-25　REST 接口文档

Swagger 是一个规范和完整的服务框架，在 Spring 框架中可以方便地使用此框架进行开发。在 Spring Boot 应用中只需要引入“springfox-boot-starter”依赖库即可直接使用此框架。下面来观察实现步骤。

> **提示：Swagger 官方 GitHub 地址。**
>
> Swagger 框架现在的最新版本是 3.x，该版本可以更加方便地与 Spring Boot 进行整合。同时在该版本对应的 GitHub 中已经明确给出了 Swagger 框架的快速整合应用。

（1）【microcloud 项目】为 provider-dept-8001 子模块添加 Swagger 3.x 依赖支持。

```
project('provider-dept-8001') {                    // 子模块
   dependencies {                                  // 重复依赖，略
      implementation('io.springfox:springfox-boot-starter:3.0.0')
   }
}
```

（2）【provider-dept-8001 子模块】引入 Swagger 依赖之后，程序就会自动查找当前项目中的所有接口，并采用默认的方式生成接口文档。开发者只需要进行微服务的启动，就可以通过/swagger-ui/路径对文档进行访问，如图 2-26 所示。

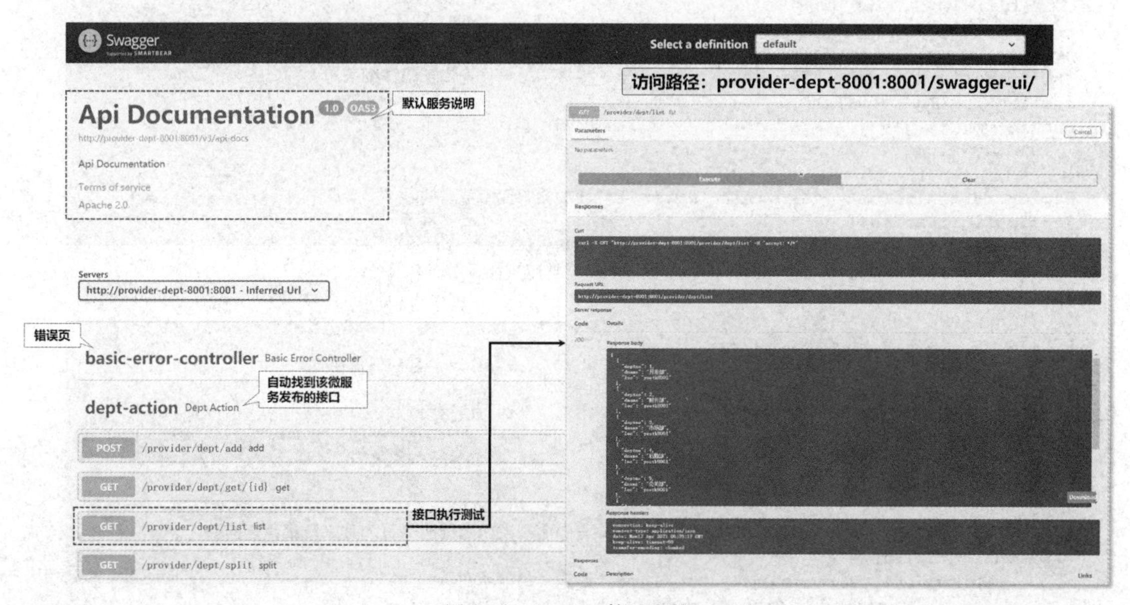

图 2-26　Swagger 接口文档

2.2.1 REST 接口描述

视频名称 0209_【理解】REST 接口描述

视频简介 Swagger 给出的接口文档可以由用户进行更加详细的设置。本视频介绍如何基于 Swagger 配置以 Swagger 注解在代码中实现接口的完整描述。

Swagger 在默认情况下仅列出接口，而无法进行接口的详细描述。如果想得到更加详细且完善的接口文档，就需要由开发者手动配置，配置包括两个部分：微服务的主要作用、微服务中接口的详细描述。下面通过具体的操作介绍如何实现。

（1）【provider-dept-8001 子模块】创建一个 Swagger 配置类并编写详细的微服务描述。

```
package com.yootk.provider.config;
@Configuration
public class SwaggerConfig {
   private ApiInfo getApiInfo() {                          // 获取接口描述信息
      return new ApiInfoBuilder().title("【沐言科技】部门微服务")
            .description("实现部门数据的统一管理，包括：增加、查询、列表显示")
            .termsOfServiceUrl("http://www.yootk.com")
            .contact(new Contact("李兴华", "edu.yootk.com", "2273627816@qq.com"))
            .license("沐言科技 - 授权管理").version("1.0").build();
   }
   @Bean
   public Docket getDocket() {                             // Swagger配置
      return new Docket(DocumentationType.SWAGGER_2)       // 文档类型
            .apiInfo(this.getApiInfo()).select()           // 接口描述
            .apis(RequestHandlerSelectors
            .basePackage("com.yootk.provider.action"))     // 扫描包
            .paths(PathSelectors.any()).build();
   }
}
```

（2）【provider-dept-8001 子模块】修改 DeptAction 控制器类，利用 Swagger 注解进行接口的详细说明。

```
package com.yootk.provider.action;
@RestController                                            // REST控制器
@RequestMapping("/provider/dept/*")                        // 父路径
public class DeptAction {
   @Autowired
   private IDeptService deptService;                       // 注入业务接口实例
   @ApiOperation(value = "部门查询", notes = "根据部门编号查询部门详细信息")
   @GetMapping("get/{id}")                                 // 子路径
   public Object get(@PathVariable("id") long id) {
      return this.deptService.get(id);                     // 部门信息加载
   }
   @ApiOperation(value = "部门增加", notes = "增加新的部门信息")
   @ApiImplicitParams({
         @ApiImplicitParam(name = "deptDTO", required = true,
               dataType = "DeptDTO", value = "部门传输对象实例")
   })
   @PostMapping("add")
   public Object add(@RequestBody DeptDTO deptDTO) {
      return this.deptService.add(deptDTO);                // 部门信息添加
   }
   @ApiOperation(value = "部门列表", notes = "得到全部部门信息")
   @GetMapping("list")                                     // 子路径
   public Object list() {
      return this.deptService.list();                      // 部门信息列表
   }
   @ApiOperation(value = "部门分页显示", notes = "实现部门数据的分页加载以及模糊查询")
   @ApiImplicitParams({
         @ApiImplicitParam(name = "cp", value = "当前所在页",
               required = true, dataType = "int"),
         @ApiImplicitParam(name = "ls", value = "每页显示数据行数",
               required = true, dataType = "int"),
```

```
        @ApiImplicitParam(name = "col", value = "模糊查询列",
                required = true, dataType = "String"),
        @ApiImplicitParam(name = "kw", value = "模糊查询关键字",
                required = true, dataType = "String")
    })
    @GetMapping("split") // 子路径
    public Object split(int cp, int ls, String col, String kw) {
        return this.deptService.split(cp, ls, col, kw); // 部门信息分页
    }
}
```

（3）【浏览器】通过浏览器访问“/swagger-ui/”路径，可以得到图 2-27 所示的详细文档结构。

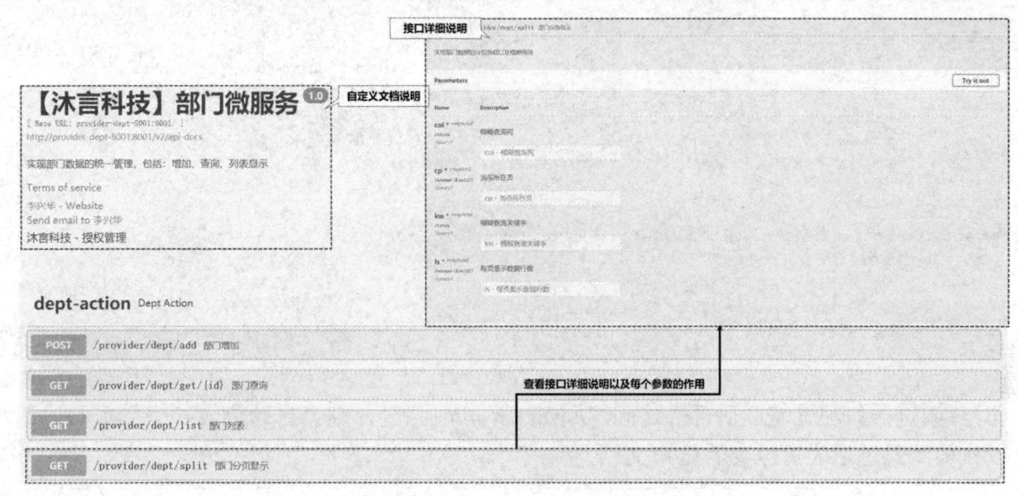

图 2-27　Swagger 文档结构

2.2.2　Swagger 安全配置

Swagger 安全配置

视频名称　0210_【理解】Swagger 安全配置

视频简介　Swagger 文档最终需要暴露给远程消费端，而为了保护文档的安全应该进行有效的安全认证管理。本视频介绍如何结合 Spring Security 开发框架采用静态认证信息的方式实现 Swagger 的安全保护。

Swagger 准确地描述了每一个微服务的接口信息，这些接口信息只应该为需要的用户服务。然而在默认情况下，由于没有任何安全认证措施，因此所有的开发者都可以浏览 Swagger 文档信息，这必然会造成安全隐患。所以，必须对 Swagger 文档进行保护，如图 2-28 所示。

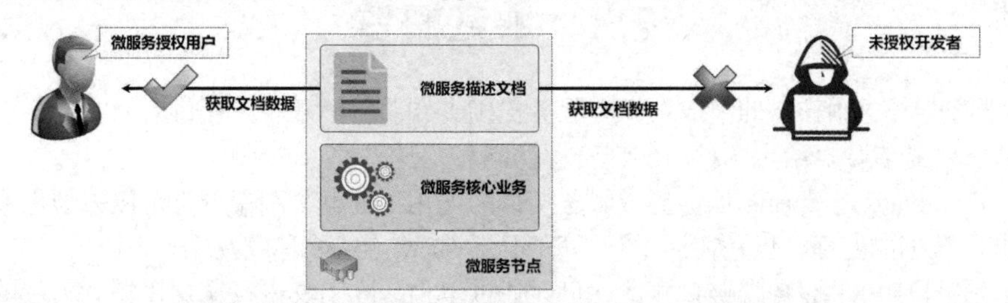

图 2-28　Swagger 文档保护

如果想保护 Swagger 文档资源，使其不被恶意访问，就需要进行有效的认证与授权管理。可以通过 Spring Security 进行安全管理，具体的实现步骤如下。

（1）【microcloud 项目】修改 build.gradle 配置文件，增加 spring-boot-starter-security 依赖配置。

```
project('provider-dept-8001') {                                    // 子模块
    dependencies {                                                  // 重复依赖，略
        implementation('org.springframework.boot:spring-boot-starter-security')
```

```
    }
}
```

（2）【provider-dept-8001】编写一个 Spring Security 安全配置类，添加一个固定账户 swagger/yootk。

```
package com.yootk.provider.config;
@Configuration
public class SwaggerWebSecurityConfig extends WebSecurityConfigurerAdapter {
    private static final String DEFAULT_PASSWORD =              // 密码明文：yootk
        "{bcrypt}$2a$10$FRy.2kILiVDf3MPTzQZcbeIvJQ.7Ic0QBJfDnx4kknsWyFH/QP/xi";
    @Bean
    public PasswordEncoder getPasswordEncoder() {               // 编码器
        return PasswordEncoderFactories.createDelegatingPasswordEncoder();
    }
    @Override
    protected void configure(AuthenticationManagerBuilder auth) throws Exception {
        auth.inMemoryAuthentication()                           // 固定认证信息
                .withUser("swagger")                            // 用户名
                .password(DEFAULT_PASSWORD)                     // 登录密码
                .roles("USER", "ADMIN");                        // 角色
    }
    @Override
    protected void configure(HttpSecurity http) throws Exception {
        http.authorizeRequests()                                // 授权请求
                .antMatchers("/swagger-ui/**", "/v2/api-docs").hasRole("ADMIN") // 检查路径
                .and().httpBasic().and().formLogin()            // 表单登录
                .permitAll().and().csrf().disable();            // 关闭CSRF
    }
}
```

（3）【浏览器】配置完成后再次访问“/swagger-ui/”路径，则会自动跳转到登录页面，用户在认证授权检测通过后才可以查看接口文档。

提问：部门微服务是否需要安全管理？

以上仅对 Swagger 组件的访问路径进行了安全控制，但是在当前给出的部门微服务中还存在大量的部门操作接口，这些接口难道不需要安全管理吗？

回答：后续将基于 JWT 处理。

微服务的安全管理并不是简单的资源接口的控制，而是需要结合一系列的微服务管理套件来实现的，所以相关技术要等待核心知识讲解完成后再引入。需要提醒读者的是，早期的 Spring Cloud 采用 OAuth2 协议实现安全管理，但是其实现存在结构复杂、性能低、不稳定的特点，所以后文在讲解微服务安全管理时将基于 JWT 进行介绍。

2.3　本章概览

1．微服务架构之中核心的组成单元有服务提供者和服务消费者。提供者负责提供资源以及资源访问接口，消费者依据给定的接口实现资源访问。

2．RestTemplate 是 Spring 提供的微服务接口的调用工具类，利用该类可以方便地实现请求的发送，以响应数据的处理，但是在调用时需要明确知晓节点的名称以及端口。

3．为了便于 REST 服务端接口测试，可以通过 Postman 工具进行调用模拟。后续的课程讲解中将会大量使用此工具进行接口的模拟调用。

4．Swagger 是一个在 REST 接口开发中被广泛使用的文档组件，通过该组件可以方便地实现接口文档的定义。

5．Spring Cloud 基于 Spring Boot 框架进行开发，所以可以方便地整合 Spring 所提供的各类组件。

6．微服务可以通过 Spring Security 进行保护，但是考虑到 Spring Security 的性能问题以及开发的复杂度问题，本书不推荐使用。后文会讲解如何基于 JWT 实现微服务认证保护。

第3章

Nacos 注册中心

本章学习目标

1. 掌握注册中心在微服务管理中的作用，并理解Nacos注册中心的工作原理；
2. 掌握Nacos程序的编译以及应用部署方法；
3. 掌握Nacos整合MySQL数据库实现配置持久化存储的方法；
4. 掌握Nacos程序开发技术，可以基于Nacos实现配置数据的程序控制；
5. 掌握微服务注册以及注册中心的安全管理方法。

利用微服务可以有效地实现大型业务开发，同时也必然会带来更多的服务节点。为了便于节点的统一管理与维护，就需要采用注册中心。本章将为读者讲解Nacos注册中心的配置以及使用。

3.1 Nacos 服务搭建

注册中心简介

视频名称 0301_【掌握】注册中心简介

视频简介 微服务的设计需要引入大量的服务节点，而为了便于这些节点的管理与维护，就需要采用注册中心。本视频为读者讲解微服务模型管理中的问题以及注册中心的作用。

基于REST实现的微服务架构，内部都是以HTTP实现数据交互处理的，因此消费端可以直接进行服务端的调用，而这样的调用过程需要明确地知道服务节点的地址和端口，如图3-1所示。在现实的生产环境中，由于网络环境的多变性以及服务的可维护性，这种消费端直接调用的模式会遇到严重的问题。一旦服务提供端节点地址发生改变，就需要及时地更新消费端，否则将无法获取到正确的服务数据。

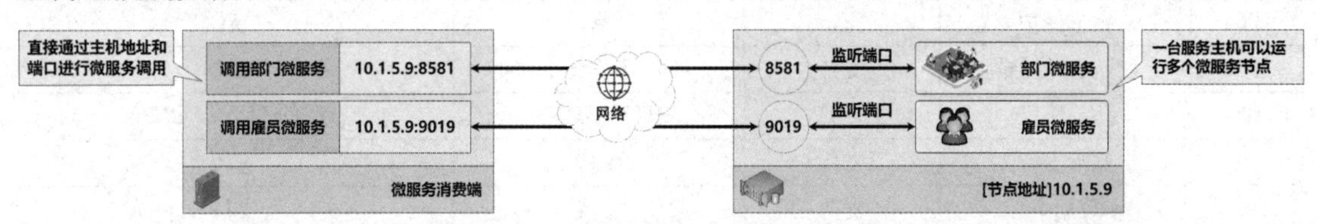

图3-1 微服务调用

为了更加合理地实现所有微服务节点的管理，就需要提供统一的注册中心，每一个微服务启动后都自动向注册中心保存各自的配置信息，而后消费端可以根据指定的微服务的名称通过注册中心来查找其可用的节点信息，从而实现正确的调用，如图3-2所示。这样即便微服务的节点数据发生了改变（地址修改、节点注销、动态扩容），只要注册中心存在，就不会影响服务的正确调用。

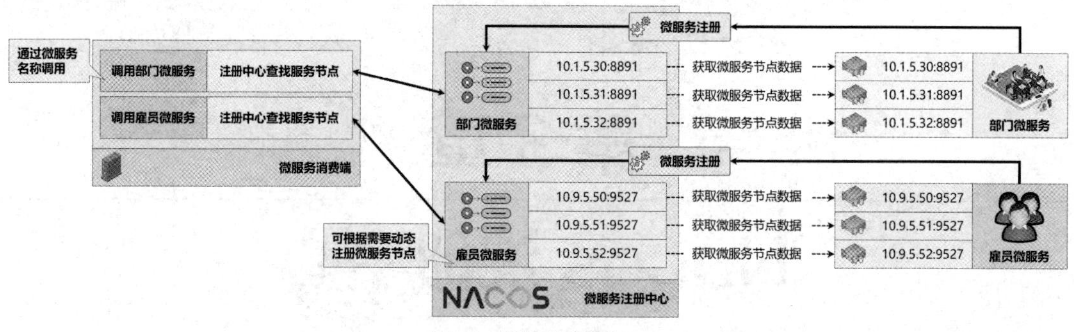

图 3-2　微服务注册中心

提示：CAP 原则。

计算机专家 Eric Brewer（布鲁尔）于 2000 年在 ACM 分布式计算原理专题讨论会（Principles Of Distributed Computing，PODC）中提出，分布式系统设计要考虑 3 个核心要素。

① **一致性（Consistency）**：同一时刻、同一请求的实例返回的结果相同。要求所有的数据具有强一致性（Strong Consistency）。

② **可用性（Availability）**：所有实例的读、写请求在一定时间内可以得到正确的响应。

③ **分区容错性（Partition Tolerance）**：在网络异常（挖断光缆、设备故障、死机等）的情况下，系统仍能提供正常的服务。

以上 3 个核心要素就是 CAP 原则（又称 CAP 定理），但是这 3 个核心要素不可能同时满足，所以分布式系统设计要考虑的是在满足 P（分区容错性）的前提下选择 C（一致性）还是 A（可用性），即 CP 或 AP。

注册中心的本质在于节点数据信息的存储，但是如果想开发出可用的注册中心，则必须遵从 CAP 原则，这样开发与维护的成本较高，所以在 Spring Cloud 开发中一般会存在 3 种注册中心：Eureka、Consul、Nacos。这 3 种注册中心的特点如表 3-1 所示。

表 3-1　Spring Cloud 注册中心

序号	比较项	Eureka	Consul	Nacos
01	开发组织	Netflix	Hashicorp	Alibaba
02	开发语言	Java	Go	Java
03	CAP 支持	AP	CP	CP + AP
04	健康检查	客户端心跳	TCP / HTTP / gRPC	TCP / HTTP / MySQL / 客户端心跳
05	负载均衡策略	Ribbon	Fabio	权重/元数据/选择器
06	雪崩保护	有	有	无
07	实例自动注销	支持	不支持	支持
08	访问协议	HTTP	HTTP / DNS	HTTP / DNS / gRPC
09	监听支持	支持	支持	支持
10	多数据中心	支持	支持	支持
11	跨注册中心同步	不支持	不支持	支持
12	Dubbo 集成	不支持	不支持	支持
13	Kubernates 集成	支持	不支持	支持

通过表 3-1 所列出的各种注册中心的特点可以发现，Nacos 注册中心的支持较为完善，也更加符合现代开发的技术特点。同时其又在阿里巴巴公司内部提供了多年的稳定服务，且依然处于更新的状态，因此是本书讲解的重点。

3.1.1 Nacos 技术架构

Nacos 技术架构

视频名称 0302_【掌握】Nacos 技术架构

视频简介 Nacos 是阿里巴巴公司推出的重要服务组件。本视频为读者讲解 Nacos 的服务架构整体组成，同时分析 Nacos 中实现 CP 机制的 Distro 和 Raft 算法的作用。

Nacos 是由阿里巴巴公司提供的一款注册与配置中心组件，除了可以方便地实现微服务的注册之外，还可以快速地实现动态服务发现、服务配置、服务元数据以及流量管理。它可以帮助用户在云服务时代更好地构建、交付、管理自己的微服务平台，更快地复用和组合业务服务，更快地实现商业创新的价值，从而为用户赢得市场。Nacos 的技术架构如图 3-3 所示。

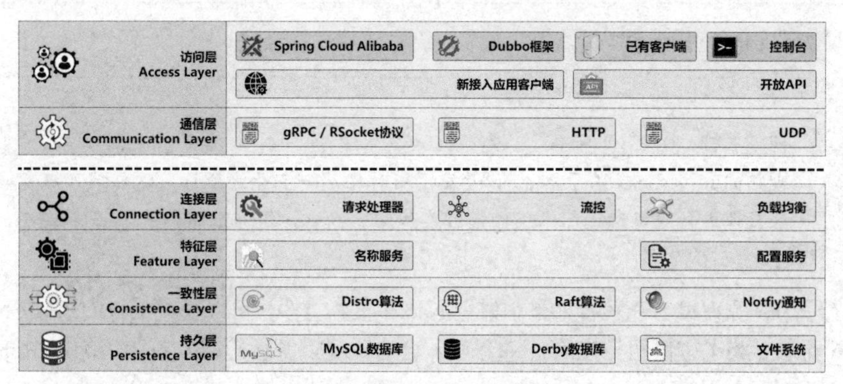

图 3-3 Nacos 技术架构

Nacos 可以很好地支持 Spring Cloud Alibaba 和 Dubbo 开发的服务管理功能，同时其他程序也可以使用 Nacos 提供的开放 API 实现服务数据处理操作。客户端在进行服务访问时，可以直接通过服务名称进行实例查找，所有的配置数据也可以被持久化存储管理。

Nacos 实现了 CAP 原则中的 CP 原则与 AP 原则，而在实现上主要采用了 Distro（阿里私有协议）和 Raft（分布式共识）两种算法，其中 Distro 算法提供了 AP 支持，而 Raft 算法提供了 CP 支持。Raft 算法采用类似于 Paxos 算法的机制，这样可以保证集群中会存在一个领导者以及多个跟随者，由领导者负责发出数据更新的指令，而后所有的跟随者可以实现数据的一致性处理，如图 3-4 所示。

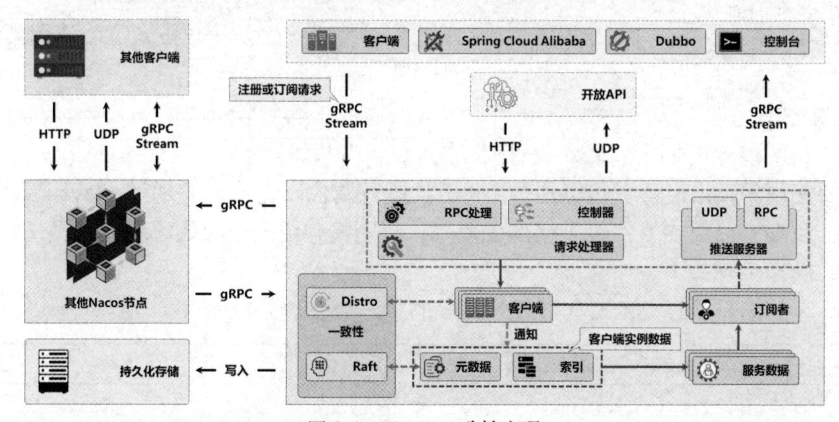

图 3-4 Nacos 一致性实现

提示：本书基于 Nacos 2.x 讲解。

截至本书编写时，Nacos 的最新版本为 2.0，虽然该版本现在可能存在一些问题，但是通过阿里巴巴公司给出的 Nacos 发展规划（见图 3-5）可以清楚地发现 Nacos 后续的发展潜力。

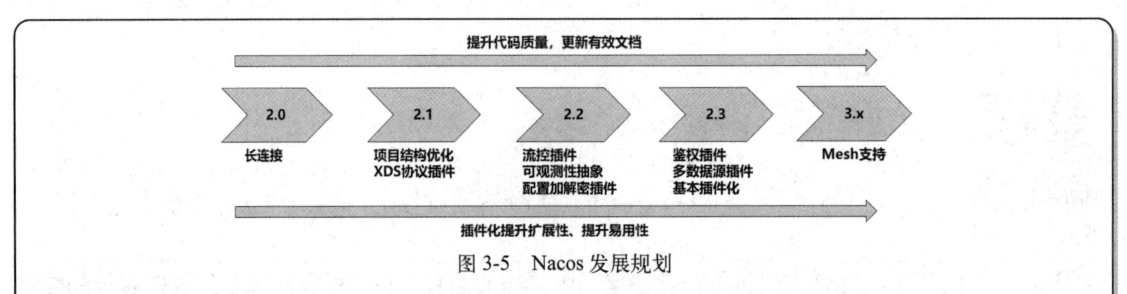

图 3-5　Nacos 发展规划

据官方测试结果，Nacos 2.x 的性能比 Nacos 1.x 提升约 10 倍，同时解决了各种长连接所带来的性能损耗问题，但是 2.x 毕竟是新版本，还存在不稳定的问题等。读者如果需要，也可以自行使用 Nacos 1.x，使用步骤与 Nacos 2.x 没有太大的差别。

3.1.2　获取 Nacos 应用组件

获取 Nacos 应用组件

视频名称　0303_【掌握】获取 Nacos 应用组件

视频简介　Nacos 基于 Maven 开发，同时又是一个开源项目，这样就可以直接依据 Nacos 源代码进行服务编译。本视频通过实例演示 Nacos 组件编译以及 Windows 下的组件使用。

Nacos 是一个开源项目，开发者如果想使用 Nacos 服务组件，可以通过 GitHub 获取 Nacos 的源代码，如图 3-6 所示。为了方便用户使用，Nacos 2.x 会自动提供源代码以及打包后的应用部署组件包。

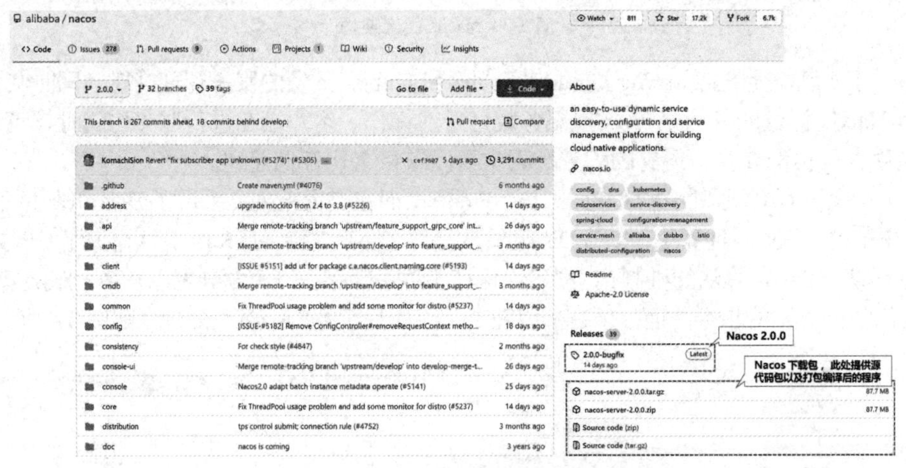

图 3-6　Nacos 代码托管

读者若要使用 Nacos 组件，可以通过 Git 工具进行源代码克隆，并通过 Maven 工具进行编译，也可以下载已经编译完成的组件包直接部署使用。此处将介绍通过源代码编译方式获取 Nacos 服务组件，具体的操作步骤如下。

（1）【Git 客户端】通过 Git 工具下载 Nacos 的指定分支（2.0.0）。

```
git clone -b 2.0.0 https://github.com/alibaba/nacos.git
```

程序执行结果

```
Cloning into 'nacos'...
remote: Enumerating objects: 990, done.
remote: Counting objects: 100% (990/990), done.
remote: Compressing objects: 100% (519/519), done.
remote: Total 76105 (delta 233), reused 880 (delta 191), pack-reused 75115
Receiving objects: 100% (76105/76105), 40.35 MiB | 5.53 MiB/s, done.
Resolving deltas: 100% (32836/32836), done.
```

（2）【Maven 工具】进入 Nacos 源代码所在目录，随后通过 Maven 工具进行项目打包编译。

```
mvn -Prelease-nacos -Dmaven.test.skip=true clean install -U
```

程序执行结果	BUILD SUCCESS

（3）【Nacos 源代码目录】通过 Maven 执行 Nacos 项目打包后，会在“${NACOS 源代码目录}/distribution\target”子目录中生成 nacos-server-2.0.0.zip 和 nacos-server-2.0.0.tar.gz 程序包。在 Windows 系统中可以直接将.zip 文件解压缩，此处将其解压缩到 h:\service 目录中。

（4）【命令行】通过命令行进入 nacos 应用所在目录，输入以下命令启动 Nacos 应用。应用启动后可以得到图 3-7 所示的 Nacos 启动信息。

```
startup.cmd -m standalone
```

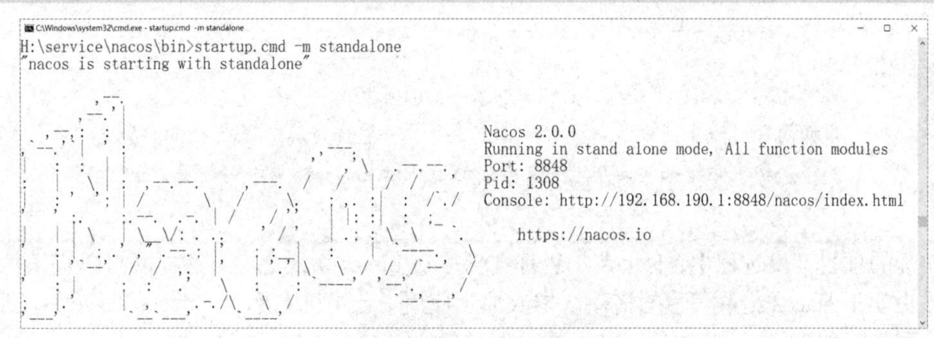

图 3-7 Nacos 启动信息

（5）【浏览器】Nacos 启动后会自动占用 8848 端口，开发者可以通过浏览器进行服务访问，如图 3-8 所示。

图 3-8 Nacos 控制台

提问：Nacos 为什么是内部系统？

在 Nacos 登录界面上会显示“内部系统，不可暴露到公网”，是什么意思？难道注册中心不能在互联网上使用吗？

回答：微服务应该工作在内网中。

微服务是通过项目业务处理设计出来的，属于项目的内部逻辑，这样的内部逻辑是应该暴露在公网上的。注册中心的出现是为了便于业务节点的管理，为了安全，注册中心只能够工作在内网中，如图 3-9 所示。

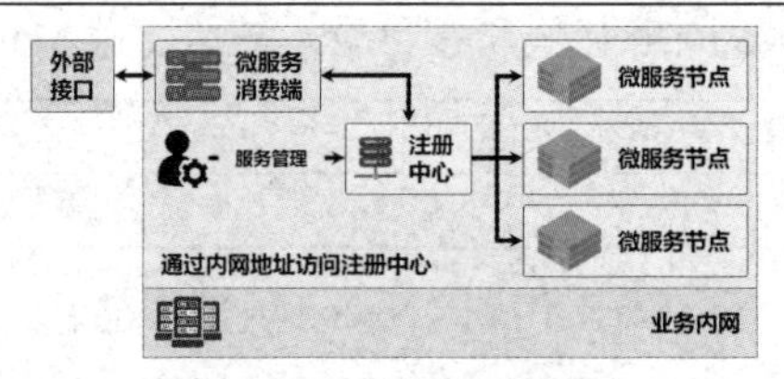

图 3-9 内网访问注册中心

内网管理员可以进行注册中心的维护，而所有的业务整合及服务接口暴露都应该由微服务消费端处理。

3.1.3 Linux 部署 Nacos 服务

Linux 部署 Nacos 服务

视频名称 0304_【掌握】Linux 部署 Nacos 服务

视频简介 为了保证注册中心的稳定运行，一般会将其部署在 Linux 系统之中。本视频将基于虚拟机介绍如何实现 Linux 系统的运行以及 Nacos 服务的部署。

在实际的项目开发过程中，Nacos 一般都会部署在 Linux 系统之中，这样既可以提供稳定的服务环境，也便于 Nacos 集群环境的搭建。本次演示使用一台主机名称为“nacos-server”的服务器，具体的实现步骤如下。

（1）【本地系统】修改 hosts 配置文件，添加 Linux 主机名称。

```
192.168.190.158nacos-server
```

（2）【本地系统】将 Nacos 打包形成的 nacos-server-2.0.0.tar.gz 程序包上传到 Linux 系统的/var/ftp 目录之中。

（3）【nacos-server 主机】将 Nacos 组件解压缩到/usr/local 目录之中。

```
tar xzvf /var/ftp/nacos-server-2.0.0.tar.gz -C /usr/local/
```

（4）【nacos-server 主机】启动 Nacos 服务。

```
bash -f /usr/local/nacos/bin/startup.sh -m standalone
```

命令执行后会出现 Nacos 启动的相关信息，同时会将所有的输出日志保存在/usr/local/nacos/logs/start.out 文件之中。运维人员可以通过此日志文件观察 Nacos 的启动信息。

> **提示：Nacos 脚本文件提示错误。**
>
> 在 Linux 系统中启动或关闭 Nacos 服务（startup.sh 与 shutdown.sh 脚本）时，有可能会出现文件格式不匹配造成服务无法启动的问题。读者可以按照如下方式确定并解决问题。
>
> ① 通过 vi 命令打开 startup.sh 脚本文件。
>
> ```
> vi /usr/local/nacos/bin/startup.sh
> ```
>
> ② 查看当前文件结构，可以发现当前的文件结构为“dos”。
>
> ```
> :set ff
> ```
>
> 程序执行结果
>
> ```
> fileformat=dos
> ```
>
> ③ 将文件结构修改为“unix”。
>
> ```
> :set ff=unix
> ```
>
> 修改完成后，再次查看当前文件结构，文件结构变为“unix”。退出 vim 编辑器后即可成功启动 Nacos 服务。

（5）【nacos-server 主机】Nacos 是基于 Java 开发的应用，可以直接查看当前主机中的 Java 进程端口占用信息。

```
netstat -nptl | grep java
```

程序执行结果	tcp6 0 0 :::7848 :::* LISTEN 1834/java tcp6 0 0 :::8848 :::* LISTEN 1834/java tcp6 0 0 :::9848 :::* LISTEN 1834/java tcp6 0 0 :::9849 :::* LISTEN 1834/java

通过此时的执行结果可以发现，Nacos 2.x 服务进程启动后会占用 4 个端口：7848、8848、9848、9849，这 4 个端口的描述如表 3-2 所示。

表 3-2 Nacos 2.x 服务端口

序号	端口	描述
01	7848	实现 Nacos 集群通信、一致性选举、心跳检测等功能
02	8848	Nacos 主端口，对外提供服务的 HTTP 端口
03	9848	客户端 gRPC 请求服务端端口，用于客户端向服务端发起连接和请求，该端口的配置为“主端口（默认为 8848）+ 1000 偏移量”
04	9849	服务端 gRPC 请求服务端端口，用于服务间同步等，该端口的配置为“主端口 + 1001 偏移量”

（6）【nacos-server 主机】修改防火墙规则。

配置访问端口	firewall-cmd --zone=public --add-port=8848/tcp --permanent firewall-cmd --zone=public --add-port=7848/tcp --permanent firewall-cmd --zone=public --add-port=9848/tcp --permanent firewall-cmd --zone=public --add-port=9849/tcp --permanent
配置重新加载	firewall-cmd --reload

（7）【浏览器】通过浏览器访问 Nacos 控制台，访问地址为 nacos-server:8848/nacos，随后输入默认账户信息即可访问。

提示：Nacos 1.x 只使用 8848 端口。

Nacos 1.x 基于 HTTP 实现服务管理，所以在注册中心启动后只会占用一个 8848 端口，这样开发者就可以通过一个端口实现 Nacos 管理、服务注册与服务发现，如图 3-10 所示。

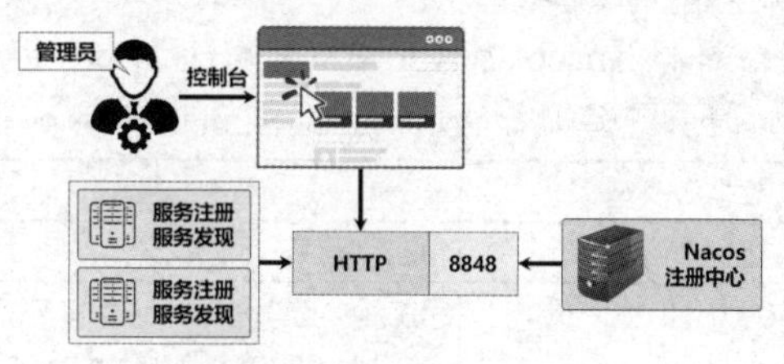

图 3-10 Nacos 1.x 服务端口

Nacos 2.x 为了提高服务性能，使用了 gRPC，这样管理端口依然为 8848（通过 HTTP 访问），但是微服务的注册与发现占用的是 9848 端口，如图 3-11 所示。

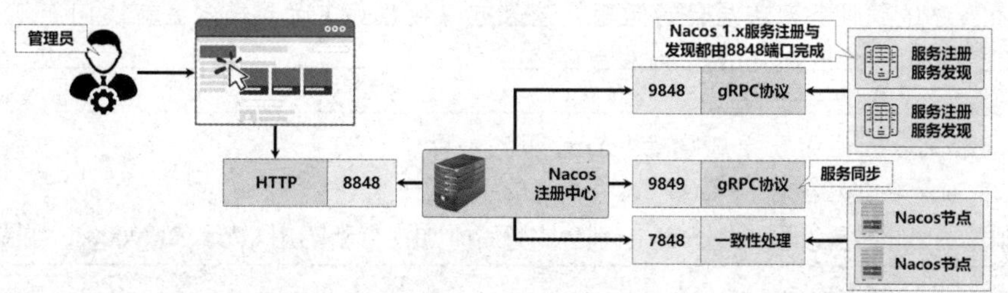

图 3-11 Nacos 2.x 服务端口

Nacos 2.x 还使用 7848 端口作为一致性处理端口，同时使用 9849 端口实现服务同步，这些操作在后文讲解 Nacos 集群时再进行分析。

（8）【nacos-server 主机】关闭 Nacos 服务。

```
bash -f /usr/local/nacos/bin/shutdown.sh
```

3.1.4　Nacos 整合 MySQL 存储

Nacos 整合 MySQL 存储

视频名称　0305_【掌握】Nacos 整合 MySQL 存储

视频简介　Nacos 依靠关系数据库实现配置数据的存储与同步。本视频为读者讲解 Nacos 与 MySQL 数据库存储的整合。

Nacos 在进行服务数据存储时会将之分为两类数据信息：一类是微服务的注册数据；另一类是相关的服务配置数据。而配置数据在开发中一般不会轻易改变，这样就可以将其存储在 MySQL 数据库之中，如图 3-12 所示。随后所有的微服务的注册数据都会基于配置的结构实现管理。

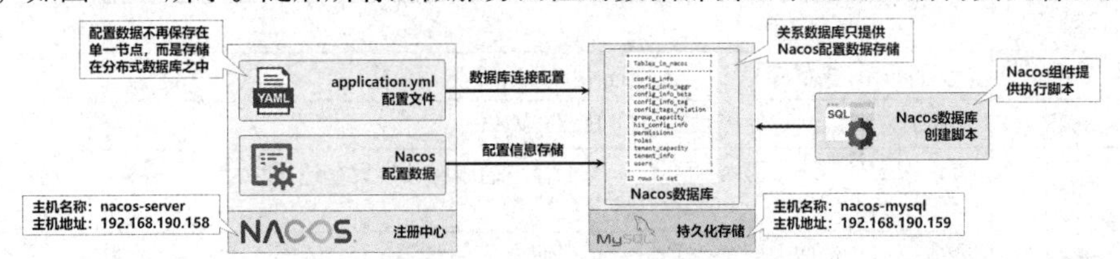

图 3-12　Nacos 整合 MySQL 存储

Nacos 安装目录中有 conf/nacos-mysql.sql 数据库脚本文件，该脚本文件提供了 Nacos 存储所需要的表结构，开发者必须手动创建数据库，而后修改 Nacos 的相关配置文件才可以实现存储。具体的配置步骤如下。

（1）【nacos-mysql 主机】启动 MySQL 数据库。

```
service mysqld start;
```

（2）【MySQL 客户端】启动 MySQL 客户端，创建 Nacos 数据库。

```
CREATE DATABASE nacos CHARACTER SET UTF8 ;
USE nacos;
```

（3）【MySQL 客户端】执行 Nacos 提供的数据库创建脚本。

```
source /usr/local/nacos/conf/nacos-mysql.sql
```

（4）【nacos-server 主机】修改 Nacos 组件中的配置文件。

```
vi /usr/local/nacos/conf/application.properties;
```

数据源类型	spring.datasource.platform=mysql
数据节点数量	db.num=1
数据库连接地址	db.url.0=jdbc:mysql://192.168.190.159:3306/nacos?characterEncoding=utf8&connectTimeout=1000&socketTimeout=3000&autoReconnect=true&useUnicode=true&useSSL=false&serverTimezone=UTC
数据库连接用户名	db.user.0=root
数据库连接密码	db.password.0=mysqladmin

（5）【nacos-server 主机】重新启动 Nacos 服务进程。

```
bash -f /usr/local/nacos/bin/startup.sh -m standalone
```

（6）【浏览器】此时重新登录 Nacos 控制台，随后添加一个新的用户 muyan/yootk，如图 3-13 所示。

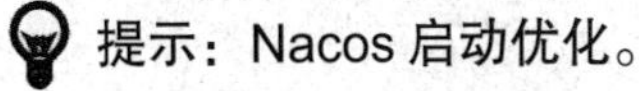

提示：Nacos 启动优化。

Nacos 基于 JVM 运行。如果需要进行 JVM 启动参数优化，可以修改“startup.sh”配置文件配置服务器的启动参数，已有的 JVM 内存配置如下。

```
JAVA_OPT="${JAVA_OPT} -server -Xms2g -Xmx2g -Xmn1g -XX:MetaspaceSize=128m -XX:
MaxMetaspaceSize=320m"
```

具体的参数设置可以根据实际应用服务器的配置进行调整，修改后重新启动 Nacos 服务即可生效。

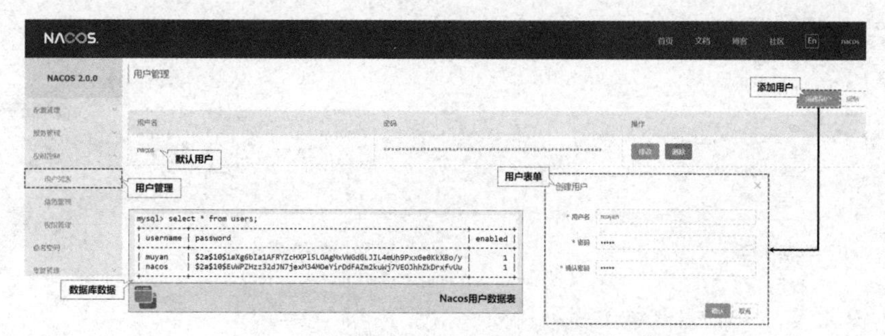

图 3-13 创建 Nacos 用户

3.2 Nacos 领域模型

Nacos 领域模型

视频名称 0306_【掌握】Nacos 领域模型

视频简介 Nacos 为了便于注册与配置数据的管理，提出了完整的领域模型。本视频将为读者讲解 Nacos 领域模型的各个组成部分，并分析其设计的目的。

由于项目应用环境的复杂度不同，同时考虑到各类数据的有效存储管理，Nacos 为数据的归类存储提供了不同的模型结构，分为命名空间（NameSpace）、分组（Group）、服务（Service）、集群（Cluster）、实例（Instance）和数据 ID（DataId），这些模型的作用如表 3-3 所示。

表 3-3 Nacos 领域模型

序号	模型	描述
01	命名空间（NameSpace）	对不同的应用环境进行隔离，如开发环境、测试环境、生产环境
02	分组（Group）	将若干个服务或配置集归为一组，推荐命名格式为“产品名_分组名称”
03	服务（Service）	指具体的某种服务，如部门微服务
04	集群（Cluster）	一个逻辑概念，可以区分不同的服务节点所处的网络环境
05	实例（Instance）	保存具体的微服务注册数据，每一个服务节点对应一个实例数据
06	数据 ID（DataId）	自定义的配置数据集，推荐命名格式为“包名称.类名称”（小写字母）

通过表 3-3 可以发现，在一个 Nacos 注册中心中，每一个产品都可能拆分为若干个项目模块，这样就可以通过分组来进行管理。最终一个项目模块会包含若干个微服务，而一个微服务又会有不同的集群环境及应用节点，这样就可以得到图 3-14 所示的 Nacos 领域模型。

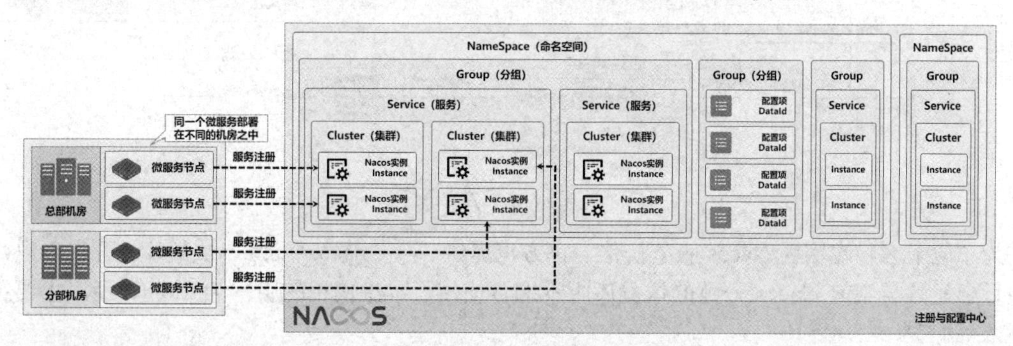

图 3-14 Nacos 领域模型

在使用 Nacos 进行项目开发前最重要的是创建一个新的命名空间，来实现不同的产品管理。本次演示将直接通过 Nacos 控制台创建一个名为“muyan”的命名空间，如图 3-15 所示。

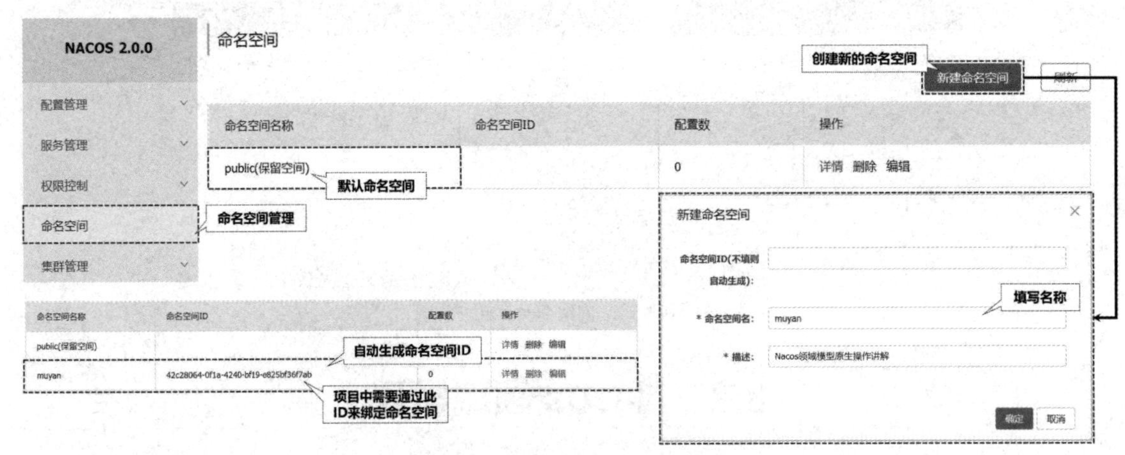

图 3-15 创建 Nacos 命名空间

提示：Nacos 数据管理。

在 Nacos 中，要想进行完整的数据操作，必须使用 NameSpace + Group 作为数据的唯一标记，不同分组无法互相访问数据。为了便于用户操作，Nacos 也提供了默认的管理组合，用户进行数据操作时没有配置 NameSpace 则使用默认的 public，没有配置 Group 则使用默认的 DEFAULT_GROUP。

3.2.1 配置数据管理

配置数据管理

视频名称 0307_【理解】配置数据管理

视频简介 Nacos 支持配置数据的管理，可以利用 Nacos 给出的原生 API，实现配置数据的发布、修改、删除、监听的相关操作。本视频为读者讲解如何基于 Nacos 客户端工具包实现 Nacos 配置项的管理。

Nacos 提供配置中心的功能，这样开发者就可以通过 Nacos 客户端实现配置数据集的发布、获取、更新、删除等常规操作。同时，相关的客户端也可以通过 Nacos 的指定领域模型（命名空间、分组、数据 ID 等）获取数据，或者定义一个监听程序及时获取每一次更新后的数据内容，如图 3-16 所示。考虑到实际开发之中配置数据保存的多样性，Nacos 支持 TXT、JSON、XML、YML、属性等格式的配置数据存储，以满足不同的应用开发要求。

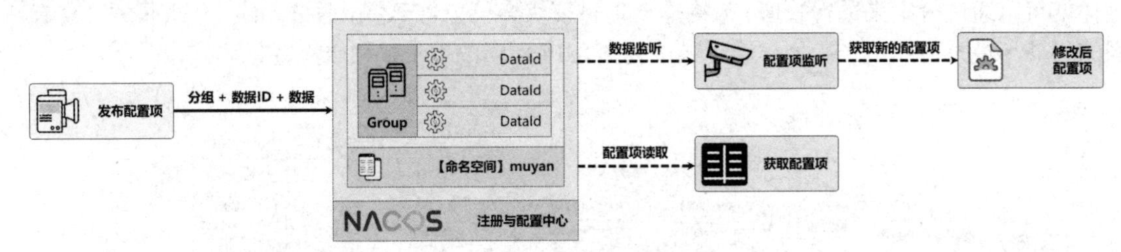

图 3-16 Nacos 配置数据管理

为了便于客户端实现 Nacos 服务操作，官方也提供一个“nacos-client”依赖库，开发者只需要在项目中配置与当前 Nacos 一致的依赖库版本即可使用。本次使用的版本为 2.0.0，所以需要进行如下依赖配置。

```
implementation('com.alibaba.nacos:nacos-client:2.0.0')        // NacosClient依赖
```

为便于用户实现配置数据操作，“nacos-client”提供了 ConfigService 操作接口，开发者利用表 3-4 列出的方法就可实现对配置数据的操作。

表 3-4 ConfigService 接口方法

序号	方法	类型	描述
01	public String getConfig(String dataId, String group, long timeoutMs) throws NacosException	普通	获取配置项
02	public String getConfigAndSignListener(String dataId, String group, long timeoutMs, Listener listener) throws NacosException	普通	获取配置项并进行单次更新监听
03	public void addListener(String dataId, String group, Listener listener) throws NacosException	普通	增加配置项监听器
04	public boolean publishConfig(String dataId, String group, String content) throws NacosException	普通	发布配置项
05	public boolean removeConfig(String dataId, String group) throws NacosException	普通	删除配置项
06	public void removeListener(String dataId, String group, Listener listener)	普通	删除配置项监听器
07	public String getServerStatus()	普通	获取服务状态
08	public void shutDown() throws NacosException	普通	关闭资源服务器

每一个 ConfigService 接口实例都可以直接实现 Nacos 配置数据的操作，而要想获取相应接口实例，则可以通过 NacosFactory 工厂类所提供的 createConfigService()方法实现，如图 3-17 所示。在调用该方法时需要传入配置操作所需的 Nacos 属性内容（如 Nacos 主机名称、命名空间等），这样就可以连接到指定的服务器，并在指定命名空间中进行操作（如果不设置命名空间则使用默认命名空间）。

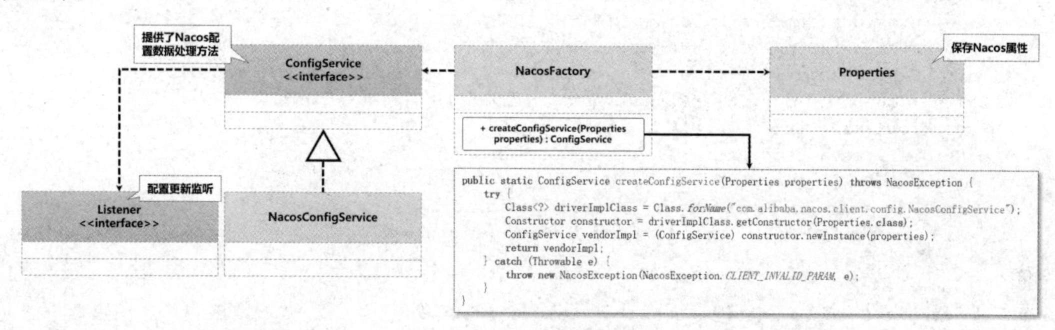

图 3-17 ConfigService 类结构

范例：向 Nacos 发布配置数据

```
package com.yootk.nacos;
public class PublishNacosConfig {
    public static final String NACOS_SERVER = "nacos-server:8848";
    public static final String DATA_ID = "com.yootk.nacos.microcloud.config";// 配置ID
    public static final String GROUP = "MICROCLOUD_GROUP";                   // 分组
    public static final String NAMESPACE = "42c28064-0f1a-4240-bf19-e825bf36f7ab";
    public static void main(String[] args) throws Exception {
        String content = "edu.yootk.com";                                    // 配置内容
        Properties properties = new Properties();                            // Nacos配置
        properties.put(PropertyKeyConst.SERVER_ADDR, NACOS_SERVER);          // 服务地址
        properties.put(PropertyKeyConst.NAMESPACE, NAMESPACE);               // 命名空间
        ConfigService configService = NacosFactory.createConfigService(properties);
        boolean isOk = configService.publishConfig(DATA_ID, GROUP, content); // 配置发布
        System.out.println(isOk ? "Nacos配置发布成功！" : "Nacos配置发布失败！");
    }
}
```

程序执行结果	Nacos配置发布成功！

本程序向 nacos-server 发布了一个文本配置项，配置项发布成功后会返回 true，同时 Nacos 控制台显示图 3-18 所示的配置列表。如果此时重复进行相同的“Group”和“Data Id”的设置，则会

使用新的内容替换旧的内容。

图 3-18　在 Nacos 控制台查看配置项

配置项发布之后，其他 Nacos 客户端可以获取此配置项，也可以利用 Listener 接口绑定数据的更新监听，这样当配置项的内容发生改变时可以及时获取最新的配置项。

范例：获取并监听 Nacos 配置项

```
package com.yootk.nacos;
public class NacosConfigListener {
    public static final String NACOS_SERVER = "nacos-server:8848";          // 服务地址
    public static final String DATA_ID = "com.yootk.nacos.microcloud.config"; // 配置ID
    public static final String GROUP = "MICROCLOUD_GROUP";                    // 分组
    public static final String NAMESPACE = "42c28064-0f1a-4240-bf19-e825bf36f7ab";
    public static void main(String[] args) throws Exception {
        Properties properties = new Properties();                            // Nacos属性配置
        properties.put(PropertyKeyConst.SERVER_ADDR, NACOS_SERVER);          // 服务地址
        properties.put(PropertyKeyConst.NAMESPACE, NAMESPACE);               // 命名空间
        ConfigService configService = NacosFactory.createConfigService(properties);
        String content = configService.getConfig(DATA_ID, GROUP, 5000);     // 获取配置内容
        System.out.println("【配置项】" + content);                          // 输出内容
        configService.addListener(DATA_ID, GROUP, new Listener() {           // 修改监听
            @Override
            public void receiveConfigInfo(String configInfo) {               // 数据修改触发此方法
                System.err.println("【接收配置】" + configInfo);             // 输出新配置项
            }
            @Override
            public Executor getExecutor() { return null; }
        });
        TimeUnit.SECONDS.sleep(Long.MAX_VALUE);                              // 主进程不关闭
    }
}
```

程序执行结果	【配置项】edu.yootk.com（获取原始配置项） 【接收配置】沐言科技：www.yootk.com（配置项更新后接收更新数据）

本程序通过 ConfigService 接口提供的 getConfig()方法获取了指定的配置项，同时为了可以及时获取最新的配置项，利用 addListener()方法添加了一个监听器，这样每当配置项修改时都会触发该监听器中的 receiveConfigInfo()方法。本程序通过 Nacos 控制台实现配置项更新，如图 3-19 所示，更新后 Nacos 客户端会接收新的配置项。

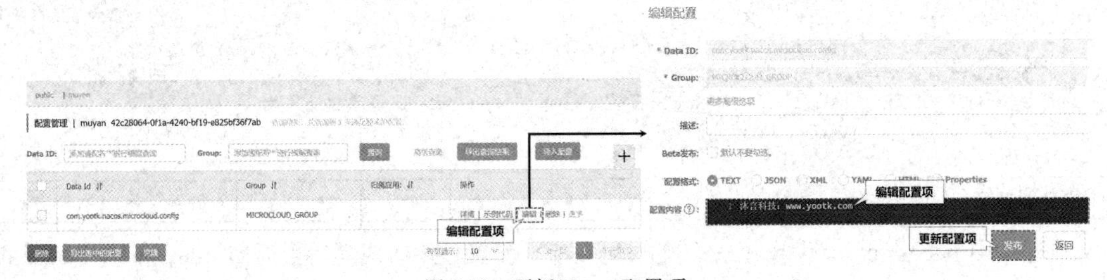

图 3-19　更新 Nacos 配置项

3.2.2 实例数据管理

实例数据管理

视频名称 0308_【理解】实例数据管理

视频简介 Nacos 提供了动态实例管理支持，这样可以及时地获取每一个实例节点的状态。本视频模拟集群节点的应用环境，介绍如何实现多个节点实例的配置，并通过具体的操作讲解 Nacos 节点配置监听的实现。

Nacos 作为注册中心，可以方便地管理微服务中的所有集群节点数据（主机地址、端口），每当有新的节点启动或者已有节点下线时都可以及时实现节点数据更新，如图 3-20 所示。其他的外部客户端也可以通过该注册中心的指定服务名称获取相应服务对应的所有节点数据，并根据需要实现节点服务的调用。

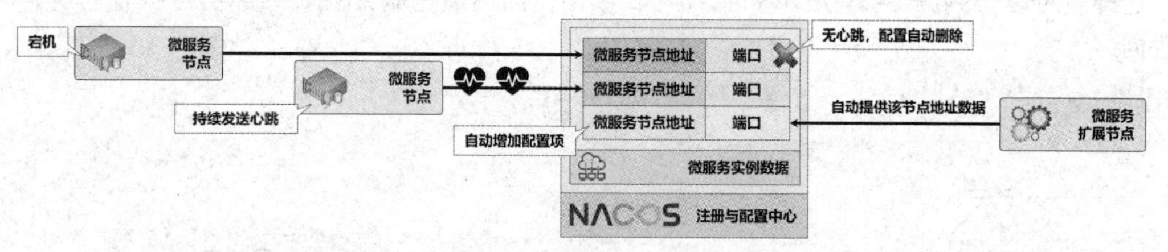

图 3-20 Nacos 实例管理

在 Nacos 客户端中可以通过 NamingService 接口实现所有服务的注册、节点数据和订阅监听管理。表 3-5 给出了该接口的常用方法，在进行实例操作时需要明确设置命名空间、服务名称、地址、端口、分组、集群名称。

表 3-5 NamingService 接口常用方法

序号	方法	类型	描述
01	public void registerInstance(String serviceName, String groupName, String ip, int port, String clusterName) throws NacosException;	普通	向 Nacos 中注册实例数据
02	public void deregisterInstance(String serviceName, String groupName, String ip, int port, String clusterName) throws NacosException;	普通	取消 Nacos 中注册的实例数据
03	public List<Instance> getAllInstances(String serviceName, String groupName) throws NacosException	普通	获取指定服务名称的实例数据
04	public List<Instance> selectInstances(String serviceName, String groupName, boolean healthy) throws NacosException	普通	根据实例健康状态获取指定服务名称的实例数据
05	public void subscribe(String serviceName, String groupName, EventListener listener) throws NacosException	普通	为实例创建订阅者，在实例更新时可以及时获取实例数据
06	public void unsubscribe(String serviceName, String groupName, EventListener listener) throws NacosException	普通	取消实例订阅者
07	public List<ServiceInfo> getSubscribeServices() throws NacosException	普通	获取所有实例订阅者数据

NamingService 是一个应用接口，在 Nacos 客户端中需要通过 NamingFactory 类中的 createNamingService()方法并根据传入的 Nacos 配置属性获取该接口实例，这样就可以实现实例操作。下面的代码可实现实例注册的操作。

范例：向 Nacos 服务器注册实例

```
package com.yootk.nacos;
public class RegisterNacosInstance {
    public static final String NACOS_SERVER = "nacos-server:8848";    // 服务地址
    public static final String NAMESPACE = "42c28064-0f1a-4240-bf19-e825bf36f7ab";
    public static void main(String[] args) throws Exception {
        Properties properties = new Properties();                      // Nacos属性配置
```

```
        properties.put(PropertyKeyConst.SERVER_ADDR, NACOS_SERVER);   // 服务地址
        properties.put(PropertyKeyConst.NAMESPACE, NAMESPACE);        // 命名空间
        NamingService naming = NamingFactory.createNamingService(properties);
        // 相同的服务可以部署在不同的节点之中，依靠集群区分不同的服务机房
        naming.registerInstance("micro.provider.dept", "MICROCLOUD_GROUP",
                "192.168.9.19", 8888, "DeptProviderCluster");
        TimeUnit.SECONDS.sleep(Long.MAX_VALUE);                       // 服务关闭实例会消失
    }
}
```

添加 Nacos 中的实例数据时需要配置命名空间，而后在服务注册时需要明确向 Nacos 传递要注册的服务名（micro.provider.dept）、分组名称（MICROCLOUD_GROUP）、实例主机（192.168.9.19）、实例端口（8888）、集群名称（DeptProviderCluster），同时该程序进程必须持续处于启动状态。这样就可以得到图 3-21 所示的服务列表数据，而后通过服务列表的详情可以查看所有的实例数据。

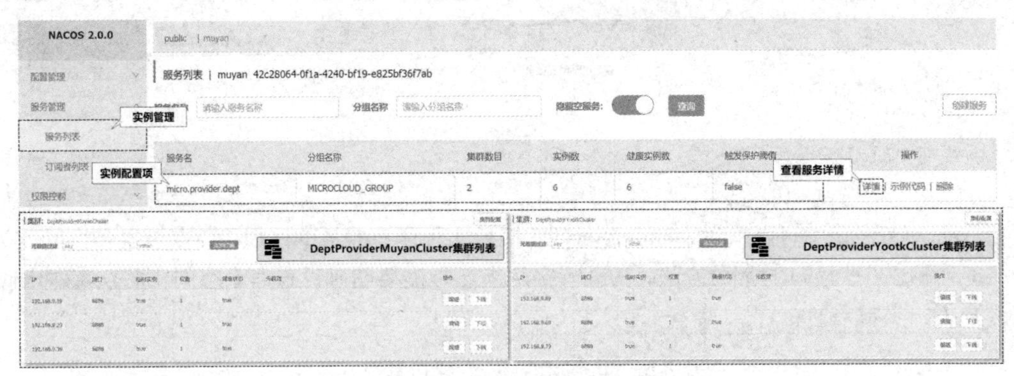

图 3-21　Nacos 服务列表

提示：多实例模拟。

为了便于读者理解，本范例分别创建了 6 个结构相同的程序，而后为不同的注册实例配置不同的主机地址，同时将这些实例分配到不同的集群（Cluster）之中，这样才可以得到图 3-21 所示的信息。读者也可以在这些实例启动后随意关闭一些实例，可以发现 Nacos 会自动进行实例数据更新。

所有在 Nacos 中注册的实例数据最终是需要被客户端获取的，这样客户端才可以根据所获取的实例数据来实现相关节点的服务调用。下面的代码将通过 NamingService 接口获取指定服务的所有实例数据。

范例：获取全部实例

```
package com.yootk.nacos;
public class NacosInstanceList {
    public static final String NACOS_SERVER = "nacos-server:8848";   // 服务地址
    public static final String NAMESPACE = "42c28064-0f1a-4240-bf19-e825bf36f7ab";
    public static void main(String[] args) throws Exception {
        Properties properties = new Properties();                    // Nacos属性配置
        properties.put(PropertyKeyConst.SERVER_ADDR, NACOS_SERVER);  // 服务地址
        properties.put(PropertyKeyConst.NAMESPACE, NAMESPACE);       // 命名空间
        NamingService naming = NamingFactory.createNamingService(properties);
        List<Instance> instances = naming.getAllInstances(
                "micro.provider.dept", "MICROCLOUD_GROUP");          // 获取实例列表
        for (Instance instance : instances) {                        // 实例迭代
            System.err.println(instance);                            // 输出实例数据
        }
    }
}
```

<table>
<tr><td rowspan="11">程序执行结果（随机实例数据）</td><td rowspan="11"><pre>Instance{
  instanceId='null',
  ip='192.168.9.19',
  port=8888,
  weight=1.0,
  healthy=true,
  enabled=true,
  ephemeral=true,
  clusterName='DeptProviderMuyanCluster',
  serviceName='MICROCLOUD_GROUP@@micro.provider.dept',
  metadata={}}</pre></td><td>实例数据（toString()输出）</td></tr>
<tr><td>实例 ID</td></tr>
<tr><td>节点注册地址</td></tr>
<tr><td>节点注册端口</td></tr>
<tr><td>节点权重</td></tr>
<tr><td>节点健康</td></tr>
<tr><td>节点启用</td></tr>
<tr><td>瞬时节点</td></tr>
<tr><td>集群名称</td></tr>
<tr><td>服务名称</td></tr>
<tr><td>实例元数据</td></tr>
</table>

本程序根据命名空间、服务名称（micro.provider.dept）以及分组名称（MICROCLOUD_GROUP）获取了该服务中的全部实例，随后通过迭代获取了每一个实例信息，这样就可以根据实例信息来实现服务主机的访问。

所有的节点都可以动态地进行扩充或删除，这样就需要对指定微服务的节点状态进行监听。Nacos 客户端提供了订阅者模式，通过 EventListener 接口即可获得动态实例的数据。

范例：实例订阅监听

```
package com.yootk.nacos;
public class NacosInstanceListener {
    public static final String NACOS_SERVER = "nacos-server:8848";            // 服务地址
    public static final String NAMESPACE = "42c28064-0f1a-4240-bf19-e825bf36f7ab";
    public static void main(String[] args) throws Exception {
        Properties properties = new Properties();                              // Nacos属性配置
        properties.put(PropertyKeyConst.SERVER_ADDR, NACOS_SERVER);            // 服务地址
        properties.put(PropertyKeyConst.NAMESPACE, NAMESPACE);                 // 命名空间
        NamingService naming = NamingFactory.createNamingService(properties);
        naming.subscribe("micro.provider.dept", "MICROCLOUD_GROUP",
          event -> {                                                           // 监听指定服务
            if (event instanceof NamingEvent) {                                // 事件类型
                System.out.println(((NamingEvent) event).getServiceName());    // 服务名称
                System.out.println(((NamingEvent) event).getInstances());      // 服务实例
            }
        });
        TimeUnit.SECONDS.sleep(Long.MAX_VALUE);                                // 持续监听
    }
}
```

本程序启动后，每当服务的实例数据发生改变时，都会产生 NamingEvent 事件类，并且可以通过该事件类的对象获取当前的全部实例数据（包括实例的主机名称、IP 地址等注册实例时提交的数据项）。

3.2.3 REST 访问配置

REST 访问配置

视频名称 0309_【掌握】REST 访问配置

视频简介 为了便于更多的平台与 Nacos 整合，Nacos 提供了 REST 接口，可以直接基于 HTTP 的形式实现 Nacos 数据操作。本视频结合 Nacos 提供的官方文档手册，通过具体的实例演示常用接口的使用。

除了使用“nacos-client”依赖库实现 Nacos 数据管理之外，也可以利用 Nacos 提供的开放平台基于 REST 接口的形式实现数据的处理操作，如图 3-22 所示。这样就可以使用任何异构平台实现 Nacos 数据操作。

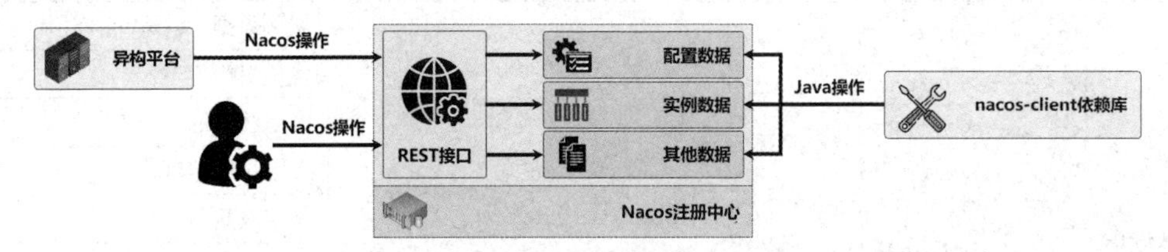

图 3-22　Nacos 开放平台

Nacos 给出的官方文档已经详细地列出了 Nacos 提供的所有开放 REST 接口路径、请求模式和请求参数等。表 3-6 为读者列出了常用的接口信息。

表 3-6　常用 REST 接口信息

序号	分类	请求类型	地址	描述
01	配置项	POST	/nacos/v1/cs/configs	新增配置项
02		GET	/nacos/v1/cs/configs	获取配置项
03		DELETE	/nacos/v1/cs/configs	删除配置项
04		POST	/nacos/v1/cs/configs/listener	监听配置项
05	实例	POST	/nacos/v1/ns/instance	发布实例
06		GET	/nacos/v1/ns/instance	获取实例
07		DELETE	/nacos/v1/ns/instance	删除实例
08		GET	/nacos/v1/ns/instance/list	查询指定服务下的实例
09		PUT	/nacos/v1/ns/instance/beat	发送实例心跳数据
10	命名空间	POST	/nacos/v1/console/namespaces	创建命名空间
11		PUT	/nacos/v1/console/namespaces	修改命名空间
12		GET	/nacos/v1/console/namespaces	命名空间列表
13		DELETE	/nacos/v1/console/namespaces	删除命名空间
14	管理	GET	/nacos/v1/ns/service/list	查询指定命名空间与分组下的服务数据
15		GET	/nacos/v1/ns/operator/metrics	查询当前服务数据指标
16		GET	/nacos/v1/ns/operator/servers	查看当前集群服务列表

表 3-6 包含 Nacos 的大部分处理操作。为便于理解，下面通过几个具体的实例进行说明。

（1）【POST 请求】创建命名空间。

```
http://nacos-server:8848/nacos/v1/console/namespaces
    ?customNamespaceId=&namespaceName=yootk&namespaceDesc=yootk.com
```

（2）【POST 请求】新增配置项。

```
http://nacos-server:8848/nacos/v1/cs/configs?dataId=com.yootk.config&group=MICROCLOUD_GROUP
    &content=edu.yootk.com&namespace=42c28064-0f1a-4240-bf19-e825bf36f7ab
```

（3）【POST 请求】发布实例。

```
http://nacos-server:8848/nacos/v1/ns/instance?serviceName=micro.provider.emp&groupName=MICROCLOUD_GROUP
    &namespaceId=42c28064-0f1a-4240-bf19-e825bf36f7ab&ip=192.168.2.17&port=8080&cluster=EmpCluster
```

（4）【GET 请求】查询指定服务下的实例。

```
http://nacos-server:8848/nacos/v1/ns/instance/list?serviceName=micro.provider.emp
    &namespaceId=42c28064-0f1a-4240-bf19-e825bf36f7ab&groupName=MICROCLOUD_GROUP
```

（5）【DELETE】删除实例。

```
http://nacos-server:8848/nacos/v1/ns/instance?serviceName=micro.provider.emp&cluster=EmpCluster&port=8080
    &namespaceId=42c28064-0f1a-4240-bf19-e825bf36f7ab&groupName=MICROCLOUD_GROUP&ip=192.168.2.17
```

3.3 微服务注册

微服务注册

视频名称 0310_【掌握】微服务注册

视频简介 Spring Cloud 中的微服务需要在 Nacos 中进行统一的注册管理。本视频通过具体的操作介绍如何修改部门微服务并实现 Nacos 服务的注册操作。

Spring Cloud Alibaba 提供了对 Nacos 注册中心的自动配置支持，开发者只需要在微服务的提供端提供 Nacos 发现（spring-cloud-starter-alibaba-nacos-discovery）与配置（spring-cloud-starter-alibaba-nacos-config）依赖。而 Nacos 配置依赖启动优先级较高，所以就需要在项目中提供 bootstrap.yml 配置文件，并在此配置文件中定义 Nacos 相关连接信息，随后再编写 application.yml 配置文件定义 Nacos 发现配置，如图 3-23 所示。

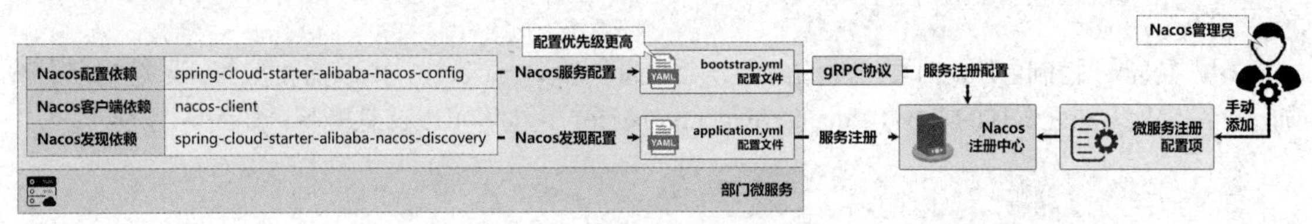

图 3-23 Nacos 服务注册

在当前环境下，由于 Spring Cloud Alibaba 暂时没有版本变更，因此就会出现默认情况下 Spring Cloud 相关的 Nacos 依赖库使用的“nacos-client”版本过低的问题，此时需要手动删除旧版本的 Nacos 客户端依赖，随后再手动引入新的 2.0.0 版本依赖。

提示：关于 Nacos 注册依赖版本。

如果开发者现在使用的是 Nacos 1.x 注册中心，则只需要在项目中配置 Nacos 发现依赖，随后编写 application.yml 配置文件即可实现服务注册。但是在使用 Nacos 2.x 注册中心时，由于其核心架构发生了改变，因此就需要额外配置 Nacos 配置依赖。未来 Spring Cloud Alibaba 在进行版本更新时，可能会解决相应的 Nacos 依赖版本问题。但是截至本书编写时，此依赖暂时没有太大变更，所以就需要开发者手动排除旧版本的依赖库。

另外需要提醒读者的是，在微服务中进行注册中心配置时，所使用的依然是 8848 端口，微服务会自动找到其对应的 9848（8848 + 1000 偏移量）端口并通过 gRPC 协议注册。

在进行 Nacos 服务注册时需要通过 bootstrap.yml 配置文件定义 Nacos 访问信息，同时也需要在 application.yml 中配置 Nacos 服务发现信息，才能够实现正确的微服务注册。下面通过具体的操作步骤进行演示。

（1）【microcloud 项目】修改 build.gradle 配置文件，为 provider-dept-8001 子模块添加 Nacos 相关依赖。

```
project('provider-dept-8001') {                                          // 子模块
    dependencies {                                                       // 重复依赖，略
        implementation('com.alibaba.cloud:' +
                'spring-cloud-starter-alibaba-nacos-discovery') {        // Nacos依赖
            exclude group: 'com.alibaba.nacos', module: 'nacos-client'   // 删除nacos-client
        }
        implementation('com.alibaba.cloud:' +
                'spring-cloud-starter-alibaba-nacos-config') {           // Nacos配置
            exclude group: 'com.alibaba.nacos', module: 'nacos-client'   // 删除nacos-client
        }
```

```
        implementation('com.alibaba.nacos:nacos-client:2.0.0')     // NacosClient依赖
    }
}
```

（2）【provider-dept-8001 子模块】在 src/main/resources 目录中创建 bootstrap.yml 配置文件，在该文件中填写要连接的 Nacos 注册中心的相关配置。

```
spring:                                              # Spring配置
  cloud:                                             # Spring Cloud配置
    nacos:                                           # Nacos注册中心
      config:                                        # Nacos配置
        server-addr: nacos-server:8848               # 服务地址
```

（3）【provider-dept-8001 子模块】修改 application.yml 配置文件，添加 Nacos 相关配置。

```
spring:
  application:                                       # 应用配置
    name: dept.provider                              # 应用名称
  cloud:                                             # Spring Cloud配置
    nacos:                                           # Nacos注册中心
      discovery:                                     # 发现服务
        service: ${spring.application.name}          # Nacos服务名称
        server-addr: nacos-server:8848               # 服务地址
```

（4）【Nacos 控制台】此时要注册的微服务名称为"dept.provider"，所以需要在 Nacos 控制台中添加一个配置项，该配置项的 ID 为"dept.provider.properties"，内容可以随意填写，如图 3-24 所示。

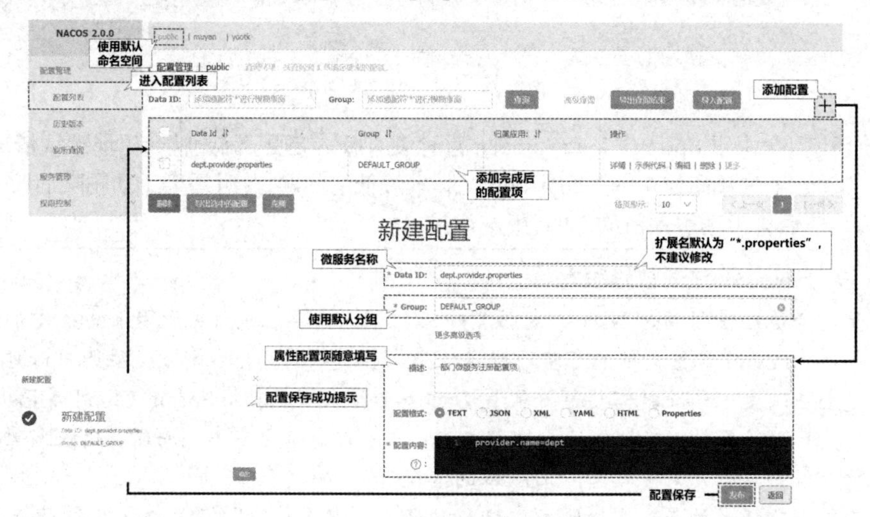

图 3-24　添加微服务配置项

（5）【Nacos 控制台】服务启动后可以查看 Nacos 控制台，而后可以看见图 3-25 所示的界面。

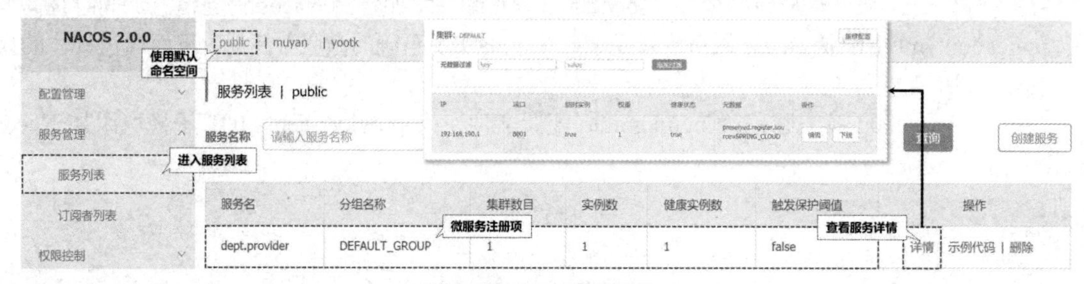

图 3-25　Nacos 服务列表

> **注意：Nacos 2.x 基于 gRPC 协议实现服务管理。**
>
> Nacos 1.x 基于 HTTP 实现服务管理，而在 Nacos 2.x 中 Nacos 控制台和服务管理各自使用不同的实现协议，即 Nacos 控制台使用 8848 端口（HTTP），而服务管理使用 9848 端口（gRPC 协议）。

gRPC 是一个高性能、开源和通用的 RPC 框架，面向服务端和移动端，适合于长连接的服务处理。同时 gRPC 基于 HTTP 2 设计，采用二进制实现数据传输，在完全兼容 HTTP 1.1 基础之上，进一步减少了网络延迟所带来的性能损耗，所以 Nacos 2.x 的处理性能要比 Nacos 1.x 的更高。

在进行 Web 程序开发过程中，客户端如果想实现页面渲染，则需要通过服务器加载所需资源，这样就有可能出现向服务器端发出多次请求的问题。在 HTTP 1 中超过限制数目的请求会被阻塞，这样就造成了传输性能下降。而 HTTP 2 采用了多路复用（Multiplexing）模型，允许一个客户端发起多次“请求-响应”，这样就很容易实现单个 TCP 连接下的多流（传输单位：二进制帧）并行传输，实现双向数据交互，如图 3-26 所示。

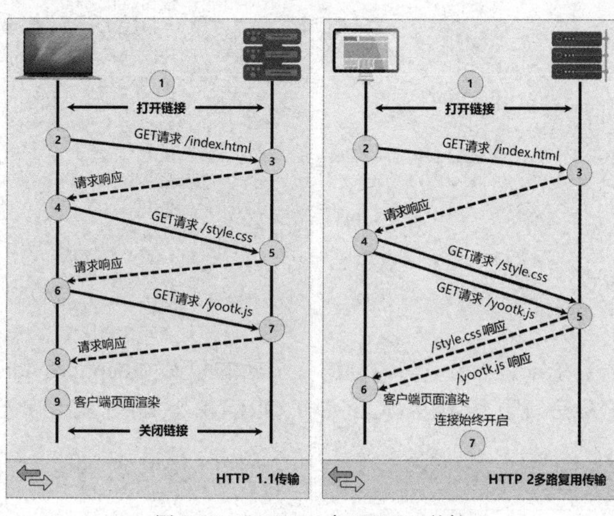

图 3-26　HTTP 1.1 与 HTTP 2 传输

3.3.1　配置 Nacos 注册信息

配置 Nacos 注册信息

视频名称　0311_【掌握】配置 Nacos 注册信息

视频简介　Nacos 为了规范化服务管理，提供了完整的领域模型，所以微服务的注册与发布也需要进行归类管理。本视频介绍对部门微服务进一步改造，实现注册数据的有效管理。

微服务在进行注册时，如果未经过任何规范化管理，则所有的配置信息都会采用默认内容。但是这样的处理形式并不便于微服务的规范化管理，这样就需要对当前的微服务做进一步的管理配置，具体的实现步骤如下。

（1）【Nacos 控制台】创建一个新的命名空间“dev”，并记录好此命名空间的 ID，如图 3-27 所示。

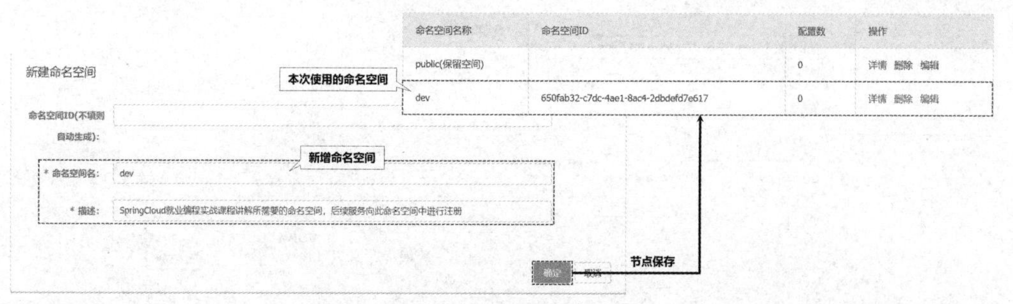

图 3-27　新建命名空间

（2）【provider-dept-8001 子模块】修改 bootstrap.yml 配置文件，添加 Nacos 领域模型。

```
spring:                                                  # Spring配置
  cloud:                                                 # Spring Cloud配置
    nacos:                                               # Nacos注册中心
      config:                                            # Nacos配置
        server-addr: nacos-server:8848                   # 服务地址
        namespace: 650fab32-c7dc-4ae1-8ac4-2dbdefd7e617  # 命名空间ID
        group: MICROCLOUD_GROUP                          # 组名称
        cluster-name: YootkCluster                       # 集群名称
```

（3）【provider-dept-8001 子模块】修改 application.yml 配置文件，添加 Nacos 领域模型定义。

```
spring:                                                  # Spring配置
  application:                                           # 应用配置
    name: dept.provider                                  # 应用名称
  cloud:                                                 # Spring Cloud配置
    nacos:                                               # Nacos注册中心
      discovery:                                         # 发现服务
        service: ${spring.application.name}              # Nacos服务名称
        server-addr: nacos-server:8848                   # 服务地址
        namespace: 650fab32-c7dc-4ae1-8ac4-2dbdefd7e617  # 命名空间ID
        group: MICROCLOUD_GROUP                          # 组名称
        cluster-name: YootkCluster                       # 集群名称
        metadata:                                        # 元数据
          version: 1.0                                   # 自定义数据项
          company: 沐言科技                               # 自定义数据项
          url: www.yootk.com                             # 自定义数据项
          author: 李兴华                                  # 自定义数据项
```

（4）【Nacos 控制台】在 dev 命名空间中创建一个新的配置项 dept.provider.properties，在设置此配置项时所使用的组名称必须为“MICROCLOUD_GROUP”，如图 3-28 所示。

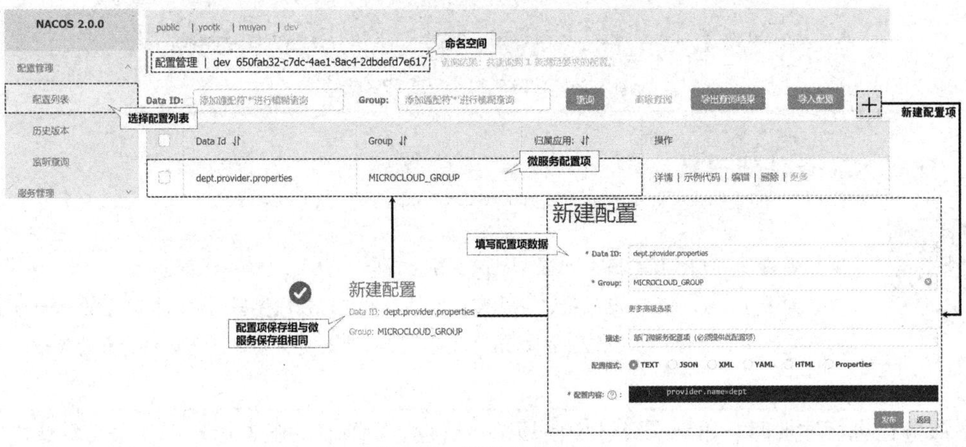

图 3-28　微服务配置项

（5）【Nacos 控制台】重新启动微服务之后，就可以在 Nacos 中发现相关的配置信息，如图 3-29 所示。

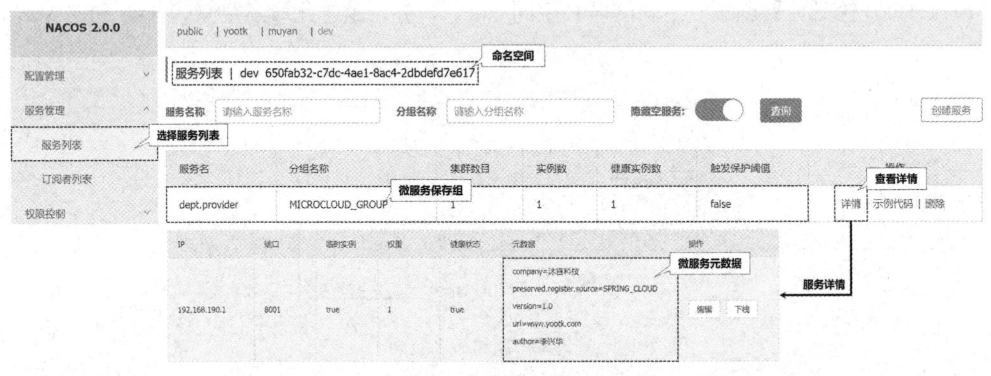

图 3-29　微服务配置信息

3.3.2 Nacos 安全注册

Nacos 安全注册

视频名称 0312_【掌握】Nacos 安全注册

视频简介 Nacos 是整个微服务的核心组件，里面保存有重要的服务实例数据，所以必须保证微服务注册的安全。本视频讲解 Nacos 安全配置启用以及微服务的安全注册。

此时的程序已经成功地实现了微服务向 Nacos 的实例注册管理，但是由于当前的 Nacos 服务中并没有启用任何安全机制，因此任何微服务只要填写了正确的服务地址就都可以进行服务注册，也都可以通过注册中心获取所有的实例信息。这样极不安全，需要进行 Nacos 安全管理，如图 3-30 所示。

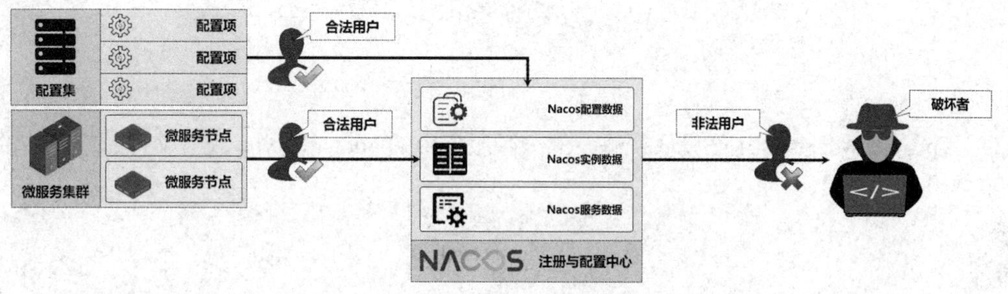

图 3-30 Nacos 安全管理

虽然 Nacos 控制台已经提供了认证访问的安全保护，但是在默认情况下此安全认证并未启动，所以需要开发者手动修改 Nacos 配置。在安全保护启用后，微服务要提供正确的账户信息才可以实现注册。下面通过具体的步骤实现。

（1）【nacos-server 主机】修改 Nacos 配置文件。

```
vi /usr/local/nacos/conf/application.properties
nacos.core.auth.enabled=true                              # 启用Nacos安全认证
nacos.core.auth.enable.userAgentAuthWhite=false           # 关闭使用user-agent判断服务端请求并放行鉴权的功能
nacos.core.auth.server.identity.key=muyan                 # 自定义身份识别Key（配置用户名）
nacos.core.auth.server.identity.value=yootk               # 自定义身份识别Value（配置密码）
```

（2）【nacos-server 主机】Nacos 配置项修改完成后需要重新启动当前的 Nacos 服务。

关闭 Nacos 服务	`bash -f /usr/local/nacos/bin/shutdown.sh`
启动 Nacos 服务	`bash -f /usr/local/nacos/bin/startup.sh -m standalone`

（3）【Nacos 控制台】创建一个新的系统角色 ROLE_MUYAN，如图 3-31 所示。

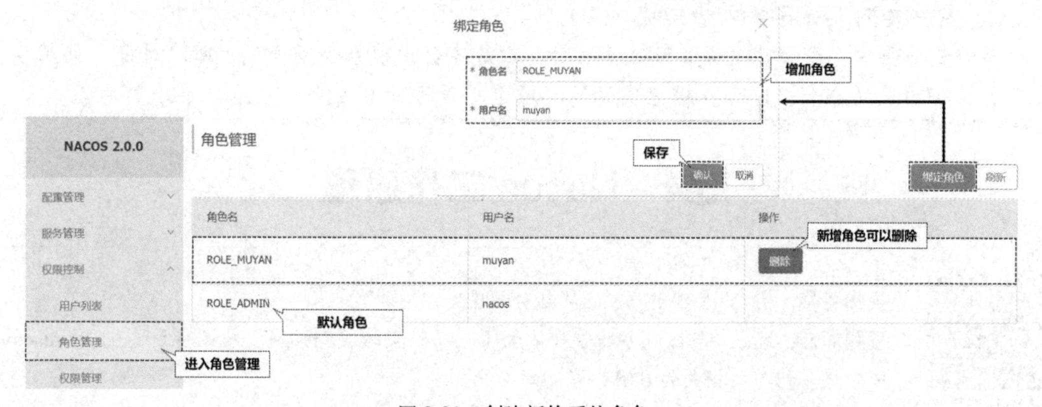

图 3-31 创建新的系统角色

（4）【Nacos 控制台】为新增的 ROLE_MUYAN 角色配置 dev 和 public 命名空间权限，如图 3-32 所示。

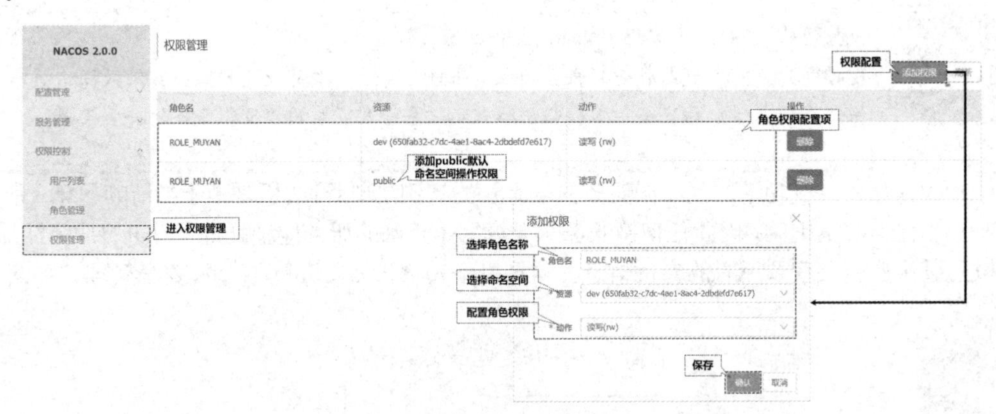

图 3-32　为角色配置权限

（5）【provider-dept-8001 子模块】修改部门微服务中的 bootstrap.yml 配置文件，在已有的配置项上增加用户配置项。

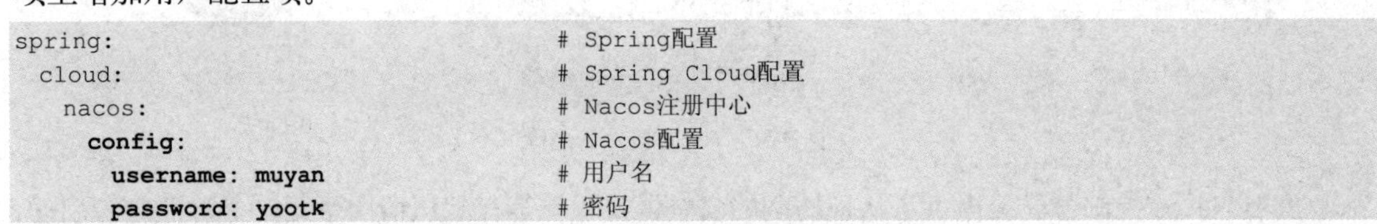

```
spring:                                          # Spring配置
  cloud:                                         # Spring Cloud配置
    nacos:                                       # Nacos注册中心
      config:                                    # Nacos配置
        username: muyan                          # 用户名
        password: yootk                          # 密码
```

（6）【provider-dept-8001 子模块】修改部门微服务中的 application.yml 配置文件，在已有的配置项上增加用户配置项。

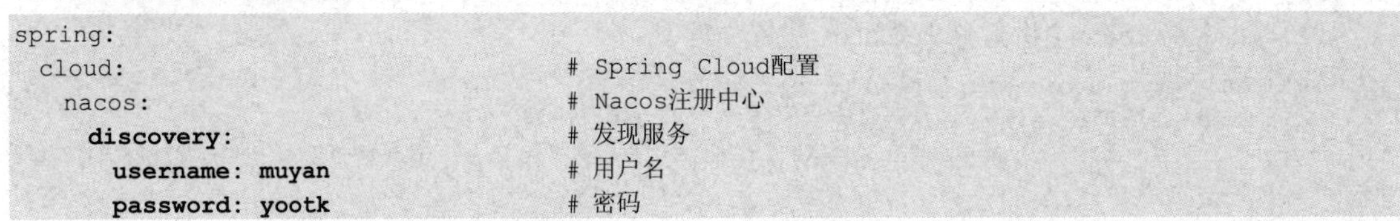

```
spring:
  cloud:                                         # Spring Cloud配置
    nacos:                                       # Nacos注册中心
      discovery:                                 # 发现服务
        username: muyan                          # 用户名
        password: yootk                          # 密码
```

此时在部门微服务中配置的用户已经被分配了正确的权限，所以当前微服务启动后可以正确通过 Nacos 认证保护并实现正确的服务注册。

提示：建议使用 nacos 用户。

虽然以上的配置解决了“muyan/yootk”用户的注册问题，但是 Nacos 的用户管理还存在不完善的情况，这些不属于本书的讨论范围。读者在发现无法进行自定义用户注册时，可采用默认的 nacos 用户实现（nacos 拥有管理员权限）。

另外需要再次提醒读者的是，Nacos 注册中心是不会出现在公网上的，所以对是否使用安全配置并没有明确的要求，即可以根据实际的环境选择是否开启安全配置。

3.4　Nacos 工作原理

Nacos 自动配置

视频名称　0313_【理解】Nacos 自动配置

视频简介　Nacos 与微服务的整合依赖于自动配置的实现机制。本视频通过 nacos-discover 依赖库介绍与之相关的自动配置类的功能。

如果想将 Spring Cloud 项目接入 Nacos 进行管理，就需要在项目中配置“spring-cloud-alibaba-nacos-discovery”依赖库，而该组件在与当前的 Spring Cloud 项目整合时，通过 spring.factories 配置文件定义了多个启动配置类，这一点可以通过该配置文件的源代码观察到。

范例：spring-cloud-alibaba-nacos-discovery/META-INF/spring.factories 配置源代码

```
org.springframework.boot.autoconfigure.EnableAutoConfiguration=\
  com.alibaba.cloud.nacos.discovery.NacosDiscoveryAutoConfiguration,\
  com.alibaba.cloud.nacos.ribbon.RibbonNacosAutoConfiguration,\
  com.alibaba.cloud.nacos.endpoint.NacosDiscoveryEndpointAutoConfiguration,\
  com.alibaba.cloud.nacos.registry.NacosServiceRegistryAutoConfiguration,\
  com.alibaba.cloud.nacos.discovery.NacosDiscoveryClientConfiguration,\
  com.alibaba.cloud.nacos.discovery.reactive
                    .NacosReactiveDiscoveryClientConfiguration,\
  com.alibaba.cloud.nacos.discovery.configclient.NacosConfigServerAutoConfiguration
org.springframework.cloud.bootstrap.BootstrapConfiguration=\
  com.alibaba.cloud.nacos.discovery.configclient
                    .NacosDiscoveryClientConfigServiceBootstrapConfiguration
```

此时可以发现，该配置文件中有两个配置加载类的定义：EnableAutoConfiguration（与 application.yml 配置文件中定义关联）和 BootstrapConfiguration（与 bootstrap.yml 配置文件中定义关联）。这些相关配置类的结构关系如图 3-33 所示。

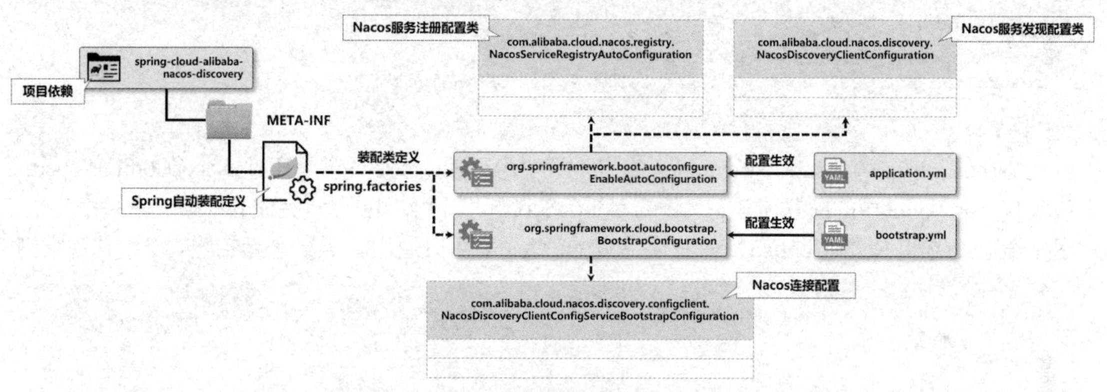

图 3-33　相关配置类的结构关系

除此之外，使用 Nacos 服务开发还需要 spring-cloud-starter-alibaba-nacos-config 模块依赖。该模块主要通过 bootstrap.yml 中的配置项实现 Nacos 的连接管理。该模块中配置的 spring.factories 文件的核心源代码如下。

范例：spring-cloud-starter-alibaba-nacos-config/META-INF/spring.factories 核心源代码

```
org.springframework.cloud.bootstrap.BootstrapConfiguration=\
          com.alibaba.cloud.nacos.NacosConfigBootstrapConfiguration
```

如果想理解 Nacos 服务注册的工作原理，则主要需研究两个 Nacos 相关依赖模块中所提供的 3 个自动配置类（NacosConfigBootstrapConfiguration、NacosServiceRegistryAutoConfiguration、NacosDiscoveryAutoConfiguration）的使用结构，本节将主要对这 3 个配置类进行分析。

3.4.1　NacosConfigBootstrapConfiguration

NacosConfig
Bootstrap
Configuration

视频名称　0314_【理解】NacosConfigBootstrapConfiguration

视频简介　Nacos 客户端可以通过 ConfigService 接口实现 Nacos 领域模型的操作，而此接口对象的实例创建是由 spring-cloud-starter-alibaba-nacos-config 依赖所提供的自动配置类实现的。本视频为读者分析 NacosConfigBootstrapConfiguration 配置类的源代码。

在 Spring Cloud 项目中使用 spring-cloud-starter-alibaba-nacos-config 依赖时，需要在项目中配置 bootstrap.yml 文件，同时在该文件中要配置当前系统中所需要使用的 Nacos 相关属性内容。最终的微服务注册实现本质上是依靠 ConfigService 接口实现的，而此接口对象的实例化处理是通过 NacosConfigBootstrapConfiguration 配置类完成的。

范例：NacosConfigBootstrapConfiguration 类源代码

```
package com.alibaba.cloud.nacos;
@Configuration(proxyBeanMethods = false)
@ConditionalOnProperty(name = "spring.cloud.nacos.config.enabled",
      matchIfMissing = true)
public class NacosConfigBootstrapConfiguration {
   @Bean
   @ConditionalOnMissingBean
   public NacosConfigProperties nacosConfigProperties() {     // Nacos配置属性
      return new NacosConfigProperties();
   }
   @Bean
   @ConditionalOnMissingBean
   public NacosConfigManager nacosConfigManager(              // Nacos配置管理类
         NacosConfigProperties nacosConfigProperties) {       // Nacos配置属性
      return new NacosConfigManager(nacosConfigProperties);   // 获取Nacos配置类实例
   }
   @Bean
   public NacosPropertySourceLocator nacosPropertySourceLocator(
         NacosConfigManager nacosConfigManager) {             // Nacos属性定位处理
      return new NacosPropertySourceLocator(nacosConfigManager);
   }
}
```

在 NacosConfigBootstrapConfiguration 自动配置类中，最为重要的就是通过 NacosConfigProperties 类获取 bootstrap.yml 中的配置项，而具体的 ConfigService 接口实例创建（完成 Nacos 服务的相关开发处理操作），则是由 NacosConfigManager 类完成的，这一点可以通过图 3-34 所示的配置类结构观察到。

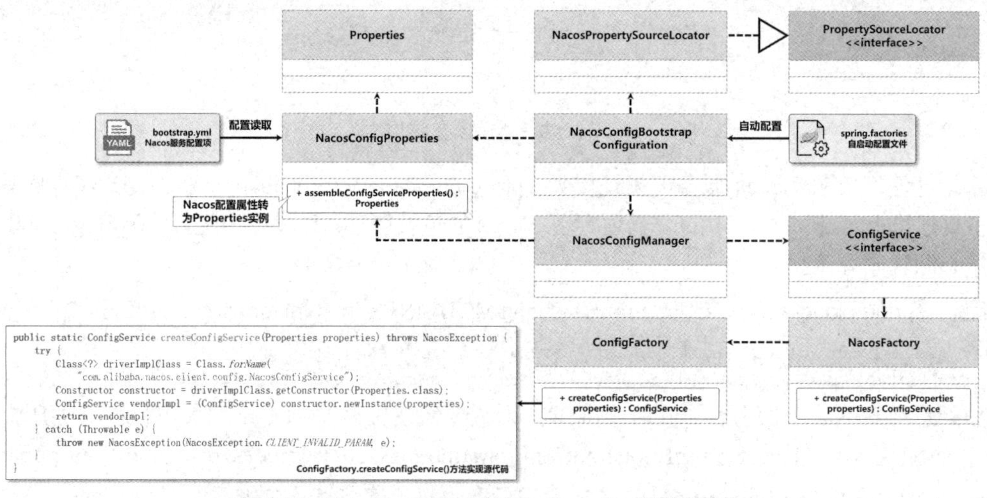

图 3-34　NacosConfigBootstrapConfiguration 配置类结构

3.4.2　@EnableDiscoveryClient 注解

@EnableDiscoveryClient
注解

视频名称　0315_【理解】@EnableDiscoveryClient 注解

视频简介　服务与注册中心的整合需要通过@EnableDiscoveryClient 注解完成。本视频为读者分析该注解的组成结构，并根据其组成结构分析导入处理类的实现机制。

Spring Cloud 服务端与客户端在进行 Nacos 整合处理时，都需要在程序的启动类中使用@EnableDiscoveryClient 注解进行配置，这样才可以触发 Nacos 相关配置类进行操作。首先观察该注解的源代码。

范例：@EnableDiscoveryClient 注解源代码

```
package org.springframework.cloud.client.discovery;
@Target(ElementType.TYPE)
@Retention(RetentionPolicy.RUNTIME)
@Documented
@Inherited
@Import(EnableDiscoveryClientImportSelector.class)          // 导入处理类
public @interface EnableDiscoveryClient {
  boolean autoRegister() default true;                       // 如果为true则自动实现服务注册
}
```

可以发现@EnableDiscoveryClient 注解中只有一个 autoRegister 属性，而是否采用自动配置主要取决于该属性配置的内容，如果为 true 则表示允许自动注册。该注解定义处使用了EnableDiscoveryClientImportSelector 导入选择器。下面继续打开该类的源代码进行观察。

范例：EnableDiscoveryClientImportSelector 类源代码

```
package org.springframework.cloud.client.discovery;
@Order(Ordered.LOWEST_PRECEDENCE - 100)                                   // 最低优先级
public class EnableDiscoveryClientImportSelector
     extends SpringFactoryImportSelector<EnableDiscoveryClient> {
  @Override
  public String[] selectImports(AnnotationMetadata metadata) {
     String[] imports = super.selectImports(metadata);
     AnnotationAttributes attributes = AnnotationAttributes.fromMap(
          metadata.getAnnotationAttributes(getAnnotationClass().getName(), true));
     boolean autoRegister = attributes.getBoolean("autoRegister");      // 属性解析
     if (autoRegister) {                                                 // 自动注册状态开启
        List<String> importsList = new ArrayList<>(Arrays.asList(imports));
        importsList.add("org.springframework.cloud.client.serviceregistry."+
              "AutoServiceRegistrationConfiguration");                  // 服务自动注册配置类
        imports = importsList.toArray(new String[0]);
     }
     else {                                                              // 未开启自动配置
        Environment env = getEnvironment();
        if (ConfigurableEnvironment.class.isInstance(env)) {
          ConfigurableEnvironment configEnv = (ConfigurableEnvironment) env;
          LinkedHashMap<String, Object> map = new LinkedHashMap<>();
          // 自动注册未开启时，会将Spring Cloud配置的属性项对应的内容设置为false
          map.put("spring.cloud.service-registry.auto-registration.enabled", false);
          MapPropertySource propertySource = new MapPropertySource(
               "springCloudDiscoveryClient", map);
          configEnv.getPropertySources().addLast(propertySource);
        }
     }
     return imports;
  }
}
```

EnableDiscoveryClientImportSelector 导入处理类主要针对@EnableDiscoveryClient 中配置的 autoRegister 属性状态进行判断：如果该属性内容为 true 则启用自动注册，此时会导入AutoServiceRegistrationConfiguration 类并进行 Bean 实例注册，实现结构如图 3-35 所示；而如果autoRegister 属性内容为空，则自动将当前系统中的相关配置属性修改为“false”。

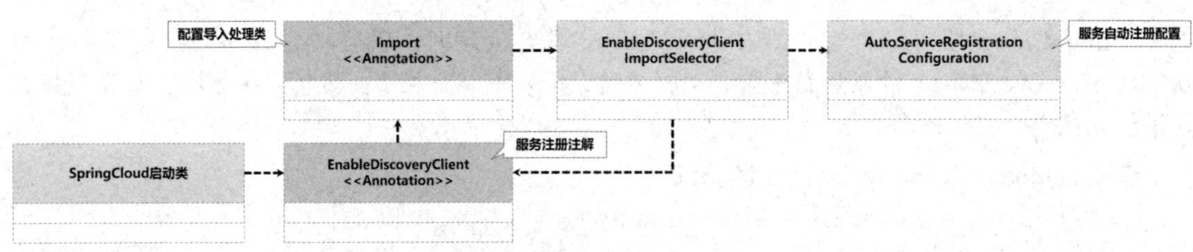

图 3-35　EnableDiscoveryClientImportSelector 实现结构

3.4.3　NacosServiceRegistryAutoConfiguration

NacosService
RegistryAuto
Configuration

视频名称　0316_【理解】NacosServiceRegistryAutoConfiguration

视频简介　为了便于服务的注册管理，Spring Cloud 提供了统一的处理标准。本视频通过 Nacos 提供的 NacosServiceRegistryAutoConfiguration 类的源代码分析服务注册流程。

在项目引入 Nacos 依赖库之后，就可以通过 NacosServiceRegistryAutoConfiguration 类实现服务的自动注册配置操作。而在进行 Nacos 服务自动注册时，实际上是创建 NacosServiceRegistry（服务注册）、NacosRegistration（Nacos 注册实例信息）、NacosAutoServiceRegistration（服务自动注册）3 个 Bean 的实例，如图 3-36 所示。

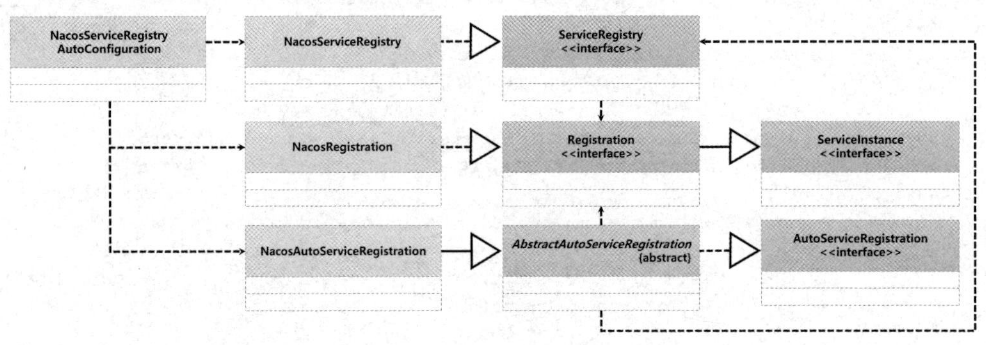

图 3-36　NacosServiceRegistryAutoConfiguration 关联结构

范例：NacosServiceRegistryAutoConfiguration 类源代码

```
package com.alibaba.cloud.nacos.registry;
@Configuration(proxyBeanMethods = false)
@EnableConfigurationProperties
@ConditionalOnNacosDiscoveryEnabled
@ConditionalOnProperty(
    value = "spring.cloud.service-registry.auto-registration.enabled",
    matchIfMissing = true)
@AutoConfigureAfter({ AutoServiceRegistrationConfiguration.class,
    AutoServiceRegistrationAutoConfiguration.class,
    NacosDiscoveryAutoConfiguration.class })
public class NacosServiceRegistryAutoConfiguration {
   @Bean
   public NacosServiceRegistry nacosServiceRegistry(
        NacosDiscoveryProperties nacosDiscoveryProperties) {
      return new NacosServiceRegistry(nacosDiscoveryProperties);          // 服务注册
   }
   @Bean
   @ConditionalOnBean(AutoServiceRegistrationProperties.class)
   public NacosRegistration nacosRegistration(
        NacosDiscoveryProperties nacosDiscoveryProperties,
        ApplicationContext context) {                                     // 服务实例数据
      return new NacosRegistration(nacosDiscoveryProperties, context);
```

```
  }
  @Bean
  @ConditionalOnBean(AutoServiceRegistrationProperties.class)
  public NacosAutoServiceRegistration nacosAutoServiceRegistration(
        NacosServiceRegistry registry,
        AutoServiceRegistrationProperties autoServiceRegistrationProperties,
        NacosRegistration registration) {                          // 服务自动注册
     return new NacosAutoServiceRegistration(registry,
           autoServiceRegistrationProperties, registration);
  }
}
```

通过以上源代码可以清楚地发现，每一个 Bean 对象实例化时都需要读取 application.yml 所配置的属性内容，而最终实现 Nacos 服务注册的类只有 NacosServiceRegistry 工具类。下面观察此类的核心源代码。

范例：NacosServiceRegistry 类核心源代码

```
package com.alibaba.cloud.nacos.registry;
public class NacosServiceRegistry implements ServiceRegistry<Registration> {
  // 保存application.yml中配置的Nacos服务主机的相关配置项
  private final NacosDiscoveryProperties nacosDiscoveryProperties;
  private final NamingService namingService;                       // 服务注册实现类
  public NacosServiceRegistry(NacosDiscoveryProperties nacosDiscoveryProperties) {
     this.nacosDiscoveryProperties = nacosDiscoveryProperties;
     this.namingService = nacosDiscoveryProperties.namingServiceInstance();
  }
  @Override
  public void register(Registration registration) {                // 注册服务实例
     if (StringUtils.isEmpty(registration.getServiceId())) {
        return;
     }
     String serviceId = registration.getServiceId();               // 获取注册名称
     String group = nacosDiscoveryProperties.getGroup();           // 获取注册分组
     Instance instance = getNacosInstanceFromRegistration(registration);
     try {// 通过NamingService业务处理类实现服务实例的注册
        namingService.registerInstance(serviceId, group, instance);
     } catch (Exception e) { }
  }
  @Override
  public Object getStatus(Registration registration) {             // 获取实例状态
     String serviceName = registration.getServiceId();             // 获取实例ID
     try {
        List<Instance> instances = nacosDiscoveryProperties.namingServiceInstance()
              .getAllInstances(serviceName);
        for (Instance instance : instances) {
           if (instance.getIp().equalsIgnoreCase(nacosDiscoveryProperties.getIp())
                 && instance.getPort() == nacosDiscoveryProperties.getPort()) {
              return instance.isEnabled() ? "UP" : "DOWN";         // 返回实例状态
           }
        }
     }
     catch (Exception e) { }
     return null;
  }
  private Instance getNacosInstanceFromRegistration(Registration registration) {
     Instance instance = new Instance();                           // 获取实例对象
     instance.setIp(registration.getHost());                       // 实例地址
     instance.setPort(registration.getPort());                     // 实例端口
     instance.setWeight(nacosDiscoveryProperties.getWeight());     // 实例权重
     instance.setClusterName(nacosDiscoveryProperties.getClusterName()); // 集群名称
     instance.setMetadata(registration.getMetadata());             // 元数据
     return instance;
  }
}
```

通过服务注册类的功能可以发现，用户在 application.yml 中配置的与 Nacos 相关的属性内容都会通过 Instance 对象实例进行保存，而真正实现服务注册的是由 Nacos 提供的 NamingService 业务处理接口。此时可以根据图 3-37 所示的结构进行分析。

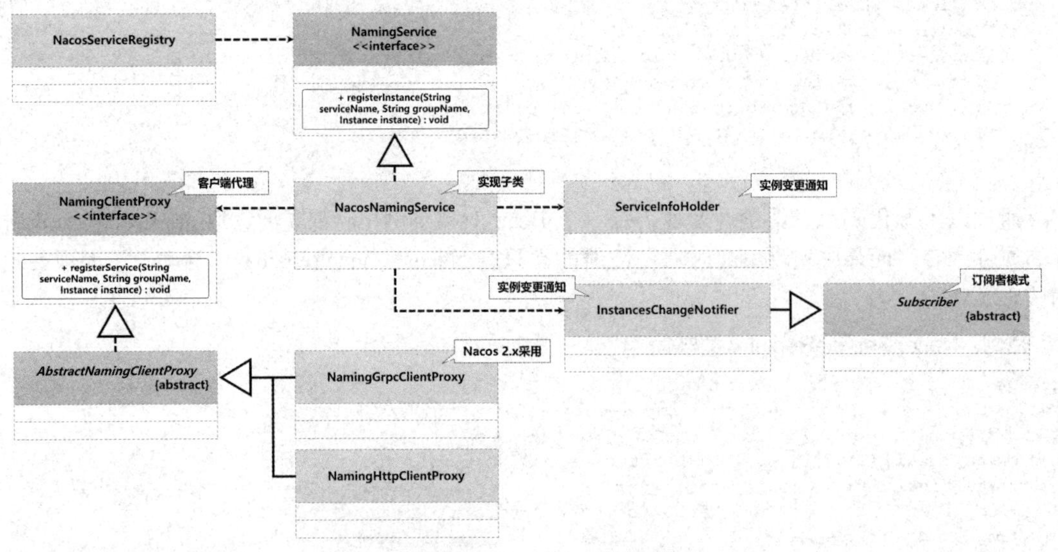

图 3-37　NamingService 实现结构

通过图 3-37 所示的结构可以清楚地发现，NacosNamingService 实现子类将通过 NamingClientProxy 客户端代理来实现 Nacos 服务注册，而服务注册所采用的协议包括 HTTP 与 gRPC 两种。由于此时所使用的 Nacos 版本为 2.x，因此打开 NamingGrpcClientProxy 类的核心源代码进行观察。

范例：NamingGrpcClientProxy 类核心源代码

```
package com.alibaba.nacos.client.naming.remote.gprc;
public class NamingGrpcClientProxy extends AbstractNamingClientProxy {   // gRPC实现
    @Override
    public void registerService(String serviceName, String groupName,
                Instance instance) throws NacosException {
        InstanceRequest request = new InstanceRequest(namespaceId, serviceName, groupName,
            NamingRemoteConstants.REGISTER_INSTANCE, instance);          // 创建注册实例数据
        requestToServer(request, Response.class);                        // 服务注册
        namingGrpcConnectionEventListener.cacheInstanceForRedo(
          serviceName, groupName, instance);
    }
    private <T extends Response> T requestToServer(AbstractNamingRequest request,
          Class<T> responseClass) throws NacosException {
        try {
            request.putAllHeader(getSecurityHeaders());                  // 设置头信息
            request.putAllHeader(getSpasHeaders(
                    NamingUtils.getGroupedNameOptional(request.getServiceName(),
                            request.getGroupName())));                   // 设置实例数据
            Response response =
                    requestTimeout < 0 ? rpcClient.request(request) :
                    rpcClient.request(request, requestTimeout);          // 请求响应
            if (ResponseCode.SUCCESS.getCode() != response.getResultCode()) {
                throw new NacosException(response.getErrorCode(), response.getMessage());
            }
            if (responseClass.isAssignableFrom(response.getClass())) {
                return (T) response;
            }
        } catch (Exception e) {
            throw new NacosException(NacosException.SERVER_ERROR, "Request … ", e);
        }
        throw new NacosException(NacosException.SERVER_ERROR, "Server return …");
    }
}
```

通过以上源代码可以发现，Nacos 服务注册时会将所有配置的注册数据包装在请求中发送到 Nacos 服务器，随后根据响应的状态判断成功或失败，如果失败则根据不同的类型抛出相应的异常。

3.4.4 NacosDiscoveryAutoConfiguration

NacosDiscovery Auto Configuration

视频名称 0317_【理解】NacosDiscoveryAutoConfiguration

视频简介 Nacos 服务除了注册支持之外还包含发现处理功能，Spring Cloud 就提供了 NacosDiscoveryAutoConfiguration 发现自动配置类。本视频通过源代码为读者分析 NacosDiscoveryAutoConfiguration 类的实现结构，并讲解 NacosServiceDiscovery 工具类所提供的服务发现方法的作用。

消费端程序需要通过 Nacos 获取全部注册的提供端微服务信息，所以 Spring Cloud 在整合 Nacos 服务时，为了便于所有服务实例数据的获取，提供了 NacosDiscoveryAutoConfiguration 自动配置类，此类的源代码如下。

范例：NacosDiscoveryAutoConfiguration 类源代码

```
package com.alibaba.cloud.nacos.discovery;
@Configuration(proxyBeanMethods = false)
@ConditionalOnDiscoveryEnabled
@ConditionalOnNacosDiscoveryEnabled
public class NacosDiscoveryAutoConfiguration {
   @Bean
   @ConditionalOnMissingBean
   public NacosDiscoveryProperties nacosProperties() {          // Nacos属性配置
      return new NacosDiscoveryProperties();
   }
   @Bean
   @ConditionalOnMissingBean
   public NacosServiceDiscovery nacosServiceDiscovery(
         NacosDiscoveryProperties discoveryProperties) {        // Nacos服务发现工具类
      return new NacosServiceDiscovery(discoveryProperties);
   }
}
```

通过此类的源代码可以发现，所有在 application.yml 中配置的 Nacos 相关属性都会注入 NacosDiscoveryProperties 对象实例，而后在此配置类中会根据 NacosDiscoveryProperties 提供的配置属性实例化 NacosServiceDiscovery 对象。这样就可以基于此对象实例根据 serviceId 或实例信息进行 Nacos 服务查询，实现结构如图 3-38 所示。

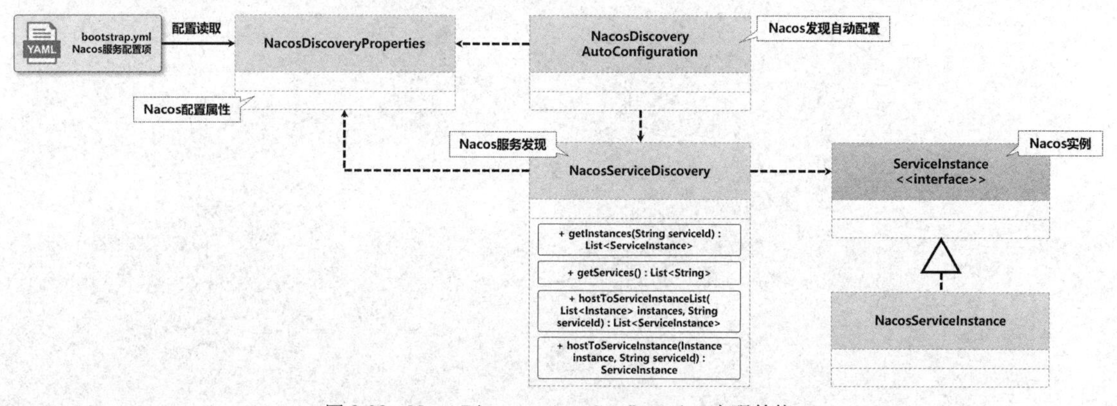

图 3-38 NacosDiscoveryAutoConfiguration 实现结构

3.5 本章概览

1．微服务的消费端在明确地知道提供端服务地址的情况下才允许访问，但这样会造成配置的维护困难，所以在微服务架构中需要通过注册中心来实现微服务的节点管理。

2．Spring Cloud 支持的注册中心包括 Eureka（闭源且不再维护）、Consul、Nacos。

3．Nacos 实现了 CAP 原则中的“CP + AP”，而且 Nacos 2.x 提供了更高的服务性能。

4．Nacos 领域模型由大到小分别为命名空间（NameSpace）、分组（Group）、服务（Service）、集群（Cluster）、实例（Instance）和数据 ID（DataId）。

5．正确的 Nacos 配置或实例发布必须按照领域模型进行归类管理。

6．Spring Cloud Alibaba 通过“spring-cloud-starter-alibaba-nacos-discovery”依赖库即可实现 Nacos 服务自动注册。

第 4 章

Spring Cloud 集群服务

本章学习目标

1. 掌握 Nacos 服务集群的搭建方法，并可以使用 Nginx 实现多 Nacos 节点代理访问；
2. 掌握 HAProxy 的作用，并可以基于 HAProxy 实现 Nacos 注册服务代理访问；
3. 掌握客户端获取 Nacos 服务注册信息的操作方法，并可以采用随机算法实现微服务调用；
4. 掌握 Ribbon 组件的使用方法，并可以使用 Ribbon 实现负载均衡以及节点优先调用配置；
5. 掌握 Feign 转换组件的使用方法，并可以使用 Feign 将 REST 服务映射为接口调度；
6. 了解 Hystrix 的主要作用与实现，并可以实现微服务调用监控处理。

集群是高性能架构之中的重要技术手段，而为了获得良好的服务性能，除了提供足够多的服务端节点之外，还需要提供注册中心的集群应用，随后再基于合适的负载均衡算法充分发挥出集群设计的优势。本章将为读者综合讲解微服务集群、Nacos 服务集群、Ribbon 负载均衡以及 Feign 转换映射的概念与应用实现。

4.1 微服务集群

微服务集群简介

视频名称 0401_【掌握】微服务集群简介

视频简介 为了提高服务的处理性能，生产环境中会大量采用集群设计。本视频为读者分析微服务集群设计的意义，同时讲解本次要搭建的服务集群架构。

为了提高微服务的处理性能，要在多个节点上进行相同的微服务部署，这样就会形成一个个微服务集群。微服务如果需要被消费端访问，就必须通过 Nacos 注册中心进行注册管理。而随着微服务集群中的节点数量增加，注册中心的访问量也会增大。所以为了保证注册中心的服务查找性能，就需要进行服务中心的集群搭建。这样对于一个微服务架构来讲就需要两个基础的集群架构：服务集群（按照功能划分）、注册中心集群，如图 4-1 所示。

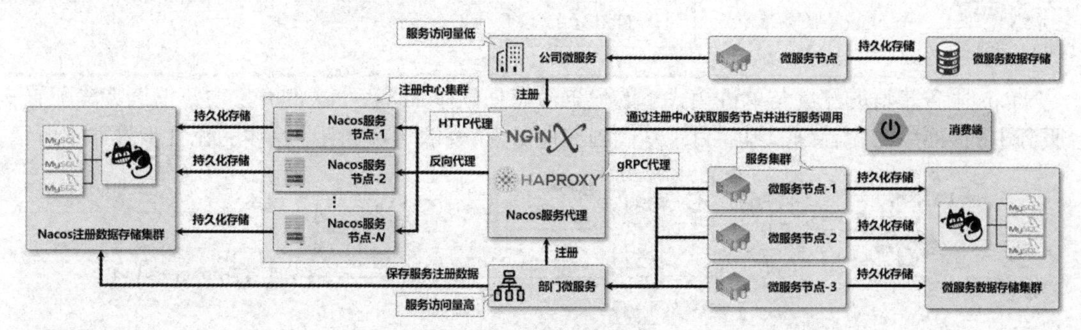

图 4-1 微服务集群架构

提示：本书暂不讨论 MySQL 集群。

本书主要围绕微服务自身的技术特点进行讲解，对 MySQL 数据库存储将采用单实例的形式完成。读者如果对此部分知识不熟悉，可以阅读本套丛书中有关架构的内容。

正常的微服务部署需要提供多台服务器。为了便于读者理解，本节将通过本地系统进程与虚拟机进行模拟，其中所涉及的主机信息以及相关的服务内容的定义如表 4-1 所示，节点分布如图 4-2 所示。

表 4-1　微服务集群架构

序号	主机名称	主机 IP 地址	端口	主机服务	认证信息
01	nacos-cluster-a	192.168.190.151	8848	Nacos	nacos / nacos
02	nacos-cluster-b	192.168.190.152	8848	Nacos	nacos / nacos
03	nacos-cluster-c	192.168.190.153	8848	Nacos	nacos / nacos
04	nacos-proxy	192.168.190.155	8848、9848	Nginx、HAProxy	-
05	nacos-mysql	192.168.190.159	3306	MySQL	mysql / mysqladmin
06	provider-dept-8001	127.0.0.1	8001	Spring Cloud 微服务	-
07	provider-dept-8002	127.0.0.1	8002	Spring Cloud 微服务	-
08	provider-dept-8003	127.0.0.1	8003	Spring Cloud 微服务	-
09	consumer-springboot-80	127.0.0.1	80	Spring Cloud 消费端	-

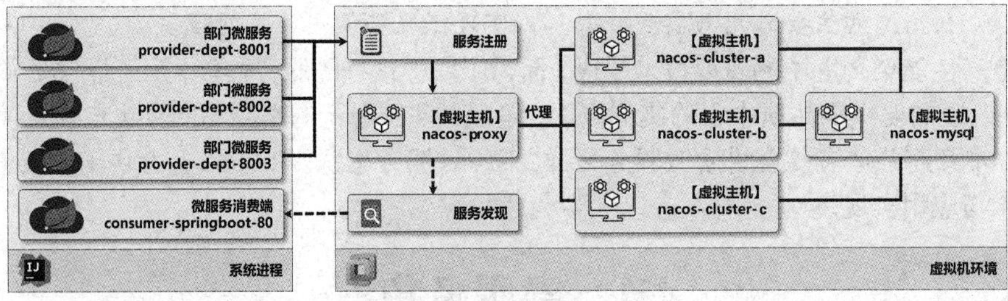

图 4-2　架构节点分布

提示：删除已有配置数据。

本节基于虚拟机实现 Nacos 集群配置，这样可以通过已有的 Nacos 虚拟机进行复制，而后实现集群配置。由于前面是采用单实例模式实现 Nacos 服务启动的，因此在进行集群配置之前请先删除 Nacos 主目录下的数据目录（命令：rm -rf /usr/local/nacos/{data,logs}），再进行虚拟机复制。同时也建议重新执行数据库脚本，使数据库的内容恢复到初始状态。

4.1.1　Nacos 服务集群

视频名称　0402_【掌握】Nacos 服务集群

视频简介　Nacos 内部直接提供了服务集群的配置支持。本视频将介绍通过虚拟机创建 3 个 Nacos 服务节点，并通过 Nacos 配置实现 3 个节点的集群关联。

Nacos 服务集群的配置与单机模式下的配置没有任何区别，在本次的开发中建议读者先配置一台服务主机，而后将配置好的组件目录复制到其他节点，如图 4-3 所示。具体的配置步骤如下所示。

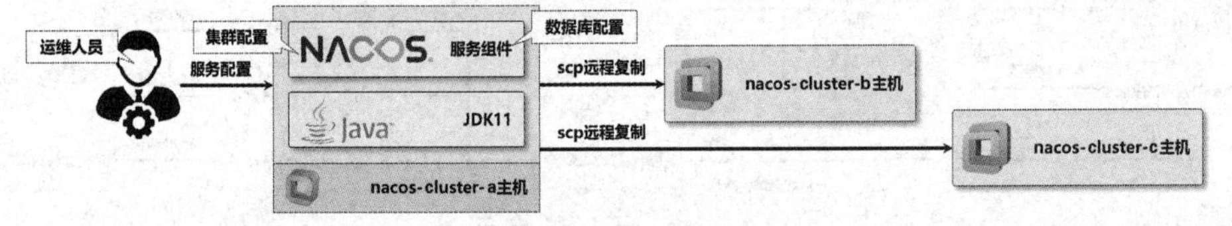

图 4-3　Nacos 服务集群配置

（1）【nacos-cluster-*主机】打开主机网卡配置文件，修改各自的 IP 地址。

```
vi /etc/sysconfig/network-scripts/ifcfg-ens33
```

nacos-cluster-a 主机 IP 地址修改	IPADDR=192.168.190.151
nacos-cluster-b 主机 IP 地址修改	IPADDR=192.168.190.152
nacos-cluster-c 主机 IP 地址修改	IPADDR=192.168.190.153

（2）【nacos-cluster-*主机】为了便于区分，建议修改每台虚拟主机中的主机名称。

```
vi /etc/hostname
```

nacos-cluster-a 主机 IP 修改	nacos-cluster-a
nacos-cluster-b 主机 IP 修改	nacos-cluster-b
nacos-cluster-c 主机 IP 修改	nacos-cluster-c

（3）【nacos-cluster-*主机】重新启动当前的虚拟机，使新的主机名称配置生效。

```
reboot
```

（4）【nacos-cluster-a 主机】为便于后续维护，修改 hosts 主机文件定义所需使用的全部主机。

```
vi /etc/hosts
192.168.190.151 nacos-cluster-a
192.168.190.152 nacos-cluster-b
192.168.190.153 nacos-cluster-c
```

（5）【nacos-cluster-a 主机】将当前主机的 hosts 配置文件复制到其他两台主机。

复制到 nacos-cluster-b 主机	scp /etc/hosts nacos-cluster-b:/etc/
复制到 nacos-cluster-c 主机	scp /etc/hosts nacos-cluster-c:/etc/

（6）【nacos-cluster-a 主机】按照前面讲解的方法配置 Nacos 服务集群，同时开启所有与 Nacos 有关的接口访问权限。

配置访问端口	firewall-cmd --zone=public --add-port=8848/tcp --permanent firewall-cmd --zone=public --add-port=7848/tcp --permanent firewall-cmd --zone=public --add-port=9848/tcp --permanent firewall-cmd --zone=public --add-port=9849/tcp --permanent
配置重新加载	firewall-cmd --reload

（7）【nacos-cluster-a 主机】Nacos 提供了集群的配置文件模板，开发者可以直接复制此模板文件定义集群主机。

```
cp /usr/local/nacos/conf/cluster.conf.example /usr/local/nacos/conf/cluster.conf
```

（8）【nacos-cluster-a 主机】编辑 Nacos 服务集群配置文件。

```
vi /usr/local/nacos/conf/cluster.conf
nacos-cluster-a:8848
nacos-cluster-b:8848
nacos-cluster-c:8848
```

（9）【nacos-cluster-a 主机】将当前配置完成的 cluster.conf 配置文件复制到其他 Nacos 节点。

复制到 nacos-cluster-b 主机	scp -r /usr/local/nacos/conf/cluster.conf nacos-cluster-b:/usr/local/nacos/conf
复制到 nacos-cluster-c 主机	scp -r /usr/local/nacos/conf/cluster.conf nacos-cluster-c:/usr/local/nacos/conf

（10）【nacos-cluster-*主机】在默认情况下 Nacos 服务集群只能通过 IP 地址进行配置，但是为了维护方便，建议采用主机名称进行配置，而这就需要在 Nacos 启动文件中追加主机 IP 地址，打开 Nacos 启动文件。

```
vi /usr/local/nacos/bin/startup.sh
```

nacos-cluster-a 主机	`JAVA_OPT="${JAVA_OPT} -Dnacos.server.ip=nacos-cluster-a"# 【新增项】主机IP地址`
nacos-cluster-b 主机	`JAVA_OPT="${JAVA_OPT} -Dnacos.server.ip=nacos-cluster-b"`
nacos-cluster-c 主机	`JAVA_OPT="${JAVA_OPT} -Dnacos.server.ip=nacos-cluster-c"`

（11）【nacos-cluster-*主机】启动所有节点的 Nacos 服务进程。

```
bash -f /usr/local/nacos/bin/startup.sh
```

后台日志输出	INFO The server IP list of Nacos is [192.168.190.151:8848, 192.168.190.152:8848, 192.168.190.153:8848]

（12）【本地系统】为便于后续访问，修改当前的开发系统的 hosts 主机文件，完整配置内容如下。

虚拟机映射	本地映射
192.168.190.151 nacos-cluster-a 192.168.190.152 nacos-cluster-b 192.168.190.153 nacos-cluster-c 192.168.190.155 nacos-proxy	127.0.0.1　provider-dept-8001 127.0.0.1　provider-dept-8002 127.0.0.1　provider-dept-8003 127.0.0.1　consumer-springboot-80

（13）【浏览器】打开客户端浏览器，访问集群中的任意 Nacos 节点，可见图 4-4 所示的界面。

图 4-4　Nacos 节点列表

提示：解决 Nacos 节点下线问题。

在查看 Nacos 节点数据或服务管理数据时，如果出现了“server is DOWN now, please try again later!”页面提示信息，那么一般都是 Nacos 服务地址发生改变所造成的。此时只需要删除所有 Nacos 主目录中的 data 目录，而后重新启动 Nacos 服务进程即可解决。

4.1.2　Nacos 控制台代理

Nacos 控制台代理

视频名称　0403_【掌握】Nacos 控制台代理

视频简介　Nacos 提供集群节点，而所有的节点需要一个统一的地址才可以对外提供集群服务。本视频将介绍通过 Nginx 实现 Nacos 集群代理配置。

Nacos 集群一旦启用，开发者就可以对该集群任意节点进行访问，以登录 Nacos 控制台，而所做的全部配置都会统一保存到 MySQL 数据库之中。为了便于 Nacos 集群节点中的控制台统一管理，就需要提供一个统一的代理地址，如图 4-5 所示。这样不仅可以有效地保护所有的 Nacos 节点，也便于运维人员管理。

Nacos 控制台基于 HTTP 访问，所以可以通过 Nginx 来实现所有的节点代理访问。Nginx 是一种高性能的 HTTP 和反向代理 Web 服务器，同时也提供了良好的负载均衡支持，读者可以按照如下步骤进行实现。

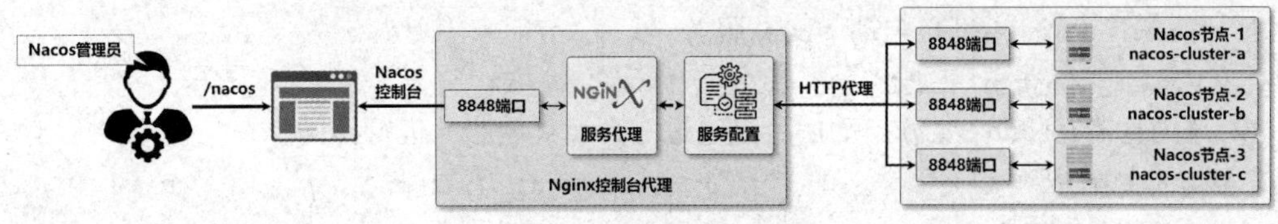

图 4-5 Nacos 控制台代理

（1）【nacos-proxy 主机】下载 Nginx 程序包并将其保存在/var/ftp 目录之中。

```
wget http://nginx.org/download/nginx-1.19.10.tar.gz
```

（2）【nacos-proxy 主机】将得到的 Nginx 源代码程序包解压缩到/usr/local/src 目录之中。

```
tar xzvf /var/ftp/nginx-1.19.10.tar.gz -C /usr/local/src/
```

（3）【nacos-proxy 主机】创建 Nginx 编译后的保存目录以及相关子目录。

```
mkdir -p /usr/local/nginx/{logs,conf,sbin}
```

（4）【nacos-proxy 主机】进入 Nginx 源代码目录。

```
cd /usr/local/src/nginx-1.19.10/
```

（5）【nacos-proxy 主机】为系统安装 Nginx 编译所需的依赖库。

```
yum -y install pcre-devel openssl openssl-devel
```

（6）【nacos-proxy 主机】对当前的 Nginx 源代码进行编译配置。

```
./configure --prefix=/usr/local/nginx/ --sbin-path=/usr/local/nginx/sbin/ --with-http_ssl_module \
--conf-path=/usr/local/nginx/conf/nginx.conf --pid-path=/usr/local/nginx/logs/nginx.pid \
--error-log-path=/usr/local/nginx/logs/error.log
--http-log-path=/usr/local/nginx/logs/access.log --with-http_v2_module
```

（7）【nacos-proxy 主机】启动 Nginx 编译与安装操作。

```
make && make install
```

（8）【nacos-proxy 主机】修改 Nginx 配置文件，追加 Nacos 集群主机。可以根据需要配置不同的访问权重。

<table>
<tr><td>打开配置文件</td><td><pre>vi /usr/local/nginx/conf/nginx.conf</pre></td></tr>
<tr><td>定义集群节点</td><td><pre>upstream nacoscluster {
	server 192.168.190.151:8848 weight=3 ;
	server 192.168.190.152:8848 weight=1 ;
	server 192.168.190.153:8848 weight=2 ;
}</pre></td></tr>
<tr><td>配置集群访问</td><td><pre>server {
	listen 8848;
	server_name localhost;
	location / {
		proxy_pass http://nacoscluster;
		root /nacos/;
	}
}</pre></td></tr>
</table>

（9）【nacos-proxy 主机】修改防火墙，添加 8848 访问规则。

配置访问端口	`firewall-cmd --zone=public --add-port=8848/tcp --permanent`
配置重新加载	`firewall-cmd --reload`

（10）【nacos-proxy 主机】查看当前的 Nginx 配置是否正确。

<table>
<tr><td colspan="2"><pre>/usr/local/nginx/sbin/nginx -t</pre></td></tr>
<tr><td>配置检测结果</td><td><pre>nginx: the configuration file /usr/local/nginx/conf/nginx.conf syntax is ok
nginx: configuration file /usr/local/nginx/conf/nginx.conf test is successful</pre></td></tr>
</table>

（11）【nacos-proxy 主机】启动 Nginx 服务。

```
/usr/local/nginx/sbin/nginx;
```

（12）【浏览器】Nginx 服务启动后可以通过 Nginx 代理服务进行访问。

```
nacos-proxy:8848/nacos;
```

4.1.3　gRPC 注册服务代理

gRPC 注册服务代理

视频名称　0404_【掌握】gRPC 注册服务代理

视频简介　为便于服务注册与消费发现处理，可以对 gRPC 的代理进行统一管理。本视频为读者讲解如何基于 HAProxy 代理组件实现 Nacos 节点访问配置。

所有的微服务在向 Nacos 注册中心进行服务注册后才可以被消费端访问，所以在 Spring Cloud Alibaba 套件之中，开发者可以直接在 Nacos 注册地址处填写全部可用的 Nacos 节点，随后可以根据需要自动地选择合适的节点进行服务注册。

范例：微服务向 Nacos 集群注册

```
spring:                                         # Spring配置
  cloud:                                        # Spring Cloud配置
    nacos:                                      # Nacos注册中心
      config:                                   # 配置服务
        server-addr: nacos-cluster-a:8848, nacos-cluster-b:8848, nacos-cluster-c:8848
```

虽然此时可以实现服务注册，但是考虑到 Nacos 集群节点的安全管理、节点扩充以及服务性能的问题，最佳的做法是直接通过代理节点来进行 Nacos 集群访问，如图 4-6 所示。

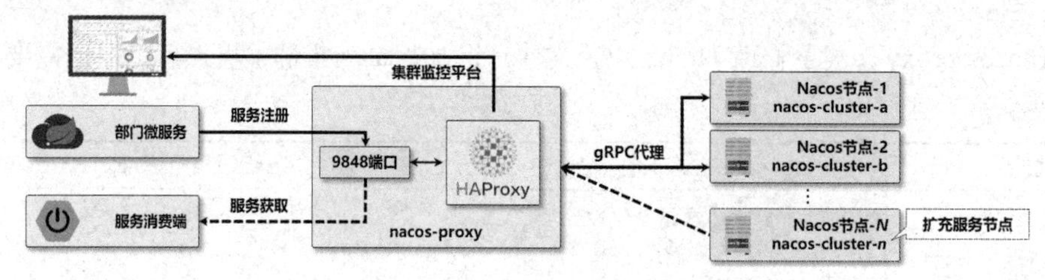

图 4-6　Nacos 服务注册与获取

Nacos 2.x 为了提高服务注册与发现管理的性能，采用了 gRPC 协议，所以在进行服务代理时就可以利用 HAProxy 代理组件来实现 Nacos 集群管理。HAProxy 是一款高可用组件，可以有效地实现集群服务节点的负载均衡以及基于 TCP（第四层）和 HTTP（第七层）应用的代理软件。开发者可以通过 HAProxy 官方站点免费获取该组件，如图 4-7 所示。

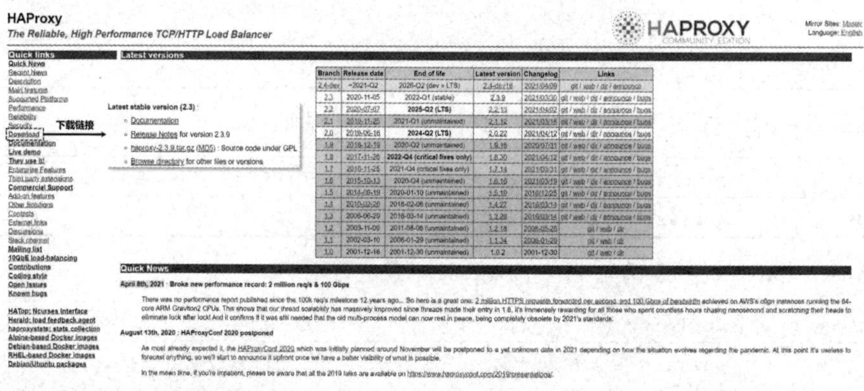

图 4-7　HAProxy 首页

提问：为什么不使用 Nginx 代理？

前面已经通过 Nginx 实现了 8848 端口集群管理，为什么此处不继续使用 Nginx 代理却额外增加了一个 HAProxy 组件？

回答：Nginx 对 gRPC 支持有限。

Nginx 从 1.13.10 版本开始支持 gRPC 协议，但是其在进行代理的过程中会遇到 HTTP 转换问题(接收的 HTTP 2 代理后变为 HTTP 1.1)，导致最终的代理失败，而且 Nginx 暂时也没有完善 gRPC 代理支持的计划。这样只能够通过 HAProxy 工具实现 gRPC（HTTP 2）代理，而 HAProxy 除了提供代理功能之外，也提供集群监控的功能，这一点会更加便于服务管理和服务状态监控。

本次使用的 HAProxy 版本为 2.3.9。HAProxy 用 C 语言开发，下载后获得的只是 HAProxy 源代码程序包，这样就需要在程序编译与安装后再进行服务配置，具体的操作步骤如下。

（1）【nacos-proxy 主机】通过 wget 命令下载 HAProxy 源代码程序包，也可以通过 FTP 客户端工具上传已经下载的 HAProxy 程序包。需要注意的是，此时程序包保存的父路径为/var/ftp/。

```
wget http://www.haproxy.org/download/2.3/src/haproxy-2.3.9.tar.gz
```

（2）【nacos-proxy 主机】将下载的 HAProxy 源代码包解压缩到/usr/local/src/目录。

```
tar xzvf /var/ftp/haproxy-2.3.9.tar.gz -C /usr/local/src/
```

（3）【nacos-proxy 主机】为了程序编译方便，建议先进入 HAProxy 解压缩后的源代码目录。

```
cd /usr/local/src/haproxy-2.3.9/
```

（4）【nacos-proxy 主机】编译 HAProxy 源代码，并设置编译前缀。

```
make TARGET=custom ARCH=x86_64 PREFIX=/usr/local/haproxy
```

（5）【nacos-proxy 主机】HAProxy 源代码编译成功后进行组件安装操作。

```
make install PREFIX=/usr/local/haproxy
```

（6）【nacos-proxy 主机】创建 haproxy.cfg 配置文件。

```
vi /usr/local/haproxy/haproxy.cfg
```

（7）【nacos-proxy 主机】在 haproxy.cfg 配置文件中需要配置默认环境、集群节点等信息，具体配置项如下。

全局配置	global # 全局配置 log 127.0.0.1 local0 # 启用日志 nbproc 1 # 监控进程个数 maxconnrate 300 # 进程每秒所能创建的最大连接数 maxcomprate 300 # 压缩速率 maxsessrate 500 # 进程每秒能创建的会话数量 chroot /usr/local/haproxy # HAProxy部署路径 pidfile /usr/local/haproxy/haproxy.pid # pid文件存储路径 maxconn 30000 # 进程所能接收的最大并发连接数 user haproxy # 启动用户名 group haproxy # 启动用户组 daemon # 后台模式运行 stats socket /usr/local/haproxy/stats # 开启统计Socket
默认配置	defaults # 默认配置 mode tcp # TCP处理模式 log global # 全局日志配置 option dontlognull # 不记录健康日志信息 option redispatch # 允许重新分配会话 option http-use-htx # 启用HTTP 2 option logasap # 传输大文件时提前记录日志 Option tcplog # 日志类别 retries 3 # 失败重试次数

默认配置

```
        timeout queue             1m                  # 队列超时
        timeout connect           5m                  # 连接超时
        timeout client            5m                  # 客户端超时
        timeout server            5m                  # 服务端超时
        timeout http-keep-alive   100s                # 保持HTTP连接
        timeout check             10s                 # 超时检查
```

管理控制台

```
listen admin_stats                                    # 管理控制台
        stats                     enable              # 启用管理控制台
        bind                      0.0.0.0:9999        # 监控端口设置
        mode                      http                # 管理控制台模式
        log                       global              # 日志配置
        maxconn                   10                  # 最大连接数量
        stats uri                 /admin              # 登录监控子路径配置
        stats realm               welcome\ Haproxy    # 登录提示信息
        stats auth                admin:admin         # 监控的账号密码
        stats admin               if TRUE             # 启用管理员模式
        option                    httplog             # HTTP日志记录
        stats refresh             30s                 # 监控刷新时间
        stats                     hide-version        # 隐藏页面版本号
```

代理集群配置

```
frontend nacos_cluster                                # 代理集群配置(名称自定义)
        bind                      :9848               # 代理端口
        mode                      tcp                 # 代理模式
        log                       global              # 日志配置
        maxconn                   8000                # 最大连接数
        default_backend           nacos_cluster_nodes # 代理节点名称(名称自定义)
```

集群节点配置

```
backend nacos_cluster_nodes                           # 集群节点配置(名称自定义)
        mode                      tcp                 # 代理模式
        server  nacos-a nacos-cluster-a:9848 check    # 集群节点
        server  nacos-b nacos-cluster-b:9848 check    # 集群节点
        server  nacos-c nacos-cluster-c:9848 check    # 集群节点
```

（8）【nacos-proxy 主机】要使用 HAProxy 服务，需要在系统中添加一个 haproxy 的用户。

```
useradd haproxy
```

（9）【nacos-proxy 主机】启动 HAProxy 服务进程，在启动时需要通过“-f”参数设置配置文件路径。

```
/usr/local/haproxy/sbin/haproxy -f /usr/local/haproxy/haproxy.cfg
```

（10）【nacos-proxy 主机】HAProxy 启动后会自动占用 9848（Nacos 注册服务代理）端口和 9999（HAProxy 管理控制台）端口，需要在防火墙中配置访问端口。

配置访问端口

```
firewall-cmd --zone=public --add-port=9848/tcp -permanent
firewall-cmd --zone=public --add-port=9999/tcp --permanent
```

配置重新加载

```
firewall-cmd --reload
```

（11）【浏览器】启动完成后通过浏览器访问 HAProxy 控制台，就可以见到图 4-8 所示的界面。

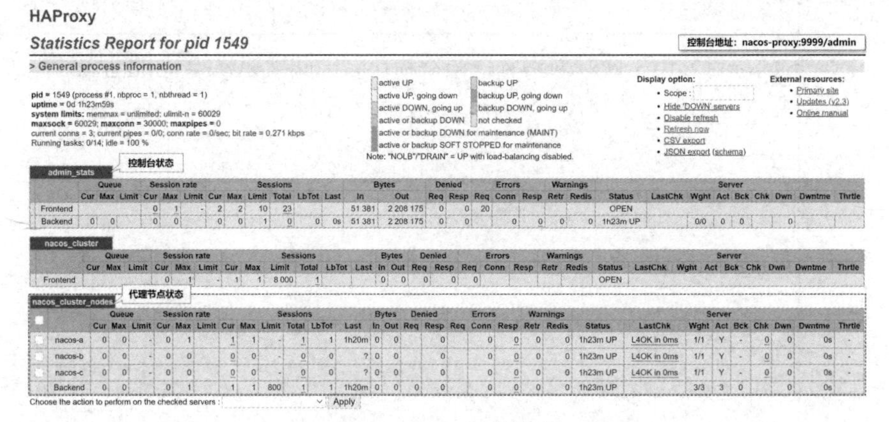

图 4-8　HAProxy 控制台

4.1.4 微服务集群注册

微服务集群注册

视频名称 0405_【掌握】微服务集群注册

视频简介 注册中心在经过集群处理后就可以保存更多的微服务注册信息。本视频介绍基于已有的部门微服务进行改造，并实现微服务集群注册。

为了提高某一个微服务的处理性能，较为常见的做法就是为其配置更多的处理节点。为了便于微服务的节点管理，必须采用相同的 Nacos 领域模型进行服务注册，这样消费端才可以通过指定的领域模型获取微服务数据。

本次的服务集群继续使用前面的部门微服务实现，在一台主机上模拟出 3 个部门微服务的节点（提供不同的监听端口）。而为了便于读者观察微服务节点数据的调用情况，本节会分别创建 3 个不同的数据库进行存储，实现架构如图 4-9 所示。具体的实现步骤如下。

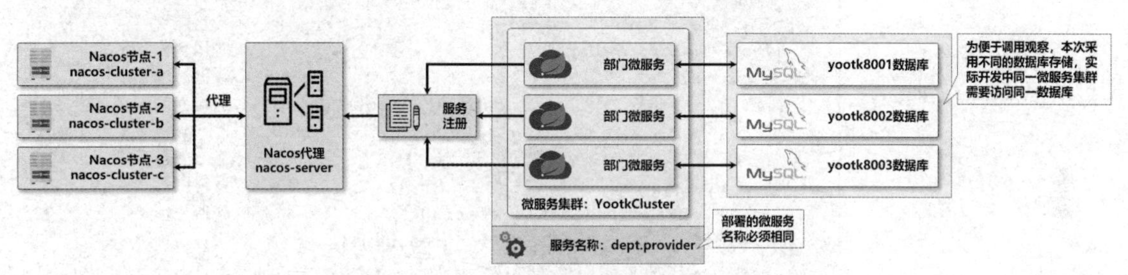

图 4-9 微服务集群注册

（1）【MySQL 数据库】分别创建 yootk8002 和 yootk8003 两个新的数据库，此时 3 个数据库所保存的数据表结构与内容相同（loc 字段内容不同，为当前各自数据库的名称，用于实现调用标记）。

创建 yootk8002 数据库	`CREATE DATABASE yootk8002 CHARACTER SET UTF8;`
创建 yootk8003 数据库	`CREATE DATABASE yootk8003 CHARACTER SET UTF8;`

（2）【microcloud 项目】复制 provider-dept-8001 子模块分别为 provider-dept-8002 和 provider-dept-8003，如果这些复制模块需要被 Gradle 项目管理，则需要同时修改 settings.gradle 配置文件，追加模块名称。

`include 'provider-dept-8001'`	已有的微服务模块
`include 'provider-dept-8002'`	复制后的微服务模块
`include 'provider-dept-8003'`	复制后的微服务模块

（3）【microcloud 项目】由于此时新增加了两个模块，因此需要修改 build.gradle 配置文件，定义这两个新模块所需要的依赖，这些依赖与 provider-dept-8001 依赖相同。

```
project('provider-dept-8002') {                    // 新增子模块
   dependencies { // 与“provider-dept-8001”依赖相同 }
}
project('provider-dept-8003') {                    // 新增子模块
   dependencies { // 与“provider-dept-8001”依赖相同 }
}
```

（4）【provider-dept-*子模块】修改项目中的 application.yml 配置文件（端口、Nacos 注册地址、数据库连接地址）。为了让读者更好地理解 Nacos 领域模型，本次将部门微服务定义在两个不同的集群中，即 YootkCluster、MuyanCluster。

配置 provider-dept-8001

```
server:                                                  # 服务端配置
  port: 8001                                             # 8001端口
```

```
spring:
  application:                                           # 应用配置
    name: dept.provider                                  # 应用名称
  datasource:                                            # 数据源配置
    url: jdbc:mysql://localhost:3306/yootk8001           # 连接地址
  cloud:                                                 # Cloud配置
    nacos:                                               # Nacos注册中心
      discovery:                                         # 发现服务
        server-addr: nacos-server:80                     # 注册地址
        service: ${spring.application.name}              # 服务名称
        cluster-name: YootkCluster                       # 集群名称
```

配置 provider-dept-8002

```
server:                                                  # 服务端配置
  port: 8002                                             # 8002端口
```

```
spring:
  application:                                           # 应用配置
    name: dept.provider                                  # 应用名称
  datasource:                                            # 数据源配置
    url: jdbc:mysql://localhost:3306/yootk8002           # 连接地址
  cloud:                                                 # Cloud配置
    nacos:                                               # Nacos注册中心
      discovery:                                         # 发现服务
        server-addr: nacos-server:80                     # 注册地址
        service: ${spring.application.name}              # 服务名称
        cluster-name: YootkCluster                       # 集群名称
```

配置 provider-dept-8003

```
server:                                                  # 服务端配置
  port: 8003                                             # 8003端口
```

```
spring:
  application:                                           # 应用配置
    name: dept.provider                                  # 应用名称
  datasource:                                            # 数据源配置
    url: jdbc:mysql://localhost:3306/yootk8003           # 连接地址
  cloud:                                                 # Cloud配置
    nacos:                                               # Nacos注册中心
      discovery:                                         # 发现服务
        server-addr: nacos-server:80                     # 注册地址
        service: ${spring.application.name}              # 服务名称
        cluster-name: MuyanCluster                       # 集群名称
```

（5）【provider-dept-*子模块】在每一个模块的启动类上追加 Nacos 发现服务注解。

```
package com.yootk.provider;
@SpringBootApplication
@EnableDiscoveryClient                          // Nacos发现服务注解
public class StartProviderDept800XApplication {
   public static void main(String[] args) {}
}
```

（6）【provider-dept-*子模块】分别启动当前的 3 个微服务程序，分别在各自的端口监听，同时这 3 个微服务也会自动注册到 Nacos 集群之中，并依据服务名称进行实例管理。

（7）【Nacos 控制台】访问 Nacos 控制台，观察当前的微服务注册信息，结果如图 4-10 所示。

提示：解决 Nacos 服务无法注册问题。

在微服务启动时，如果出现“Connection is unregistered”或“Client not connected,current status:STARTING.”错误，则都是因为无法访问 9848 端口（gRPC 无法通信）。此时需要排查代理服务器是否已经正确开启了 9848 端口。

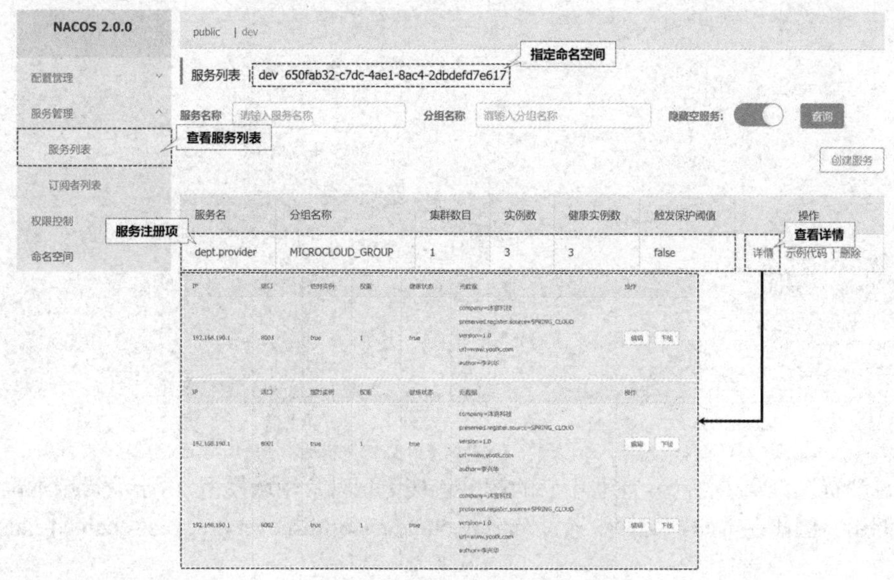

图 4-10　查看微服务注册信息

4.1.5　客户端服务访问

视频名称　0406_【掌握】客户端服务访问

视频简介　Nacos 除实现了微服务数据的管理之外，还提供客户端的服务发现管理支持，这样就可以通过服务的信息配置找到相关的服务节点进行调用。本视频将介绍实现消费端的服务注册以及服务调用处理。

此时所有的微服务已经成功注册到 Nacos 集群之中，下面需要编写消费端，而后该消费端设置与微服务相同的领域模型数据（NameSpaceID、Group、ServiceName）即可实现微服务数据调用，如图 4-11 所示。

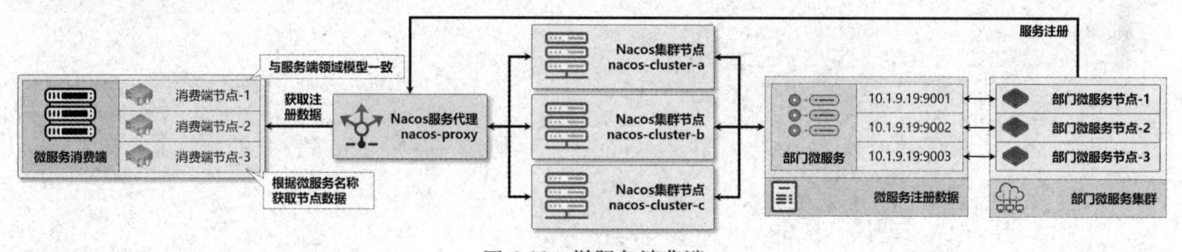

图 4-11　微服务消费端

客户端在实现服务查找时需要在项目中配置“spring-cloud-starter-alibaba-nacos-discovery”依赖库，而后利用该依赖库所提供的 DiscoveryClient 实例，即可根据指定的微服务名称实现所有相关服务节点的获取。为便于理解，下面的程序将对消费端程序进行改造，同时将基于随机算法实现服务端节点调用。具体实现步骤如下。

（1）【microcloud 项目】修改 build.gradle 配置文件，为消费端引入 Nacos 依赖库。

```
project('consumer-springboot-80') {                          // 子模块
    dependencies {                                           // 配置子模块依赖
        implementation(project(':common-api'))               // 引入子模块
        implementation('com.alibaba.cloud:spring-cloud-starter-alibaba-nacos-discovery')
    }
}
```

（2）【consumer-springboot-80 子模块】修改 application.yml 配置文件定义中与 Nacos 访问的相

关配置项。

```
server:                                                    # 服务端配置
  port: 80                                                 # 80端口监听
spring:                                                    # Spring配置
  application:                                             # 应用配置
    name: consumer                                         # 应用名称
  cloud:                                                   # Spring Cloud配置
    nacos:                                                 # Nacos注册中心
      discovery:                                           # 发现服务
        username: nacos                                    # 用户名
        password: nacos                                    # 密码
        service: ${spring.application.name}                # 服务名称
        server-addr: nacos-proxy:8848                      # 服务地址
        namespace: 650fab32-c7dc-4ae1-8ac4-2dbdefd7e617    # 命名空间
        group: MICROCLOUD_GROUP                            # 组名称
        register-enabled: false                            # 取消服务注册
```

此时在消费端设置了 Nacos 注册中心的相关连接信息以及领域模型，由于消费端不再需要被其他服务所调用，因此在本配置文件中通过 spring.cloud.nacos.discovery.register-enabled=false 来取消服务注册。

（3）【consumer-springboot-80 子模块】消费端在进行服务调用时需要通过注册中心获取指定微服务的全部节点，为了便于服务调用，可以编写一个随机返回一个节点的工具类。

```
package com.yootk.consumer.util;
@Component
@Slf4j
public class RandomAccessUtil {
    @Autowired
    private DiscoveryClient discoveryClient;                              // 发现服务
    /**
     * 根据指定的微服务名称随机获取一个微服务的节点实例
     * @param serviceName 调用的微服务名称
     * @param uri 微服务调用子路径（服务功能）
     * @return 随机获取一个完整的微服务调用地址
     */
    public String getTargetUrl(String serviceName, String uri) {
        List<ServiceInstance> instances = this.discoveryClient
                .getInstances(serviceName);                               // 根据服务名称获取实例列表
        if (instances.size() == 0) {                                      // 没有返回任何实例数据
            throw new RuntimeException("Nacos服务名称ID，无法获取服务信息。"); // 抛出调用异常
        }
        List<String> collect = instances.stream()
              .map(instance -> instance.getUri().toString() + uri)
              .collect(Collectors.toList());                              // 拼凑服务地址
        int num = ThreadLocalRandom.current().nextInt(collect.size());    // 随机算法
        String targetURL = collect.get(num);                              // 随机获取一个地址
        log.info("获取Nacos服务注册地址：{}", targetURL);                    // 日志输出
        return targetURL;                                                 // 返回调用地址
    }
}
```

（4）【consumer-springboot-80 子模块】修改 DeptConsumerAction 程序类，基于 RandomAccessUtil 工具类实现接口调用。

```
package com.yootk.consumer.action;
@RestController                                                            // REST响应
@RequestMapping("/consumer/dept/*")                                        // 父路径
public class DeptConsumerAction {
    public static final String DEPT_ADD_URL = "/provider/dept/add";        // 部门增加地址
    public static final String DEPT_GET_URL = "/provider/dept/get/";       // 部门查询地址
    public static final String DEPT_LIST_URL = "/provider/dept/list";      // 部门列表地址
```

```
    public static final String DEPT_SPLIT_URL = "/provider/dept/split";  // 分页列表地址
    public static final String SERVICE_ID = "dept.provider";             // 服务ID
    @Autowired
    private RestTemplate restTemplate;                                     // 注入对象实例
    @Autowired
    private RandomAccessUtil randomAccessService;                          // 服务工具实例
    @GetMapping("add")
    public Object addDept(DeptDTO dept) {                                  // 增加部门
        String deptAddUrl = this.randomAccessService.getTargetUrl(
                SERVICE_ID, DEPT_ADD_URL);
        return this.restTemplate.postForObject(deptAddUrl, dept, Boolean.class);
    }
    @GetMapping("get")
    public Object getDept(Long deptno) {                                   // 查询部门
        String deptGetUrl = this.randomAccessService.getTargetUrl(
                SERVICE_ID, DEPT_GET_URL + deptno);
        return this.restTemplate.getForObject(deptGetUrl, DeptDTO.class);
    }
    @GetMapping("list")
    public Object listDept() {                                             // 部门列表
        String deptListUrl = this.randomAccessService.getTargetUrl(
                SERVICE_ID, DEPT_LIST_URL);
        return this.restTemplate.getForObject(deptListUrl, List.class);
    }
    @GetMapping("split")
    public Object splitDept(int cp, int ls, String col, String kw) {      // 分页列表
        String deptSplitUrl = this.randomAccessService.getTargetUrl(SERVICE_ID,
          DEPT_SPLIT_URL + "?cp=" + cp + "&ls=" + ls + "&col=" + col + "&kw=" + kw);
        return this.restTemplate.getForObject(deptSplitUrl, Map.class);
    }
}
```

（5）【consumer-springboot-80 子模块】修改消费端启动程序类，添加发现服务注解。

```
package com.yootk.consumer;
@SpringBootApplication
@EnableDiscoveryClient                                                     // 启用发现服务
public class StartConsumerApplication {
    public static void main(String[] args) {
        SpringApplication.run(StringConsumerApplication.class, args);     // 服务启动
    }
}
```

修改完成后，可以通过 Postman 工具访问消费端提供的 REST 接口，多次调用接口后可以发现会随机分配到不同的部门微服务节点实现数据操作。

4.1.6 CP 与 AP 模式切换

CP与AP模式切换

视频名称 0407_【掌握】CP 与 AP 模式切换

视频简介 Nacos 服务集群提供了数据一致性的同步管理。本视频为读者分析 Nacos 对 CAP 原则的实现，同时分析 CP 与 AP 实现算法，以及与之相关的源代码。

引入 Nacos 服务集群可以提高服务注册与发现的处理性能，然而在实际的 Nacos 运行过程之中，每一个 Nacos 数据节点都只会保存自己的配置数据。由于集群中的每个节点都需要满足数据的修改与获取的功能需求，因此需要对集群中的节点数据进行自动同步处理，如图 4-12 所示，即某一个节点数据发生改变时将同步到其他节点。

Nacos 服务集群会包含众多数据节点，这样所有的数据可以分别保存在不同的数据节点之中，一旦某一个节点发生故障，应该可以及时满足分区容错性（Partition Tolerance）原则（简称“P 原

则”），即可以通过其他非故障节点获取所需要的数据，如图 4-13 所示。

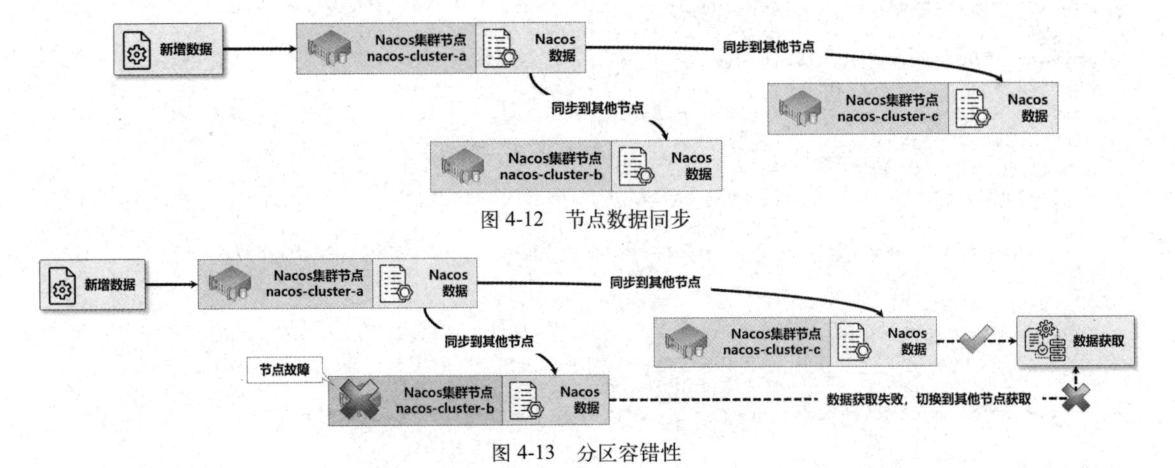

图 4-12　节点数据同步

图 4-13　分区容错性

虽然通过若干个节点可以实现数据的分区存储，但是这若干个节点之间毕竟有数据同步的需要，所以在进行数据处理时就存在两种实现原则：一致性（Consistency）原则（简称“C 原则”）、可用性（Availability）原则（简称“A 原则”）。CP 原则与 AP 原则的具体概念如下。

1. CP 原则：一致性原则 + 分区容错性原则

CP 原则属于强一致性原则，要求所有节点可以查询到的数据随时随刻保持一致（同步中的数据不可查询），即若干个节点形成一个逻辑的共享区域，如图 4-14 所示。某一个节点更新的数据会立即同步到其他数据节点之中，数据同步完成后才能返回成功的结果。但是在实际的运行过程中网络故障在所难免，如果此时若干个服务节点之间无法通信，就会出现错误，即牺牲了可用性原则（A 原则）。

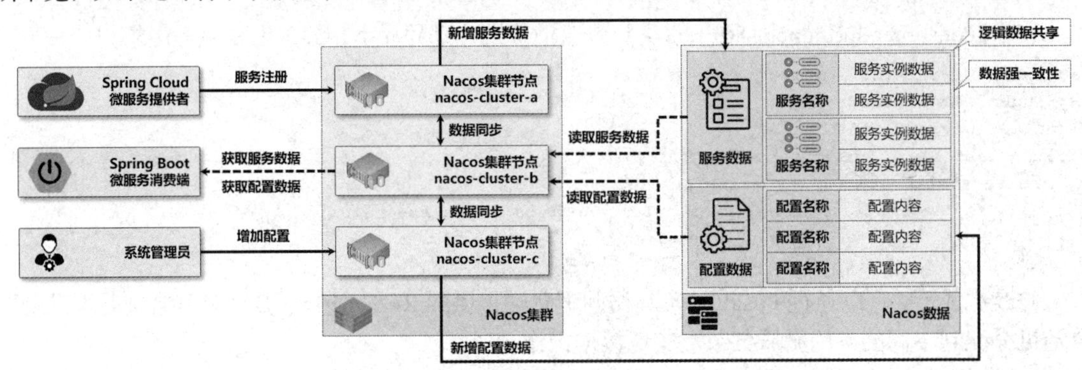

图 4-14　强一致性原则

在 Nacos 中的 CP 原则实现依靠的是 Raft 算法，Raft 将一致性算法分为几个部分，包括领导选取（Leader Selection）、日志复制（Log Replication）、安全（Safety），并且使用更强的一致性来减少必须考虑的状态。Raft 算法将服务端节点划分为 3 种不同的状态（或者称为角色）。

- **领导者（Leader）**：负责客户端交互及日志复制。在同一时刻的集群环境中最多存在一个 Leader。
- **跟随者（Follower）**：被动请求的节点，跟随 Leader 实现数据同步。
- **候选人（Candidate）**：一个临时的角色，只存在于 Leader 选举阶段。某个 Follower 要想成为 Leader，就需要发出投票请求，同时自己变为 Candidate，如果选举成功则变为 Leader，否则退回为 Follower。

在 Raft 算法中，所有的数据更新指令全部由 Leader 发出，更新命令全部保存在状态机（节点更新命令）之中，这样所有的 Follower 会依据状态机的顺序执行更新命令。当 Leader 不可用时，所有的 Follower 会变为 Candidate，并由一个节点发起投票。在投票通过后该节点成为新的 Leader，而此时其他 Candidate 将变为 Follower 继续实现数据同步处理。操作流程如图 4-15 所示。

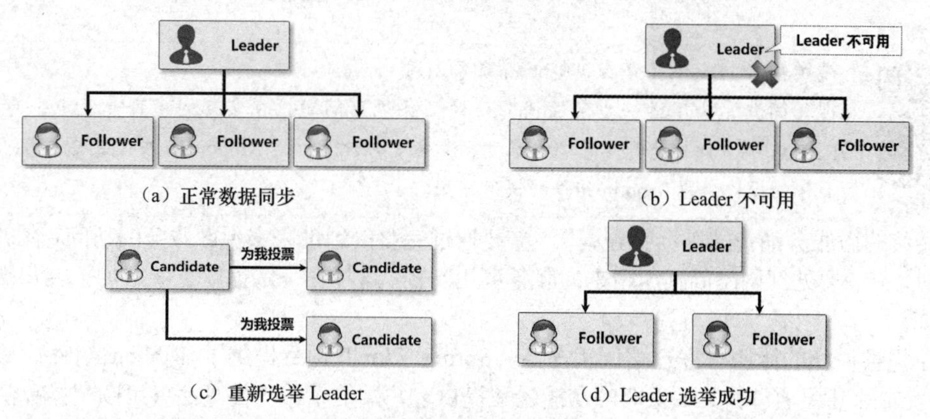

（a）正常数据同步　（b）Leader 不可用

（c）重新选举 Leader　（d）Leader 选举成功

图 4-15　Raft 节点状态

2. AP 原则：可用性原则 + 分区容错性原则

AP 原则属于弱一致性原则，集群中只要有存活的节点，发送来的所有请求就都可以得到正确的响应。在数据同步处理操作中，即便某些节点没有成功实现数据同步，也返回成功，这样就牺牲了一致性原则。

在 Nacos 中，AP 原则主要依靠 Distro 协议实现，该协议是阿里巴巴公司的私有协议，是一种定位临时数据的一致性协议。在该协议中不需要把数据存储到磁盘或者数据库，因为临时数据通常会和服务器保持一个会话，只要该会话存在，数据就不会消失。在 Distro 协议中，服务端接收数据后，服务端的负责节点实现数据写入后返回，而后台会将这些数据异步发送给其他节点。

提示：CP 与 AP 具体应用。

Spring Cloud 早期使用的 Eureka 注册中心采用 AP 模式，所以会出现数据更新不及时的问题（在一定时间内不同步）。而较为常见的 ZooKeeper 组件采用 CP 模式，所以在使用 ZooKeeper 时，更新任意一个节点数据后，其他节点可以立即获取该数据内容。

综上所述，当开发者选择 CP 模式时，可满足数据一致性，但是一旦出现网络异常就会导致服务停止。如果选择 AP 模式（Nacos 默认模式），即便存在网络异常，系统也可以提供服务，但是会出现节点数据不一致的情况。为了便于开发者使用，Nacos 提供了如下的 REST 接口，便于使用者实现 CP 模式与 AP 模式的切换。

切换为 CP 模式	`curl -X PUT "nacos-proxy:8848/nacos/v1/ns/operator/switches?entry=serverMode&value=CP&username=nacos&password=nacos"`
切换为 AP 模式	`curl -X PUT "nacos-proxy:8848/nacos/v1/ns/operator/switches?entry=serverMode&value=AP&username=nacos&password=nacos"`

根据大量的工程经验可以发现，集群服务的可用性要比集群服务的一致性更加重要，集群允许存在一定的数据同步时差，只要保证在一定的时间内达到数据的一致性，即所谓的最终一致性即可。

提示：Base 方案与 CAP 原则。

Base 是基本可用（Basically Available）、软状态（Soft State）和最终一致性（Eventual Consistency）这 3 个短语的简写，是对 CAP 原则的扩展。Base 方案允许系统集群在一段时间内存在数据不一致的情况，但是在规定的时间后数据必须实现最终一致性，这样就可以在满足 AP 原则的同时也在一定程度上符合 CP 原则。

4.2　Ribbon 负载均衡

视频名称　0408_【掌握】Ribbon 服务调用

视频简介　为了进一步解决消费端的服务调用问题，在实际的项目开发中，可以基于 Ribbon 组件简化消费端的调用难度。本视频为读者介绍 Ribbon 组件的主要作用，并通过具体的操作介绍 Ribbon 组件的快速应用。

在传统的微服务消费端进行服务调用，需要通过完整的地址来实现微服务的访问；在进行微服务调用时，又需要处理网络问题所带来的服务调用问题。这样在一定程度上就增加了客户端调用与维护难度。

为了提供便捷的微服务客户端调用支持，Spring Cloud 项目提供了 Ribbon 组件。该组件是 Netflix 发布的开源项目，其最重要的功能就是提供客户端的负载均衡算法（可以根据算法随机实现不同的服务端节点调用）。除此之外 Ribbon 组件还提供一系列完善的服务调用配置项，如连接超时、失败重试、访问权重、优先级调用等，在使用时将之与 RestTemplate 结合即可通过微服务的名称实现服务调用，如图 4-16 所示。

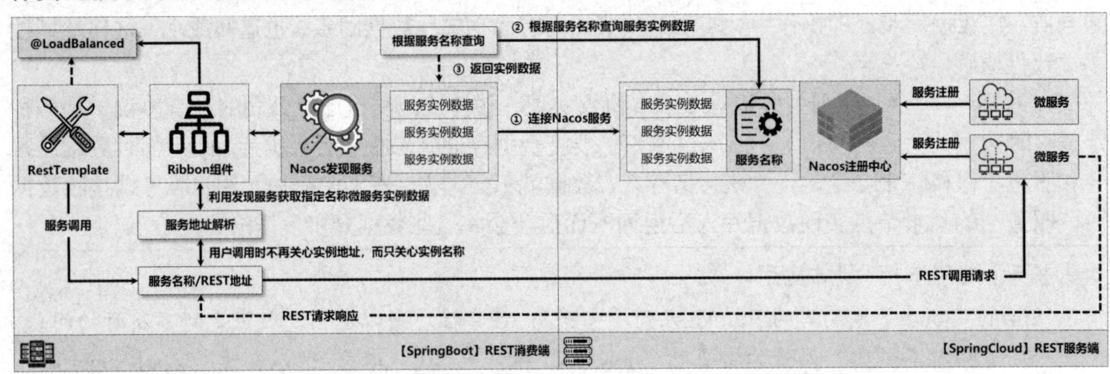

图 4-16　Ribbon 组件作用

Spring Cloud 为 Ribbon 提供了“org.springframework.cloud:spring-cloud-netflix-ribbon”专属依赖库（Nacos 发现服务中自动提供此依赖），只需要通过一个@LoadBalanced 注解来实现服务调用。下面通过具体的步骤来演示 Ribbon 组件的基本使用方法。

（1）【consumer-springboot-80 子模块】修改 RestTemplateConfig 配置类，在进行 RestTemplate 注册时使用 Ribbon 组件提供的@LoadBalanced 注解进行配置。

```
package com.yootk.consumer.config;
@Configuration                                          // 配置类
public class RestTemplateConfig {
    @Autowired                                          // 注入拦截器
    private MicroServiceLoadBalancerInterceptor interceptor;
    @Bean                                               // Bean注册
    @LoadBalanced                                       // Ribbon提供注解
    public RestTemplate getRestTemplate() {             // 获取RestTemplate实例
        RestTemplate restTemplate = new RestTemplate();
        restTemplate.setInterceptors(Collections.singletonList(
                this.interceptor));                     // 添加拦截器
        return restTemplate;
    }
}
```

（2）【consumer-springboot-80 子模块】修改 DeptConsumerAction 控制器类，在此类中，通过微服务名称，即可由 Ribbon 组件根据其内部使用的负载均衡算法动态获取一个微服务实例，以实现目标微服务调用。

```
package com.yootk.consumer.action;
@RestController                                          // REST响应
@RequestMapping("/consumer/dept/*")                      // 父路径
public class DeptConsumerAction {
    // 此时直接根据微服务名称即可实现微服务调用，Ribbon内部默认提供负载均衡算法
    public static final String DEPT_ADD_URL =
        "http://dept.provider/provider/dept/add";        // 部门增加REST地址
    public static final String DEPT_GET_URL =
        "http://dept.provider/provider/dept/get/";       // 部门查询REST地址
    public static final String DEPT_LIST_URL =
        "http://dept.provider/provider/dept/list";       // 部门列表REST地址
    public static final String DEPT_SPLIT_URL =
        "http://dept.provider/provider/dept/split";      // 部门分页列表REST地址
    @Autowired
    private RestTemplate restTemplate;                   // 注入对象实例
    @GetMapping("add")
    public Object addDept(DeptDTO dept) {                // 增加部门
        return this.restTemplate.postForObject(DEPT_ADD_URL, dept, Boolean.class);
    }
    @GetMapping("get")
    public Object getDept(Long deptno) {                 // 查询部门
        return this.restTemplate.getForObject(DEPT_GET_URL + deptno, DeptDTO.class);
    }
    @GetMapping("list")
    public Object listDept() {                           // 部门列表
        return this.restTemplate.getForObject(DEPT_LIST_URL, List.class);
    }
    @GetMapping("split")
    public Object splitDept(int cp, int ls, String col, String kw) { // 分页列表
        return this.restTemplate.getForObject(DEPT_SPLIT_URL + "?cp=" + cp + "&ls=" +
                ls + "&col=" + col + "&kw=" + kw, Map.class);
    }
}
```

此处代码中的消费端依然采用 RestTemplate 模板类，并根据注册的服务名称实现微服务的调用。在用户第一次进行微服务调用时，消费端会自动抓取 Nacos 上的服务配置数据，而后才会进行服务调用，同时可以在当前应用中得到如下日志信息。

```
c.n.l.DynamicServerListLoadBalancer      : DynamicServerListLoadBalancer for client dept.provider
initialized: DynamicServerListLoadBalancer:{
 NFLoadBalancer:name=dept.provider,current list of Servers=[169.254.19.173:8001, 169.254.19.173:8003,
169.254.19.173:8002], ...]
},Server stats: [
    [Server:169.254.19.173:8001;   Zone:UNKNOWN;  Total Requests:0;  ...]
,   [Server:169.254.19.173:8003;   Zone:UNKNOWN;  Total Requests:0;  ...]
,   [Server:169.254.19.173:8002;   Zone:UNKNOWN;  Total Requests:0;  ...]
]}ServerList:com.alibaba.cloud.nacos.ribbon.NacosServerList@73ff3c28
```

通过该日志信息可以发现，虽然服务调用是通过微服务名称进行的，但是其内部是直接基于 Ribbon 组件实现服务名称与实例地址之间的转换处理的。而当一个服务由多个节点组成时，也可以由 Ribbon 组件所提供的负载均衡算法随机抽取一个实例进行调用，如图 4-17 所示。由于此时的消费端只与服务名称关联，因此即便微服务集群存在节点变化，也不会对调用有任何影响。

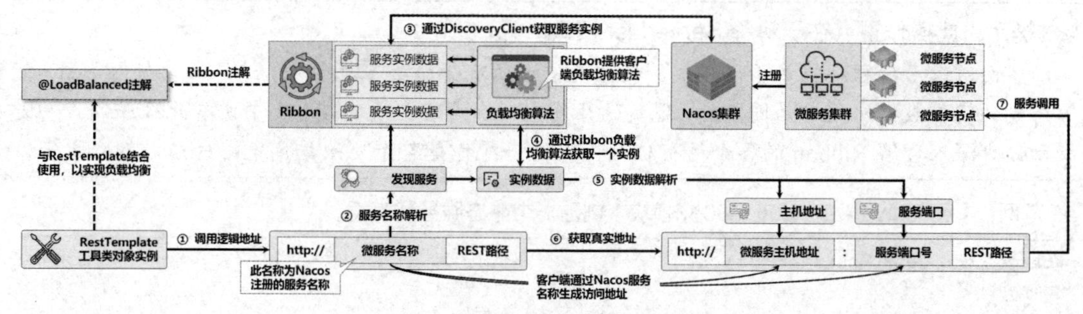

图 4-17　Ribbon 组件消费端调用

提示：开启 Ribbon“饿汉式”加载模式。

默认情况下 Ribbon 中的数据采用的是“懒加载”的模式，在第一次请求时会首先通过 Nacos 获取服务的实例数据，而后才会进行服务调用，所以执行速度会相对较慢。在第二次访问时，由于实例数据已经加载完成，因此执行速度就会快很多。如果想解决这个问题，则可以开启“饿汉式”加载模式，在消费端微服务启动时就加载所有相关的 Nacos 实例数据，这一操作可以通过修改 application.yml 配置文件实现。

范例：开启“饿汉式”加载模式

```
ribbon:                          # Ribbon配置
  eager-load:                    # 饿汉式加载
    enabled: true                # 模式开启
    clients: dept.provider       # 服务名称
```

程序重新启动时会自动获得“dept.provider”的所有注册的实例数据，这样在进行微服务调用时第一次就可以得到较好的性能。需要注意的是，此时的配置仅对“dept.provider”微服务提供了“饿汉式”加载，如果需要配置多个微服务的“饿汉式”加载，则可以对多个微服务名称采用“,”进行分隔。

4.2.1　ServerList 实例列表

ServerList 实例列表

视频名称　0409_【理解】ServerList 实例列表

视频简介　消费端通过 Ribbon 实现的微服务调用中需要及时进行 Nacos 数据更新，所以提供了 ServerList 内容保存接口。本视频为读者分析 ServerList 接口的主要使用方法，同时基于 NacosServerList 类的源代码进行 Nacos 服务数据获取分析。

客户端通过 Ribbon 进行微服务调用时，首先要通过 Nacos 注册中心来获取全部的微服务实例列表，而这些 Nacos 数据就会统一保存在 ServerList 接口实例之中，如图 4-18 所示。消费端在进行微服务调用时，会通过此实例列表获取一个实例地址，最终实现微服务的调用。

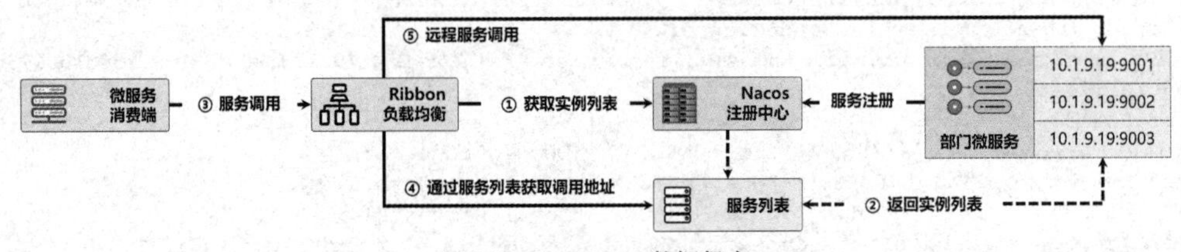

图 4-18　ServetList 数据保存

com.netflix.loadbalancer.ServerList 是由 Ribbon 提供的一个服务接口，该接口包含一个 List 集合，里面保存当前微服务的所有实例数据，该数据可以由开发者设置静态内容，或者动态获取。为便于读者理解，下面通过一个静态列表的形式为读者演示 ServerList 接口的作用。

提示：此操作重点在于解释 ServerList 接口。

下面所演示的程序将基于静态实例的方式进行所有微服务节点管理，这样不管如何更新，能够得到的实例数据永远都是固定的，现实的开发中不会存在此类需要。本书讲解此程序的目的是帮助读者深入理解 Ribbon 的各个运行机制，而这一操作是整个负载均衡策略的组成部分之一。

范例：【consumer-springboot-80 子模块】创建一个静态服务列表类

```
package com.yootk.consumer.ribbon;
@Component                              // Bean注册
@Slf4j                                  // 日志输出
```

```
public class DefaultServerList implements ServerList<Server> {   // 自定义服务列表
    @Override
    public List<Server> getInitialListOfServers() {              // 初始化服务列表
        return null;
    }
    @Override
    public List<Server> getUpdatedListOfServers() {              // 更新后的服务列表
        // 此时的实例数据可以通过DiscoveryClient类来动态获取，会定时刷新
        List<Server> allServers = new ArrayList<>();             // 实例集合
        allServers.add(new Server("127.0.0.1", 8001));           // 静态数据
        allServers.add(new Server("127.0.0.1", 8002));           // 静态数据
        log.info("更新服务列表：{}", allServers);                 // 日志输出
        return allServers;
    }
}
```

后台日志输出	c.y.consumer.ribbon.DefaultServerList : 更新服务列表：[127.0.0.1:8001, 127.0.0.1:8002]

此时的程序将所有的微服务调用固定在了两个实例之中，这样不管如何调用都只会集中在两个固定的实例上。同时通过后台的日志也可以发现，Ribbon 会持续进行服务列表的更新方法调用，以便及时获取具体的实例数据。

为便于用户开发，Ribbon 内部提供了 ConfigurationBasedServerList 子类（配置静态服务列表），以及 NacosServerList 子类（动态获取 Nacos 服务列表），继承结构如图 4-19 所示。Ribbon 会根据当前的系统环境配置选择相应的子类实现服务数据存储。

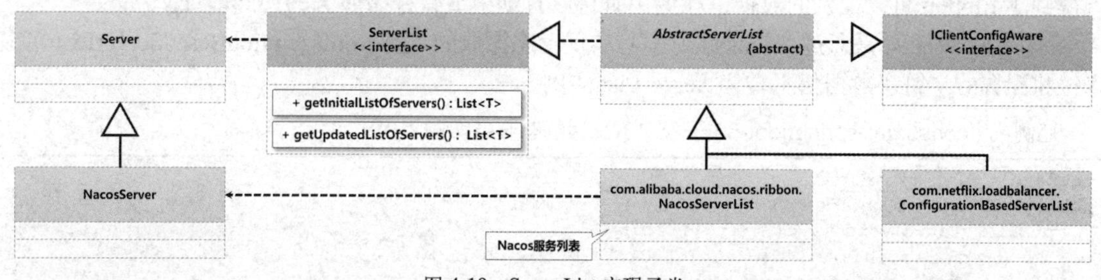

图 4-19 ServerList 实现子类

这是基于 Nacos 注册中心实现的服务管理，所有的服务数据都需要通过 Nacos 发现服务来动态查询。为便于读者理解，下面通过 NacosServerList 类源代码进行分析。

范例：NacosServerList 类源代码

```
package com.alibaba.cloud.nacos.ribbon;
public class NacosServerList extends AbstractServerList<NacosServer> {
   private NacosDiscoveryProperties discoveryProperties; // Nacos发现服务属性
   private String serviceId;                             // 调用服务ID
   public NacosServerList(NacosDiscoveryProperties discoveryProperties) {
      this.discoveryProperties = discoveryProperties;
   }
   @Override
   public List<NacosServer> getInitialListOfServers() {   // 返回初始化列表
      return getServers();
   }
   @Override
   public List<NacosServer> getUpdatedListOfServers() {   // 返回更新后列表
      return getServers();
   }
   private List<NacosServer> getServers() {               // 获取服务列表
      try {
         String group = discoveryProperties.getGroup();
         List<Instance> instances = discoveryProperties.namingServiceInstance()
              .selectInstances(serviceId, group, true);   // 服务查询
         return instancesToServerList(instances);         // Instance转NacosServer
      }
```

```
        catch (Exception e) {
            throw new IllegalStateException(
                    "Can not get service instances from nacos, serviceId=" + serviceId, e);
        }
    }
    private List<NacosServer> instancesToServerList(List<Instance> instances) {
        List<NacosServer> result = new ArrayList<>();              // 目标集合
        if (CollectionUtils.isEmpty(instances)) {                  // 集合个数为空
            return result;
        }
        for (Instance instance : instances) {                      // Instance集合迭代
            result.add(new NacosServer(instance));                 // 保存NacosServer集合
        }
        return result;
    }
    public String getServiceId() {                                 // 返回服务ID
        return serviceId;
    }
    @Override
    public void initWithNiwsConfig(IClientConfig iClientConfig) {
        this.serviceId = iClientConfig.getClientName();           // 根据配置返回服务名称
    }
}
```

通过以上源代码可以发现，在进行服务列表获取时，都是通过 Nacos 实现服务数据查询的。由于 Nacos 以 Instance 实例的形式返回微服务注册信息，因此需要将其转为 NacosServer 实例。这样就得到了 Ribbon 可以识别的数据，在调用时可以自动生成目标微服务的访问地址。

NacosServerList 是通过 Nacos 注册中心实现动态抓取的，而 ConfigurationBasedServerList 可以直接将配置的字符串数据信息转为 ServerList 实例。

范例：【consumer-springboot-80 子模块】配置静态服务实例

```
package com.yootk.test;
public class TestConfigurationBasedServerList {
    public static void main(String[] args) {
        String instants = "10.9.19.1:8001,10.9.19.2:8001,10.9.19.3:8001";     // 实例数据
        DefaultClientConfigImpl clientConfig = new DefaultClientConfigImpl();  // 客户端配置
        clientConfig.set(CommonClientConfigKey.ListOfServers, instants);      // 设置实例属性
        ConfigurationBasedServerList baseList = new ConfigurationBasedServerList();
        baseList.initWithNiwsConfig(clientConfig);                            // 初始化配置
        System.out.println(baseList.getUpdatedListOfServers());               // 获取服务列表
    }
}
```

程序执行结果	[10.9.19.1:8001, 10.9.19.2:8001, 10.9.19.3:8001]

此时的程序通过客户端的配置类实现了所有实例数据的配置，这样就可以通过 ConfigurationBasedServerList 提供的方法将相应的数据转换为 List<Server>数据信息。此时所采用的类结构关系如图 4-20 所示。

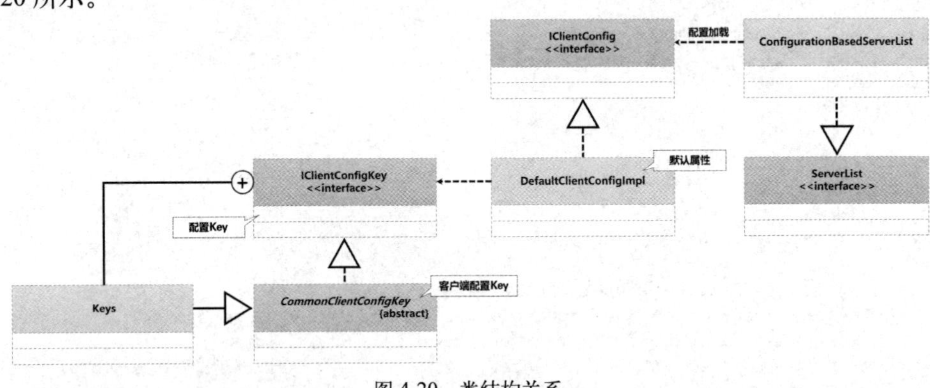

图 4-20　类结构关系

提示：IClientConfig 接口主要实现 Ribbon 环境配置。

在后续的 Ribbon 讲解中会存在大量的 IClientConfig 接口应用。该接口可以获取一些默认的消费端的配置。Ribbon 只为此接口提供了一个 DefaultClientConfigImpl 子类，该子类中存在一些默认的环境属性内容。这些属性的配置构成了整个 Ribbon 负载均衡机制，后文会有详细的讲解。

4.2.2 ILoadBalancer 负载均衡器

ILoadBalancer 负载均衡器

视频名称 0410_【理解】ILoadBalancer 负载均衡器

视频简介 在进行微服务集群构建后，通过 Ribbon 可以获取全部的服务实例数据，那么此时需要通过一定的算法进行调度，以避免某一个节点服务量过大。本视频为读者分析 Ribbon 负载均衡器的接口标准，同时讲解内部负载均衡器的实现与配置。

在进行微服务创建时往往要预估实际的并发访问量来进行服务节点的配置，而在服务集群配置完成后，又需要做到负载均衡，即不能让一个或几个节点的访问量过大，而应该让所有的节点分摊访问的负载压力。此时就需要通过 Ribbon 提供的负载均衡器来进行控制。

为了便于负载均衡器的管理，Ribbon 提供了 ILoadBalancer 接口标准，可以将所有的服务数据在此接口中进行配置，而后利用此接口提供的服务下线功能，实现服务状态的动态控制。用户需要获取某一个服务时，可以通过其内部提供的 chooseServer()方法实现服务实例获取，如图 4-21 所示。ILoadBalancer 接口定义如下。

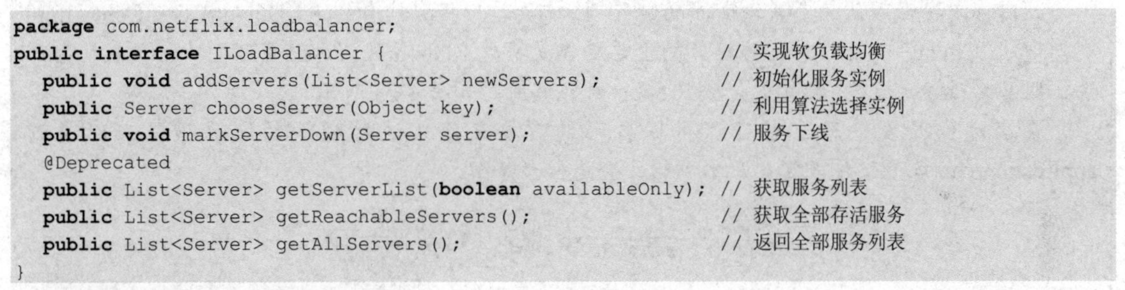

```
package com.netflix.loadbalancer;
public interface ILoadBalancer {                                       // 实现软负载均衡
  public void addServers(List<Server> newServers);                     // 初始化服务实例
  public Server chooseServer(Object key);                              // 利用算法选择实例
  public void markServerDown(Server server);                           // 服务下线
  @Deprecated
  public List<Server> getServerList(boolean availableOnly);            // 获取服务列表
  public List<Server> getReachableServers();                           // 获取全部存活服务
  public List<Server> getAllServers();                                 // 返回全部服务列表
}
```

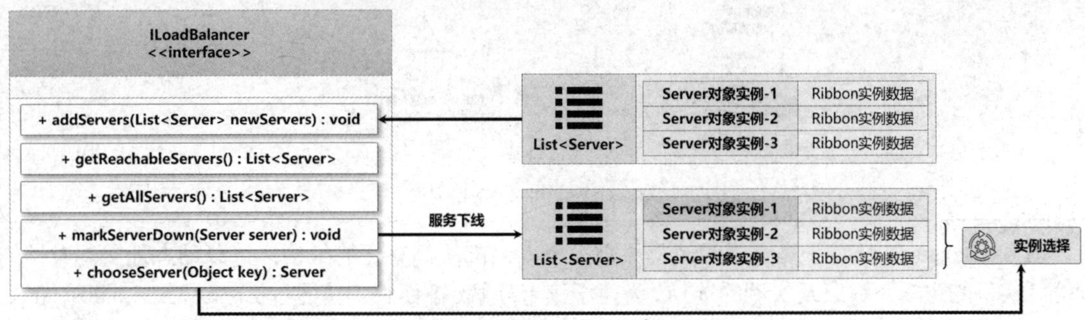

图 4-21 ILoadBanalcer 接口使用

ILoadBalancer 是一个接口标准，所以在实际的使用中需要通过接口的子类来实现具体的负载均衡器。在默认情况下 Ribbon 会使用 com.netflix.loadbalancer.ZoneAwareLoadBalancer 子类来实现负载均衡处理，而除了该子类之外，Ribbon 还提供 BaseLoadBalancer 子类、DynamicServerListLoadBalancer 子类、NoOpLoadBalancer 子类，这些子类的继承关系如图 4-22 所示。

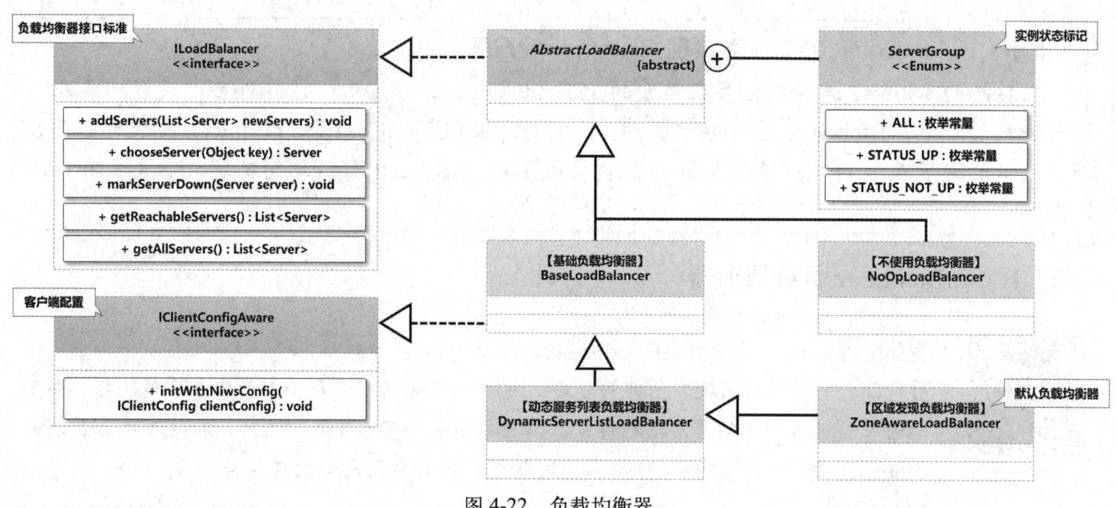

图 4-22　负载均衡器

提示：关于 Ribbon 默认配置

Ribbon 在服务处理过程中会提供许多接口，而这些接口在使用中都会提供大量的默认子类。这些子类的配置可以通过 DefaultClientConfigImpl 类查看。

范例：查看默认的负载均衡器配置

```
public static final String DEFAULT_NFLOADBALANCER_CLASSNAME = "com.netflix.loadbalancer.
ZoneAwareLoadBalancer";
```

此时可以发现默认的负载均衡器的配置项。需要特别提醒读者的是，DefaultClientConfigImpl 类实现了 IClientConfig 接口，该接口的主要功能是提供 Ribbon 默认的配置环境（如负载均衡算法、服务名称等），这些配置项会被 Ribbon 内部调用，如图 4-23 所示。

在实际的开发中用户也可以根据自己的需要来修改这些默认的配置项，可以通过 application.yml 配置文件来定义。本章后文会进一步讲解。

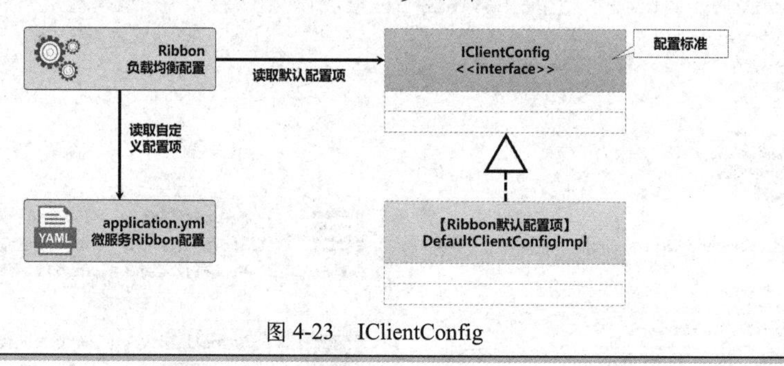

图 4-23　IClientConfig

ILoadBalancer 接口子类实现的核心原则在于对内部实例集合的维护，可以动态地实现数据的添加、实例的逻辑下线以及实例的选择。为便于读者理解，下面将直接通过 ZoneAwareLoadBalancer 演示实例数据的操作。

范例：【consumer-springboot-80 子模块】加载服务实例

```
package com.yootk.test;
public class TestLoadBalancer {
    public static void main(String[] args) {
        List<Server> serverList = new ArrayList<>();                             // 实例化集合
        serverList.add(createServer("muyan", "muyan.provider-dept", 8001));   // 保存实例
        serverList.add(createServer("muyan", "muyan.provider-dept", 8002));   // 保存实例
        serverList.add(createServer("yootk", "yootk.provider-dept", 8001));   // 保存实例
        ZoneAwareLoadBalancer loadBalancer = new ZoneAwareLoadBalancer();     // 负载均衡器
```

```
        loadBalancer.addServers(serverList);                           // 添加服务集合
        LoadBalancerStats stats = loadBalancer.getLoadBalancerStats(); // 获取统计数据
        stats.updateServerList(serverList);                            // 更新服务列表
        stats.getServerStats().keySet().forEach(currentServer -> {
            if (currentServer.getZone().equals("yootk")) {             // 查询标记
                loadBalancer.markServerDown(currentServer);            // 服务逻辑下线
            }
        });
        for (int i = 0; i < loadBalancer.getServerCount(true); i++) {
            System.out.println(loadBalancer.chooseServer(null));      // 筛选服务实例
        }
    }
    public static Server createServer(String zone, String url, int index) { // 创建实例
        Server server = new Server(url, index);                        // 创建实例
        server.setZone(zone);                                          // 设置Zone
        return server;                                                 // 返回实例
    }
}
```

程序执行结果

```
muyan.provider-dept:8001
muyan.provider-dept:8002
muyan.provider-dept:8002
```

本程序通过 serverList 集合保存了全部要操作的实例数据，而后通过 LoadBalancerStats 工具类实现了服务数据的处理，随后对区域为“yootk”的服务实例进行下线处理（此时服务数据依然存在，但是无法通过算法获取）。不管外部如何调用 chooseServer()方法，都不会选择已下线的实例。

4.2.3 ServerListUpdater 服务列表更新

ServerListUpdater 服务列表更新

视频名称 0411_【理解】ServerListUpdater 服务列表更新

视频简介 为了进行有效的微服务实例数据维护，Ribbon 提供了服务列表更新接口。本视频为读者分析 ServerListUpdater 接口的使用，以及服务更新的操作监听处理。

Ribbon 在进行微服务实例数据获取时，需要不断考虑可能存在的微服务节点变化（增加新的节点或者某些节点不可达），所以为了便于微服务的数据更新，Ribbon 提供了 ServerListUpdater 列表更新接口。只需将之与动态服务列表负载均衡器（Dynamic Server List Load Balancer）绑定，即可实现服务列表更新的周期性触发，实现结构如图 4-24 所示。

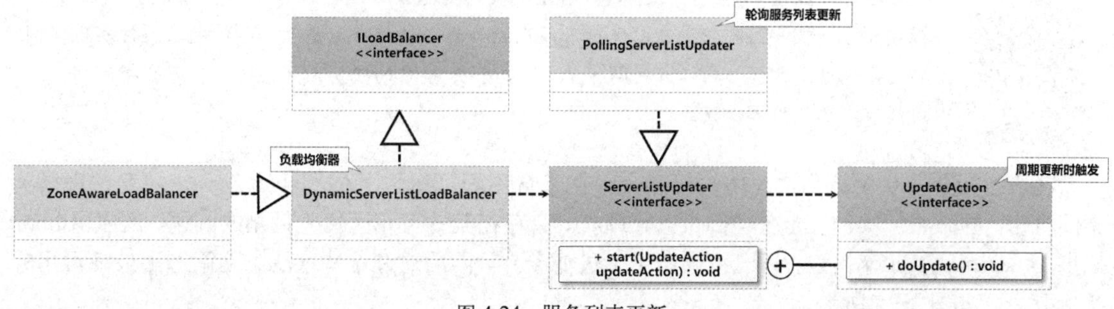

图 4-24 服务列表更新

ServerListUpdater 接口提供 start()方法，该方法会开启一个新的线程，用于实现服务更新的监听操作。而所有的更新操作（定时发出）一旦发出，都可以利用 ServerListUpdater.UpdateAction 子接口提供的 doUpdate()方法来接收。

范例：【consumer-springboot-80 子模块】实现更新监听

```
package com.yootk.test;
public class TestServerListUpdater {
    public static void main(String[] args) {
        String instants = "10.9.19.1:8001,10.9.19.2:8001,10.9.19.3:8001";     // 实例数据
        DefaultClientConfigImpl clientConfig = new DefaultClientConfigImpl();  // 配置
        clientConfig.set(CommonClientConfigKey.ListOfServers, instants);      // 实例属性
        ConfigurationBasedServerList baseList = new ConfigurationBasedServerList();
        baseList.initWithNiwsConfig(clientConfig);      // 初始化配置
        ServerListUpdater updater = new PollingServerListUpdater();           // 服务列表更新
        updater.start(new ServerListUpdater.UpdateAction() {
            @Override
            public void doUpdate() {                                           // 获取监听操作
                System.out.println("【实例更新】最后更新时间：" + updater.getLastUpdate() +
                        "、上次更新间隔时长：" + updater.getDurationSinceLastUpdateMs() +
                        "、错过更新的周期数量：" + updater.getNumberMissedCycles() +
                        "、使用线程数：" + updater.getCoreThreads());
                // 可以在此处实现服务列表的动态抓取，代码略
            }
        });
        ZoneAwareLoadBalancer loadBalancer = new ZoneAwareLoadBalancer();// 负载均衡器
        loadBalancer.setServerListImpl(baseList);                      // 设置ServerList实例
        loadBalancer.setServerListUpdater(updater);                    // 设置更新监听
        try {
            TimeUnit.SECONDS.sleep(1000);                              // 程序不退出
        } catch (InterruptedException e) {}
    }
}
```

后台日志输出	【实例更新】最后更新时间：Wed Apr 28 16:44:28 CST 2023、上次更新间隔时长：1114、错过更新的周期数量：0、使用线程数：2

此时的程序在 ZoneAwareLoadBalancer 负载均衡器中配置了一个 ServerListUpdater 接口并设置了服务更新监听，随后在 ServerListUpdater 实例中配置了 ServerListUpdater.UpdateAction 内部接口实例，这样每当任务更新时就会自动触发该实例中的 doUpdate()方法。本次处理只是输出了一些基本的更新信息，开发者可以在实际使用时通过 Nacos 发现服务实现动态的实例数据加载。

4.2.4　ServerListFilter 实例过滤器

ServerListFilter 实例过滤器

视频名称　0412_【理解】ServerListFilter 实例过滤器

视频简介　服务列表数据可以进行有效过滤处理，这样可以基于一些特定的条件实现实例的优先调度。本视频为读者讲解 ServerListFilter 接口的使用，并通过源代码分析可用区选用阈值以及可用区切换原理。

在进行微服务实例调用过程中，可能会存在来自不同可用区（Available Zone，可简单理解为不同机房）的微服务实例。而在进行服务管理时，会存在某些可用区优先调用的问题。所以 Ribbon 提供了 ServerListFilter 接口，可以设置若干个过滤条件，而后通过这些过滤条件筛选出最终可用的实例列表，如图 4-25 所示。

为了便于读者理解，下面首先通过 ZoneAffinityServerListFilter 过滤实现子类实现一个指定区域服务列表的筛选处理。需要注意的是，由于此时没有启动 Spring Boot 容器，因此相关的属性内容需要开发者自行通过特定的属性标记设置。

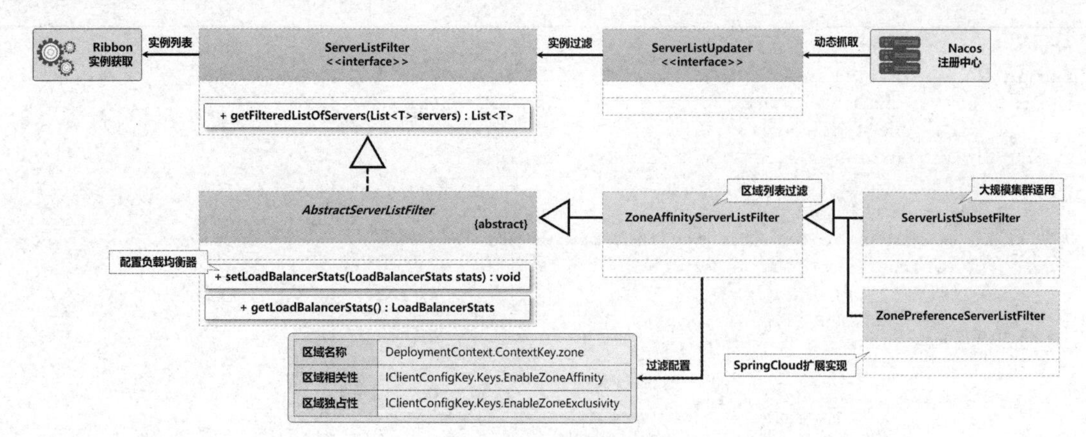

图 4-25 ServerListFilter

范例：【consumer-springboot-80 子模块】区域服务过滤

```
package com.yootk.test;
public class TestServerListFilter {
    public static void main(String[] args) {
        List<Server> serverList = new ArrayList<>();                      // 实例化集合
        serverList.add(createServer("muyan", "muyan.provider-dept", 8001)); // 保存实例
        serverList.add(createServer("muyan", "muyan.provider-dept", 8002)); // 保存实例
        serverList.add(createServer("yootk", "yootk.provider-dept", 8001)); // 保存实例
        DefaultClientConfigImpl config = new DefaultClientConfigImpl();    // 配置类
        config.set(IClientConfigKey.Keys.EnableZoneAffinity, true);         // 区域相关性
        config.set(IClientConfigKey.Keys.EnableZoneExclusivity, true);      // 区域独占性
        ConfigurationManager.getDeploymentContext().setValue(
                DeploymentContext.ContextKey.zone, "muyan");                // 设置过滤区域
        ZoneAffinityServerListFilter filter = new ZoneAffinityServerListFilter();
        filter.initWithNiwsConfig(config);                                  // 配置初始化
        System.out.println(filter.getFilteredListOfServers(serverList)); // 过滤查询
    }
    public static Server createServer(String zone, String url,
                int index) {                                                // 创建服务实例
        Server server = new Server(url, index);                             // 创建实例
        server.setZone(zone);                                               // 设置Zone
        return server;                                                      // 返回实例
    }
}
```

后台日志输出	DEBUG com.netflix.loadbalancer.ZoneAffinityServerListFilter - activeReqeustsPerServerThreshold: 0.6 DEBUG com.netflix.loadbalancer.ZoneAffinityServerListFilter - blackOutServerPercentageThreshold: 0.8 DEBUG com.netflix.loadbalancer.ZoneAffinityServerListFilter - availableServersThreshold: 2
程序执行结果	[muyan.provider-dept:8001, muyan.provider-dept:8002]

本程序通过ZoneAffinityServerListFilter过滤类筛选出了区域名称为“muyan”的所有实例数据，而在进行筛选前必须设置相关的区域配置选项，否则将无法成功实现数据过滤。

提问：如果区域无可用实例该如何？

在使用ZoneAffinityServerListFilter类进行区域实例过滤时，如果当前所选定的区域并没有具体的实例数据，那么最终将无法获取到服务列表，就会出现明明有其他区域实例却调用失败的问题，这种问题该如何解决？

回答：根据指标判断是否切换区域。

关于防止多区域情况下某一区域实例无法获取而造成服务调用问题，在区域服务列表过滤类中覆写的getFilteredListOfServers()方法中已经有了明确的定义：

```
@Override
public List<T> getFilteredListOfServers(List<T> servers) {
    if (zone != null && (zoneAffinity || zoneExclusive)&&
      servers !=null && servers.size() > 0){ // 过滤判断
        List<T> filteredServers = Lists.newArrayList(
          Iterables.filter(servers,
          this.zoneAffinityPredicate.getServerOnlyPredicate()));
        if (shouldEnableZoneAffinity(filteredServers)) {
            return filteredServers;
        } else if (zoneAffinity) {
            overrideCounter.increment();
        }
    }
    return servers;
}
```

观察此源代码，可以发现在真正进行过滤处理时有一个 shouldEnableZoneAffinity()方法，该方法的主要作用在于使用 LoadBalancerStats 提供的 getLoadPerServer()方法获取同区域实例的相关数据指标（实例数量、断路器断开数、活动请求数、实例平均负载等），而后根据一系列的算法与设置进行比对。如果条件满足，则不开启区域感知，否则开启区域感知。这样在集群出现区域性故障时，仍可使用其他区域的实例进行正常服务调用，从而实现微服务集群的高可用特性。

范例：shouldEnableZoneAffinity()源代码

```
private boolean shouldEnableZoneAffinity(List<T> filtered) {
    if (!zoneAffinity && !zoneExclusive) {   // 属性判断
        return false;
    }
    if (zoneExclusive) {                          // 属性判断
        return true;
    }
    // 获取负载均衡统计数据，用于进行后续判断
    LoadBalancerStats stats = getLoadBalancerStats();
    if (stats == null) {
        return zoneAffinity;
    } else {
        logger.debug("Determining … server list: {}", filtered);
        ZoneSnapshot snapshot = stats.getZoneSnapshot(filtered);
        double loadPerServer = snapshot.getLoadPerServer();
        int instanceCount = snapshot.getInstanceCount();
        int circuitBreakerTrippedCount =
            snapshot.getCircuitTrippedCount();    // 断路器数量
        if (((double) circuitBreakerTrippedCount) / instanceCount
          >= blackOutServerPercentageThreshold.get()  ||
          loadPerServer >= activeReqeustsPerServerThreshold.get()
          || (instanceCount - circuitBreakerTrippedCount) <
          availableServersThreshold.get()) {
            logger.debug("zoneAffinity…", …);
            return false;
        } else {
            return true;
        }
    }
}
```

在进行以上指标计算时，实际上有 3 个重要的指标计算公式，分别如下。

① 故障实例占比，默认阈值为 0.8：

```
circuitBreakerTrippedCount（断路器断开数） / instanceCount（实例数量）
    >= blackOutServerPercentageThreshold.get()（默认阈值）
```

② 实例平均负载，默认阈值为 0.6：

```
loadPerServer（实例平均负载） >= activeReqeustsPerServerThreshold.get()（默认阈值）
```

③ 可用实例数量，默认阈值为 2:

```
(instanceCount（实例数量） - circuitBreakerTrippedCount（断路器断开数））
    < availableServersThreshold.get()（默认阈值）
```

在综合判断时，若以上条件有一个不满足，就会触发区域感知功能（查询其他区域），而所有的阈值数据内容都在 ZoneAffinityServerListFilter 类初始化部分中有明确的配置。

此外，ZoneAffinityServerListFilter 类中还有一个 ServerListSubsetFilter 子类，该子类最大的特点是可以产生一个“区域感知”结果的子集列表，同时还可以自动剔除所有不健康的实例数据，适合于拥有大规模服务集群的系统环境。

4.2.5 IPing 存活检查

IPing 存活检查

视频名称 0413_【掌握】IPing 存活检查

视频简介 为了便于节点存活状态的检查，Ribbon 提供了 IPing 接口。本视频为读者分析 IPing 接口的主要组成结构，同时通过源代码方式分析 Ping 操作的伪实现意义。

IPing 接口提供一个 isAlive()方法，用于判断微服务的状态是否存活，并且基于 Iping 接口提供了若干个实现子类，如图 4-26 所示。在默认情况下使用的是 DummyPing 子类，这一点可以通过 DefaultClientConfigImpl 子类源代码查询到：

```
public static final String DEFAULT_NFLOADBALANCER_PING_CLASSNAME =
        "com.netflix.loadbalancer.DummyPing";
```

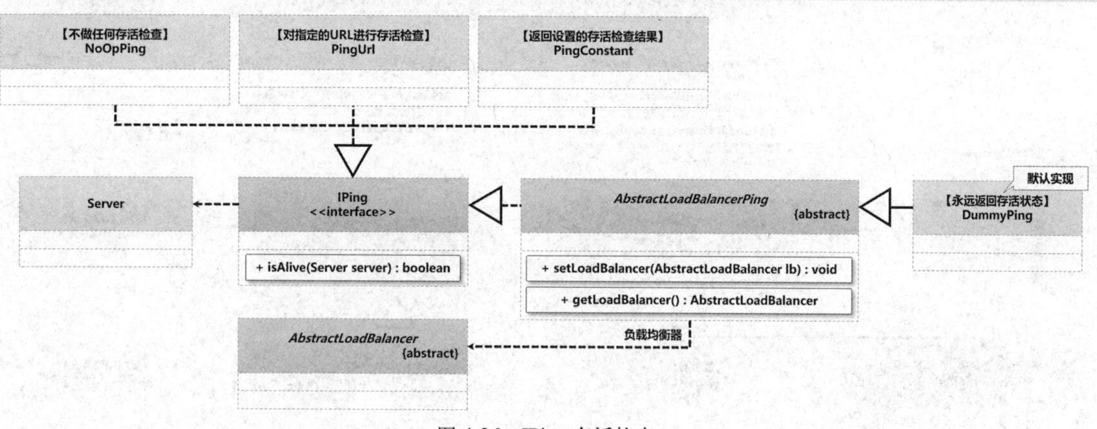

图 4-26 IPing 存活检查

范例：URL 存活检查

```
package com.yootk.test;
public class TestPing {
    public static void main(String[] args) {
        String host = "provider-dept-8001";     // 主机名称
        String uri = "/provider/dept/list";     // 子路径
        IPing ping = new PingUrl(false, uri);      // false表示使用HTTP
        Server server = new Server(host, 8001);    // 创建实例
        System.out.println(ping.isAlive(server)); // 存活检查
    }
}
```

程序执行结果	true

由于此时已经开启了部门微服务的节点，因此在进行存活检查时，返回的结果为 true。又由于此时没有使用 HTTPS 安全访问，因此在实例化 PingUrl 类时将“isSecure”参数内容设置为 false。

提示：Ping 操作并非重点。

读者如果打开 DummyPing 类实现的源代码，可以发现此类中提供的 isAlive()方法返回的内容为 true，即节点永远都是存活状态。

而如果继续查看其他的 IPing 接口的实现子类（非 PingUrl），也可以发现几乎都没有实质上的存活检查。这是因为在基于 Nacos 的处理中，可以基于间隔任务实现动态的服务列表获取，且最终进行调用时也提供失败重试的支持，这样就没有必要实现具体的 Ping 处理。

4.2.6　IRule 负载均衡算法

视频名称　0414_【掌握】IRule 负载均衡算法

视频简介　Ribbon 可以帮助用户隐藏注册中心的实例发现处理，同时又提供一系列的负载均衡算法，帮助用户更合理地实现微服务集群调用。本视频将为读者讲解 Ribbon 所提供的负载均衡算法以及算法的使用配置。

消费端调用微服务时，可以通过 Ribbon 实现服务实例查询，并可以依据 Ribbon 所提供的负载均衡算法实现指定微服务的实例调用。而为了便于负载均衡算法的管理，Ribbon 提供了 IRule 规则接口，所有的负载均衡算法都必须实现此接口，如图 4-27 所示。消费端在进行微服务调用时，可以根据此接口提供的 choose()方法并根据所设置的负载均衡器（ILoadBalancer 接口实例）实现服务实例的获取。

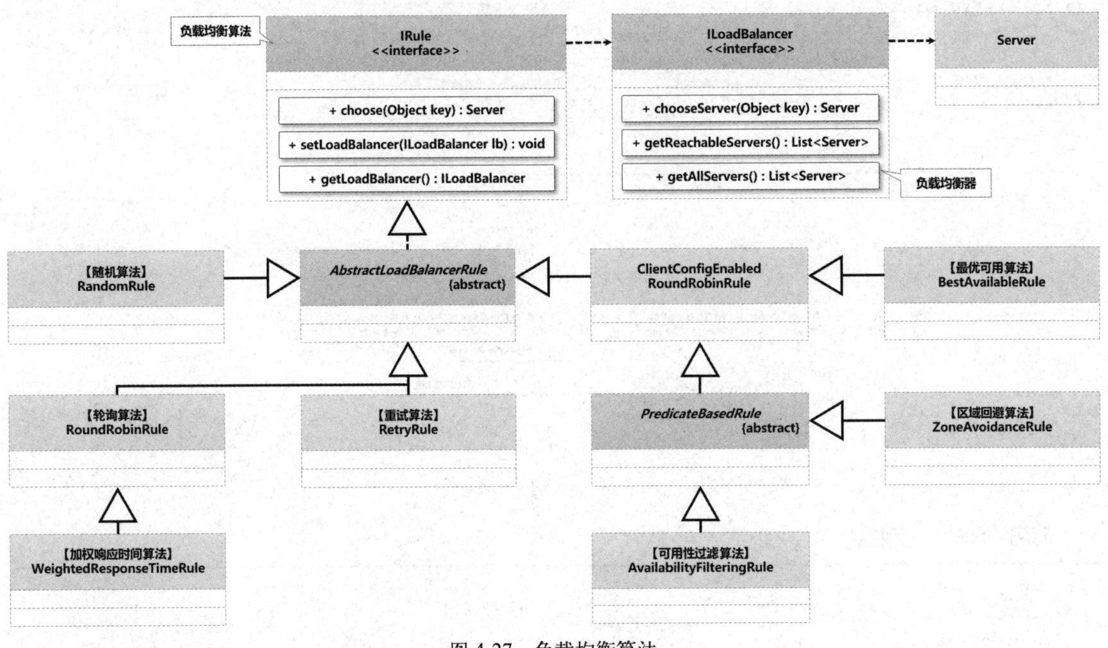

图 4-27　负载均衡算法

提问：Ribbon 默认的负载均衡算法是什么？

在使用 Ribbon 实现的服务调用过程中，每一次消费端调用都会访问不同的服务器端，所以其一定采用了默认的负载均衡算法。请问在默认情况下所使用的是哪一种负载均衡算法？

回答：AvailabilityFilteringRule 为默认算法。

Ribbon 为了便于用户使用，默认配置了一个 AvailabilityFilteringRule 负载均衡算法。这一点可以通过 DefaultClientConfigImpl 子类观察到。

范例：查看默认的负载均衡算法

```
public static final String DEFAULT_NFLOADBALANCER_RULE_CLASSNAME =
"com.netflix.loadbalancer.AvailabilityFilteringRule";
```

通过 DefaultClientConfigImpl 子类可以发现，当前所使用的 Ribbon 组件版本的默认负载均衡算法规则为 AvailabilityFilteringRule。而随着版本的不同，也会有不同的默认负载均衡算法，这一点可以根据对应的源代码类进行查询。

通过图 4-27 可以发现，Ribbon 提供了若干种负载均衡算法，开发者可以根据实际的微服务调用需要来进行选择。这些算法的具体作用如表 4-2 所示。

表 4-2　负载均衡算法

序号	负载均衡算法	描述
01	AvailabilityFilteringRule	可用性过滤算法，会取消一直连接失败的服务端
02	BestAvailableRule	最优可用资源算法，选择一个请求并发量最小的实例
03	RandomRule	随机算法，随机选择一个实例
04	RetryRule	重试算法，如果获取失败，则在指定时间内重试
05	RoundRobinRule	轮询算法，会依据节点的顺序依次调用
06	WeightedResponseTimeRule	加权响应时间算法，根据响应时间增加权重，响应时间越长权重越小，实例被选中的概率也会降低
07	ZoneAvoidanceRule	区域回避算法，综合判断服务所在区域的性能以及服务的可用性，而后选择一个实例，在没有区域的环境下采用类似轮询的机制选择实例

如果开发者需要修改当前所使用的负载均衡算法，那么可以直接定义一个具体的 Ribbon 配置类，而后向 Spring 容器手动注入 IRule 接口实例即可生效，可以采用如下配置步骤实现。

注意：Ribbon 配置类不要与主应用重合。

如果要进行 Ribbon 配置类的定义，则需要在类定义时使用@Configuragion 注解进行标记，同时所配置的算法对象也需要使用@Bean 进行注册，但是这个配置类不能直接出现在当前的主应用（包含@SpringBootApplication 注解的启动类）扫描包之中，否则就会被主应用中的@RibbonClients 共享，引发问题。

正确的做法是，在主应用扫描包之外进行 Ribbon 配置类的定义，而后在应用主类中通过 Ribbon 提供的@RibbonClient 注解进行引入，如图 4-28 所示。

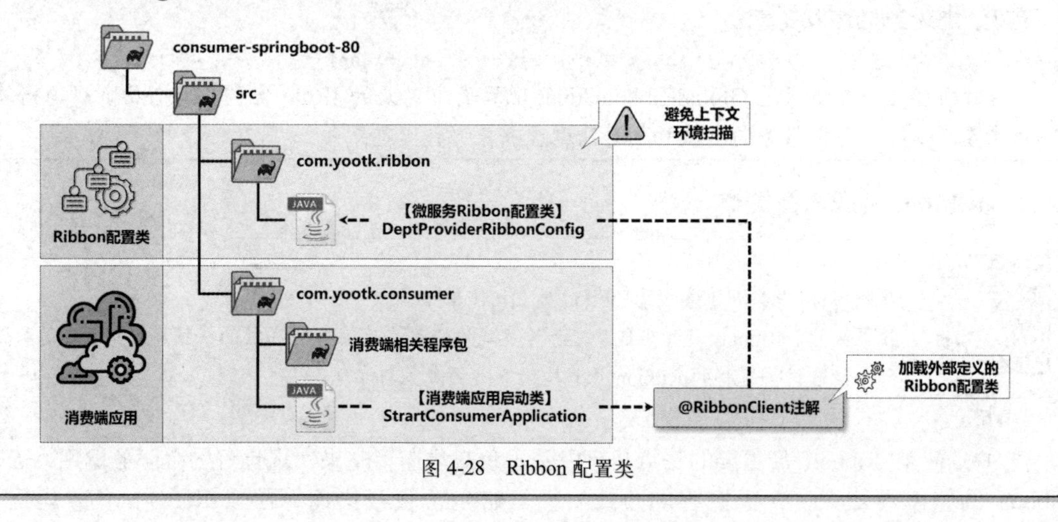

图 4-28　Ribbon 配置类

（1）【consumer-springboot-80 子模块】在主应用之外的包中创建一个 Ribbon 配置类，并自定义负载均衡算法。

```
package muyan.yootk.config.ribbon;              // 与主应用扫描包不同
@Configuration                                  // 配置类
public class DeptProviderRibbonConfig {
    @Bean                                       // Bean注册
    public IRule ribbonRule() {
        return new RandomRule();                // 随机读取
    }
}
```

（2）【consumer-springboot-80 子模块】在消费端主应用中引用负载均衡算法的配置 Bean。

```
package com.yootk.consumer;                     // 主应用扫描包
@SpringBootApplication
@EnableDiscoveryClient                          // 启用发现服务
@RibbonClient(name = "dept.provider", configuration =
        DeptProviderRibbonConfig.class)         // 为部门微服务添加负载均衡算法
public class StartConsumerApplication {
    public static void main(String[] args) {
        SpringApplication.run(StartConsumerApplication.class, args); // 服务启动
    }
}
```

应用启动主类中使用了@RibbonClient 注解为 dept.provider 微服务配置了一个随机调用的算法，随后通过服务最终调用可发现此时没有任何调用规律可循，所有的调用完全随机。

提示：结合@RibbonClients 注解配置。

在消费端进行 Ribbon 配置时，也可以通过@RibbonClients 注解管理多个@RibbonClient 注解的配置，此时只需要通过数组的形式定义@RibbonClient 配置，代码如下。

范例：通过@RibbonClients 配置

```
@RibbonClients(                  // Ribbon配置类
   value = {                     // 针对不同的微服务配置负载均衡算法
      @RibbonClient(name = "dept.provider",
          configuration = DeptProviderRibbonConfig.class)}
)
```

另外需要提醒读者的是，如果现在希望配置全局的负载均衡算法，则可以直接在应用主类中使用@RibbonClients 注解配置。

范例：定义全局负载均衡算法

```
@RibbonClients(defaultConfiguration = GlobalRibbonClient.class)
```

此时项目之中可以通过 GlobalRibbonClient 配置类中定义的 IRule 实例实现全局负载均衡算法的配置，而后具备微服务的负载均衡算法配置将失效。

4.2.7　Ribbon 负载均衡策略

Ribbon 负载均衡策略

视频名称　0415_【掌握】Ribbon 负载均衡策略

视频简介　Ribbon 通过其自身完善的负载均衡策略可以实现高效的微服务调用管理。本视频为读者讲解完整的 Ribbon 负载均衡策略的配置与实现。

客户端使用 Ribbon 最重要的一点是可以基于负载均衡策略来实现良好的微服务调用，而在 Ribbon 的整体设计中，一套完善的负载均衡策略包含负载均衡算法（IRule）、负载均衡器（ILoadBalancer）、服务列表（ServerList）等核心组成结构，如图 4-29 所示。

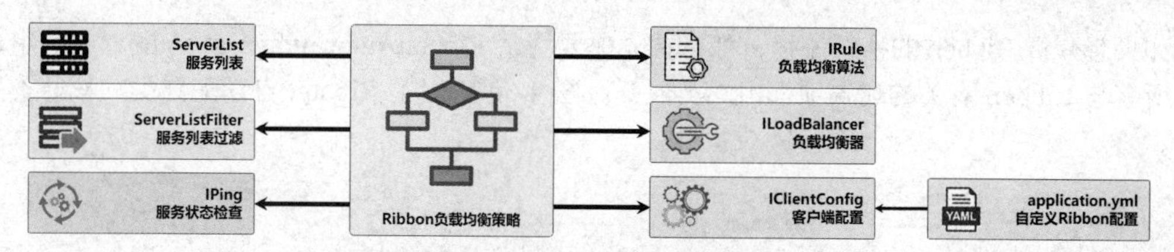

图 4-29 Ribbon 负载均衡策略

这些结构在 Ribbon 中有大量的接口实现类，同时在 DefaultClientConfigImpl 子类中也都定义了相关的默认配置项。使用者如果需要去改变这些默认配置，则可以通过表 4-3 所示的配置项实现。

表 4-3 Ribbon 负载均衡策略

序号	配置项	描述
01	NFLoadBalancerClassName	ILoadBalancer 为使用的微服务配置负载均衡器实现类
02	NFLoadBalancerRuleClassName	IRule 为使用的微服务配置负载均衡算法实现类
03	NFLoadBalancerPingClassName	IPing 为使用的微服务配置存活检查实现类
04	NIWSServerListClassName	ServerList 为使用的微服务配置服务列表实现类
05	NIWSServerListFilterClassName	ServiceListFilter 为使用的微服务配置服务列表过滤实现类

范例：【consumer-springboot-80 子模块】为部门微服务配置负载均衡策略，修改 application.yml 配置文件

```
dept.provider:                                                    # 微服务配置
  ribbon:                                                         # Ribbon负载均衡策略
    NFLoadBalancerClassName:
        com.netflix.loadbalancer.DynamicServerListLoadBalancer    # 负载均衡器
    NFLoadBalancerRuleClassName: com.netflix.loadbalancer.RandomRule # 负载均衡算法
```

此时的配置中通过“服务名称.ribbon.负载均衡策略”的形式对“dept.provider”微服务的调用配置了负载均衡器和负载均衡算法，而除了这些已有的功能类之外，还可以根据实际的要求进行相关接口的实现扩充。

 注意：不要与 Bean 冲突。

在本次进行 Ribbon 配置时，建议删除消费端主类中配置的@RibbonClient 注解，如果未删除，则对于负载均衡策略也存在配置优先级：application.yml 配置策略 > Bean 配置策略。

在 Spring Cloud 中，建议开发者通过 application.yml 进行负载均衡策略配置，但是在一些奇特且无法解释的环境下，配置文件定义均衡策略可能无法生效，所以此时开发者只能选择配置类的方式定义。

4.2.8 Ribbon 执行分析

Ribbon 执行分析

视频名称 0416_【理解】Ribbon 执行分析

视频简介 Ribbon 是基于 Starter 组件的形式自动配置到系统之中的。本视频为读者分析 Ribbon 配置加载，并结合 ILoadBalancer、ServerList、IRule 等及核心处理逻辑分析 Ribbon 源代码中的启动流程。

Spring Cloud 中，当需要消费端进行服务调用时一般都需要使用“spring-cloud-netflix-ribbon”依赖库，而此依赖库中有“spring.factories”配置文件，同时该文件中明确定义了 Ribbon 的自动配置类：

```
org.springframework.boot.autoconfigure.EnableAutoConfiguration=\
     org.springframework.cloud.netflix.ribbon.RibbonAutoConfiguration
```

如果想分析 Ribbon 的执行流程，那么首先需要清楚 RibbonAutoConfiguration 配置类的组成，因为所有与 Ribbon 有关的配置项都由此类展开。图 4-30 给出了 Ribbon 自动配置类的核心类关联结构。

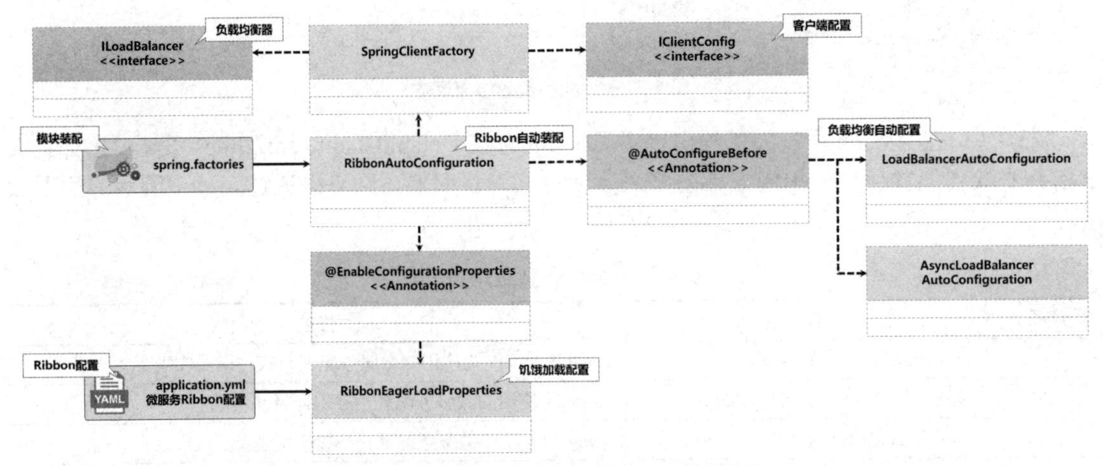

图 4-30　Ribbon 自动配置类的核心类关联结构

范例：【spring-cloud-netflix-ribbon 依赖库】RibbonAutoConfiguration 核心源代码

```
package org.springframework.cloud.netflix.ribbon;
@Configuration
@Conditional(RibbonAutoConfiguration.RibbonClassesConditions.class)
@RibbonClients
@AutoConfigureAfter(
    name = "org.springframework.cloud.netflix.eureka.EurekaClientAutoConfiguration")
@AutoConfigureBefore({ LoadBalancerAutoConfiguration.class,
    AsyncLoadBalancerAutoConfiguration.class })
@EnableConfigurationProperties({ RibbonEagerLoadProperties.class,
    ServerIntrospectorProperties.class })
public class RibbonAutoConfiguration {
  @Autowired
  private RibbonEagerLoadProperties ribbonEagerLoadProperties;  // Ribbon饥饿配置
  @Bean
  public SpringClientFactory springClientFactory() {
    SpringClientFactory factory = new SpringClientFactory();    // Spring客户端工厂
    factory.setConfigurations(this.configurations);
    return factory;
  }
  @Bean
  @ConditionalOnMissingBean(LoadBalancerClient.class)
  public LoadBalancerClient loadBalancerClient() {              // 负载均衡客户端
    return new RibbonLoadBalancerClient(springClientFactory());// 对象实例化
  }
}
```

通过以上 RibbonAutoConfiguration 类核心源代码可以清楚地发现，在 Ribbon 中使用了@AutoConfigureBefore 与@EnableConfigurationProperties 两个重要的注解。其中@EnableConfigurationProperties 主要实现了 application.yml 关于“饿汉式”加载配置的属性接收，而@AutoConfigureBefore 注解则直接设置了 LoadBalancerAutoConfiguration 配置类的定义，这个类也是实现整个 Ribbon 负载均衡策略的核心配置类。这些类的关联结构如图 4-31 所示。

通过图 4-31 所给出的结构可以清楚地发现，所有的 Ribbon 负载均衡都是以 HTTP 拦截器的形式存在的，即用户发送的请求被请求拦截器拦截，而后在此拦截器中通过配置的负载均衡策略来实现 Ribbon 服务调用。

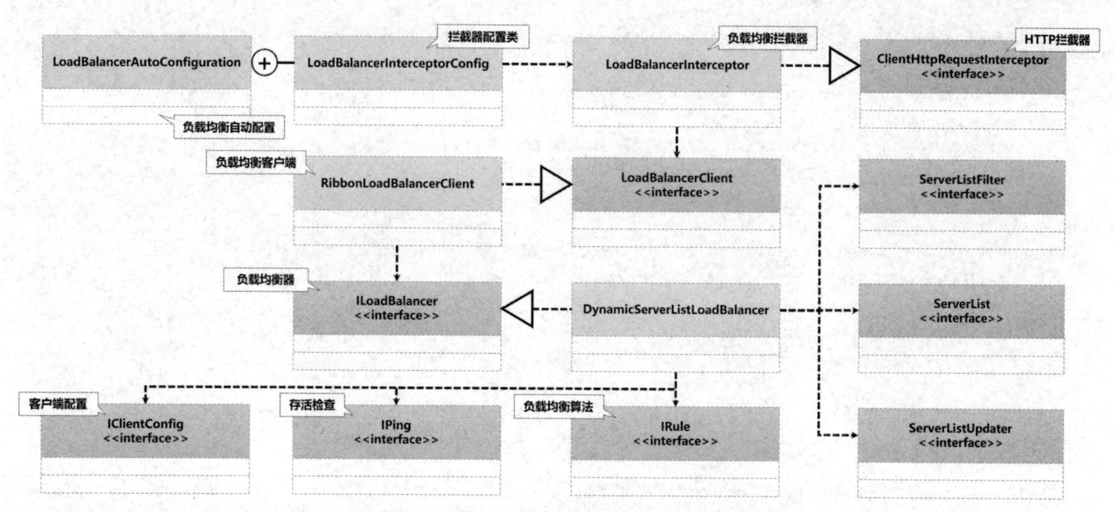

图 4-31 RibbonAutoConfiguration 类关联结构

范例：【spring-cloud-netflix-ribbon 依赖库】RibbonLoadBalancerClient 类核心源代码

```
package org.springframework.cloud.netflix.ribbon;
public class RibbonLoadBalancerClient implements LoadBalancerClient {
   private SpringClientFactory clientFactory;                    // Spring客户端工厂类
   public RibbonLoadBalancerClient(SpringClientFactory clientFactory) {
      this.clientFactory = clientFactory;                        // 保存工厂类实例
   }
   // 为系统构造一个完整的“主机:端口”请求路径
   public URI reconstructURI(ServiceInstance instance, URI original) {
      String serviceId = instance.getServiceId();                // 获取服务ID
      RibbonLoadBalancerContext context = this.clientFactory
           .getLoadBalancerContext(serviceId);
      URI uri;
      Server server;
      if (instance instanceof RibbonServer) {
         RibbonServer ribbonServer = (RibbonServer) instance;
         server = ribbonServer.getServer();                     // 获取服务实例
         uri = updateToSecureConnectionIfNeeded(original, ribbonServer);
      }
      else {
         server = new Server(instance.getScheme(), instance.getHost(),
              instance.getPort());
         IClientConfig clientConfig = clientFactory.getClientConfig(serviceId);
         ServerIntrospector serverIntrospector = serverIntrospector(serviceId);
         uri = updateToSecureConnectionIfNeeded(original, clientConfig,
              serverIntrospector, server);
      }
      return context.reconstructURIWithServer(server, uri);
   }
   public ServiceInstance choose(String serviceId, Object hint) {   // 实例选择
      Server server = getServer(getLoadBalancer(serviceId), hint);   // 获取一个实例
      if (server == null) {
         return null;
      }
      return new RibbonServer(serviceId, server, isSecure(server, serviceId),
        serverIntrospector(serviceId).getMetadata(server));          // 返回Ribbon服务实例
   }
   // 通过指定的服务ID进行服务调用
   public <T> T execute(String serviceId, LoadBalancerRequest<T> request, Object hint)
         throws IOException {
      ILoadBalancer loadBalancer = getLoadBalancer(serviceId);       // 负载均衡器
      Server server = getServer(loadBalancer, hint);                 // 服务获取
      if (server == null) {
```

```
            throw new IllegalStateException("No instances available for " + serviceId);
        }
        RibbonServer ribbonServer = new RibbonServer(serviceId, server,
            isSecure(server, serviceId),
            serverIntrospector(serviceId).getMetadata(server));        // 获取Ribbon实例
        return execute(serviceId, ribbonServer, request);             // 调用重载方法
    }
    @Override
    public <T> T execute(String serviceId, ServiceInstance serviceInstance,
            LoadBalancerRequest<T> request) throws IOException {
        Server server = null;
        if (serviceInstance instanceof RibbonServer) {
            server = ((RibbonServer) serviceInstance).getServer();
        }
        if (server == null) {
            throw new IllegalStateException("No instances available for " + serviceId);
        }
        RibbonLoadBalancerContext context = this.clientFactory
                .getLoadBalancerContext(serviceId);                    // 获取实例
        RibbonStatsRecorder statsRecorder = new RibbonStatsRecorder(context, server);
        try {
            T returnVal = request.apply(serviceInstance);
            statsRecorder.recordStats(returnVal);                     // 状态记录
            return returnVal;
        } catch (Exception ex) {}
        return null;
    }
    protected Server getServer(ILoadBalancer loadBalancer, Object hint) {
        if (loadBalancer == null) {
            return null;
        } // 通过负载均衡器选择一个调用实例
        return loadBalancer.chooseServer(hint != null ? hint : "default");
    }
    protected ILoadBalancer getLoadBalancer(String serviceId) {   // 获取负载均衡器
        return this.clientFactory.getLoadBalancer(serviceId);
    }
}
```

在 RibbonLoadBalancerClient 类中最为重要的一个方法为 execute()，该方法可以根据指定的负载均衡策略执行服务调用。在进行实例获取时通过 ILoadBalancer 接口定义的 chooseServer()实现一个可用实例的获取，而根据 Ribbon 的默认配置设置，此时所使用的是 ZoneAwareLoadBalancer 子类，根据配置的区域进行实例查找。

4.3　自定义 Ribbon 负载均衡算法

在 Spring Cloud Alibaba 服务套件中，Nacos 注册中心给了开发者很大的发挥空间。除了 Ribbon 自身所提供的负载均衡算法之外，用户也可以使用 IRule 接口（或者直接使用 AbstractLoadBalancerRule 抽象子类）来根据实际的情况定义需要的负载均衡算法。这样就可以实现权重调度算法、集群优先级调度算法以及元数据版本控制调度算法，可以对集群服务进行更加有效的调度管理。

4.3.1　Nacos 权重优先调度

Nacos 权重优先调度

视频名称　0417_【理解】Nacos 权重优先调度

视频简介　基于 Nacos 服务注册实例支持权重的配置，可以基于权重实现不同实例的调度。本视频为读者讲解微服务权重的作用以及具体实现。

为了提供更高效的微服务处理，需要根据当前不同的运行环境进行节点的扩充。这样一来在一个庞大的微服务集群中，因服务实例上线的时间及并发量的不同，就可能存在不同硬件配置的服务主机，如图 4-32 所示。

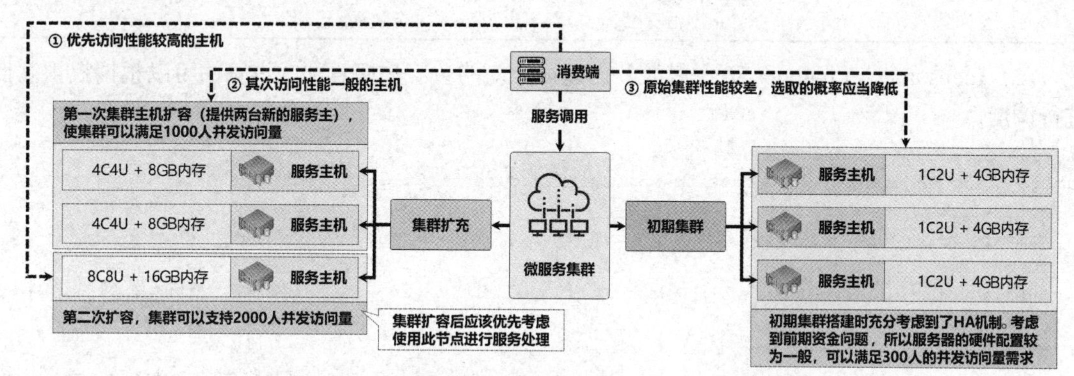

图 4-32　微服务集群扩容

此时微服务集群中的主机就会由于硬件的配置而存在处理性能高低不一的情况，而考虑到集群的稳定性，应该首先分配性能较高的服务主机，而后降低初期集群服务器的访问概率。要想实现这样的选择，可以为不同的微服务实例配置不同的访问权重，这样通过一定的 Ribbon 算法实现高权重主机的优先调度。为了简化用户的权重配置，所有在 Nacos 中注册的微服务都可以动态地进行权重的调整，如图 4-33 所示。

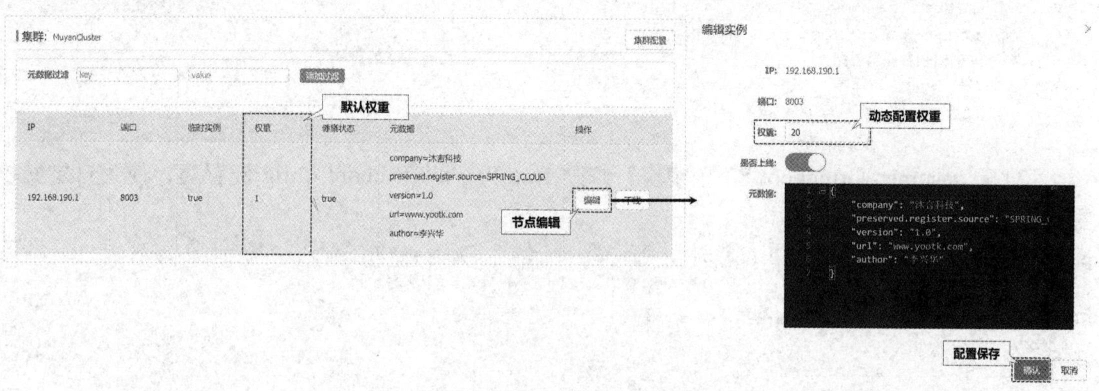

图 4-33　在 Nacos 控制台动态配置微服务权重

除了可以在 Nacos 控制台进行权重配置之外，开发者也可以在微服务中进行权重的配置。而要想实现基于权重实现服务调度，则还需要开发者扩充一个新的 Ribbon 负载均衡算法。下面通过具体的步骤介绍如何实现这一操作。

（1）【provider-dept-*子模块】修改每一个微服务提供端注册配置，在注册时为不同的微服务设置不同的权重。

provider-dept-8001 配置

```
spring:                              # Spring配置
  cloud:                             # Spring Cloud配置
    nacos:                           # Nacos注册中心
      discovery:                     # 发现服务
        weight: 10                   # 服务权重
```

provider-dept-8002 配置

```
spring:                              # Spring配置
  cloud:                             # Spring Cloud配置
    nacos:                           # Nacos注册中心
      discovery:                     # 发现服务
        weight: 50                   # 服务权重
```

<table>
<tr><td>provider-dept-8003 配置</td><td><pre>spring: # Spring配置
  cloud: # Spring Cloud配置
    nacos: # Nacos注册中心
      discovery: # 发现服务
        weight: 90   # 服务权重</pre></td></tr>
</table>

（2）【consumer-springboot-80 子模块】创建一个新的负载均衡算法，该算法可以根据权重高低进行调度。

```
package muyan.yootk.loadbalancer.rule;
public class NacosWeightRule extends AbstractLoadBalancerRule {      // Nacos权重规则
   @Autowired
   private NacosDiscoveryProperties nacosDiscoveryProperties ;       // Nacos发现服务
   @Override
   public void initWithNiwsConfig(IClientConfig clientConfig) {}     // 配置数据初始化
   @Override
   public Server choose(Object key) {                                // 服务选择
      try {                                                          // Ribbon入口
         BaseLoadBalancer loadBalancer = (BaseLoadBalancer)
               super.getLoadBalancer();                              // 获取负载均衡器
         String name = loadBalancer.getName();                       // 调用微服务名称
         NamingService namingService = nacosDiscoveryProperties
               .namingServiceInstance();                             // 命名发现
         // Nacos客户端自动提供基于权重实现的负载均衡调度算法，直接查询实例即可
         Instance instance = namingService.selectOneHealthyInstance(name,
               this.nacosDiscoveryProperties.getGroup());
         return new NacosServer(instance);                           // 返回Nacos实例
      } catch (NacosException e) {
         return null;
      }
   }
}
```

（3）【consumer-springboot-80 子模块】修改 DeptProviderRibbonConfig 配置类，使用自定义权重均衡算法。

```
package muyan.yootk.config.ribbon;                   // 与主应用扫描包不同
@Configuration                                       // 配置类
public class DeptProviderRibbonConfig {
   @Bean                                             // Bean注册
   public IRule ribbonRule() {
      return new NacosWeightRule();                  // 随机读取
   }
}
```

（4）【consumer-springboot-80 子模块】在启动类中定义此配置类。

```
@RibbonClient(name = "dept.provider",
configuration = DeptProviderRibbonConfig.class)      // 负载均衡策略
public class StartConsumerApplication {}
```

此时为 3 个微服务的节点分别配置了不同的权重，这样在消费端进行微服务调用时，权重大的节点会被 Nacos 消费端优先调用（并非绝对优先，只是与其他节点相比更容易被选中）。

4.3.2　Nacos 集群优先调度

Nacos 集群优先调度

视频名称　0418_【理解】Nacos 集群优先调度

视频简介　在 Nacos 领域模型中可以基于集群实现不同机房服务节点的管理，这样也可以基于集群名称实现消费端优先调用。本视频为读者分析集群优先调度的意义，并通过代码介绍该负载均衡算法的实现。

一个庞大的微服务集群在设计时必然要充分地考虑到 HA 机制。即便微服务有很多服务节点，也不应该将其统一部署在一个机房之中，而应在不同的机房中设置若干个服务节点，每个机房的节点都是一个 Nacos 集群，而后依靠 Nacos 提供的领域模型实现逻辑上的绑定，如图 4-34 所示。

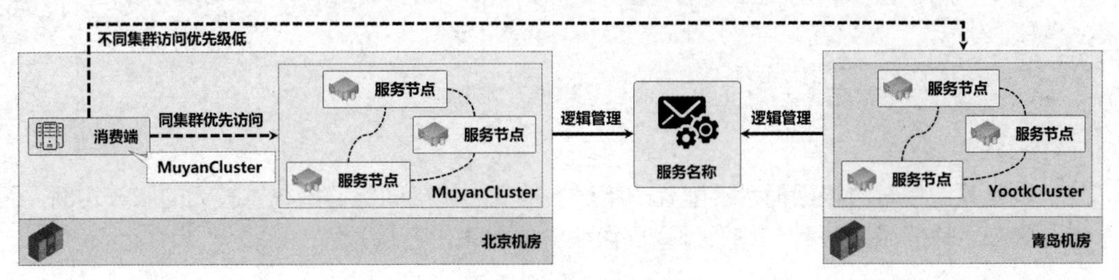

图 4-34 服务分布式部署

如果此时微服务消费端部署在“北京机房”之中（集群名称为“MuyanCluster”），就应该优先访问同集群中的服务实例，从而得到最佳的处理性能；如果当前机房没有可用实例，再去访问其他集群。那么此时就可以自定义一个 Ribbon 负载均衡算法，实现步骤如下。

（1）【consumer-springboot-80 子模块】定义集群优先调度的负载均衡算法。

```
package muyan.yootk.loadbalancer.rule;
public class NacosClusterWeightRule extends AbstractLoadBalancerRule {
   @Autowired
   private NacosDiscoveryProperties nacosDiscoveryProperties;       // Nacos发现服务
   @Override
   public Server choose(Object key) {                                // 服务选择
      BaseLoadBalancer loadBalancer = (BaseLoadBalancer) super.getLoadBalancer();
      String name = loadBalancer.getName();                          // 请求的微服务名称
      NamingService namingService = this.nacosDiscoveryProperties
            .namingServiceInstance();                                // 获取名称服务
      try {
         List<Instance> instances = namingService
               .selectInstances(name, this.nacosDiscoveryProperties.getGroup(),
                     true);                                          // 找到指定服务的全部实例
         List<Instance> clusterInstances = instances.stream()
               .filter(instance -> Objects.equals(instance.getClusterName(),
                     this.nacosDiscoveryProperties.getClusterName())) // 实例筛选
               .collect(Collectors.toList());
         List<Instance> instancesChoose = null;                      // 候选实例
         if (CollectionUtils.isEmpty(clusterInstances)) {            // 没有集群匹配成功
            instancesChoose = instances;                             // 全部实例
         } else {                                                    // 集群名称匹配成功
            instancesChoose = clusterInstances;                      // 集群实例
         }
         Instance selectedInstance = ExtendBalancer
               .getHostByRandomWeight2(instancesChoose);             // 基于权重负载均衡算法返回实例
         return new NacosServer(selectedInstance);                   // 返回实例
      } catch (NacosException e) {}
      return null;
   }
   @Override
   public void initWithNiwsConfig(IClientConfig clientConfig) {}    // 配置数据初始化
}
```

（2）【consumer-springboot-80 子模块】修改消费端的 application.yml 配置优先访问的集群名称。

```
spring:                                       # Spring配置
  cloud:                                      # Spring Cloud配置
    nacos:                                    # Nacos注册中心
      discovery:                              # 发现服务
        cluster-name: MuyanCluster            # 匹配集群名称
```

（3）【consumer-springboot-80 子模块】修改 Ribbon 配置类，使用自定义负载均衡算法。

```
package muyan.yootk.config.ribbon;                   // 与主应用扫描包不同
@Configuration                                       // 配置类
public class DeptProviderRibbonConfig {
    @Bean                                            // Bean注册
    public IRule ribbonRule() {
        return new NacosClusterWeightRule();         // 集群优先读取
    }
}
```

此时微服务调用时会根据消费端配置的集群名称优先访问同集群中权重较大的服务实例，而当没有指定集群名称的微服务时，将通过全部微服务的权重进行筛选。

4.3.3 元数据优先调度

元数据优先调度

视频名称 0419_【理解】元数据优先调度

视频简介 微服务的消费端可能会有多种，而为了便于微服务的版本维护，可以基于元数据进行调度配置。本视频为读者讲解版本优先调度的意义，并通过自定义的负载均衡算法介绍同一版本微服务的优先调度实现。

在一个完整的微服务集群中，需要不断地进行服务实现的版本升级。如果此时只有一个消费端，那么只需要同步进行消费端的版本更新即可实现微服务的正确调用。但是如果一个微服务有多个不同的消费端，并且有些消费端不能及时更新，就需要基于版本编号的形式实现服务实例的优先调度，如图 4-35 所示。

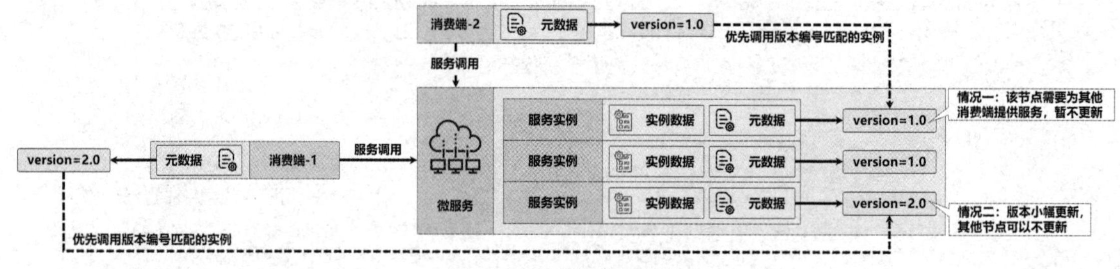

图 4-35 元数据优先调度

为了便于这种烦琐的微服务调用管理，开发者可以基于元数据的形式进行服务版本配置，消费端在进行服务调用时，可以将之与自身所配置的元数据版本进行匹配，实现指定版本微服务的优先调度。下面通过具体的步骤实现这一操作。

（1）【provider-dept-8001 子模块】修改部门元数据中的版本编号，使其版本编号为 2.0（其余微服务版本不变）。

```
spring:                                    # Spring配置
  cloud:                                   # Spring Cloud配置
    nacos:                                 # Nacos注册中心
      discovery:                           # 发现服务
        metadata:                          # 元数据
          version: 2.0                     # 自定义数据项
```

（2）【consumer-springboot-80 子模块】修改消费端的元数据，也设置一个版本的元数据信息，这样当服务端有相同版本时，可以优先调用指定节点的服务。

```
spring:                                    # Spring配置
  cloud:                                   # Spring Cloud配置
    nacos:                                 # Nacos注册中心
      discovery:                           # 发现服务
        metadata:                          # 元数据
          version: 2.0                     # 优先匹配2.0版本微服务
```

（3）【consumer-springboot-80 子模块】创建一个新的负载均衡算法类实现版本匹配。

```
package muyan.yootk.loadbalancer.rule;
public class NacosVersionRule extends AbstractLoadBalancerRule {
    @Autowired
    private NacosDiscoveryProperties nacosDiscoveryProperties ;
    @Override
    public void initWithNiwsConfig(IClientConfig clientConfig) {}            // 配置数据初始化
    @Override
    public Server choose(Object key) {
        BaseLoadBalancer loadBalancer = (BaseLoadBalancer) super.getLoadBalancer();
        String name = loadBalancer.getName();                  // 请求微服务名称
        NamingService namingService = this.nacosDiscoveryProperties
                .namingServiceInstance();                       // 命名实例
        try {
            List<Instance> instances = namingService
                    .selectInstances(name, this.nacosDiscoveryProperties.getGroup(),
                            true);                 // 获取指定名称的所有实例
            List<Instance> metadataVersionMatchInstances = instances.stream()
                    .filter(instance -> Objects.equals(this.nacosDiscoveryProperties
                            .getMetadata().get("version"),
                                    instance.getMetadata().get("version")))   // 版本过滤
                    .collect(Collectors.toList());
            List<Instance> instancesToBeanChoose = null;    // 候选实例
            if (CollectionUtils.isEmpty(metadataVersionMatchInstances)) {      // 未选中实例
                instancesToBeanChoose = instances;           // 全部实例
            } else {
                instancesToBeanChoose = metadataVersionMatchInstances;        // 匹配实例
            }
            Instance selectedInstance = ExtendBalancer.getHostByRandomWeight2(
                    instancesToBeanChoose);                     // 基于权重负载均衡算法返回实例
            return new NacosServer(selectedInstance);       // 返回实例
        } catch (NacosException e) {}
        return null;
    }
}
```

（4）【consumer-springboot-80 子模块】修改 Ribbon 配置类。

```
package muyan.yootk.config.ribbon;                          // 与主应用扫描包不同
@Configuration                                               // 配置类
public class DeptProviderRibbonConfig {
    @Bean                                                    // Bean注册
    public IRule ribbonRule() {
        return new NacosVersionRule();                       // 版本匹配优先读取
    }
}
```

此时由于消费端在元数据中配置了 version 属性，因此在进行服务调用时，会将之与微服务实例的版本编号进行匹配，匹配成功后再基于权重进行实例调用。如果没有匹配版本，则会根据权重调用从全部实例中获取一个实例进行调用。

4.4 Feign 接口转换

Feign 接口转换

视频名称 0420_【掌握】Feign 接口转换

视频简介 为了便于微服务的调用管理，Spring Cloud 提供了 Feign 组件，利用此组件可以实现远程 REST 接口与 Java 接口的映射访问。本视频通过一个完整的实例为读者讲解 Feign 组件的基本使用以及服务调用操作。

在进行远程接口调用时，“原始”的做法是直接依据微服务的名称找到与之相关的 REST 接口，随后通过 RestTemplate 对象实例实现服务调用，这样在每次进行服务调用时必须明确地知道服务的地址。但是这种做法并不符合传统的 RPC 实现形式，最佳的形式是客户端通过业务接口实现微服务的调用，如图 4-36 所示。

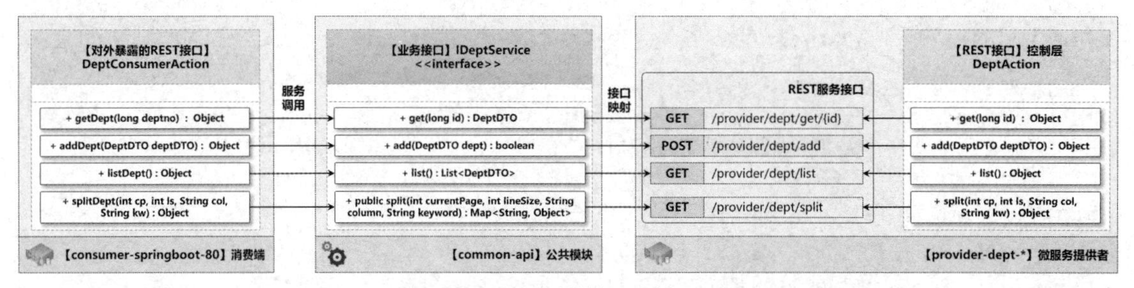

图 4-36　远程接口映射

为了便于远程 REST 接口访问，Spring Cloud 提供了 Feign 工具，利用其内部提供的注解并结合接口中的方法可实现映射处理，这样用户调用接口即可实现远程服务的调用。下面通过具体的操作步骤进行介绍。

(1)【microcloud 项目】由于业务接口统一在 common-api 子模块中定义，因此要修改 build.gradle 配置文件，为此模块添加 Feign 依赖库。

```
project('common-api') {                                          // 子模块
   dependencies {                                                // 配置子模块依赖
      compile('org.springframework.boot:spring-boot-starter-web')          // Spring Boot
      compile('org.springframework.cloud:spring-cloud-starter-openfeign')  // OpenFeign
   }
}
```

(2)【common-api 子模块】为了便于描述，本次将在 IDeptService 业务接口中引入 Feign 相关配置，实现接口映射。

```
package com.yootk.service;
@FeignClient("dept.provider")                                    // 映射微服务名称
public interface IDeptService {                                  // 部门业务接口
   @GetMapping("/provider/dept/get/{deptno}")                    // 微服务路径
   public DeptDTO get(@PathVariable("deptno") long id);
   @PostMapping("/provider/dept/add")                            // 微服务路径
   public boolean add(DeptDTO dept);
   @GetMapping("/provider/dept/list")                            // 微服务路径
   public List<DeptDTO> list();
   @GetMapping("/provider/dept/split")                           // 微服务路径
   public Map<String, Object> split(@RequestParam("cp") int currentPage,
        @RequestParam("ls") int lineSize, @RequestParam("col") String column,
        @RequestParam("kw") String keyword);
}
```

(3)【consumer-springboot-80 子模块】修改消费端 Action 程序类，此时的消费端直接通过业务接口即可实现服务调用。

```
package com.yootk.consumer.action;
@RestController                                                  // REST响应
@RequestMapping("/consumer/dept/*")                              // 父路径
public class DeptConsumerAction {
   @Autowired
   private IDeptService deptService;                             // 业务接口实例
   @GetMapping("add")
   public Object addDept(DeptDTO dept) {                         // 增加部门
      return this.deptService.add(dept);
   }
```

```
    @GetMapping("get")
    public Object getDept(Long deptno) {                              // 查询部门
        return this.deptService.get(deptno);
    }
    @GetMapping("list")
    public Object listDept() {                                        // 部门列表
        return this.deptService.list();
    }
    @GetMapping("split")
    public Object splitDept(int cp, int ls, String col, String kw) { // 分页列表
        return this.deptService.split(cp, ls, col, kw);
    }
}
```

（4）【consumer-springboot-80 子模块】修改启动类，追加 Feign 扫描配置。

```
package com.yootk.consumer;
@SpringBootApplication
@EnableDiscoveryClient                                              // 启用发现服务
@RibbonClient(name = "dept.provider",
        configuration = DeptProviderRibbonConfig.class)             // 负载均衡策略
@EnableFeignClients("com.yootk.service")                            // Feign扫描包
public class StartConsumerApplication {
    public static void main(String[] args) {
        SpringApplication.run(StartConsumerApplication.class, args); // 服务启动
    }
}
```

由于此时 IDeptService 业务接口与消费端的主应用扫描包有所不同，因此在使用@EnableFeignClients 注解时就必须明确配置扫描包的名称，否则消费端将无法实现 IDeptService 业务接口实例的生成与注册。

4.4.1 Feign 转换日志

Feign 转换日志

视频名称 0421_【掌握】Feign 转换日志

视频简介 Feign 在进行转换时实际上依然采用了 HTTP 请求与响应处理机制，而为了便于使用者观察，Feign 组件提供了日志的支持。本视频为读者讲解如何配置 Feign 组件的转换类，并分析日志的组成结构。

REST 服务是以 HTTP 为基础进行搭建的，而在使用 Feign 进行转换时，实际上封装了用户对 REST 接口的调用过程，使之可以通过 RPC 的方式实现服务调用。如果开发者需要关注这一操作的过程，可以开启 Feign 转换日志。在 Feign 中，转换日志的级别定义有 4 个级别，如表 4-4 所示。

表 4-4 Feign 日志级别

序号	日志级别	描述
01	feign.Logger.Level.NONE	不记录任何日志信息
02	feign.Logger.Level.BASIC	仅记录基本信息，如请求方法、URL 以及响应状态、执行时间等
03	feign.Logger.Level.HEADERS	在基本信息基础之上记录请求和响应头信息
04	feign.Logger.Level.FULL	记录所有请求与响应的完整信息，包括请求数据、头信息等

Feign 中的日志实现需要使用 feign.Logger 日志处理类，而后在该类中提供了一个 Level 枚举类型。为了便于读者观察，下面将采用完整日志的形式跟踪消费端与提供端之间的通信流程，具体的实现步骤如下所示。

（1）【common-api 子模块】创建一个 Feign 配置类，进行日志输出级别的配置。

```
package com.yootk.service.config;
import feign.Logger;                          // 此时导入的为Feign提供的日志组件
```

```java
public class FeignConfig {
    @Bean
    public Logger.Level level() {
        return Logger.Level.FULL;                // 输出所有请求的细节
    }
}
```

为便于读者更好地跟踪 Feign 转换过程，本次采用了全部详细日志的输出。而在实际的生产环境中，一般使用“Logger.Level.BASIC”级别即可，该级别会清楚地显示出请求与响应的操作信息。

（2）【common-api 子模块】修改 IDeptService 接口中的@FeignClient 注解，引用 FeignConfig 配置类。

```java
package com.yootk.service;
@FeignClient(value="dept.provider", configuration = FeignConfig.class)  // 配置类
public interface IDeptService { }
```

（3）【consumer-springboot-80 子模块】修改 application.yml 配置文件，追加 Feign 转换日志输出配置项。

```yaml
logging:
  level:
    com.yootk.service.IDeptService: DEBUG     # 定义接口日志级别
```

（4）【Postman】此时已经成功地实现了 Feign 日志的输出操作。为了验证当前的配置是否正确，可以通过 Postman 工具发出两个消费端请求，而后可以在后台观察到相关的日志信息。

①【GET 请求】查询部门信息。

```
consumer-springboot-80/consumer/dept/get?deptno=1
```

后台日志输出	
后台日志输出	com.yootk.service.IDeptService : [IDeptService#get] ---> GET http://dept.provider/provider/dept/get/1 HTTP/1.1 com.yootk.service.IDeptService : [IDeptService#get] ---> END HTTP (0-byte body) com.yootk.service.IDeptService : [IDeptService#get] <--- HTTP/1.1 200 (4ms) com.yootk.service.IDeptService : [IDeptService#get] cache-control: no-cache, no-store, max-age=0, must-revalidate com.yootk.service.IDeptService : [IDeptService#get] connection: keep-alive com.yootk.service.IDeptService : [IDeptService#get] content-type: application/json com.yootk.service.IDeptService : [IDeptService#get] expires: 0 com.yootk.service.IDeptService : [IDeptService#get] keep-alive: timeout=60 com.yootk.service.IDeptService : [IDeptService#get] pragma: no-cache com.yootk.service.IDeptService : [IDeptService#get] transfer-encoding: chunked com.yootk.service.IDeptService : [IDeptService#get] x-content-type-options: nosniff com.yootk.service.IDeptService : [IDeptService#get] x-frame-options: DENY com.yootk.service.IDeptService : [IDeptService#get] x-xss-protection: 1; mode=block com.yootk.service.IDeptService : [IDeptService#get] {"deptno":1,"dname":"开发部","loc":"yootk8003"} com.yootk.service.IDeptService : [IDeptService#get] <--- END HTTP (50-byte body)

②【POST 请求】增加部门数据。

```
consumer-springboot-80/consumer/dept/add?dname=沐言优拓&loc=洛阳
```

后台日志输出	
后台日志输出	com.yootk.service.IDeptService : [IDeptService#add] ---> POST http://dept.provider/provider/dept/add HTTP/1.1 com.yootk.service.IDeptService : [IDeptService#add] Content-Length: 53 com.yootk.service.IDeptService : [IDeptService#add] Content-Type: application/json com.yootk.service.IDeptService : [IDeptService#add] com.yootk.service.IDeptService : [IDeptService#add] {"deptno":null,"dname":"沐言优拓","loc":"洛阳"} com.yootk.service.IDeptService : [IDeptService#add] ---> END HTTP (53-byte body) com.yootk.service.IDeptService : [IDeptService#add] <--- HTTP/1.1 200 (28ms) com.yootk.service.IDeptService : [IDeptService#add]

后台日志输出

```
                                        cache-control: no-cache, no-store,
                                        max-age=0, must-revalidate
com.yootk.service.IDeptService : [IDeptService#add] connection: keep-alive
com.yootk.service.IDeptService : [IDeptService#add] content-type: application/json
com.yootk.service.IDeptService : [IDeptService#add] expires: 0
com.yootk.service.IDeptService : [IDeptService#add] keep-alive: timeout=60
com.yootk.service.IDeptService : [IDeptService#add] pragma: no-cache
com.yootk.service.IDeptService : [IDeptService#add] transfer-encoding: chunked
com.yootk.service.IDeptService : [IDeptService#add] x-content-type-options: nosniff
com.yootk.service.IDeptService : [IDeptService#add] x-frame-options: DENY
com.yootk.service.IDeptService : [IDeptService#add] x-xss-protection: 1; mode=block
com.yootk.service.IDeptService : [IDeptService#add] true
com.yootk.service.IDeptService : [IDeptService#add] <--- END HTTP (4-byte body)
```

4.4.2 Feign 连接池

Feign 连接池

视频名称 0422_【掌握】Feign 连接池

视频简介 Feign 服务通信基于 HTTP 完成，而为了得到最佳的处理性能，需要在项目中进行服务的配置。本视频为读者讲解 Feign 的相关配置以及实现意义。

在使用 Feign 组件实现远程微服务调用时，全部的处理操作都是基于 HTTP 完成的，而 HTTP 是基于 TCP 开发的。HTTP 采用了无状态的处理形式，这样一来用户每一次发出请求都需要进行开启连接与连接关闭操作，如图 4-37 所示。这样的处理形式最终一定会导致严重的性能损耗。

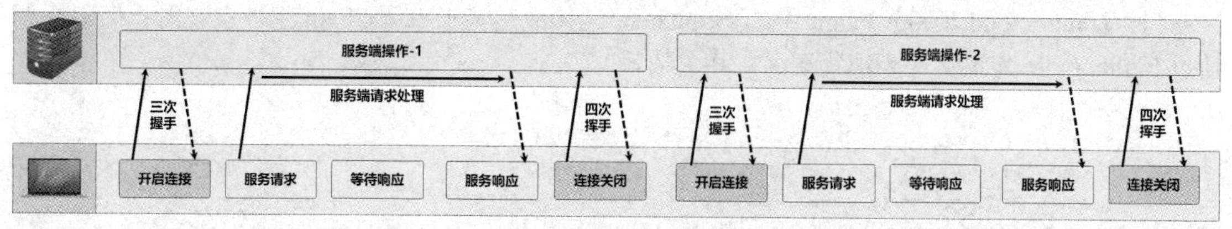

图 4-37 HTTP 的反复开启连接与连接关闭

HTTP 1.0 采用的是全双工协议，为了创建可靠的连接，要在连接建立与连接断开时采用“三次握手”与“四次挥手”的处理机制，这样在每次进行请求和响应时都会造成大量的资源消耗。为了弥补这样的缺陷，最佳的做法是采用持久化连接的形式来实现 Socket 连接复用，这样用户只需要创建一次连接，就可以实现多次的请求与响应处理，如图 4-38 所示。

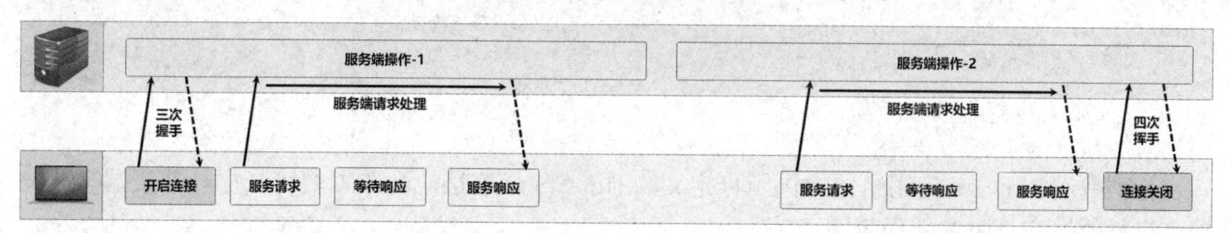

图 4-38 HTTP 持久化连接

提示：HTTP 1.0 的“Keep-Alive”机制。

从 1996 年开始，为了提高 HTTP 1.0 的处理性能，很多支持 HTTP 1.0 的浏览器与服务器都进行了协议的扩展，提出了“Keep-Alive”机制（HTTP 1.1 规范没有对此机制的描述，该机制只是作为一种应用设计被保留下来），只需要在每次请求的头信息中加上“Connection:Keep-Alive”参数即可。服务端如果需要让这条连接保持打开的状态，就会在响应头信息中包含同样的头信息，客户端会根据这个头信息是否存在来进行是否要关闭连接的判断。

HTTP 1.1 为了提高 Socket 复用性采用了多路复用的设计原则，这样可以让一个连接为不同的用户提供服务，从而实现服务端的服务处理性能。客户端如果想实现持久化连接，则可以基于 HttpClient 组件提供的连接池来进行管理，以实现同一条 TCP 链路上的连接复用，如图 4-39 所示。

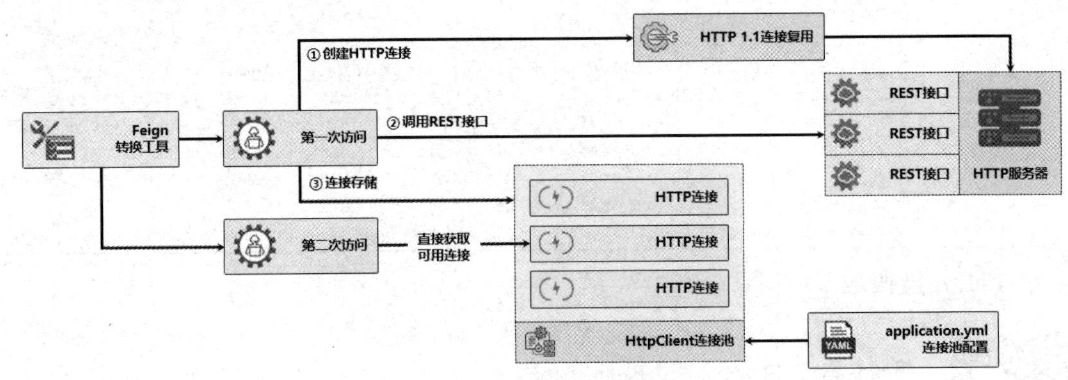

图 4-39　HttpClient 连接池

HttpClient 连接池的核心流程在于：第一次进行服务端访问时才需要创建服务端连接，服务调用完成后对应的连接并不关闭，而是直接将之归还到连接池，这样在下一次访问时通过 HttpClient 连接池即可获取一个可用的连接。考虑到连接池的占用过多问题，还需要进行定时清理操作，将已经失效或不再使用的连接释放。下面通过具体的操作步骤介绍如何在 Feign 组件中使用 HttpClient 连接池。

（1）【microcloud 项目】Feign 中的 HttpClient 连接池是基于 HttpClient 组件实现的，所以需要通过 build.gradle 配置文件为消费端添加相关依赖。

```
project('consumer-springboot-80') {                  // 子模块
    dependencies {                                    // 配置子模块依赖
        implementation('org.apache.httpcomponents:httpclient:4.5.13')// HttpClient
        implementation('io.github.openfeign:feign-httpclient:11.1')   // HttpClient
        implementation(project(':common-api')) // 子模块
        implementation('com.alibaba.cloud:spring-cloud-starter-alibaba-nacos-discovery')
    }
}
```

（2）【consumer-springboot-80 子模块】修改 application.yml 配置文件，添加 HttpClient 连接池配置。

```
feign:                                          # Feign配置
  httpclient:                                   # HttpClient连接池配置
    enabled: true                               # 使用HttpClient处理机制
    max-connections: 200                        # Feign的最大连接数
    max-connections-per-route: 50               # Feign单个路径的最大连接数
    connection-timeout: 2000                    # 连接超时
```

为消费端的 Feign 应用组件提供了自定义的 HttpClient 连接池配置，就可以提升消费端与服务端在进行数据交互时的处理效率。

4.4.3　数据压缩传输

数据压缩传输

视频名称　0423_【理解】数据压缩传输

视频简介　为了高效地实现数据的传输处理，Feign 组件提供了数据压缩的处理支持。本视频为读者讲解如何在 Feign 组件中实现 HTTP 数据压缩传输。

为了进一步提升数据传输的性能，Feign 组件提供了请求与响应数据的 GZIP 处理操作。这样当传输的数据量较大时就可以自动实现请求与响应数据的压缩与解压缩处理，如图 4-40 所示。在 Feign 中对数据的压缩可以直接通过 application.yml 配置实现，而在配置时需要同时启用请求和响应数据的压缩处理。

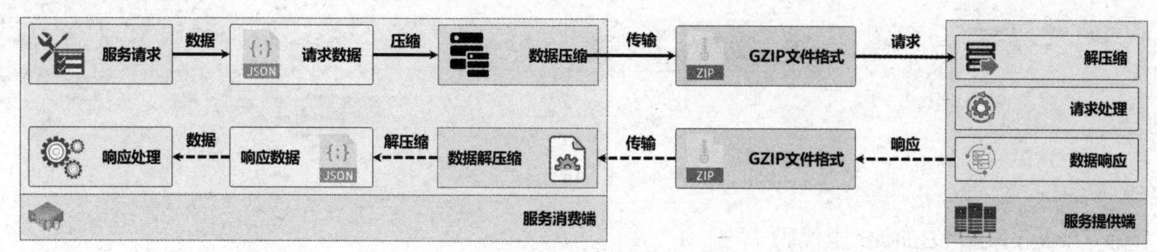

图 4-40　数据压缩传输

提示：GZIP 是压缩的简称。

GZIP（GUN Zip）主要用来描述文件的压缩格式，由 Jean-loup Gailly 和 Mark Adler 创建，主要在 UNIX 系统中使用。在 HTTP 上的 GZIP 编码是一种用来改进 Web 应用程序性能的技术，可以基于压缩结构高效实现网络传输（文本数据内容可以压缩到原始文本大小的约 40%），现在大部分的服务器都有 GZIP 支持。

范例：【consumer-springboot-80 子模块】Feign 压缩传输

```
server:                                    # 服务端配置
 port: 80                                  # 80端口监听
 compression:                              # Spring Boot压缩配置
  enabled: true                            # 启用压缩
  mime-types: application/json,application/xml,text/html,text/xml,text/plain # 压缩MIME
feign:
 compression:                              # GZIP压缩配置
  request:                                 # 请求配置
   enabled: true                           # 启用压缩
   mime-types: text/xml,application/xml,application/json   # 配置压缩支持的MIME
   min-request-size: 512                   # 压缩数据大小的最小阈值
  response:                                # 响应配置
   enabled: true                           # 启用压缩
```

后台日志输出	com.yootk.service.IDeptService : [IDeptService#list] Accept-Encoding: gzip

本配置中对文本、XML、JSON 数据全部采用了压缩处理，而根据此时后台日志的输出也可以清楚地发现当前数据传输的类型为 GZIP 类型。

4.5　Feign 核心源代码分析

Feign 工作原理

视频名称　0424_【理解】Feign 工作原理

视频简介　Feign 组件为了便于接口映射管理，提供了一系列处理步骤。本视频为读者讲解 Feign 组件的工作原理，通过源代码分析原生 Feign 组件的构建，并浅析在 Spring Cloud 组件中 Feign 组件的构建与管理。

Feign 组件本质上封装了 HTTP 请求与响应处理过程，使得整个处理机制更加符合面向对象的设计形式，但是最终的接口实现子类必定在远程服务端，所以消费端在通过接口执行远程方法时，就需要基于动态代理机制来生成接口实现子类，最终实现服务调用。其工作原理如图 4-41 所示。

Feign 原始开发包为了便于所有 Feign 组件的实例化管理，提供了一个 Feign.Builder 内部构建器类，利用此类可以直接获取所有的核心接口，该类的部分源代码如下所示。

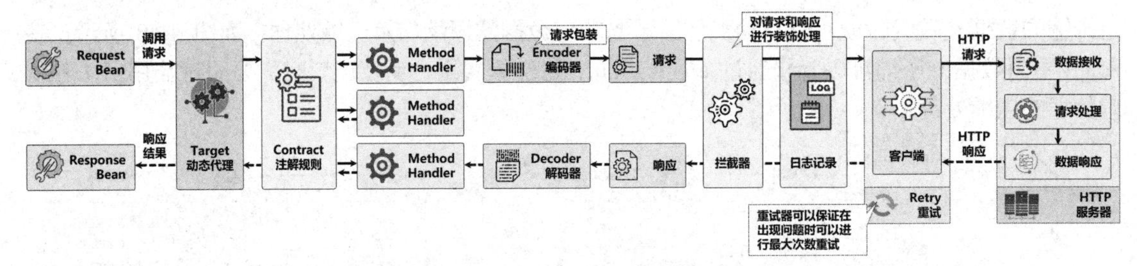

图 4-41　Feign 工作原理

范例：feign.Feign.Builder 源代码

```
package feign;                                             // Feign组件原生包
public abstract class Feign {
  public static Builder builder() {                        // 获取Feign.Builder实例
    return new Builder();
  }
  public abstract <T> T newInstance(Target<T> target);
  public static class Builder {
    private final List<RequestInterceptor> requestInterceptors =
        new ArrayList<RequestInterceptor>();               // 拦截器
    private Logger.Level logLevel = Logger.Level.NONE;     // 日志级别
    private Contract contract = new Contract.Default();    // 注解规则
    private Client client = new Client.Default(null, null);  // 客户端
    private Retryer retryer = new Retryer.Default();       // 重试器
    private Logger logger = new NoOpLogger();              // 日志
    private Encoder encoder = new Encoder.Default();       // 编码器
    private Decoder decoder = new Decoder.Default();       // 解码器
    private QueryMapEncoder queryMapEncoder = new FieldQueryMapEncoder();
    private ErrorDecoder errorDecoder = new ErrorDecoder.Default();
    private Options options = new Options();
    private InvocationHandlerFactory invocationHandlerFactory =
        new InvocationHandlerFactory.Default();            // 动态代理工厂类
    private boolean decode404;
    private boolean closeAfterDecode = true;
    private ExceptionPropagationPolicy propagationPolicy = NONE;
  }
}
```

通过如上源代码可以发现，在 Feign 组件中所有操作结构的实例都可以通过“Feign.Builder”来获取，这样就可以得到图 4-42 所示的类结构。

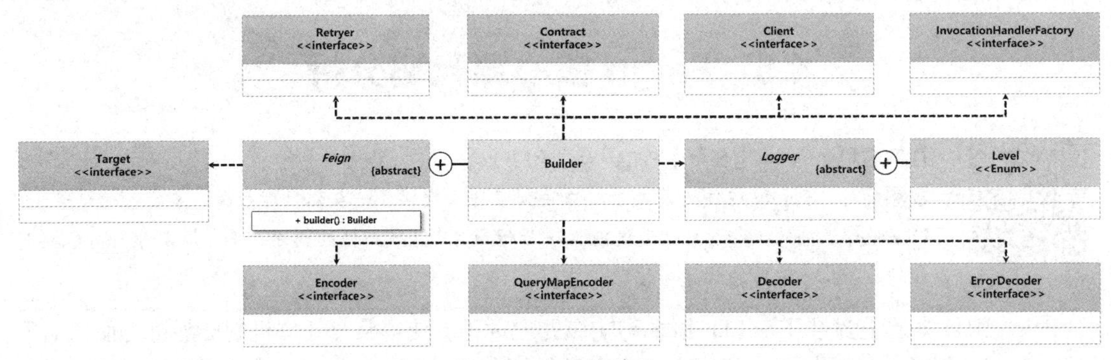

图 4-42　Feign 类结构

Spring Cloud 整合 Feign 组件之后，修改了 Feign 组件的构造过程，为了便于自动配置的实现，它提供了一系列 AutoConfig 配置类，这样就可以实现 Feign 组件的配置处理。这些都在“spring-cloud-openfeign-core”模块提供的自启动配置文件中进行了定义。

范例：spring.factories 源代码

```
org.springframework.boot.autoconfigure.EnableAutoConfiguration=\
org.springframework.cloud.openfeign.ribbon.FeignRibbonClientAutoConfiguration,\
org.springframework.cloud.openfeign.hateoas.FeignHalAutoConfiguration,\
org.springframework.cloud.openfeign.FeignAutoConfiguration,\
org.springframework.cloud.openfeign.encoding.FeignAcceptGzipEncodingAutoConfiguration,\
org.springframework.cloud.openfeign.encoding.FeignContentGzipEncodingAutoConfiguration,\
org.springframework.cloud.openfeign.loadbalancer.FeignLoadBalancerAutoConfiguration
```

通过此处定义的配置文件可以清楚地发现，Feign 组件在使用时需要与 Ribbon、GZIP、负载均衡等相关的组件整合。后文将通过具体的操作分析这些启动类的处理机制。

提示：spring-cloud-starter-openfeign 依赖。

用户在进行依赖配置时，使用了 spring-cloud-starter-openfeign 启动依赖库，但是需要注意的是，真正的 Feign 核心配置都在“spring-cloud-openfeign-core”依赖库中定义，这一点可以通过观察源代码明确。

范例：【spring-cloud-starter-openfeign 依赖】观察依赖配置文件源代码

```
<dependency>
        <groupId>org.springframework.cloud</groupId>
        <artifactId>spring-cloud-openfeign-core</artifactId>
        <version>2.2.2.RELEASE</version>
        <scope>compile</scope>
</dependency>
```

后文主要分析“spring-cloud-openfeign-core”依赖库中提供的程序。

4.5.1 FeignAutoConfiguration

FeignAuto Configuration

视频名称 0425_【理解】FeignAutoConfiguration

视频简介 本视频通过 Starter 组件的执行流程分析 FeignAutoConfiguration 配置类的组成，并依据组成分析 Feign 中的动态代理生成机制以及 HTTP 客户端的使用选择。

Feign 在进行组件依赖时主要依靠核心类 FeignAutoConfiguration，所有在 application.yml 中定义的 Feign 的配置项都会通过 FeignClientProperties（基本配置）与 FeignHttpClientProperties（HTTP 配置）类进行保存，而后在 Feign 自动配置类的内部会生成 FeignContext 上下文环境，并且依据此上下文的环境自动生成接口实现类，最终在 Spring 容器中进行注册。FeignAutoConfiguration 类的核心源代码如下。

范例：【spring-cloud-openfeign-core 依赖】FeignAutoConfiguration 类的核心源代码

```
package org.springframework.cloud.openfeign;
@Configuration(proxyBeanMethods = false)
@ConditionalOnClass(Feign.class)                          // Feign原生功能类
@EnableConfigurationProperties({ FeignClientProperties.class,
      FeignHttpClientProperties.class })                  // Feign配置属性
@Import(DefaultGzipDecoderConfiguration.class)            // 压缩解码配置
public class FeignAutoConfiguration {
   @Autowired(required = false)
   private List<FeignClientSpecification> configurations = new ArrayList<>();
   @Bean
   public FeignContext feignContext() {                   // Feign上下文
      FeignContext context = new FeignContext();
      context.setConfigurations(this.configurations); // Feign配置
      return context;
   }
   @Configuration(proxyBeanMethods = false)
```

```
  @ConditionalOnClass(name = "feign.hystrix.HystrixFeign")
  protected static class HystrixFeignTargeterConfiguration {
     @Bean
     @ConditionalOnMissingBean
     public Targeter feignTargeter() {                    // 生成代理类
        return new HystrixTargeter();                     // 生成Hystrix代理类
     }
  }
  @Configuration(proxyBeanMethods = false)
  @ConditionalOnMissingClass("feign.hystrix.HystrixFeign")
  protected static class DefaultFeignTargeterConfiguration {
     @Bean
     @ConditionalOnMissingBean
     public Targeter feignTargeter() {                    // 生成代理类
        return new DefaultTargeter();                     // 生成默认代理类
     }
  }
  @Configuration(proxyBeanMethods = false)
  @ConditionalOnClass(ApacheHttpClient.class)             // HttpClient实现
  @ConditionalOnMissingClass("com.netflix.loadbalancer.ILoadBalancer")
  @ConditionalOnMissingBean(CloseableHttpClient.class)
  @ConditionalOnProperty(value = "feign.httpclient.enabled", matchIfMissing = true)
  protected static class HttpClientFeignConfiguration {}
  @Configuration(proxyBeanMethods = false)
  @ConditionalOnClass(OkHttpClient.class)                 // OkHttpClient实现
  @ConditionalOnMissingClass("com.netflix.loadbalancer.ILoadBalancer")
  @ConditionalOnMissingBean(okhttp3.OkHttpClient.class)
  @ConditionalOnProperty("feign.okhttp.enabled")
  protected static class OkHttpFeignConfiguration {}
}
```

通过以上配置类中的“@ConditionalOnClass(Feign.**class**)”代码，可以清楚地发现，如果要进行 Feign 自动配置，则必须要有“feign.Feign”原生 Feign 功能类。而后在该类中有 Feign 构建类，就可以根据此构建类进行 Feign 各个组件的构建管理，同时在该类中会根据当前环境配置不同的动态代理实现类。其结构如图 4-43 所示。

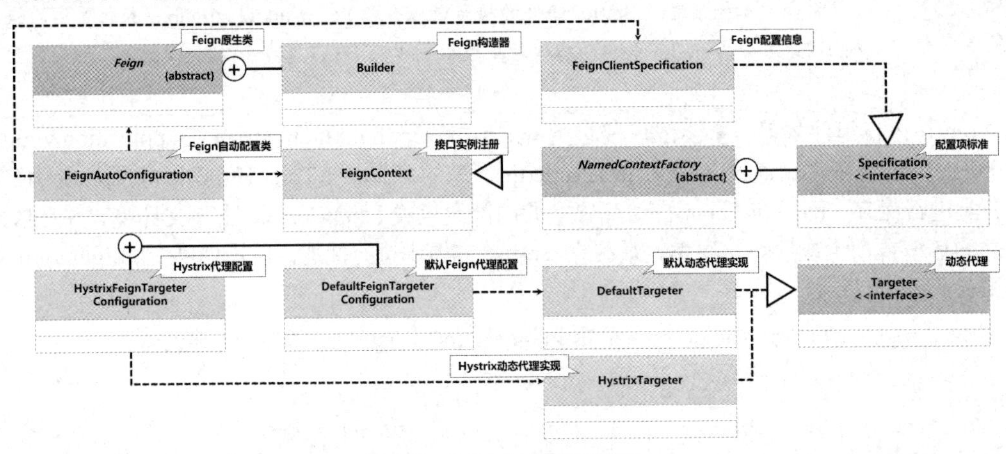

图 4-43　Feign 自动配置类结构

4.5.2　FeignRibbonClientAutoConfiguration

FeignRibbon
ClientAuto
Configuration

视频名称　0426_【理解】FeignRibbonClientAutoConfiguration

视频简介　Feign 工作在消费端，所以在设计时充分地利用了 Ribbon 处理机制实现服务端负载均衡策略。本视频为读者分析 FeignRibbonClientAutoConfiguration 配置类的源代码结构及其与 Ribbon 标准的关联。

在 Spring Cloud 中消费端与服务端之间的调用主要通过 Ribbon 负载均衡实现，所以在引入 Feign 依赖时就会自动进行 Ribbon 依赖的引入。这样在使用 Feign 调用服务之前就需要对 Ribbon 客户端进行自动配置，所以 Feign 组件内部提供了 FeignRibbonClientAutoConfiguration 自动配置类，该类的源代码如下。

范例：【spring-cloud-openfeign-core 依赖】FeignRibbonClientAutoConfiguration 源代码

```
package org.springframework.cloud.openfeign.ribbon;
@ConditionalOnClass({ ILoadBalancer.class, Feign.class })
@ConditionalOnProperty(value = "spring.cloud.loadbalancer.ribbon.enabled",
    matchIfMissing = true)                                      // 默认启用Ribbon负载均衡
@Configuration(proxyBeanMethods = false)
@AutoConfigureBefore(FeignAutoConfiguration.class)              // Feign自动配置之后处理
@EnableConfigurationProperties({ FeignHttpClientProperties.class })
@Import({ HttpClientFeignLoadBalancedConfiguration.class,
    OkHttpFeignLoadBalancedConfiguration.class,
    DefaultFeignLoadBalancedConfiguration.class })
public class FeignRibbonClientAutoConfiguration {
  @Bean
  @Primary
  @ConditionalOnMissingBean
  @ConditionalOnMissingClass("org.springframework.retry.support.RetryTemplate")
  public CachingSpringLoadBalancerFactory cachingLBClientFactory(
      SpringClientFactory factory) {
    return new CachingSpringLoadBalancerFactory(factory);       // 负载均衡工厂类
  }
  @Bean
  @Primary
  @ConditionalOnMissingBean
  @ConditionalOnClass(name = "org.springframework.retry.support.RetryTemplate")
  public CachingSpringLoadBalancerFactory retryabeCachingLBClientFactory(
      SpringClientFactory factory, LoadBalancedRetryFactory retryFactory) {
    return new CachingSpringLoadBalancerFactory(factory, retryFactory);
  }
  @Bean
  @ConditionalOnMissingBean
  public Request.Options feignRequestOptions() {
    return LoadBalancerFeignClient.DEFAULT_OPTIONS;
  }
}
```

通过当前的源代码可以发现，在进行 Ribbon 初始化时会获得一个 CachingSpringLoadBalancerFactory 对象实例，而在获取该类实例时需要传入 SpringClientFactory 对象实例（这一点与 Ribbon 的实现结构类似）。此时所使用的类结构如图 4-44 所示。

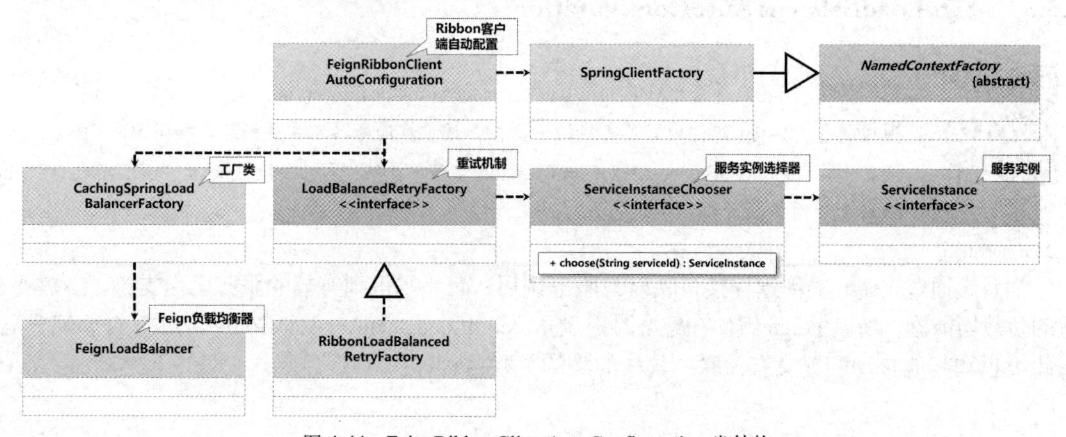

图 4-44　FeignRibbonClientAutoConfiguration 类结构

此程序的实现关键在于 CachingSpringLoadBalancerFactory 工厂类，该工厂类可以采用缓存的方式实现工厂加载数据的存储。下面打开 CachingSpringLoadBalancerFactory 程序类的源代码进行观察。

范例：【spring-cloud-openfeign-core 依赖】CachingSpringLoadBalancerFactory 源代码

```
package org.springframework.cloud.openfeign.ribbon;
public class CachingSpringLoadBalancerFactory {
   protected final SpringClientFactory factory;
   protected LoadBalancedRetryFactory loadBalancedRetryFactory = null;
   private volatile Map<String, FeignLoadBalancer> cache =
               new ConcurrentReferenceHashMap<>();          // 缓存配置
   public CachingSpringLoadBalancerFactory(SpringClientFactory factory) {
      this.factory = factory;
   }
   public CachingSpringLoadBalancerFactory(SpringClientFactory factory,
         LoadBalancedRetryFactory loadBalancedRetryPolicyFactory) {
      this.factory = factory;
      this.loadBalancedRetryFactory = loadBalancedRetryPolicyFactory;
   }
   public FeignLoadBalancer create(String clientName) {
      FeignLoadBalancer client = this.cache.get(clientName); // 获取缓存数据
      if (client != null) {                                   // 缓存数据不为空
         return client;                                       // 返回缓存数据
      }
      // 获取IClientConfig、ILoadBalancer接口对象实例
      IClientConfig config = this.factory.getClientConfig(clientName);
      ILoadBalancer lb = this.factory.getLoadBalancer(clientName);
      ServerIntrospector serverIntrospector = this.factory.getInstance(clientName,
            ServerIntrospector.class);
      client = this.loadBalancedRetryFactory != null
            ? new RetryableFeignLoadBalancer(lb, config, serverIntrospector,
               this.loadBalancedRetryFactory)                 // 创建Feign负载均衡器
            : new FeignLoadBalancer(lb, config, serverIntrospector);
      this.cache.put(clientName, client);                     // 负载均衡器缓存
      return client;
   }
}
```

该类的主要功能是根据客户端的名称创建一个 FeignLoadBalancer 对象实例，随后通过 LoadBalancedRetryFactory 重试工厂类实现服务调用。同时，为了提高处理效率，其会在每次创建 FeignLoadBalancer 实例后将所有的相关配置信息保存在 Map 集合之中（基于 Map 集合实现一个配置的缓存）。

4.5.3　FeignLoadBalancerAutoConfiguration

FeignLoad BalancerAuto Configuration

视频名称　0427_【了解】FeignLoadBalancerAutoConfiguration

视频简介　Feign 组件内置了负载均衡器，而该负载均衡器又需要稳定的服务调用。本视频为读者分析 FeignLoadBalancerAutoConfiguration 自动配置类，同时分析 BlockingLoadBalancerClient 阻塞负载均衡器的作用。

消费端通过 Feign 实现远程接口映射与服务调用，而为了得到有效的服务调用支持，就需要合适的负载均衡器，所以 Feign 组件在整合时提供了 FeignLoadBalancerAutoConfiguration 自动配置类，而此类也和其他自动配置类有关联。具体的类的关联结构如图 4-45 所示。

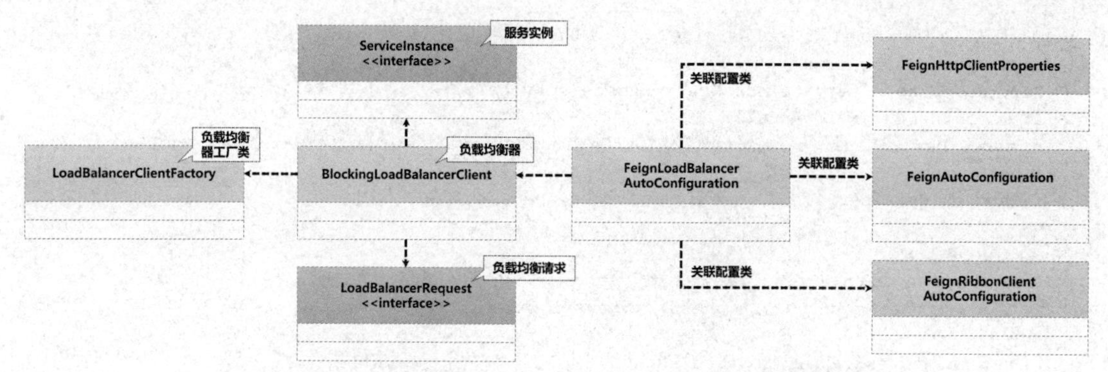

图 4-45　FeignLoadBalancerAutoConfiguration 类的关联结构

范例：【spring-cloud-openfeign-core 依赖】FeignLoadBalancerAutoConfiguration 源代码

```
package org.springframework.cloud.openfeign.loadbalancer;
@ConditionalOnClass(Feign.class)
@ConditionalOnBean(BlockingLoadBalancerClient.class)                // 存在指定Bean
@AutoConfigureBefore(FeignAutoConfiguration.class)                  // Feign配置之前
@AutoConfigureAfter(FeignRibbonClientAutoConfiguration.class)       // Ribbon配置之后
@EnableConfigurationProperties(FeignHttpClientProperties.class)     // HttpClient属性
@Configuration(proxyBeanMethods = false)
@Import({ HttpClientFeignLoadBalancerConfiguration.class,
     OkHttpFeignLoadBalancerConfiguration.class,
     DefaultFeignLoadBalancerConfiguration.class })
public class FeignLoadBalancerAutoConfiguration {}
```

通过此时源代码的定义可以发现，当前的自动配置类并没有提供任何配置方法，但是在这个类的注解定义中提供一个@ConditionalOnBean(BlockingLoadBalancerClient.class)定义。这就表示在当前 Spring 容器中必须首先存在 BlockingLoadBalancerClient 实例，该配置类才可以生效（由于此 Bean 并不存在，因此此配置类暂不生效），所以研究此启动类的重点就在于 BlockingLoadBalancerClient 阻塞负载均衡器。

提示：spring-cloud-loadbalancer 依赖。

在 2017 年时 Spring 开始尝试开发“spring-cloud-loadbalancer”项目来替代 Ribbon 组件（目的是不再受限于 Netflix 维护），一段时间后，此项目被规划到了“spring-cloud- commons”项目之中。开发者如果想观察 BlockingLoadBalancerClient 类的源代码，就需要在项目中引入指定依赖。

范例：【microcloud 项目】修改 build.gradle 依赖配置

```
implementation('org.springframework.cloud:' +
      'spring-cloud-commons:2.2.8.RELEASE')
implementation('org.springframework.cloud:' +
      'spring-cloud-loadbalancer:2.2.8.RELEASE')
```

另外需要提醒读者的是，最初的“spring-cloud-loadbalancer”项目托管在 spring-cloud-incubator（Spring Cloud 项目孵化器）中，而本书所讲解的 Spring Cloud Alibaba 就来自此孵化器。

范例：【spring-cloud-openfeign-core 依赖】BlockingLoadBalancerClient 源代码

```
package org.springframework.cloud.loadbalancer.blocking.client;
public class BlockingLoadBalancerClient implements LoadBalancerClient {
  private final LoadBalancerClientFactory loadBalancerClientFactory;
  public BlockingLoadBalancerClient(
      LoadBalancerClientFactory loadBalancerClientFactory) {
    this.loadBalancerClientFactory = loadBalancerClientFactory;
  }
  @Override
```

```
    public <T> T execute(String serviceId, LoadBalancerRequest<T> request)
            throws IOException {
        ServiceInstance serviceInstance = choose(serviceId);
        if (serviceInstance == null) {
            throw new IllegalStateException("No instances available for " + serviceId);
        }
        return execute(serviceId, serviceInstance, request);
    }
    @Override
    public <T> T execute(String serviceId, ServiceInstance serviceInstance,
            LoadBalancerRequest<T> request) throws IOException {
        try {
            return request.apply(serviceInstance);                // 请求调用
        } catch (IOException iOException) { throw iOException;
        } catch (Exception exception) {
            ReflectionUtils.rethrowRuntimeException(exception);
        }
        return null;
    }
    @Override
    public URI reconstructURI(ServiceInstance serviceInstance, URI original) {
        // 根据服务实例以及访问的URI构建一个完整的访问路径
        return LoadBalancerUriTools.reconstructURI(serviceInstance, original);
    }
    @Override
    public ServiceInstance choose(String serviceId) {          // 根据服务ID选择一个实例
        ReactiveLoadBalancer<ServiceInstance> loadBalancer = loadBalancerClientFactory
                .getInstance(serviceId);                        // 获取Reactive负载均衡器
        if (loadBalancer == null) {
            return null;
        }
        Response<ServiceInstance> loadBalancerResponse = Mono.from(loadBalancer.choose())
                .block();                                       // 阻塞模式获取响应数据
        if (loadBalancerResponse == null) {
            return null;
        }
        return loadBalancerResponse.getServer();                // 获取服务实例
    }
}
```

通过此类的源代码可以发现，该类提供了类似于 Ribbon 处理结构的负载均衡处理，也提供实例选择器、服务调用等相关处理方法，但是这个类是基于 spring-cloud-loadbalancer 均衡器实现的，而此操作暂时还未使用。

4.5.4　FeignClientsRegistrar

FeignClients Registrar

视频名称　0428_【理解】FeignClientsRegistrar

视频简介　在使用 Feign 组件时需要通过@EnableFeignClients 和@FeignClient 两个注解进行标记，而在 Feign 内部会自动进行扫描包的配置与 Bean 注册。本视频为读者分析 FeignClientsRegistrar 类源代码的实现。

在 SpringCloud 中所有的远程接口都需要通过@FeignClient 注解进行定义，而后所有被该注解声明过的接口需要通过@EnableFeignClients 注解进行扫描与 Bean 注册。在此注解中除了配置扫描包之外，也可以配置具体的类名称，这一点可以通过该注解的源代码观察到。

范例：【spring-cloud-openfeign-core 依赖】EnableFeignClients 源代码

```
package org.springframework.cloud.openfeign;
@Retention(RetentionPolicy.RUNTIME)
@Target(ElementType.TYPE)
```

```
@Documented
@Import(FeignClientsRegistrar.class)                   // 导入指定类
public @interface EnableFeignClients {
   String[] value() default {};                        // 等价于扫描包配置
   String[] basePackages() default {};                 // 扫描包配置
   Class<?>[] basePackageClasses() default {};         // 定义扫描包程序类
   Class<?>[] defaultConfiguration() default {};       // 定义配置类
   Class<?>[] clients() default {};                    // 配置@FeignClient注解类
}
```

此时可以发现该注解中有一个@Import(FeignClientsRegistrar.class)导入配置，同时导入的类型为 FeignClientsRegistrar，即该类实现了接口 Bean 注册支持。此类的关联结构如图 4-46 所示。

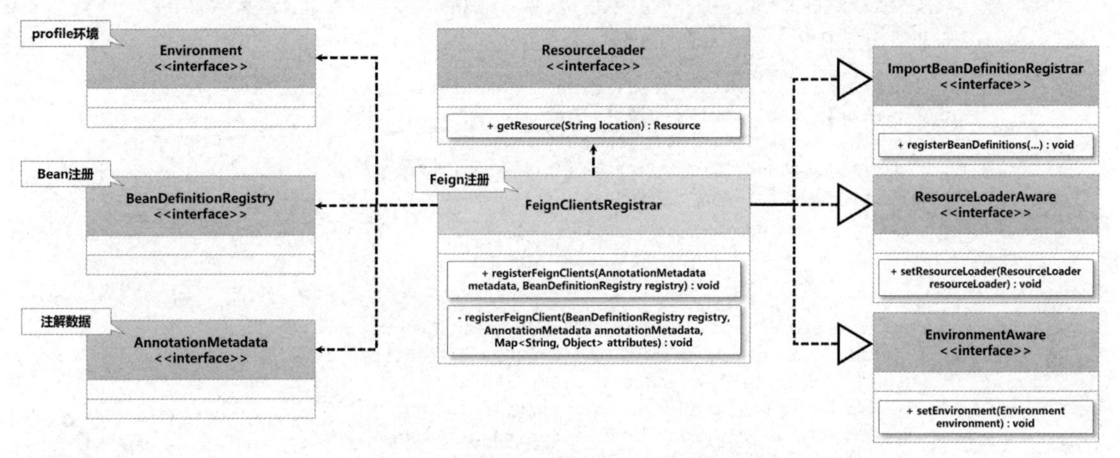

图 4-46　FeignClientsRegistrar 类的关联结构

范例：【spring-cloud-openfeign-core 依赖】FeignClientsRegistrar 核心源代码

```
package org.springframework.cloud.openfeign;
class FeignClientsRegistrar implements ImportBeanDefinitionRegistrar,
      ResourceLoaderAware, EnvironmentAware {                 // FeignClient接口注册
  public void registerFeignClients(AnnotationMetadata metadata,
      BeanDefinitionRegistry registry) {                      // 扫描包注册配置
    ClassPathScanningCandidateComponentProvider scanner = getScanner();
    scanner.setResourceLoader(this.resourceLoader);
    Set<String> basePackages;                                 // 保存所有扫描包
    Map<String, Object> attrs = metadata
        .getAnnotationAttributes(EnableFeignClients.class.getName());
    AnnotationTypeFilter annotationTypeFilter = new AnnotationTypeFilter(
        FeignClient.class);                                   // 获取指定Annotation
    final Class<?>[] clients = attrs == null ? null
        : (Class<?>[]) attrs.get("clients");
    if (clients == null || clients.length == 0) {             // 没有FeignClient
      scanner.addIncludeFilter(annotationTypeFilter);         // 定义扫描注解
      basePackages = getBasePackages(metadata);               // 扫描包
    }
    else {
      final Set<String> clientClasses = new HashSet<>();      // Feign客户端类型保存
      basePackages = new HashSet<>();
      for (Class<?> clazz : clients) {                        // 迭代每一个FeignClient
        basePackages.add(ClassUtils.getPackageName(clazz));
        clientClasses.add(clazz.getCanonicalName());
      }
      AbstractClassTestingTypeFilter filter = new AbstractClassTestingTypeFilter() {
        @Override
        protected boolean match(ClassMetadata metadata) {
          String cleaned = metadata.getClassName().replaceAll("\\$", ".");
          return clientClasses.contains(cleaned);
```

```
            }
        };
        scanner.addIncludeFilter(
            new AllTypeFilter(Arrays.asList(filter, annotationTypeFilter)));
    }
    for (String basePackage : basePackages) {
        Set<BeanDefinition> candidateComponents = scanner
            .findCandidateComponents(basePackage);         // 扫描Bean组件
        for (BeanDefinition candidateComponent : candidateComponents) {
            if (candidateComponent instanceof AnnotatedBeanDefinition) {
                AnnotatedBeanDefinition beanDefinition = (AnnotatedBeanDefinition)
                        candidateComponent;                // 注解定义在接口上
                AnnotationMetadata annotationMetadata = beanDefinition.getMetadata();
                // 根据指定的注解获取其所有的属性内容
                Map<String, Object> attributes = annotationMetadata
                    .getAnnotationAttributes(FeignClient.class.getCanonicalName());
                String name = getClientName(attributes);     // 获取服务名称
                registerClientConfiguration(registry, name,
                    attributes.get("configuration"));        // 配置注册
                registerFeignClient(registry, annotationMetadata, attributes);
            }
        }
    }
}
private void registerFeignClient(BeanDefinitionRegistry registry,
        AnnotationMetadata annotationMetadata, Map<String, Object> attributes) {
    String className = annotationMetadata.getClassName();
    BeanDefinitionBuilder definition = BeanDefinitionBuilder
        .genericBeanDefinition(FeignClientFactoryBean.class);
    // 属性验证与内容处理保存部分代码，略
    BeanDefinitionHolder holder = new BeanDefinitionHolder(beanDefinition, className,
        new String[] { alias });                            // Bean定义
    BeanDefinitionReaderUtils.registerBeanDefinition(holder, registry); // Bean注册
}
}
```

以上源代码首先针对@EnableFeignClients 注解中配置的包名称进行扫描，而后查找出所有带有@FeignClient 的接口，接下来通过 Feign 组件自带的动态代理机制生成实现类，并将此代理实现类的对象注册到 Spring 容器之中，这样就可以在使用时直接进行接口对象的依赖注入。

提示：FeignClientBuilder 构建器。

如果现在的程序中没有使用@FeignClient 注解进行配置，则会自动通过 FeignClientBuilder 类实现解析与注册处理。该类的实现结构如图 4-47 所示。

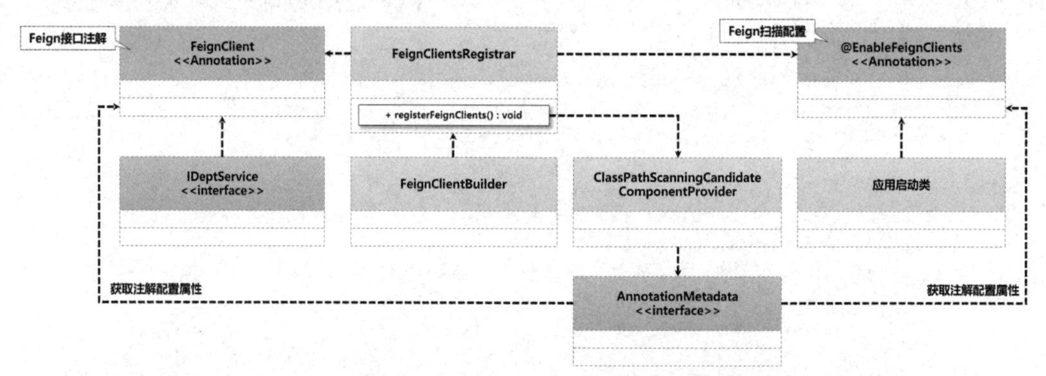

图 4-47　FeignClientBuilder 类的实现结构

该操作类采用与 FeignClientsRegistrar 类似的机制，也需要自动进行目标路径的解析，而后生成最终访问路径。这一点读者可以通过 FeignClientsRegistrar 类的源代码观察到。

4.6 Hystrix 熔断机制

视频名称 0429_【了解】Hystrix 熔断机制

视频简介 Hystrix 是由 Netflix 提供的微服务熔断处理机制。本视频为读者分析微服务调用链出现的“雪崩效应”，同时为读者讲解 Hystrix 基本概念。

在微服务的设计之中，开发者会不断地根据业务的需要进行服务的扩充，同时这些扩充的微服务又有可能要调用其他微服务，这样在整个微服务的调用过程中就会形成图 4-48 所示的调用链。

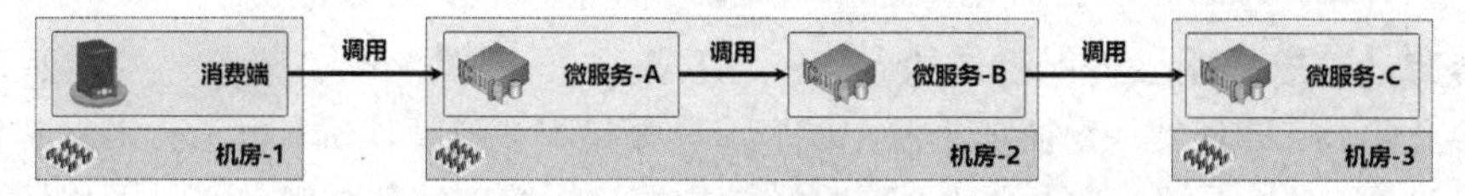

图 4-48 微服务调用链

但是在实际的生产运行过程中，有可能会因机房断电、网络不稳、并发访问量过大而出现响应速度过慢或服务不稳定，或者内部程序设计产生异常等，导致微服务不可用的情况出现。现在假设“微服务-C”出现了问题，就有可能连带着“微服务-B”出现问题，从而再导致“微服务-A”出现问题，最终影响到“消费端”正常执行。此类问题在微服务设计中称为雪崩效应，如图 4-49 所示。

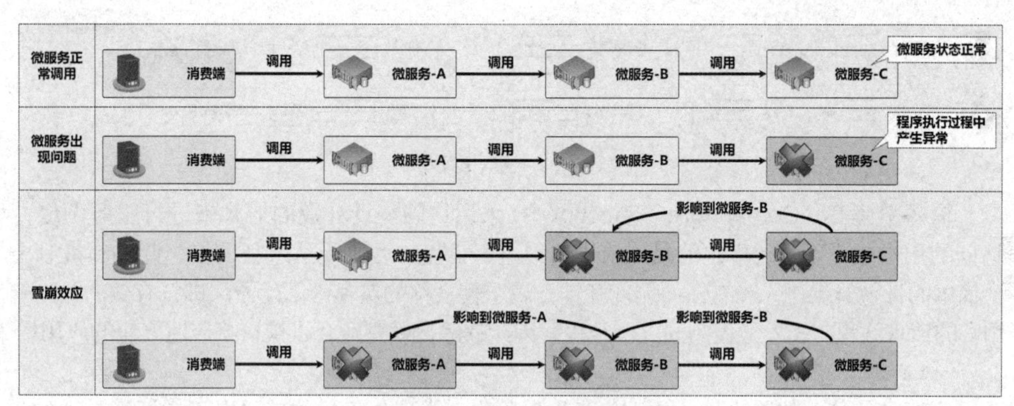

图 4-49 微服务雪崩效应

为了防止微服务出现雪崩效应，可以在微服务之中设置熔断机制，在某一个微服务出现问题之后，可以直接通过服务的降级机制进行保护。这样就不会影响到其他的微服务执行，同时也便于开发者进行错误的排查。这样的熔断机制在 Spring Cloud 中可以通过 Hystrix 组件来实现，该组件是由 Netflix 公司开源的一款容错框架，具有如下功能。

1. 断路器

当服务访问量过大，并且超过了预先配置的访问阈值时，Hystrix 会直接开启断路器，同时会拒绝后续所有的请求。在一段时间之后断路器会进入半打开状态，此时会允许一部分请求通过并进行调用尝试。如果尝试依然存在问题，则断路器继续保持开启状态；如果尝试没有问题，则断路器会进入关闭状态，如图 4-50 所示。

2. 资源隔离

Hystrix 采用了舱壁隔离模式对资源进行隔离，使多个资源彼此不会产生影响，而在 Hystrix 中资源隔离形式分为线程池隔离与信号量隔离。

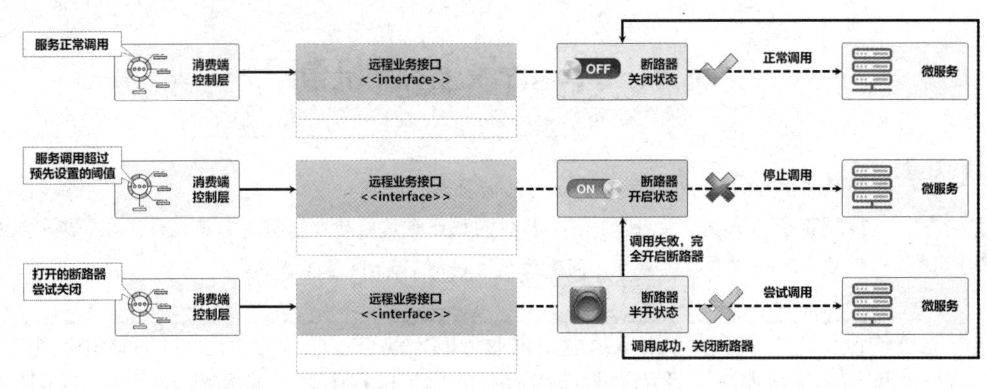

图 4-50　断路器处理

（1）线程池隔离（ThreadPool Rejection）：Hystrix 会为每一个资源创建一个线程池，这若干个线程池彼此隔离，即便某一个服务资源出问题，导致线程池资源耗尽，也不会影响其他线程池的正常工作，如图 4-51 所示。该模式适合耗时较长的接口（接口处理逻辑负载）调用场景。同时，为了进一步提升处理性能，线程池中的所有的线程并不参与实际的服务调用，真正的服务调用由转发线程处理，这样就可保证容器有足够的线程来处理新的请求。使用线程池隔离模式虽然会有一定的线程调度开销，但是在并发访问量大的情况下可以基于延迟队列实现调度处理（如等待超时、异步调用等），带来较好的并发处理性能。

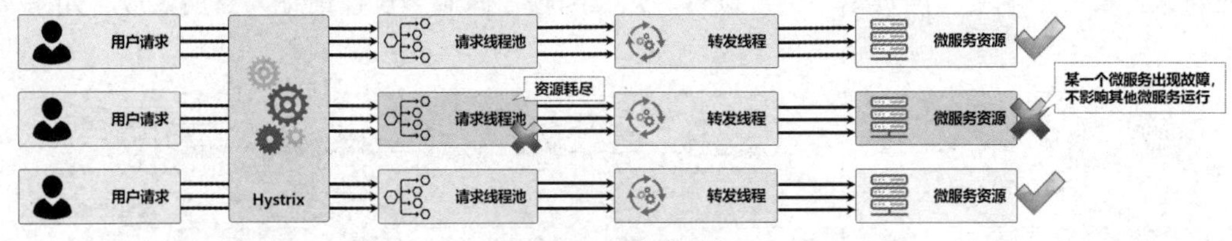

图 4-51　线程池隔离

（2）信号量隔离（Semaphore Rejection）：采用访问统计计数的形式来进行服务调度，如果当前访问的用户量小于系统规定的最大值则可以进行调度，否则无法进行调度。此隔离机制由于只是一个简单的计数器操作，没有任何调度开销，属于轻量化的实现，因此无法进行有效的队列管理。由于用户的请求线程与转发线程为同一个线程，因此该隔离模式适合于接口响应速度快的应用场景。

3. 降级回退

为了保证整体业务的稳定性，保证在某些服务节点出现故障后依然可以正常返回处理结果，就需要通过降级回退机制进行实现，一般以下几种情况会启用降级回退处理。

- 断路器打开：所有的客户端请求无法发送到服务端，所以直接返回既定的降级回退数据。
- 线程池/信号量资源不足：当前服务调度资源过大，已经无法保证正常的服务提供。
- 服务调用超时：服务处理逻辑复杂，或者与该服务有关的其他服务调度时间过长所产生的超时情况。
- 服务调用异常：目标服务主机停电或者网络故障。

由于实际项目运行中的服务调用的复杂性，因此对于服务降级回退的也有多种实现方式。

- 快速失败（Fail Fast）：最普通的执行方式，不需要重写任何降级逻辑，如果出现故障直接抛出异常。
- 无声失败（Fail Silent）：在服务降级时返回一些特定的数据，如 null、空集合或其他类似响应。
- 静态值（Static）：由开发者预先配置的数据内容，例如，部门增加微服务调用失败时直接返回 false。
- 复合数据（Stubbed）：由开发者定义的复合对象内容。

- 网络缓存（Cache）：如果服务调用失败，可以通过缓存服务查询旧数据。
- 主次策略（Primary + Secondary）：将失败回退分为主要故障（网络）与次要故障（业务）两种处理模式。

4. 请求结果缓存

Hystrix 实现了一个内部缓存机制，可以将请求结果进行缓存，对于相同的请求则会直接使用缓存而不用请求后端服务。

5. 请求合并

可以实现将一段时间内的请求合并，然后只对后端服务发送一次请求。

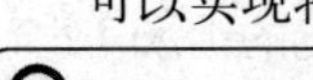

提示：Hystrix 不是 Spring Cloud Alibaba 套件。

Hystrix 最早是由 Netflix 维护的开发组件，但是并不在 Spring Cloud Alibaba 体系之中。本书讲解 Hystrix 主要是考虑到让读者知识全面。本书的第 5 章将为读者讲解 Sentinel 组件，该组件可以更好地替代 Hystrix 组件。

4.6.1 Hystrix 简介

Hystrix 简介

视频名称 0430_【了解】Hystrix 简介

视频简介 为了便于整合处理，Spring Cloud 专门提供了 Hystrix 服务整合组件。本视频通过具体的实例代码讲解 Hystrix 中的失败处理支持。

Hystrix 本身提供失败回退处理支持，为了便于读者理解 Hystrix 的基本作用，下面将介绍针对已有的部门微服务进行改进，为其追加断路器功能，这样在其出现异常后其他微服务的运行不会受影响。具体的实现步骤如下。

（1）【microcloud 项目】修改 build.gradle 配置文件，在 provider-dept-*子模块中添加 Hystrix 依赖库。

```
project('provider-dept-8001') {                                          // 子模块
    dependencies {                                                       // 重复依赖，略
        implementation('org.springframework.cloud:spring-cloud-starter-netflix-hystrix')
    }
}
```

（2）【provider-dept-*子模块】修改 DeptAction 程序类，针对部门数据 ID 查询操作配置一个失败回退处理方法。

```
package com.yootk.provider.action;
@RestController                                                          // REST控制器
@RequestMapping("provider/dept/*")                                       // 父路径
public class DeptAction {
    @Autowired
    private IdeptService deptService;                                    // 注入业务接口实例
    @ApiOperation(value = "部门查询", notes = "根据部门编号查询部门详细信息")
    @GetMapping("get/{id}")                                              // 子路径
    @HystrixCommand(fallbackMethod="getFallback")                        // 服务降级
    public Object get(@PathVariable("id") long id) {                     // 部门编号不存在产生异常
        return this.deptService.get(id);                                 // 部门信息加载
    }
    public Object getFallback(@PathVariable("id") long id) {             // 服务降级方法
        DeptDTO dept = new DeptDTO();                                    // 实例化DTO对象
        dept.setDeptno(id);                                              // 属性设置
        dept.setDname("【Fallback】部门名称");                             // 属性设置
        dept.setLoc("【Fallback】部门位置");                               // 属性设置
        return dept;                                                     // 数据响应
    }
}
```

（3）【provider-dept-*子模块】修改部门微服务的启动类，定义熔断机制启用注解。

```
package com.yootk.provider;
@SpringBootApplication
@EnableDiscoveryClient
@EnableCircuitBreaker                           // 启用熔断机制
public class StartProviderDept8001Application {
    public static void main(String[] args) {
        SpringApplication.run(StartProviderDept8001Application.class, args); // 服务启动
    }
}
```

（4）【Postman】分别启动服务端与消费端代码，随后通过 Postman 发出部门查询的需求（此时查询 ID 不存在）。

consumer-springboot-80/consumer/dept/get?deptno=19	
程序执行结果	{ "deptno": 19, "dname": "【Fallback】部门名称", "loc": "【Fallback】部门位置" }

此时部门数据库中没有编号为 19 的部门数据，所以程序会产生异常，而后会自动启用服务降级机制，利用 getFallback()服务降级方法进行请求响应，这样整个程序就不会因为产生异常而导致最终的服务调用失败。

4.6.2　Feign 失败回退

Feign 失败回退

视频名称　0431_【了解】Feign 失败回退

视频简介　服务调用都会通过 Feign 进行远程接口映射，所以 Feign 对 Hystrix 也有所支持，可以直接通过失败回退工厂类进行服务降级处理。本视频通过实例讲解如何基于 FeignClient 实现失败回退处理。

此时所实现的 Hystrix 失败回退是在 Action 类中通过@HystrixCommand 注解定义的处理方法，这样的失败回退（或称“服务降级”）处理并不便于管理。所以最佳的做法是在 Feign 远程接口中进行失败回退的配置，这就需要创建 FallbackFactory 对象实例，程序的实现结构如图 4-52 所示。具体的实现步骤如下。

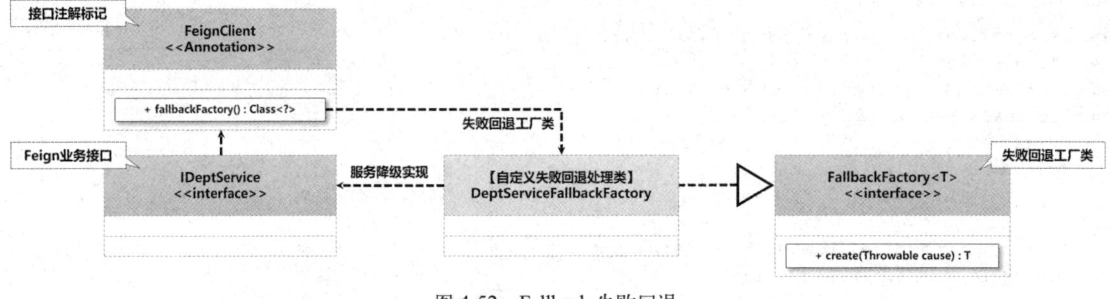

图 4-52　Fallback 失败回退

> **注意：@HystrixCommand 与 Feign 失败回退。**
>
> 在进行失败回退处理时，可以在部门控制层方法中使用@HystrixCommand，也可以基于 Feign 提供的 FallbackFactory 实现。如果两者同时出现，则根据情况优先选择。例如，在服务正常运行时，如果出现了错误，则@HystrixCommand 注解配置的失败回退方法生效；如果服务已经关闭，则可以通过 Feign 提供的失败回退来处理。

（1）【common-api 子模块】创建一个 Fallback 服务降级处理类，对 IdeptService 业务接口实现服务降级定义。

```
package com.yootk.service.fallback;
@Component                                                          // 组件注册
public class DeptServiceFallbackFactory
        implements FallbackFactory<IdeptService> {                  // 服务降级工厂类
    @Override
    public IdeptService create(Throwable cause) {                   // 创建IdeptService服务降级
        return new IdeptService() {                                 // 实例化业务接口对象
            @Override
            public DeptDTO get(long id) {                           // 服务降级处理
                DeptDTO dept = new DeptDTO();                       // 实例化DTO对象
                dept.setDeptno(id);                                 // 属性设置
                dept.setDname("【部门名称】" + cause.getMessage());  // 属性设置
                dept.setLoc("【部门位置】" + cause.getMessage());    // 属性设置
                return dept;                                        // 数据响应
            }
            @Override
            public  oolean add(DeptDTO dept) {                      // 服务降级处理
                return false;                                       // 增加失败
            }
            @Override
            public List<DeptDTO> list() {                           // 服务降级处理
                return new ArrayList<DeptDTO>();                    // 返回空集合
            }
            @Override
            public Map<String, Object> split(int currentPage, int lineSize,
                    String column, String keyword) {                // 服务降级处理
                return new HashMap<>();                             // 返回空集合
            }
        };
    }
}
```

（2）【common-api 子模块】在业务接口中定义失败回退处理类。

```
package com.yootk.service;
@FeignClient(value="dept.provider",configuration = FeignConfig.class,
    fallbackFactory = DeptServiceFallbackFactory.class)             // 服务降级配置
public interface IdeptService {}                                    // 部门业务接口
```

（3）【consumer-springboot-80 子模块】修改消费端启动主类，在该类中配置失败回退类的扫描包定义。

```
package com.yootk.consumer;
@SpringBootApplication
@EnableDiscoveryClient                                              // 启用发现服务
@RibbonClient(name = "dept.provider", configuration = DeptProviderRibbonConfig.class)
@ComponentScan({"com.yootk.service","com.yootk.consumer"})          // 配置扫描包
@EnableFeignClients("com.yootk.service")
public class StartConsumerApplication {
    public static void main(String[] args) {
        SpringApplication.run(StartConsumerApplication.class, args); // 服务启动
    }
}
```

（4）【consumer-springboot-80 子模块】修改 application.yml 配置文件，启用 Hystrix 熔断配置。

```
feign:                          # Feign配置
  hystrix:                      # Hystrix配置
    enabled: true               # 组件启用
```

（5）【Postman】对消费端进行测试，在测试时可以尝试关闭部门微服务，此时会得到如下服务降级信息。

服务访问路径	consumer-springboot-80/consumer/dept/list
程序执行结果	[] （空集合）

此时由于微服务已经关闭了，因此在消费端进行服务调用时，会触发失败回退机制，执行 DeptServiceFallbackFactory 类定义的服务处理方法。

4.6.3　HystrixDashboard

HystrixDashboard

视频名称　0432_【了解】HystrixDashboard

视频简介　Hystrix 除了提供熔断机制外，还提供 Hystrix 监控面板支持。本视频讲解 Hystrix 微服务的开启以及服务面板应用的创建。

Hystrix 可以根据微服务中提供的 HystrixCommand 方法提供执行数据（如每秒的请求数、请求成功数等），同时这些数据可以通过 HystrixDashboard 提供的面板进行可视化监控，如图 4-53 所示。这样开发者就可以实时掌握微服务的执行情况。具体的实现步骤如下。

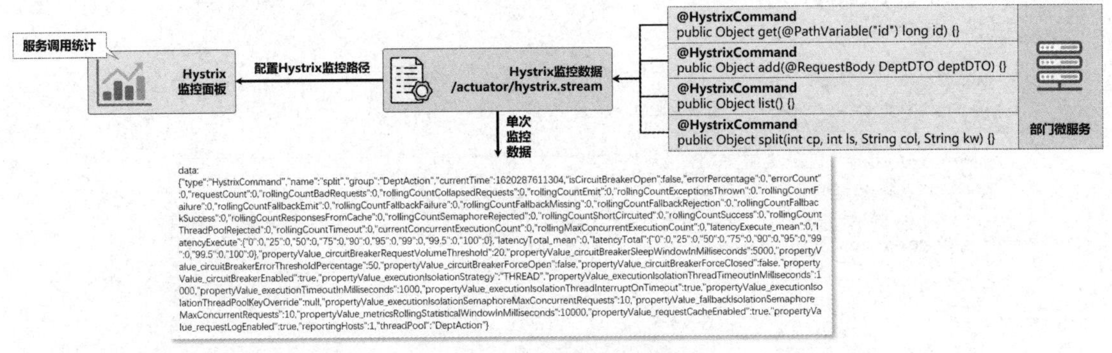

图 4-53　Hystrix 监控处理

（1）【microcloud 项目】为了便于服务创建，建立一个新的“hystrix-dashboard-8101”项目。

（2）【microcloud 项目】修改 build.gradle 配置文件，为相关子模块添加服务依赖。

```
Project('provider-dept-8001') {                 // 子模块
    dependencies {                              // 重复依赖，略
        implementation('org.springframework.boot:spring-boot-starter-actuator')
        implementation('org.springframework.cloud:spring-cloud-starter-netflix-hystrix')
    }
}
project('hystrix-dashboard-8101') {
    dependencies {                              // 配置子模块依赖
        implementation('org.springframework.boot:spring-boot-starter-web')
        implementation('org.springframework.cloud:' +
                'spring-cloud-starter-netflix-hystrix')
        implementation('org.springframework.cloud:' +
                'spring-cloud-starter-netflix-hystrix-dashboard')
    }
}
```

（3）【provider-dept-*子模块】修改 DeptAction 控制器类，为控制器中的每个方法添加@HystrixCommand 注解，这样就可以为 Hystrix 面板提供监控数据。

```
Package com.yootk.provider.action;
@RestController                                 // REST控制器
@RequestMapping("/provider/dept/*")             // 父路径
public class DeptAction {
    @HystrixCommand                             // Hystrix监控
```

```
    public Object get(@PathVariable("id") long id) { … }
    @HystrixCommand                          // Hystrix监控
    public Object add(@RequestBody DeptDTO deptDTO) { … }
    @HystrixCommand                          // Hystrix监控
    public Object list() { … }
    @HystrixCommand                          // Hystrix监控
    public Object split(int cp, int ls, String col, String kw) { … }
}
```

（4）【provider-dept-*子模块】修改 application.yml 配置文件，启用 Actuator 监控路径。

```
Management:                                  # Actuator配置
  endpoints:                                 # 访问终端
    web:                                     # Web配置
      exposure:                              # 端口暴露
        include: '*'                         # 开放全部端口
```

（5）【provider-dept-*子模块】修改程序启动类，启用 Hystrix 支持。

```
Package com.yootk.provider;
@SpringBootApplication
@EnableDiscoveryClient
@EnableCircuitBreaker                        // 启用熔断机制
@EnableHystrix                               // 启用Hystrix支持
public class StartProviderDept8001Application {}
```

（6）【浏览器】启动部门微服务。由于部门微服务已经配置了 Actuator 监控，因此可以直接通过“/actuator/hystrix.stream”获取与 Hystrix 有关的服务信息。

（7）【hystrix-dashboard-8101 子模块】修改 application.yml 配置文件，定义服务监听端口。

```
Server:                                      # 服务配置
  port: 8101                                 # 监听端口
```

（8）【hystrix-dashboard-8101 子模块】创建 Hystrix 面板启动类并添加相关注解。

```
Package com.yootk.hystrix;
@SpringBootApplication
@EnableHystrixDashboard                      // 启动HystrixDashboard
public class StartHystrixDashboardApplication {
    public static void main(String[] args) {
        SpringApplication.run(StartHystrixDashboardApplication.class, args); // 服务启动
    }
}
```

（9）【本地系统】修改 hosts 主机配置文件，添加新的服务主机名称。

```
127.0.0.1  hystrix-dashboard-8101
```

（10）【浏览器】通过浏览器访问 Hystrix 面板，访问地址为“hystrix-dashboard-8101:8101/hystrix”，而后输入要监控的微服务监控路径，即可得到监控数据，如图 4-54 所示。

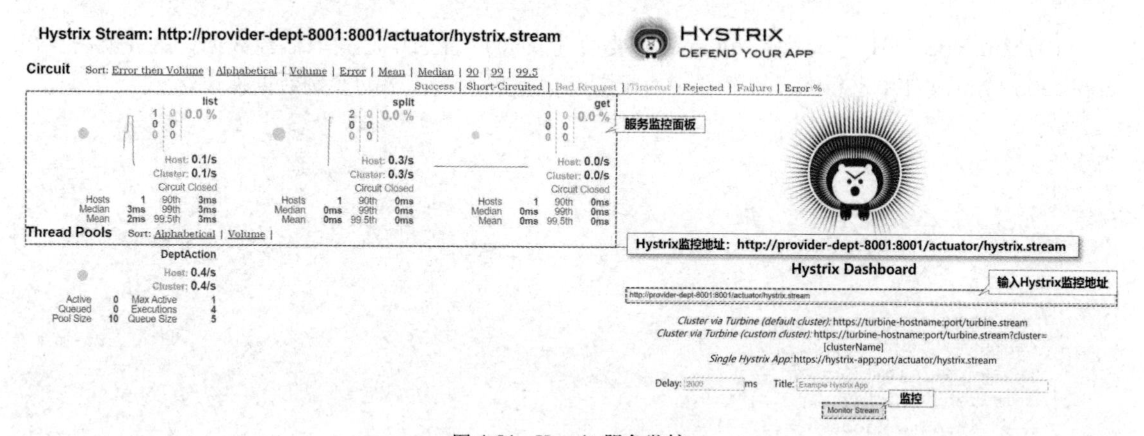

图 4-54　Hystrix 服务监控

4.6.4 Turbine 聚合监控

Turbine 聚合监控

视频名称 0433_【了解】Turbine 聚合监控

视频简介 为了进行微服务的监控统一管理，Spring Cloud 提供了 Turbine 组件。本视频为读者讲解统一监控的意义，并讲解 Turbine 的配置以及服务监控数据获取。

在直接使用 HystrixCommand 进行服务监控时，每次都需要设置完整的监控路径。这样一旦微服务节点发生改变，就无法准确地获取监控数据。所以最佳的做法是直接通过 Nacos 实现服务实例数据的抓取，而后通过一个统一的访问路径为 HystrixDashboard 提供监控数据，如图 4-55 所示。这样就可使得监控数据的维护更加方便。

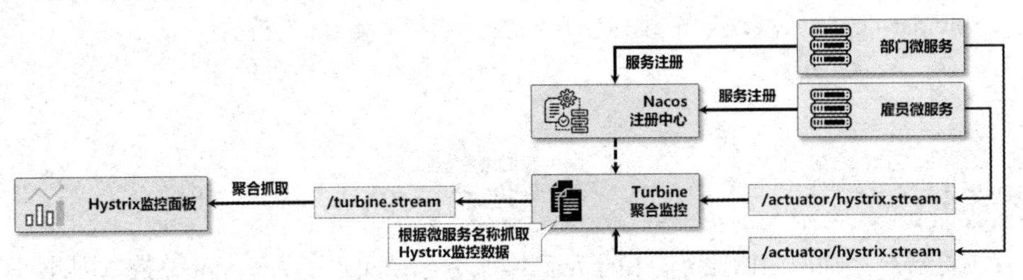

图 4-55 Turbine 聚合监控

Spring Cloud 提供了 Turbine 组件，并利用此组件来实现聚合监控管理。只要通过配置文件定义所需要监控的微服务名称，即可实现监控数据的获取。下面通过具体的步骤来介绍此操作。

（1）【microcloud 项目】创建 turbine-8201 子模块，用于实现所有微服务监控数据的聚合配置。

（2）【microcloud 项目】修改 build.gradle 配置文件，为 turbine-dashboard-8201 子模块添加相关依赖。需要注意的是，在默认情况下 Turbine 是基于 Eureka 注册中心进行整合的，所以在依赖配置时需要删除 Eureka 相关依赖。

```
project('turbine-8201') {
    dependencies {                                                      // 配置子模块依赖
        implementation('org.springframework.boot:spring-boot-starter-web')
        implementation('com.alibaba.cloud:spring-cloud-starter-alibaba-nacos-discovery')
        implementation('org.springframework.cloud:' +
                'spring-cloud-starter-netflix-turbine') {              // Nacos配置
            exclude group: 'org.springframework.cloud',
                module: 'spring-cloud-starter-netflix-eureka-client'   // 删除Eureka相关依赖
        }
    }
}
```

（3）【turbine-8201 子模块】Turbine 需要通过 Nacos 注册中心获取监控数据，这样就需要在 application.yml 文件中进行 Nacos 注册中心配置，以实现 Nacos 中的实例数据抓取。

```
server:                                                    # 服务端配置
  port: 8201                                               # 监听端口
spring:                                                    # Spring配置
  application:                                             # 应用配置
    name: turbine                                          # 应用名称
  cloud:                                                   # Spring Cloud配置
    nacos:                                                 # Nacos注册中心
      discovery:                                           # 发现服务
        username: nacos                                    # 用户名
        password: nacos                                    # 密码
        service: ${spring.application.name}                # 服务名称
        server-addr: nacos-server:8848                     # 服务地址
        namespace: 650fab32-c7dc-4ae1-8ac4-2dbdefd7e617    # 命名空间
```

```
      group: MICROCLOUD_GROUP                                    # 组名称
      register-enabled: false                                    # 不注册
      cluster-name: MuyanCluster                                 # 匹配集群名称
turbine:
  app-config: dept.provider                                      # 配置监控服务信息
  cluster-name-expression: new String("default")                 # 监控表达式，获取监控信息名称
```

（4）【turbine-8201 子模块】创建 Turbine 应用启动类，同时绑定 Turebine 启用注解。

```
package com.yootk.turbine;
@SpringBootApplication
@EnableDiscoveryClient                                             // 启用发现服务
@EnableTurbine                                                     // Turbine启用
public class StartTurbineApplication {
    public static void main(String[] args) {
        SpringApplication.run(StartTurbineApplication.class, args);  // 服务启动
    }
}
```

（5）【本地系统】修改 hosts 主机配置文件，定义 Turbine 应用访问名称。

```
127.0.0.1   turbine-8201
```

主机名称配置完成后，就可以通过“turbine-8201:8201/turbine.stream”获取所配置的集群监控数据。

（6）【浏览器】访问 Hystrix-Dashboard 提供的监控服务页面“hystrix-dashboard-8101:8101/hystrix”，随后输入 Turbine 监控地址，就可以实现所有在 Turbine 中配置的微服务监控数据的汇总，如图 4-56 所示。

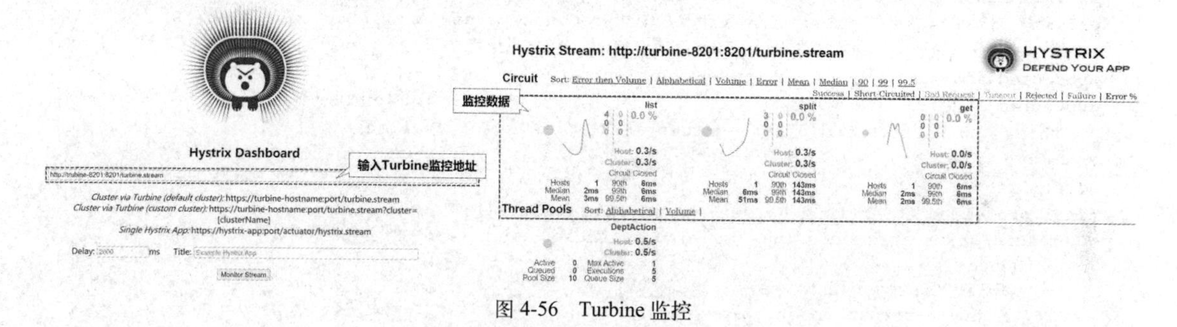

图 4-56　Turbine 监控

4.7　Hystrix 源代码分析

Hystrix 工作流程

视频名称　0434_【了解】Hystrix 工作流程

视频简介　Hystrix 所有的操作都被封装在了 HystrixCommand 对象实例之中，如果想弄清楚 Hystrix 工作流程，就需要通过 HystrixCommand 类的源代码进行观察。本视频通过 HystrixCommand 类的继承结构以及核心方法分析 Hystrix 的工作流程。

在 Hystrix 组件运行过程中，需要根据当前请求线程的情况来进行熔断以及降级回退处理，而这些操作全部被封装在 HystrixCommand 抽象类之中，此类的关联结构如图 4-57 所示。下面打开此类的源代码进行观察。

范例：【hystrix-core 依赖】HystrixCommand 源代码

```
package com.netflix.hystrix;
public abstract class HystrixCommand<R> extends AbstractCommand<R> implements
        HystrixExecutable<R>, HystrixInvokableInfo<R>, HystrixObservable<R> {
    // Hystrix处理类中提供了多个构造方法，以实现Hystrix相关属性的配置，此处只列出全参构造方法的源代码
```

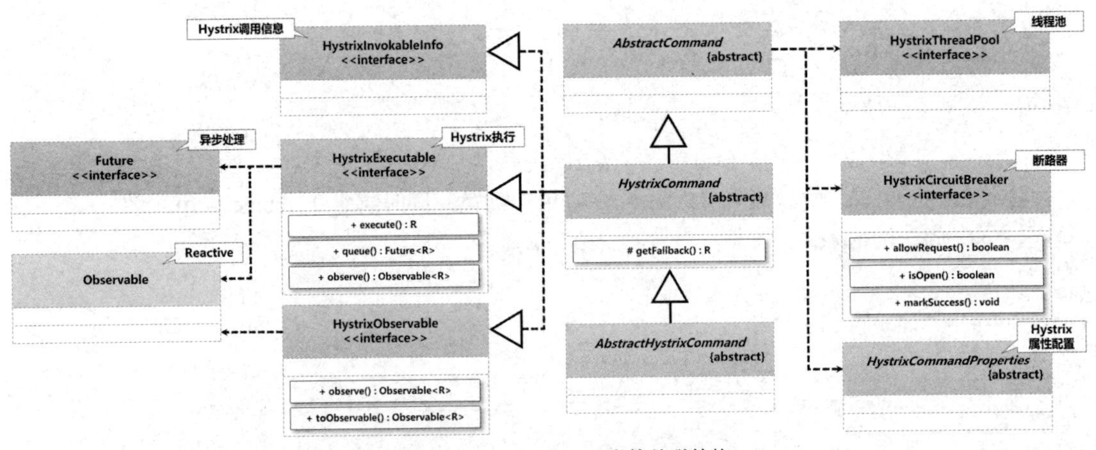

图 4-57 HystrixCommand 类的关联结构

```
    protected HystrixCommand(HystrixCommandGroupKey group,
        HystrixThreadPoolKey threadPool,
        int executionIsolationThreadTimeoutInMilliseconds) {
      super(group, null, threadPool, null, null, HystrixCommandProperties.Setter()
        .withExecutionTimeoutInMilliseconds(
                executionIsolationThreadTimeoutInMilliseconds),
        null, null, null, null, null, null);
    }
    // 原子引用，用于保存当前的执行线程
    private final AtomicReference<Thread> executionThread = new AtomicReference<Thread>();
    private final AtomicBoolean interruptOnFutureCancel = new AtomicBoolean(false);
    // 执行Hystrix处理，可能调用的是execute()或者是queue()方法
    protected abstract R run() throws Exception;
    protected R getFallback() {                              // 降级回退方法
        throw new UnsupportedOperationException("No fallback available.");
    }
    @Override
    final protected Observable<R> getExecutionObservable() {  // 执行请求
        return Observable.defer(new Func0<Observable<R>>() {  // 创建响应数据
            @Override
            public Observable<R> call() {
                try { return Observable.just(run());          // 请求发送
                } catch (Throwable ex) {  return Observable.error(ex); }
            }
        }).doOnSubscribe(new Action0() {                      // 订阅处理
            @Override
            public void call() { executionThread.set(Thread.currentThread()); }
        });
    }
    @Override
    final protected Observable<R> getFallbackObservable() {  // 降级回退处理
        return Observable.defer(new Func0<Observable<R>>() {
            @Override
            public Observable<R> call() {
                try { return Observable.just(getFallback());  // 降级方法
                } catch (Throwable ex) { return Observable.error(ex); }
            }
        });
    }
    public R execute() {                                      // 同步处理
        try { return queue().get();
        } catch (Exception e) { }
    }
    public Future<R> queue() {}                               // 异步处理
}
```

通过 HystrixCommand 类的源代码可以清楚地发现，此类在进行实例化时需要传递线程池、信号量配置、断路器等配置信息。同时该类中明确地给出了具体的调用执行方法（execute()、queue()、getFallback()），不同的调用方法对应不同的执行形式。这些执行类型全部由 ExecutionType 枚举类定义，而具体执行操作是由 CommandExecutor 类实现的，这样就可以得到图 4-58 所示的类关联结构。

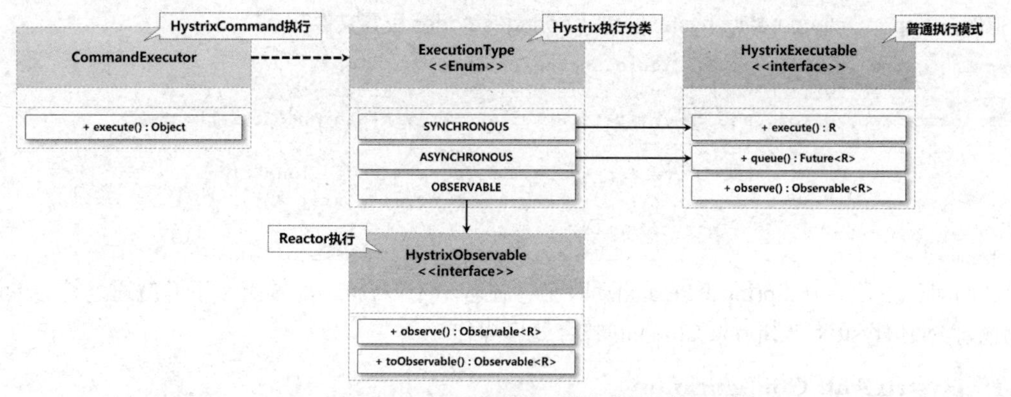

图 4-58　Hystrix 执行处理的类关联结构

范例：【hystrix-core 依赖】CommandExecutor 类源代码

```
package com.netflix.hystrix.contrib.javanica.command;
public class CommandExecutor {                // 命令执行器
   public static Object execute(HystrixInvokable invokable,
        ExecutionType executionType, MetaHolder metaHolder) throws RuntimeException {
      switch (executionType) {                // 判断执行模式
         case SYNCHRONOUS: {                  // 同步处理
            return castToExecutable(invokable, executionType).execute();
         }
         case ASYNCHRONOUS: {                 // 异步处理
            HystrixExecutable executable = castToExecutable(invokable, executionType);
            if (metaHolder.hasFallbackMethodCommand()
                  && ExecutionType.ASYNCHRONOUS == metaHolder
                     .getFallbackExecutionType()) {
               return new FutureDecorator(executable.queue());
            }
            return executable.queue();
         }
         case OBSERVABLE: {                   // 响应式处理
            HystrixObservable observable = castToObservable(invokable);
            return ObservableExecutionMode.EAGER == metaHolder
                .getObservableExecutionMode() ? observable.observe() :
                        observable.toObservable();
         }
         default:
            throw new RuntimeException("unsupported execution type: " + executionType);
      }
   }
}
```

通过此处的 CommandExecutor 源代码，结合断路器的概念，就可以得到图 4-59 所示的执行流程。

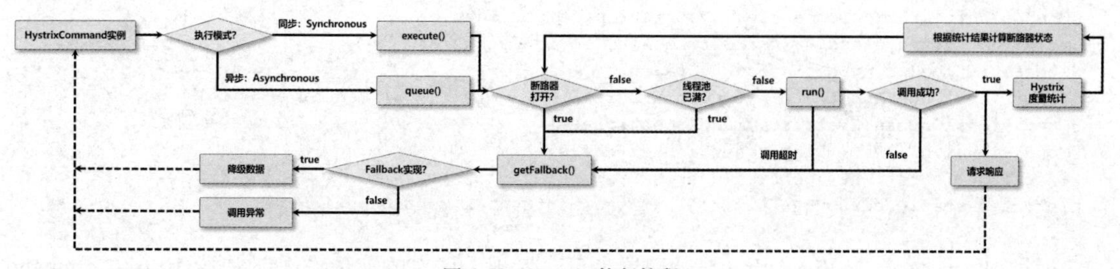

图 4-59　Hystrix 执行流程

通过图 4-59 可以发现，整个 Hystrix 的实现关键就是命令的执行以及断路器控制，所以 Spring Cloud 在整合时也提供了相应的自动配置器，这一点可以通过 spring-cloud-netflix-hystrix 依赖库所提供的 spring.factories 文件观察到。

范例：【spring-cloud-netflix-hystrix 依赖】spring.factories 配置文件

```
org.springframework.boot.autoconfigure.EnableAutoConfiguration=\
org.springframework.cloud.netflix.hystrix.HystrixAutoConfiguration,\
org.springframework.cloud.netflix.hystrix.HystrixCircuitBreakerAutoConfiguration,\
org.springframework.cloud.netflix.hystrix
                    .ReactiveHystrixCircuitBreakerAutoConfiguration,\
org.springframework.cloud.netflix.hystrix.security.HystrixSecurityAutoConfiguration
org.springframework.cloud.client.circuitbreaker.EnableCircuitBreaker=\
org.springframework.cloud.netflix.hystrix.HystrixCircuitBreakerConfiguration
```

此时可以发现，在 Spring Cloud 启动时会存在若干个 Hystrix 配置类。下面就通过核心的配置类为读者介绍 Hystrix 与 Spring Cloud 的整合处理操作。

4.7.1　HystrixAutoConfiguration

HystrixAuto
Configuration

视频名称　0435_【了解】HystrixAutoConfiguration
视频简介　Hystrix 提供了自动配置处理，而在 Hystrix 整合时需要进行 HystrixCommand 切面配置和度量配置。本视频通过 HystrixAutoConfiguration 类源代码以及对应的关联结构进行 HystrixCommand 流程分析。

Spring Cloud 在引入 Hystrix 组件时，会通过 HystrixAutoConfiguration 类进行组件的自动配置，主要是开启 Hystrix 统计数据，同时绑定 Hystrix 监控终端。

范例：【spring-cloud-netflix-hystrix 依赖】HystrixAutoConfiguration 核心源代码

```
package org.springframework.cloud.netflix.hystrix;
@Configuration(proxyBeanMethods = false)
@ConditionalOnClass({ Hystrix.class, HealthIndicator.class,
    HealthContributorAutoConfiguration.class })
@AutoConfigureAfter({ HealthContributorAutoConfiguration.class })
public class HystrixAutoConfiguration {                         // Hystrix自动配置
  @Bean
  @ConditionalOnEnabledHealthIndicator("hystrix")
  public HystrixHealthIndicator hystrixHealthIndicator() {    // Hystrix健康指示器
    return new HystrixHealthIndicator();
  }
  @Configuration(proxyBeanMethods = false)
  @ConditionalOnProperty(value = "management.metrics.binders.hystrix.enabled",
      matchIfMissing = true)
  @ConditionalOnClass({ HystrixMetricsBinder.class })
  protected static class HystrixMetricsConfiguration {}      // Hystrix度量配置
  @Configuration(proxyBeanMethods = false)
  @ConditionalOnWebApplication(type = SERVLET)                // Servlet度量采集
  @ConditionalOnBean(HystrixCommandAspect.class)              // HystrixCommand切面
  @ConditionalOnClass({ HystrixMetricsStreamServlet.class })
  @EnableConfigurationProperties(HystrixProperties.class)
  protected static class HystrixServletAutoConfiguration {}
  @Configuration(proxyBeanMethods = false)
  @ConditionalOnWebApplication(type = REACTIVE)               // Reactive度量采集
  @ConditionalOnBean(HystrixCommandAspect.class)              // HystrixCommand切面
  @ConditionalOnClass({ DispatcherHandler.class })
  @EnableConfigurationProperties(HystrixProperties.class)
  protected static class HystrixWebfluxManagementContextConfiguration {}
}
```

通过 HystrixAutoConfiguration 自动配置类可以发现，所有的度量操作都需要 HystrixCommand

支持，而这个切面处理是由 HystrixCommandAspect 实现的，这样就可以得到图 4-60 所示的类的关联结构。

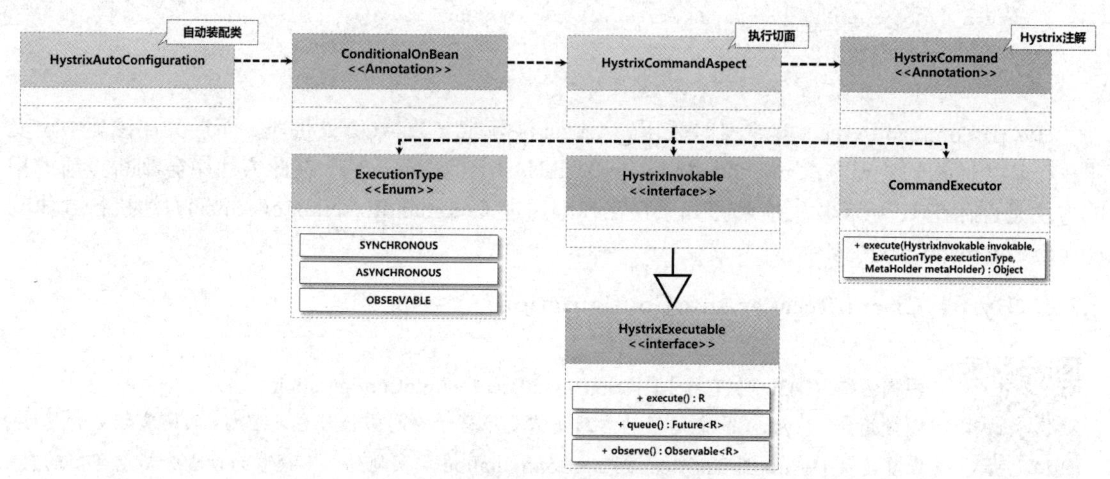

图 4-60　HystrixAutoConfiguration 类的关联结构

通过图 4-60 可以清楚地发现，整个 Hystrix 运行处理全部在 HystrixCommandAspect 切面处理类中定义，在此类中可以进行 HystrixCommand 的调用，也可以进行 Fallback 的设置。下面重点观察 HystrixCommand 调用。

范例：【spring-cloud-netflix-hystrix 依赖】HystrixCommandAspect 核心源代码

```
package com.netflix.hystrix.contrib.javanica.aop.aspectj;
@Aspect                                                     // AOP切面处理
public class HystrixCommandAspect {
   @Around("hystrixCommandAnnotationPointcut() ||
            hystrixCollapserAnnotationPointcut()")          // 环绕通知
   public Object methodsAnnotatedWithHystrixCommand(
        final ProceedingJoinPoint joinPoint) throws Throwable {
      Method method = getMethodFromTarget(joinPoint);       // 获取调用方法
      Validate.notNull(method, "failed to get method from joinPoint: %s", joinPoint);
      if (method.isAnnotationPresent(HystrixCommand.class) &&
            method.isAnnotationPresent(HystrixCollapser.class)) { // 是否存在注解
         throw new IllegalStateException("method canno…");
      }
      MetaHolderFactory metaHolderFactory = META_HOLDER_FACTORY_MAP
            .get(HystrixPointcutType.of(method));          // 获取元数据
      MetaHolder metaHolder = metaHolderFactory.create(joinPoint);
      HystrixInvokable invokable = HystrixCommandFactory
            .getInstance().create(metaHolder);              // 实例化Hystrix执行接口
      ExecutionType executionType = metaHolder.isCollapserAnnotationPresent() ?
          metaHolder.getCollapserExecutionType() :
            metaHolder.getExecutionType();                  // 获取当前Hystrix执行类型
      Object result;                                        // 保存执行结果
      try {
         if (!metaHolder.isObservable()) {                  // 采用非响应模式
            result = CommandExecutor.execute(invokable, executionType, metaHolder);
         } else {                                           // 采用响应模式
            result = executeObservable(invokable, executionType, metaHolder);
         }
      } catch (HystrixBadRequestException e) {
         throw e.getCause();
      } catch (HystrixRuntimeException e) {
         throw hystrixRuntimeExceptionToThrowable(metaHolder, e);
      }
      return result;
```

```
    }
    private Observable executeObservable(HystrixInvokable invokable,
                ExecutionType executionType, final MetaHolder metaHolder) {
        return ((Observable) CommandExecutor.execute(invokable,
                executionType, metaHolder))… ;
    }
}
```

HystrixCommandAspect 类定义时采用@Aspect 注解定义为 AOP 处理类，而后又在该类中定义了一个环绕通知的处理方法 methodsAnnotatedWithHystrixCommand()。在此方法中会判断当前调用的方法是否存在 Hystrix 相关注解，如果存在，则通过 CommandExecutor.execute()方法执行具体的 HystrixCommand 操作。

4.7.2　HystrixCircuitBreakerAutoConfiguration

HystrixCircuit BreakerAuto Configuration

视频名称　0436_【了解】HystrixCircuitBreakerAutoConfiguration

视频简介　Hystrix 中断路器的启用是需要结合特定的阈值配置处理的，而所有相关的属性配置是由 HystrixCircuitBreakerAutoConfiguration 类实现的。本视频为读者分析断路器的自动配置类的作用。

Hystrix 中断路器是最重要的组成单元之一，而断路器的整合可以通过 HystrixCircuitBreaker AutoConfiguration 配置类来实现。在此类中会通过 CircuitBreakerFactory 工厂类创建一个具体的 Hystrix 断路器实例，在创建断路器的时候会自动设置当前系统所需要的相关 Hystrix 属性内容，如 Hystrix 执行属性（状态启用、调用超时、度量采集周期等）、Hystrix 线程池配置属性（内核线程大小、工作线程大小、延迟队列大小等），类结构如图 4-61 所示。

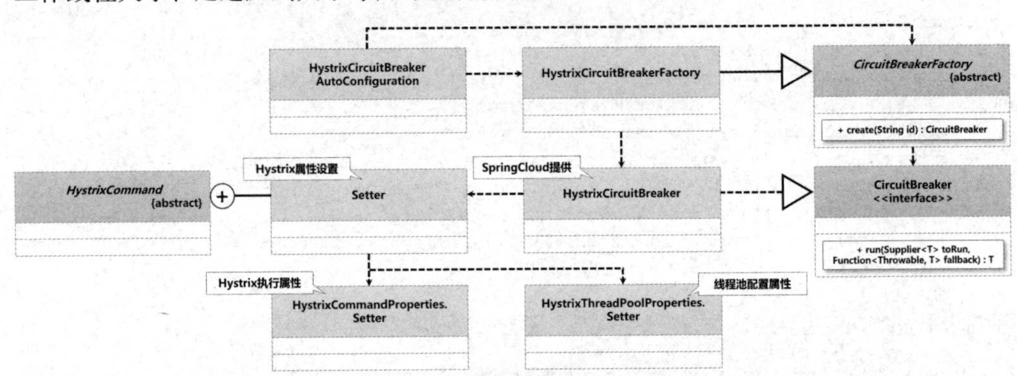

图 4-61　HystrixCircuitBreakerAutoConfiguration 类结构

范例：【spring-cloud-netflix-hystrix 依赖】HystrixCircuitBreakerAutoConfiguration 源代码

```
package org.springframework.cloud.netflix.hystrix;
@Configuration(proxyBeanMethods = false)
@ConditionalOnClass({ Hystrix.class })
@ConditionalOnProperty(name = "spring.cloud.circuitbreaker.hystrix.enabled",
    matchIfMissing = true)
public class HystrixCircuitBreakerAutoConfiguration {
    @Autowired(required = false)
    private List<Customizer<HystrixCircuitBreakerFactory>> customizers = new ArrayList<>();
    @Bean
    @ConditionalOnMissingBean(CircuitBreakerFactory.class)
    public CircuitBreakerFactory hystrixCircuitBreakerFactory() {
        HystrixCircuitBreakerFactory factory = new HystrixCircuitBreakerFactory();
        customizers.forEach(customizer -> customizer.customize(factory));
        return factory;
    }
}
```

通过该源代码可以发现，Spring 容器启动之后，会自动提供一个 CircuitBreakerFactory 工厂类的对象实例，而后就可以依据工厂类提供的 create()方法创建 CircuitBreaker 接口实例。

范例：【spring-cloud-netflix-hystrix 依赖】HystrixCircuitBreakerFactory 源代码

```
package org.springframework.cloud.netflix.hystrix;
public class HystrixCircuitBreakerFactory extends
    CircuitBreakerFactory<HystrixCommand.Setter,
              HystrixCircuitBreakerFactory.HystrixConfigBuilder> {
  public HystrixCircuitBreaker create(String id) {          // 创建Hystrix断路器
    Assert.hasText(id, "A CircuitBreaker must have an id.");
    HystrixCommand.Setter setter = getConfigurations().computeIfAbsent(id,
        defaultConfiguration);                               // 获取断路器配置属性
    return new HystrixCircuitBreaker(setter);                // 返回断路器实例
  }
}
```

通过 create()方法实现的源代码可以发现，在创建 HystrixCircuitBreaker 实例时需要通过 HystrixCommand.Setter 内部类获取相关的配置属性，而这些属性就包含断路器的相关阈值数据。

4.7.3 HystrixCircuitBreakerConfiguration

HystrixCircuit Breaker Configuration

视频名称 0437_【了解】HystrixCircuitBreakerConfiguration

视频简介 HystrixCircuitBreakerConfiguration 实现了与 Hystrix 断路器处理有关的 Bean 注册操作。本视频通过源代码为读者分析此配置器的主要作用，同时分析 Hystrix 中线程池与断路器之间的配置关联。

要执行 Hystrix 调用处理，最终肯定需要依靠 HystrixCommand 对象实例。Spring Cloud 在进行整合时，是基于 AOP 环绕通知的形式实现 HystrixCommand 方法调用的，而 HystrixCommandAspect 实例的注册就是通过 HystrixCircuitBreakerConfiguration 配置类实现的定义。首先来观察此类的源代码。

范例：【spring-cloud-netflix-hystrix 依赖】HystrixCircuitBreakerConfiguration 源代码

```
package org.springframework.cloud.netflix.hystrix;
@Configuration(proxyBeanMethods = false)
public class HystrixCircuitBreakerConfiguration {
  @Bean
  public HystrixCommandAspect hystrixCommandAspect() {   // Hystrix切面配置
    return new HystrixCommandAspect();
  }
  @Bean
  public HystrixShutdownHook hystrixShutdownHook() {     // Hystrix关闭处理
    return new HystrixShutdownHook();
  }
  private class HystrixShutdownHook implements DisposableBean {
    @Override
    public void destroy() throws Exception {
      Hystrix.reset();                                   // 清空统计数据
    }
  }
}
```

通过该源代码可以发现，此时最重要的一点是向 Spring 注册 HystrixCommandAspect 对象实例，由于所有的 Hystrix 断路器都需要通过统计数据来计算是否打开断路器，因此在每次请求的最后都需要通过 Hystrix.reset()方法及时进行资源释放（Hystrix 提供了一个 com.netflix.hystrix.Hystrix 类实现生命周期控制），此时的类结构如图 4-62 所示。

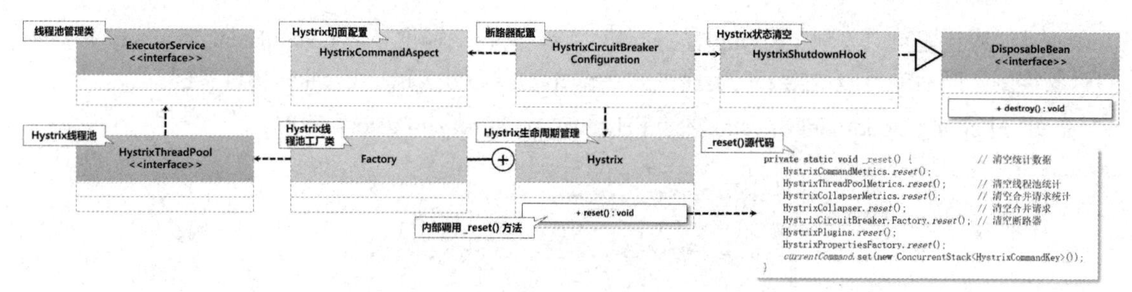

图 4-62　HystrixCircuitBreakerConfiguration 类结构

4.7.4　HystrixCircuitBreaker

HystrixCircuit
Breaker

视频名称　0438_【了解】HystrixCircuitBreaker

视频简介　断路器的最终实现操作是由 HystrixCircuitBreaker 接口定义的。本视频通过断路器的源代码为读者分析断路器的具体实现逻辑。

在实例化 HystrixCommand 对象时一般都需要传递 com.netflix.hystrix.HystrixCircuitBreaker 接口实例，而后在该接口中规定所有的断路器的相关实现。为了便于读者理解，图 4-63 给出了断路器相关实现类的结构。

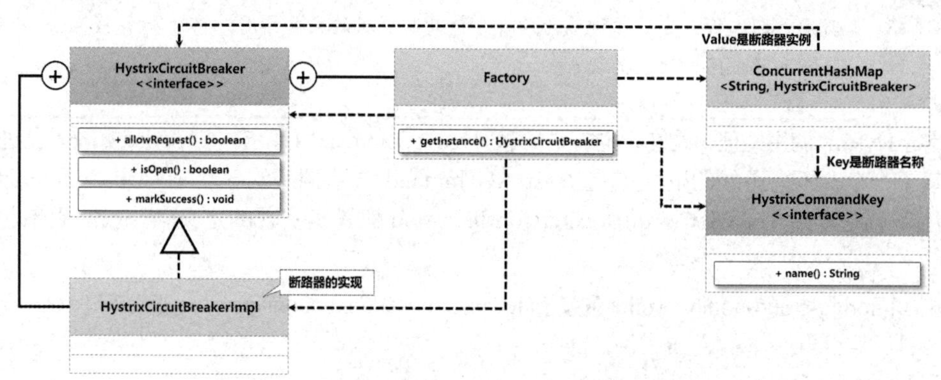

图 4-63　断路器相关实现类的结构

范例：【hystrix-core 依赖】HystrixCircuitBreakerImpl 内部类源代码

```
package com.netflix.hystrix;
public interface HystrixCircuitBreaker {
   static class HystrixCircuitBreakerImpl implements HystrixCircuitBreaker {
      private final HystrixCommandProperties properties;           // Hystrix命令属性
      private final HystrixCommandMetrics metrics;                 // Hystrix度量统计
      // 保存断路器打开或关闭的状态，默认情况下断路器状态保持关闭（false）
      private AtomicBoolean circuitOpen = new AtomicBoolean(false);
      // 断路器处于开启状态或最后尝试调用singleTest()测试的时间
      private AtomicLong circuitOpenedOrLastTestedTime = new AtomicLong();
      protected HystrixCircuitBreakerImpl(HystrixCommandKey key,
               HystrixCommandGroupKey commandGroup,
               HystrixCommandProperties properties, HystrixCommandMetrics metrics) {
         this.properties = properties;
         this.metrics = metrics;
      }
      public void markSuccess() {                                  // 断路器半开状态下执行
         if (circuitOpen.get()) {                                  // 获取断路器状态
            if (circuitOpen.compareAndSet(true, false)) {          // 修改断路器状态
               metrics.resetStream();                              // 数据重置
            }
```

```
            }
        }
        @Override
        public boolean allowRequest() {
            if (properties.circuitBreakerForceOpen().get()) {          // 默认为强制打开
                return false;                                          // 请求无法成功
            }
            if (properties.circuitBreakerForceClosed().get()) {        // 默认为强制关闭
                return true;                                           // 允许请求
            }
            return !isOpen() || allowSingleTest();                     // 阈值检测与开启访问检测
        }
        public boolean allowSingleTest() {                             // 状态检测
            long timeCircuitOpenedOrWasLastTested = circuitOpenedOrLastTestedTime.get();
            // 在断路器处于开启状态，并且已经开启了较长时间的情况下执行本操作
            if (circuitOpen.get() &&
                    System.currentTimeMillis() > timeCircuitOpenedOrWasLastTested +
                    properties.circuitBreakerSleepWindowInMilliseconds().get()) {
                if (circuitOpenedOrLastTestedTime.compareAndSet(
                    timeCircuitOpenedOrWasLastTested, System.currentTimeMillis())) {
                     return true;                                      // 允许执行状态测试
                }
            }
            return false;                                              // 其他线程在执行状态测试
        }
        @Override
        public boolean isOpen() {                                      // 是否打开
            if (circuitOpen.get()) {                                   // 当前断路器状态为打开
                return true;                                           // 返回true（打开状态）
            }
            HealthCounts health = metrics.getHealthCounts();           // 获取访问统计
            if (health.getTotalRequests() < properties
                    .circuitBreakerRequestVolumeThreshold().get()) {   // 访问阈值检测
                return false;                                          // 未超过阈值不开启断路器
            }
            if (health.getErrorPercentage() < properties.
                    circuitBreakerErrorThresholdPercentage().get()) {  // 错误阈值检测
                return false;                                          // 未超过阈值不开启断路器
            } else {                                                   // 超过阈值，开启断路器
                if (circuitOpen.compareAndSet(false, true)) {          // 开启断路器
                    circuitOpenedOrLastTestedTime.set(System.currentTimeMillis());
                    return true;                                       // 开启断路器
                } else {                                               // 预留其他可能存在的处理
                    return true;
                }
            }
        }
    }
}
```

由于断路器在执行状态下有可能被多个线程调用，因此所有断路器的状态都是基于 J.U.C（Java.Util. Concurrent）提供的原子类进行保存的，每一次进行访问请求之前都要判断当前断路器的状态，同时也需要根据阈值来选择是否要修改状态。

4.7.5 AbstractCommand

Abstract Command

视频名称 0439_【了解】AbstractCommand

视频简介 为了保证资源调度稳定性，Hystrix 提供了资源隔离机制。本视频基于此概念为读者分析 AbstractCommand 类的核心源代码。

AbstractCommand 是 HystrixCommand 的父类，也是服务调用的核心处理工具类。Hystrix 提供的资源隔离策略中有线程池隔离与信号量隔离，而这两种策略的执行最终都是由 AbstractCommand 类实现的。

范例：【hystrix-core 依赖】AbstractCommand 核心源代码

```
package com.netflix.hystrix;
abstract class AbstractCommand<R> implements
HystrixInvokableInfo<R>, HystrixObservable<R> {
   protected final HystrixCircuitBreaker circuitBreaker;          // 断路器
   protected final HystrixThreadPool threadPool;                  // 线程池
    protected final HystrixCommandExecutionHook executionHook;   // 钩子调用
   protected volatile ExecutionResult executionResult = ExecutionResult.EMPTY;
   private Observable<R> applyHystrixSemantics(final AbstractCommand<R> _cmd) {
      executionHook.onStart(_cmd);                                // 执行处理
      if (circuitBreaker.allowRequest()) {                        // 断路器判断
         // 获取执行的信号量，每一个默认的commandKey对应的最大信号量为10
         final TryableSemaphore executionSemaphore = getExecutionSemaphore();
         // 信号量释放标记
         final AtomicBoolean semaphoreHasBeenReleased = new AtomicBoolean(false);
         final Action0 singleSemaphoreRelease = new Action0() {
            @Override
            public void call() {                                  // 信号量释放逻辑
               if (semaphoreHasBeenReleased.compareAndSet(false, true)) {
                  executionSemaphore.release();                   // 信号量释放
               }
            }
         };
         final Action1<Throwable> markExceptionThrown = new Action1<Throwable>() {
            @Override
            public void call(Throwable t) {                       // 标记异常逻辑
               eventNotifier.markEvent(
                     HystrixEventType.EXCEPTION_THROWN, commandKey);
            }
         };
         if (executionSemaphore.tryAcquire()) {                   // 尝试获取信号量
            try {
               executionResult = executionResult.setInvocationStartTime(
               System.currentTimeMillis());                       // 命令执行开始时间
               return executeCommandAndObserve(_cmd)              // 命令执行
                     .doOnError(markExceptionThrown)
                     .doOnTerminate(singleSemaphoreRelease)
                     .doOnUnsubscribe(singleSemaphoreRelease);
            } catch (RuntimeException e) { return Observable.error(e); }
         } else {
            return handleSemaphoreRejectionViaFallback();         // 获取信号量失败的逻辑
         }
      } else {
         return handleShortCircuitViaFallback();                  // 执行断路逻辑
      }
   }
   private Observable<R> executeCommandAndObserve(final AbstractCommand<R> _cmd) {
      final HystrixRequestContext currentRequestContext = HystrixRequestContext
            .getContextForCurrentThread();                        // 获取请求上下文
      final Action1<R> markEmits = new Action1<R>() {
         @Override
         public void call(R r) {                                  // 发送数据之前的处理逻辑
            if (shouldOutputOnNextEvents()) {                     // 是否要报告事件
               executionResult = executionResult.addEvent(HystrixEventType.EMIT);
               eventNotifier.markEvent(HystrixEventType.EMIT, commandKey);
            }
            if (commandIsScalar()) {                              // 获取度量数据
```

```
            long latency = System.currentTimeMillis() - executionResult
                    .getStartTimestamp();                      // 获取延迟时间
            eventNotifier.markCommandExecution(getCommandKey(),
                    properties.executionIsolationStrategy().get(), (int) latency,
                    executionResult.getOrderedList());     // 命令执行
            eventNotifier.markEvent(HystrixEventType.SUCCESS, commandKey);
            executionResult = executionResult.addEvent((int) latency,
                    HystrixEventType.SUCCESS);
            circuitBreaker.markSuccess();                      // 标记成功
        }
    }
};
final Action0 markOnCompleted = new Action0() {
    @Override
    public void call() {                                       // 命令完成时的执行逻辑
        if (!commandIsScalar()) { … }
    }
};
final Func1<Throwable, Observable<R>> handleFallback =
  new Func1<Throwable, Observable<R>>() {                      // 降级回退时的执行逻辑
    @Override
    public Observable<R> call(Throwable t) { … };
final Action1<Notification<? super R>> setRequestContext =
  new Action1<Notification<? super R>>() {                     // 设置请求上下文
    @Override
    public void call(Notification<? super R> rNotification) { … }
};
Observable<R> execution;                                       // 隔离策略处理逻辑
if (properties.executionTimeoutEnabled().get()) {
    execution = executeCommandWithSpecifiedIsolation(_cmd)
            .lift(new HystrixObservableTimeoutOperator<R>(_cmd)); // 超时处理
} else {
    execution = executeCommandWithSpecifiedIsolation(_cmd);
}
return execution.doOnNext(markEmits).doOnCompleted(markOnCompleted)
        .onErrorResumeNext(handleFallback).doOnEach(setRequestContext);
}
private Observable<R> executeCommandWithSpecifiedIsolation(
      final AbstractCommand<R> _cmd) {
    if (properties.executionIsolationStrategy().get() ==
        ExecutionIsolationStrategy.THREAD) {                   // 线程池隔离策略
        // 标记线程执行（即便被拒绝，最终也通过线程执行，而不是信号量）
        return Observable.defer(new Func0<Observable<R>>() {
            @Override
            public Observable<R> call() {                      // 隔离处理逻辑
                executionResult = executionResult.setExecutionOccurred();
                if (!commandState.compareAndSet(CommandState.OBSERVABLE_CHAIN_CREATED,
                CommandState.USER_CODE_EXECUTED)) {            // 判断当前命令是否执行过
                    return Observable.error(new IllegalStateException("…"));
                }
                metrics.markCommandStart(commandKey, threadPoolKey,
                        ExecutionIsolationStrategy.THREAD);    // 上传报告
                if (isCommandTimedOut.get() == TimedOutStatus.TIMED_OUT) {// 超时
                    return Observable.error(new RuntimeException("timed out …"));
                }
                if (threadState.compareAndSet(ThreadState.NOT_USING_THREAD,
                        ThreadState.STARTED)) {                // 修改线程状态
                    // Hystrix计数，此时还没有退订，需要继续执行
                    HystrixCounters.incrementGlobalConcurrentThreads();
                    threadPool.markThreadExecution();          // 上报线程统计指标
                    // 记录正在运行的命令，将当前命令保存到ThreadLocal中，执行完后删除
                    endCurrentThreadExecutingCommand = Hystrix
```

```
                    .startCurrentThreadExecutingCommand(getCommandKey());
            executionResult = executionResult.setExecutedInThread();
            try {
                executionHook.onThreadStart(_cmd);          // 钩子状态记录
                executionHook.onRunStart(_cmd);             // 钩子状态记录
                executionHook.onExecutionStart(_cmd);       // 钩子状态记录
                return getUserExecutionObservable(_cmd); // 执行用户任务
            } catch (Throwable ex) { return Observable.error(ex); }
        } else {                    // 命令取消订阅，抛出异常
            return Observable.error(new RuntimeException("unsubscribed…"));
        }
    }
}).doOnTerminate(new Action0() {    …                   // 中断时调用
}).doOnUnsubscribe(new Action0() { …                   // 订阅取消
}).subscribeOn(threadPool.getScheduler(                 // Hystrix线程池
        new Func0<Boolean>() {
    @Override
    public Boolean call() {                             // 修改订阅线程，交由线程池执行
        return properties.executionIsolationThreadInterruptOnTimeout().get()
                && _cmd.isCommandTimedOut.get() == TimedOutStatus.TIMED_OUT;
    }
}));
} else {                                                // 信号量隔离策略
    return Observable.defer(new Func0<Observable<R>>() {
        @Override
        public Observable<R> call() {
            executionResult = executionResult.setExecutionOccurred();
            if (!commandState.compareAndSet(CommandState.OBSERVABLE_CHAIN_CREATED,
                    CommandState.USER_CODE_EXECUTED)) {          // 命令是否执行过
                return Observable.error(new IllegalStateException("execution …));
            }
            metrics.markCommandStart(commandKey, threadPoolKey,
                    ExecutionIsolationStrategy.SEMAPHORE);       // 上报统计信息
            // 记录当前正在运行的命令，将其保存在ThreadLocal中，处理完后删除
            endCurrentThreadExecutingCommand = Hystrix
                    .startCurrentThreadExecutingCommand(getCommandKey());
            try {
                executionHook.onRunStart(_cmd);                  // 钩子状态记录
                executionHook.onExecutionStart(_cmd);            // 钩子状态记录
                return getUserExecutionObservable(_cmd);         // 执行用户任务
            } catch (Throwable ex) { return Observable.error(ex); }
        }
    });
}
}
}
```

以上的源代码已经清楚地描述了关于 Hystrix 执行处理操作的流程，在每次执行前都需要判断断路器的状态，在每次执行完成后还需要对断路器的状态进行标记。在最终执行 Hystrix 命令时，会根据当前不同的隔离策略执行。每次执行完成后都会通过钩子函数进行状态数据的记录，同时也会将用户的请求线程变为转发线程进行服务资源调用。

4.8　本章概览

1．为了提高注册中心的处理性能，可以采用集群模式进行服务搭建，在 Nacos 中直接通过 cluster.conf 定义主机列表。

2．Nacos 默认情况下只能够采用 IP 地址进行集群搭建，这样的搭建方式并不适合于节点维护，所以需要手动修改 startup.sh 配置文件为其追加主机名称（“-Dnacos.server.ip=主机名称”属性设置）。

3．Nacos 注册中心提供的 Web 管理服务（默认为 8848 端口）可以通过 Nginx 进行代理访问，而服务注册（默认端口为“8848 + 1000 = 9848”）与发现需要通过 HAProxy 实现代理。

4．Nacos 集群服务支持 CP 与 AP 模式，默认情况下采用的是 AP 模式，即牺牲实时一致性，而考虑最终一致性。

5．Nacos 生产端进行服务注册时，需要配置 bootstrap.yml（nacos-config）与 application.yml（nacos-discovery）。

6．消费端可以通过 Nacos 注册中心进行服务访问，此时可以通过 DiscoveryClient 进行服务发现处理。

7．为了简化消费端的调用难度，可以使用 Ribbon 组件。Ribbon 组件的核心组成包括 ILoadBalancer、ServerList、ServerListUpdater、ServerListFilter、IPing、IRule 等。在通过 Bean 进行 Ribbon 配置时应避免将之与主应用保存在同一扫描包。

8．基于 Nacos 注册中心可以自定义负载均衡策略，可以实现权重优先调度、集群优先调度、元数据版本优先调度。

9．为了便于远程接口的调用，可以通过 Feign 组件进行远程接口映射，在 Feign 组件中默认整合了 Ribbon，可以实现负载均衡处理。

10．Feign 配置需要通过@FeignClient 注解进行，而后需要通过@EnableFeignClients 注解配置扫描包。

11．为了防止微服务集群中出现雪崩效应，可以引入 Hystrix 熔断机制。这样在出现问题时，可以通过断路器断开微服务资源访问，同时基于降级回退机制返回指定结构的数据内容。

12．Hystrix 提供了 Dashboard 监控面板服务，可以针对 Hystrix 的监控数据进行监控图形的绘制。

13．为了便于集群环境的监控，可以通过 Turbine 进行聚合监控配置。

第 5 章 Sentinel

本章学习目标

1. 掌握 Sentinel 组件的作用，并可以理解 Sentinel 与 Hystrix 的实现区别；
2. 掌握 Sentinel 监控数据的持久化存储方法，并可以基于 MySQL、ElasticSearch、Kafka 实现监控数据保存；
3. 理解 Sentinel 的工作原理与服务降级实现原理；
4. 理解 Sentinel 流控规则持久化的实现。

良好的微服务设计需要考虑流控与熔断机制，而 Spring Cloud Alibaba 组件中提供了 Sentinel 组件。本章将为读者分析 Sentinel 组件的工作原理以及与 REST 微服务的整合实现。

5.1 Sentinel 服务搭建

Sentinel 简介

视频名称 0501_【理解】Sentinel 简介

视频简介 Spring Cloud Alibaba 提供了良好的服务流控管理支持。本视频为读者讲解访问流控的作用，并介绍 Sentinel 组件的功能。

当服务在调用中面对大规模的并发请求时，如果没有进行合理的请求规划，那么会出现服务资源调用量过大导致服务响应慢而出现“假死”状态，甚至会出现服务死机导致的雪崩效应。所以一个合理的微服务架构设计必须进行有效的流量防护控制，如图 5-1 所示。利用 Sentinel 防护组件可以在高并发访问量出现的情况下实现有效的流理控制（简称流控）管理，并基于一定的规则对大规模请求进行调整，从而实现对资源的保护。

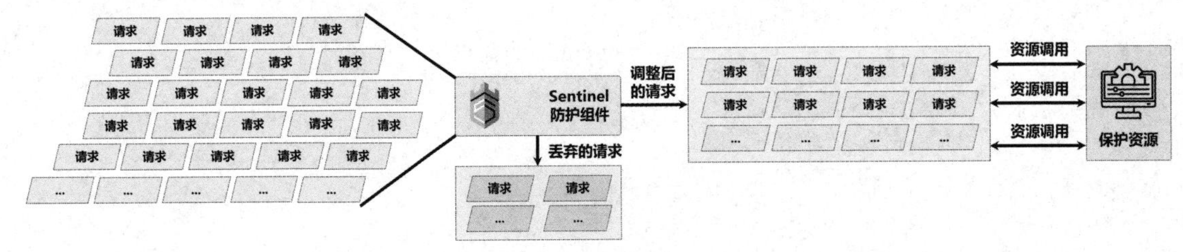

图 5-1 流量防护

Sentinel 是由阿里巴巴公司提供的一款面向分布式服务架构的高可用流量防护开源组件，也是 Spring Cloud Alibaba 之中的重要组成单元。Sentinel 组件主要以流量为切入点实现调用关系流控、流量整形、慢调用降级、系统自适应保护、热点流控等功能，如图 5-2 所示，从而帮助开发者保障微服务运行的稳定性。

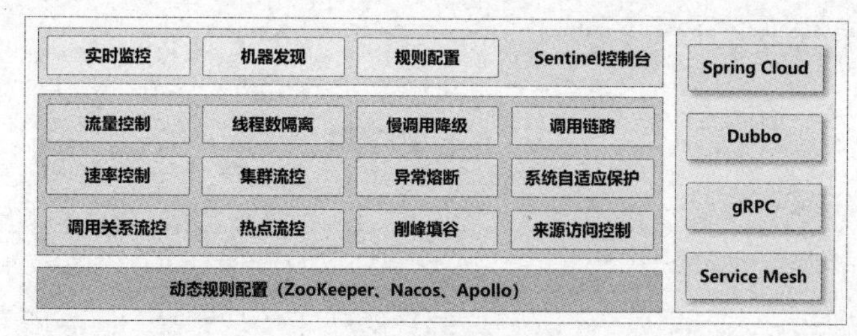

图 5-2 Sentinel 组件功能

Sentinel 组件诞生于 2012 年并且于 2018 年实现开源，最初的功能主要是入口流量控制，而后在近十年的时间里承担了阿里巴巴"双十一"大促的应用场景，使突发流量可以被控制在系统能承受的范围之内。Sentinel 组件可以方便地与 Spring Cloud、Dubbo、gRPC、Service Mesh 进行整合。为了便于服务管理，Sentinel 组件内部提供了完备的实时监控支持，可以轻松查看接入应用的流量数据。

Sentinel 与 Hystrix 在实现原则上是相同的，当检测到调用链路中某个资源出现不稳定的表现（如请求响应时间长或异常比例升高）时，则对这个资源的调用进行限制，让请求快速失败，避免影响到其他资源而导致级联故障。Hystrix 是通过线程池隔离的方式实现的，虽然可以有效地实现资源之间的彻底隔离，但是也会增加线程切换的开销。同时，过多的线程池也会导致线程数量增多。Sentinel 并没有采用这样的线程池隔离策略，而是针对微服务访问中可能存在的问题采取了不同的实现手段。

- 通过并发线程数进行限制：Sentinel 组件中通过限制资源并发线程的数量来减少不稳定资源对其他资源的影响。当某一个资源造成线程数量堆积过多时，Sentinel 会自动拒绝新的请求，待堆积线程处理完成后才开始继续接收新的请求。
- 服务慢调用和异常出现时对资源进行降级：Sentinel 可以根据响应时间和异常等不稳定的因素来实现对资源调用的熔断。当资源调用响应时间过长时，Sentinel 会自动拒绝对该资源后续的访问请求，直到过了指定的窗口时间才会重新尝试进行资源调用的恢复。

提示：Sentinel 与 Hystrix 的选择。

Sentinel（Spring Cloud Alibaba 套件）与 Hystrix（Spring Cloud Netflix 套件）是两种不同的流量控制组件，在 Spring Cloud Alibaba 套件中可以使用 Hystrix，也可以使用 Sentinel。由于 Hystrix 已经不再维护，因此建议使用 Sentinel 组件进行开发。

5.1.1 Sentinel 控制台

Sentinel 控制台

视频名称 0502_【掌握】Sentinel 控制台

视频简介 Sentinel 是开源组件，可以通过 GitHub 获取。本视频为读者讲解 Sentinel 组件的下载、组件启动以及控制台使用。

Sentinel 基于 Java 开发，同时也是免费的开源组件，开发者直接通过 GitHub 即可获取 Sentinel 源代码以及打包后的部署程序，如图 5-3 所示。截至本书编写时 Sentinel 组件的最新版本为 1.8.1，本次将基于此版本进行讲解。

提示：建议下载 Sentinel 源代码。

虽然 Sentinel 组件提供了打包后的应用程序，但是由于在随后的讲解中需要对 Sentinel 功能进行扩展，这样就需要对 Sentinel 的源代码进行修改并重新打包，因此建议读者保存 Sentinel 源代码。

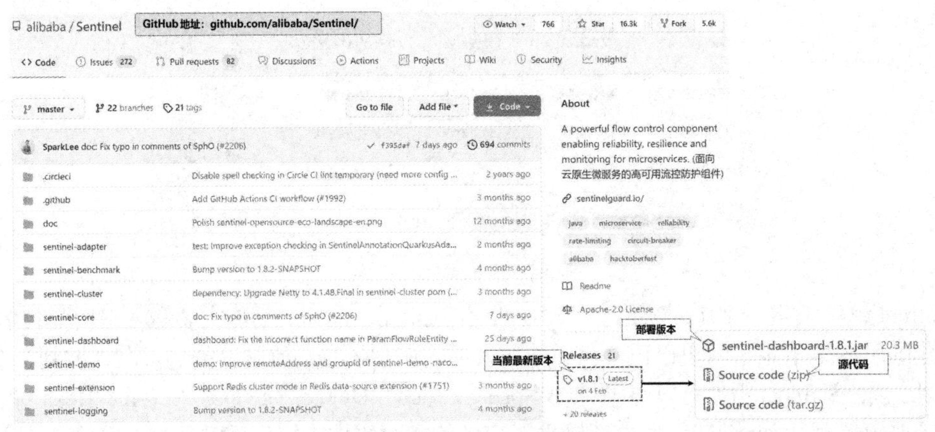

图 5-3 获取 Sentinel 组件

为了便于使用，本次直接下载 Sentinel 部署包（“sentinel-dashboard-1.8.1.jar”）进行服务的创建。由于 Sentinel 是通过 Spring Boot 开发的，因此可以直接通过命令行工具进行启动，具体的使用步骤如下。

（1）【sentinel-server 主机】下载 Sentinel 部署程序包，保存路径为/usr/local/src/sentinel-dashboard-1.8.1.jar。

（2）【sentinel-server 主机】启动 Sentinel 组件，并且在启动时设置 Sentinel 服务的监听端口、用户名以及密码信息。

```
java -Dserver.port=8888 -Dproject.name=sentinel-dashboard -Dcsp.sentinel.dashboard.server=
localhost:8888 \
    -Dsentinel.dashboard.auth.username=muyan -Dsentinel.dashboard.auth.password=yootk \
    -jar /usr/local/src/sentinel-dashboard-1.8.1.jar >/usr/local/src/sentinel.log 2>&1 &
```

因为 Sentinel 是通过 Spring Boot 开发的，所以在默认启动时会直接在控制台进行日志输出。为了使其可以在后台运行，本次将所有的日志保存在/usr/local/src/sentinel.log 日志文件之中。

（3）【sentinel-server 主机】Sentinel 服务启动之后会占用 8888（Web 控制台）与 8719（组件通信）两个端口。

配置访问端口	`firewall-cmd --zone=public --add-port=8888/tcp --permanent`
	`firewall-cmd --zone=public --add-port=8719/tcp --permanent`
配置重新加载	`firewall-cmd --reload`

（4）【本地系统】为了便于服务访问，可以在本地系统的 hosts 文件中添加主机映射配置。

```
192.168.190.168sentinel-server
```

（5）【浏览器】打开 http://sentinel-server:8888，而后就可以见到 Sentinel 控制台，如图 5-4 所示。

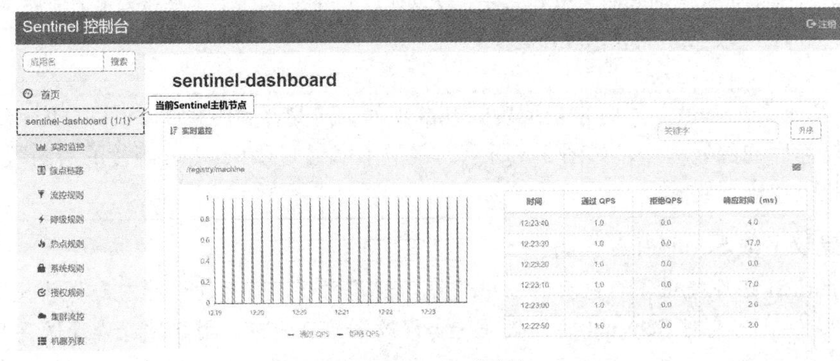

图 5-4 Sentinel 控制台

5.1.2 Sentinel 资源监控

视频名称 0503_【掌握】Sentinel 资源监控

视频简介 项目中的微服务如果需要通过 Sentinel 进行流控保护，则需要引入相关依赖并进行配置实现。本视频通过具体的操作实例，为读者讲解微服务的改造配置，并基于 AOP 切面实现 Sentinel 流量数据的获取以及流控保护控制。

Spring Cloud Alibaba 套件直接支持 Sentinel 服务的接入配置，开发者只需要导入所需要的依赖库，而后通过 application.yml 配置文件定义 Sentinel 相关服务地址，就可以自动扫描项目中的所有 REST 接口，并在微服务启动后实现自动服务注册，如图 5-5 所示。下面通过具体的操作步骤介绍如何实现服务资源与 Sentinel 组件的整合。

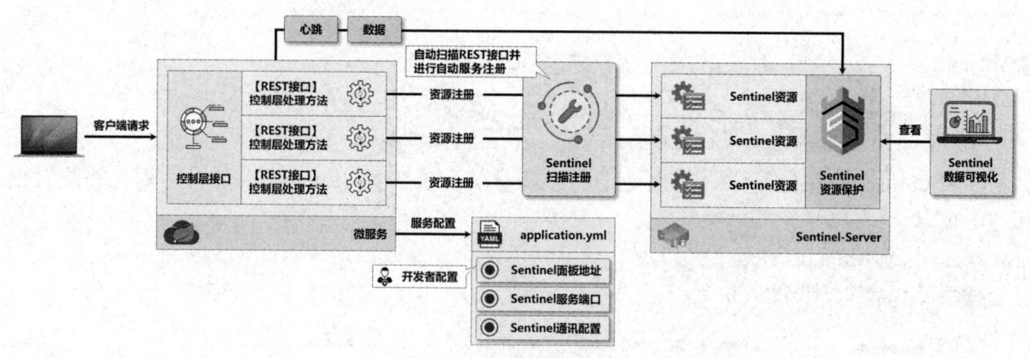

图 5-5 Sentinel 资源监控

（1）【microcloud 项目】修改 build.gradle 配置文件，为“provider-dept-*”部门微服务子模块添加 Sentinel 依赖库。

```
project('provider-dept-8002') {          // 子模块
    dependencies {                       // 重复依赖，略
        implementation('com.alibaba.cloud:spring-cloud-starter-alibaba-sentinel')
    }
}
```

在本次项目中，部门微服务一共存在 3 个子模块，所以需要在 3 个子模块中同时配置 Sentinel 依赖库，并需要取消项目中的 Hystrix 依赖。在项目中引入“spring-cloud-starter-alibaba-sentinel”依赖后就可以直接通过 application.yml 配置 Sentinel 相关的使用环境，具体的配置项如表 5-1 所示。

表 5-1 Spring Cloud 整合 Sentinel 配置项

序号	配置项	描述
01	spring.cloud.sentinel.enabled	Sentinel 配置是否生效，默认为 true
02	spring.cloud.sentinel.eager	Sentinel 控制台懒注册，默认为 false
03	spring.cloud.sentinel.transport.port	Sentinel 控制台交互端口，默认为 8719
04	spring.cloud.sentinel.transport.dashboard	Sentinel 控制台地址
05	spring.cloud.sentinel.transport.heartbeat-interval-ms	心跳发送间隔时间
06	spring.cloud.sentinel.transport.client-ip	客户端 IP 地址
07	spring.cloud.sentinel.filter.enabled	是否启用 Sentinel 过滤，默认为 true
08	spring.cloud.sentinel.filter.order	Sentinel 过滤加载顺序
09	spring.cloud.sentinel.filter.url-patterns	过滤器匹配路径
10	spring.cloud.sentinel.metric.charset	Metric 文件字符集
11	spring.cloud.sentinel.metric.file-single-size	单个 Metric 文件大小

续表

序号	配置项	描述
12	spring.cloud.sentinel.metric.file-total-count	Metric 总文件数量
13	spring.cloud.sentinel.log.dir	Sentinel 日志文件所在目录
14	spring.cloud.sentinel.log.switch-pid	Sentinel 日志文件是否带上 pid（进程 ID）
15	spring.cloud.sentinel.servlet.block-page	流控时跳转路径
16	spring.cloud.sentinel.flow.cold-factor	启动冷因子

（2）【provider-dept-*子模块】修改 application.yml 配置文件，添加 Sentinel 相关配置项。

```
spring:                                    # Spring配置
  application:                             # 应用配置
    name: dept.provider                    # 应用名称（Sentinel注册名称）
  cloud:                                   # Spring Cloud配置
    sentinel:                              # Sentinel配置
      transport:                           # 通信配置
        port: 8719                         # Sentinel通信端口
        dashboard: sentinel-server:8888    # Sentinel控制面板
```

（3）【Sentinel 控制台】微服务启动之后，就可以直接在 Sentinel 控制台看见相关的配置信息，如图 5-6 所示。

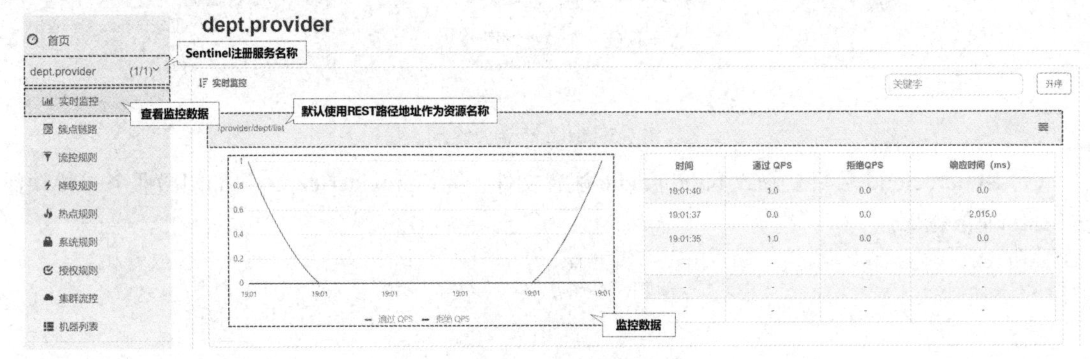

图 5-6 Sentinel 监控数据

> **提示：资源信息显示。**
>
> 在默认情况下，所有在 Sentinel 中注册的资源数据都是在第一次使用之后（此时会向 Sentinel Dashboard 发送一次心跳）才能在 Sentinel 控制台中显示。如果希望服务在启动时可以自动地进行服务注册，可以在 application.yml 中修改懒注册配置项。
>
> **范例：配置服务懒注册**
>
> ```
> spring.cloud.sentinel.eager = true                  # 服务启动时注册
> ```
>
> 配置此项后一旦微服务启动，则会自动发出心跳。在 Sentinel 控制台就能看见相关的注册信息。

5.1.3 实时监控数据

实时监控数据

视频名称 0504_【理解】实时监控数据

视频简介 Sentinel 为了与其他平台进行整合处理，提供了若干个 REST 数据接口，开发者可以直接通过该接口获取 Sentinel 中的流量数据、日志数据。本视频为读者介绍这些数据接口的名称以及作用。

Sentinel 可以实现所有服务资源的性能监控，同时为了便于这些监控服务数据的获取，又提供了一系列的 REST 接口，如图 5-7 所示。利用这些接口可以方便地实现与其他应用平台的对接。下面将为读者列出几个 REST 接口进行说明。

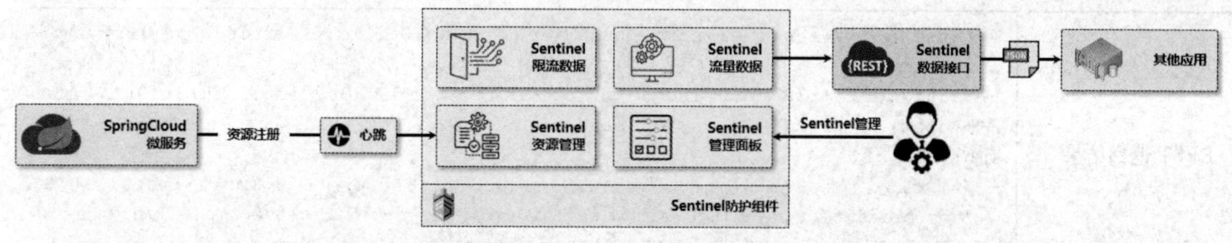

图 5-7　Sentinel 实时监控数据

（1）【REST 接口】查看当前所使用的 Sentinel 版本。

GET 请求模式	`sentinel-server:8719/version`
REST 返回结果	`1.8.1`

（2）【REST 接口】获取全部簇点列表。

GET 请求模式	`sentinel-server:8719/clusterNode`

REST 返回结果

```
[{
  "averageRt": 0,                          // 平均响应时间
  "blockQps": 0,                           // 每秒拦截请求数
  "exceptionQps": 0,                       // 产生异常的请求数
  "oneMinuteBlock": 0,                     // 每分钟拦截请求数
  "oneMinuteException": 0,                 // 每分钟产生的异常数
  "oneMinutePass": 2,                      // 每分钟通过的请求数
  "oneMinuteTotal": 2,                     // 每分钟的请求总数
  "passQps": 0,                            // 每秒通过请求数
  "resource": "/app/dept.provider/machines.json", // 资源名称
  "successQps": 0,                         // 每秒成功请求数
  "threadNum": 0,                          // 每秒线程个数
  "timestamp": 1622207613213,              // 数据产生的日期时间
  "totalQps": 0                            // 每秒总共的请求数
}, ...]
```

（3）【REST 接口】获取指定簇点信息（查询时使用的资源名称可以通过 ID 模糊匹配）。

GET 请求模式	`sentinel-server:8719/cnode?id=e`

REST 返回结果

idx	id	thread	pass	blocked	success	total	aRt	1m-pass	1m-block	1m-all	exception
13	/metric/queryTopResourceMetric.json	0	0.0	0.0	0.0	0.0	0.0	0	0	0	0.0
14	/system/rules.json	0	0.0	0.0	0.0	0.0	0.0	1	0	1	0.0
15	/version	0	0.0	0.0	0.0	0.0	0.0	0	0	0	0.0
16	/cluster/server state/sentinel-dashboard	0	0.0	0.0	0.0	0.0	0.0	1	0	1	0.0
17	/degrade/rules.json	0	0.0	0.0	0.0	0.0	0.0	1	0	1	0.0
18	/resource/machineResource.json	0	0.0	0.0	0.0	0.0	0.0	3	0	3	0.0
19	/app/dept.provider/machines.json	0	0.0	0.0	0.0	0.0	0.0	3	0	3	0.0
20	/v1/flow/rules	0	0.0	0.0	0.0	0.0	0.0	2	0	2	0.0
21	/app/sentinel-dashboard/machines.json	0	0.0	0.0	0.0	0.0	0.0	5	0	5	0.0
22	/authority/rules	0	0.0	0.0	0.0	0.0	0.0	1	0	1	0.0
23	/registry/machine	0	0.0	0.0	0.0	0.0	0.0	18	0	18	0.0
24	/app/briefinfos.json	0	0.0	0.0	0.0	0.0	0.0	0	0	0	0.0

（4）【REST 接口】链路监控。

GET 请求模式	`sentinel-server:8719/tree`

REST 返回结果

```
EntranceNode: machine-root(t:0 pq:1.0 bq:0.0 tq:1.0 rt:1.0 prq:1.0 1mp:17 1mb:0
1mt:17)
-EntranceNode: sentinel_web_servlet_context(t:0 pq:1.0 bq:0.0 tq:1.0 rt:1.0
prq:1.0 1mp:17 1mb:0 1mt:17)
--/app/dept.provider/machines.json(t:0 pq:0.0 bq:0.0 tq:0.0 rt:0.0 prq:0.0
1mp:0 1mb:0 1mt:0)
--/version(t:0 pq:0.0 bq:0.0 tq:0.0 rt:0.0 prq:0.0 1mp:0 1mb:0 1mt:0)
--/registry/machine(t:0 pq:1.0 bq:0.0 tq:1.0 rt:1.0 prq:1.0 1mp:17 1mb:0 1mt:17)
--/(t:0 pq:0.0 bq:0.0 tq:0.0 rt:0.0 prq:0.0 1mp:0 1mb:0 1mt:0)
--/metric/queryTopResourceMetric.json(t:0 pq:0.0 bq:0.0 tq:0.0 rt:0.0 prq:0.0
1mp:0 1mb:0 1mt:0)
```

<table>
<tr><td>REST 返回结果</td><td>--/assets/img/sentinel-logo.png(t:0 pq:0.0 bq:0.0 tq:0.0 rt:0.0 prq:0.0 1mp:0 1mb:0 1mt:0)
--/auth/login(t:0 pq:0.0 bq:0.0 tq:0.0 rt:0.0 prq:0.0 1mp:0 1mb:0 1mt:0)
--/resource/machineResource.json(t:0 pq:0.0 bq:0.0 tq:0.0 rt:0.0 prq:0.0 1mp:0 1mb:0 1mt:0)
--/v1/flow/rules(t:0 pq:0.0 bq:0.0 tq:0.0 rt:0.0 prq:0.0 1mp:0 1mb:0 1mt:0)
--/app/briefinfos.json(t:0 pq:0.0 bq:0.0 tq:0.0 rt:0.0 prq:0.0 1mp:0 1mb:0 1mt:0)
-EntranceNode: sentinel_default_context(t:0 pq:0.0 bq:0.0 tq:0.0 rt:0.0 prq:0.0 1mp:0 1mb:0 1mt:0)</td></tr>
<tr><td>响应参数标记</td><td>1. t: threadNum（线程个数）
2. pq: passQps（每秒通过的访问请求数量）
3. bq: blockedQps（每秒阻断的访问请求数量）
4. tq: totalQps（总共的访问请求数量）
5. rt: averageRt（平均响应时间）
6. prq: passRequestQps（通过的请求数量）
7. 1mp: 1m-passed（每分钟通过的请求数量）
8. 1mb: 1m-blocked（每分钟阻断的请求数量）
9. 1mt: 1m-total（每分钟通过的请求总量）</td></tr>
</table>

（5）【REST 接口】查询数据日志。

<table>
<tr><td>GET 请求模式</td><td>sentinel-server:8719/metric?id=dept&maxLines=10&
startTime=1622358000000&endTime=1622372400000</td></tr>
<tr><td>REST 返回结果</td><td>1622371245000|/resource/machineResource.json|1|0|1|0|178|0|0|1
1622371245000|/app/dept.provider/machines.json|2|0|2|0|6|0|0|1
1622371245000|/v1/flow/rules|1|0|1|0|17|0|0|1
1622371245000|__total_inbound_traffic__|4|0|4|0|52|0|0|0</td></tr>
</table>

5.2 Sentinel 流控保护

Sentinel 流控保护

视频名称 0505_【掌握】Sentinel 流控保护

视频简介 Sentinel 进行微服务资源保护时，提供了多种流控规则。本视频为读者分析基础流控规则的使用，并通过具体的操作演示服务流控以及内置 Fallback 执行。

微服务接入 Sentinel 组件之后，所有被调用过的方法都可以通过 Sentinel 控制台提供的“簇点链路”进行查看，如图 5-8 所示。在该界面中除了可以查看资源的访问数据之外，还可以进行各种流量控制操作，如流控、降级、热点、授权。开发者可以直接通过此界面根据自身的需要设置资源保护规则。

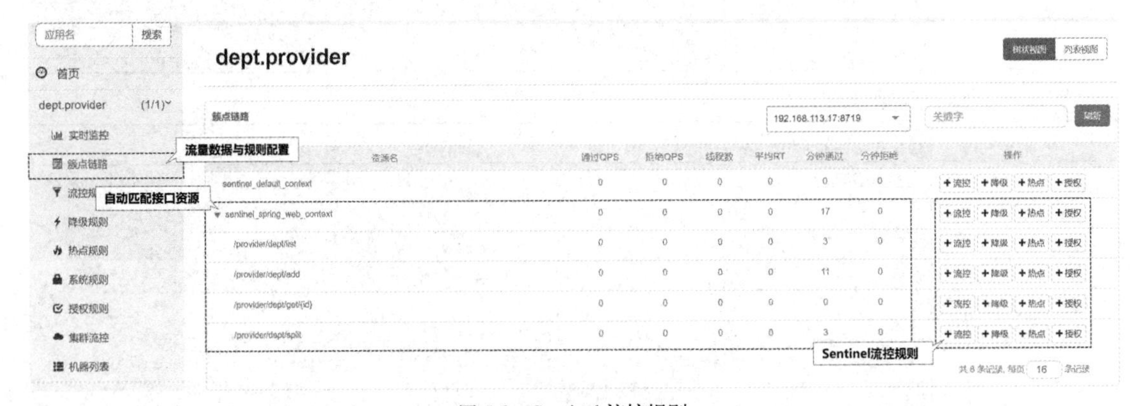

图 5-8 Sentinel 流控规则

1. 流控规则

流量控制（Flow Control）主要是监控保护资源流量的 QPS（Query Per Second，每秒查询率）或并发线程数量，当指定的监控指标达到阈值时对流量进行控制，以避免系统被瞬时流量高峰冲垮，从而保障服务资源的高可用性。开发者可以直接通过相关选项为指定的资源添加流控规则，如图 5-9 所示。

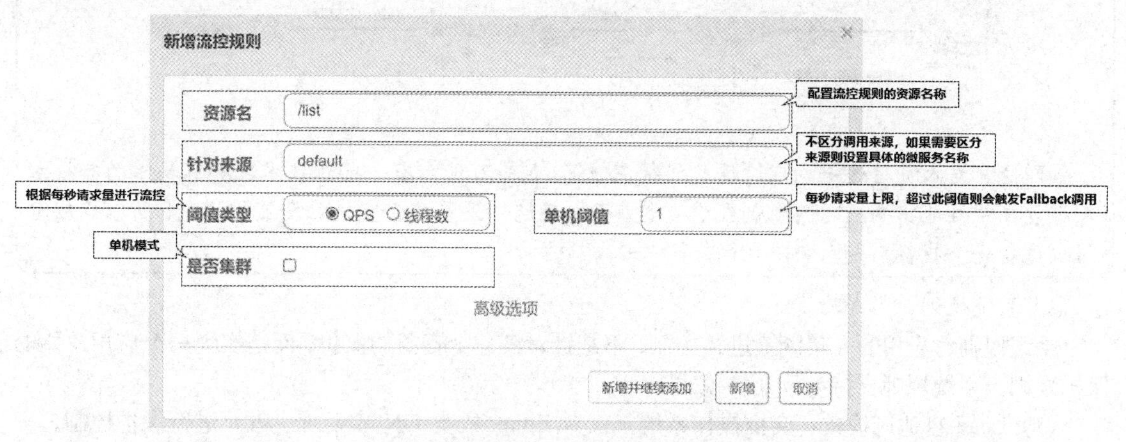

图 5-9 新增流控规则

在开启流量控制时需要明确设置流控的阈值，只要超过该阈值就会触发具体的流控处理操作。而 Sentinel 之中对流控有 3 种处理方式（需要开启"高级选项"才可以配置），具体的使用特点如下。

（1）快速失败：默认流控效果，访问量超过规定阈值后，新的请求会被立即拒绝。

（2）Warm Up：采用"预热/冷启动"方式，当访问量瞬间激增时让通过的流量缓慢增加，给冷系统预热缓冲。

（3）排队等待：当出现间隔性流量激增时，会根据请求通过的间隔时间让所有的请求匀速通过（漏桶算法）。

2. 降级规则

降级主要是指当微服务调用链中某个资源出现不稳定状态时（如响应时间过长、产生异常），为避免其他资源调用而导致的级联错误，在某个特定的时长内实现资源熔断处理。可以在图 5-10 所示的界面进行降级规则配置。

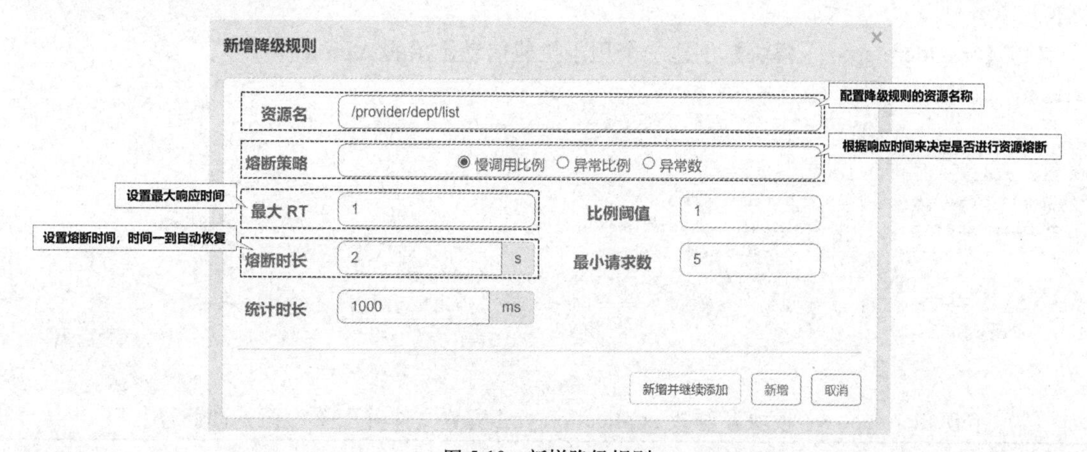

图 5-10 新增降级规则

提示：RT 是人为标准。

RT（Response Time，响应时间）是指系统对请求做出响应的时间限制，但是这个限制的标准并不是固定的。现在假设有一个业务请求处理，该业务需要整合若干个微服务，如图 5-11 所示。由于网络延时的问题，系统的响应时间为 3 毫秒。

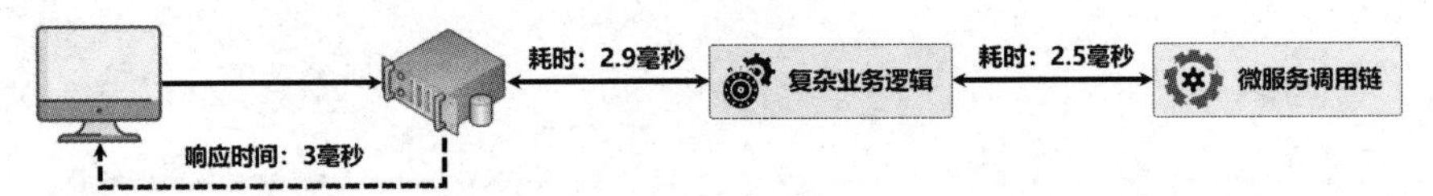

图 5-11 服务响应

即便开发人员对这一响应速度已经很满意了，但是可能会有一些对技术理解有局限的非技术人员要强制性地将响应时间限制在 1 毫秒，而一旦这样设置就会直接触发流控规则，所以响应时间仅仅是一个指标，是人为决定的。

3. 系统规则

系统规则对应用的入口流量进行控制，不是针对某一个资源的保护，而是实现一个应用总体的保护规则。系统规则支持如下几种模式。

（1）Load 自适应模式：交由操作系统（仅对 Linux 或类 UNIX 系统生效）进行保护控制，一般的参考值为“系统硬件的 CPU 内核数量 × 2.5”。

（2）平均响应时间：当单台主机上的所有入口流量的平均 RT 达到阈值时触发系统保护。

（3）并发线程数：当单台主机上的所有入口流量达到并发线程阈值时触发系统保护。

（4）入口 QPS：当单台主机上的所有入口流量达到阈值时触发系统保护。

（5）CPU 使用率：当系统 CPU 使用率超过阈值时触发系统保护。

5.2.1 自定义流控错误页

视频名称 0506_【理解】自定义流控错误页

视频简介 客户端请求一旦触发 Sentinel 资源保护规则便会进行错误信息的显示，开发者可以根据需要自定义错误信息。本视频将介绍基于 application.yml 配置实现统一拦截错误的显示以及拦截处理路径的配置。

在默认情况下，一旦用户的请求触发 Sentinel 保护规则，客户端就会输出默认拦截提示信息“Blocked by Sentinel (flow limiting)”。开发者如果有需要也可以自定义拦截信息提示，实现步骤如下。

（1）【provider-dept-*子模块】创建一个用于拦截信息显示的 Action 程序类。

```
package com.yootk.provider.action;
@RestController
@RequestMapping("/errors/*")
public class BlockAction {
    @RequestMapping("block_handler")
    public static Object globalBlockHandler() {                          // 部门增加拦截
        Map<String, Object> map = new HashMap<>();                       // 响应结果
        map.put("status", HttpServletResponse.SC_BAD_REQUEST);           // 操作状态
        map.put("message", "Blocked by Sentinel (flow limiting)");      // 保存失败信息
        return map;                                 // 数据响应
    }
}
```

（2）【provider-dept-*子模块】修改 application.yml 配置文件并配置拦截处理路径。

```
spring:                                  # Spring配置
  cloud:                                 # Spring Cloud配置
    sentinel:                            # Sentinel配置
```

	`transport:` # 通信配置 `  port: 8719` # Sentinel通信端口 `  dashboard: sentinel-server:8888` # Sentinel控制面板 `block-page: /errors/block_handler` # 拦截处理路径

拦截返回数据	`{`
	`  "message": "Blocked by Sentinel (flow limiting)",`
	`  "status": 400`
	`}`

配置完成后，当用户请求触发 Sentinel 保护规则时，程序将会跳转到“/errors/block_handler”路径显示相关拦截信息。

5.2.2 失败回退

失败回退

视频名称 0507_【理解】失败回退

视频简介 Sentinel 在与 Spring Cloud 整合时提供了默认的失败回退处理支持，但是考虑到定制化的开发要求，开发者也可以通过@SentinelResource 注解实现自定义失败回退操作。本视频通过具体的实例讲解自定义资源失败回退开发实现。

默认情况下，一旦触发 Sentinel 保护规则，程序就会使用默认的失败回退方法进行处理。如果用户有需要，也可以针对某些特定的操作方法自定义失败回退处理，此时就需要在 REST 方法实现中使用@SentinelResource 注解进行配置。该注解所定义的属性如表 5-2 所示。

表 5-2　@SentinelResource 注解定义的属性

序号	属性	类型	描述
01	value	String	资源名称，不能为空
02	entryType	EntryType	资源调用类型，默认为 EntryType.OUT
03	resourceType	int	资源分类
04	blockHandler	String	拦截产生后的处理方法名称
05	blockHandlerClass	Class<?>[]	拦截产生后的处理类
06	fallback	String	失败回退处理方法名称
07	defaultFallback	String	通用的失败回退处理方法名称
08	fallbackClass	Class<?>[]	失败回退处理类
09	exceptionsToTrace()	Class<? extends Throwable>[]	需要跟踪的异常类型
10	exceptionsToIgnore	Class<? extends Throwable>[]	忽略的异常类型

如果想在 Sentinel 控制台获取接口的流量统计信息并实现有效的流量控制，则还需要在项目中配置切面处理结构。Sentinel 组件中提供了 SentinelResourceAspect 来实现此操作，如图 5-12 所示。

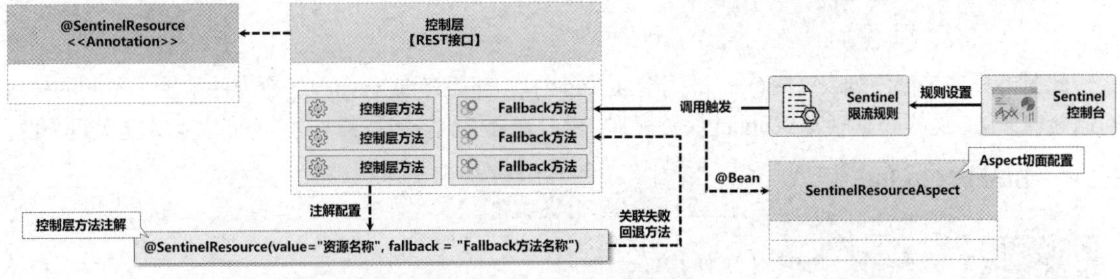

图 5-12　自定义失败回退处理

（1）【provider-dept-*子模块】创建 Sentinel 资源切面配置类，而后才可以在服务降级时自动调用失败回退方法。

```
package com.yootk.provider.config;
@Configuration
// 开启AOP注解模式（CGLib实现），通过AOP框架暴露代理对象（exposeProxy = true）
@EnableAspectJAutoProxy(exposeProxy = true, proxyTargetClass = true)
public class SentinelAOPConfig {
    @Bean
    public SentinelResourceAspect getSentinelResourceAspect() {      // Sentinel资源切面
        return new SentinelResourceAspect();
    }
}
```

（2）【provider-dept-*子模块】修改 DeptAction 程序类，在每一个 REST 接口方法中使用 @SentinelResource 定义注册的资源名称和失败回退处理方法名称。

```
package com.yootk.provider.action;
@RestController                                                      // REST控制器
@RequestMapping("/provider/dept/*")                                  // 父路径
public class DeptAction {
    @Autowired
    private IDeptService deptService;                                // 注入业务接口实例
    @GetMapping("get/{id}")                                          // 子路径
    @SentinelResource(value="/dept_get", fallback = "getFallback")   // Sentinel保护
    public Object get(@PathVariable("id") long id) {
        return this.deptService.get(id);                             // 部门信息加载
    }
    @PostMapping("add")
    @SentinelResource(value="/dept_add", fallback = "addFallback")   // Sentinel保护
    public Object add(@RequestBody DeptDTO deptDTO) {
        return this.deptService.add(deptDTO);                        // 部门信息添加
    }
    @GetMapping("list")                                              // 子路径
    @SentinelResource(value="/dept_list", fallback = "listFallback") // Sentinel保护
    public Object list() {
        return this.deptService.list();                              // 部门信息列表
    }
    @GetMapping("split")                                             // 子路径
    public Object split(int cp, int ls, String col, String kw) {
        return this.deptService.split(cp, ls, col, kw);              // 部门信息分页
    }
    public Object getFallback(@PathVariable("id") long id) {         // 失败回退方法
        DeptDTO dept = new DeptDTO();                                // 实例化DTO对象
        dept.setDeptno(id);                                          // 属性设置
        dept.setDname("【Fallback】部门名称");                         // 属性设置
        dept.setLoc("【Fallback】部门位置");                           // 属性设置
        return dept;                                                 // 数据响应
    }
    public Object addFallback(@RequestBody DeptDTO deptDTO) {        // 失败回退方法
        return false;
    }
    public Object listFallback() {                                   // 失败回退方法
        return new ArrayList<Dept>();
    }
}
```

此时的程序为部门列表（list()）、部门增加（add()）、部门查询（get()）3 个方法配置了不同的失败回退处理方法。这样一旦触发 Sentinel 保护规则，程序就会根据资源找到与之匹配的失败回退处理方法。

5.2.3　BlockHandler

BlockHandler

视频名称　0508_【理解】BlockHandler

视频简介　为了便于失败回退的统一管理，Sentinel 提供了 BlockHandler 支持。本视频为读者讲解 BlockHandler 操作的实现结构，并介绍如何实现与控制层方法的关联配置。

控制层接口处理方法使用@SentinelResource注解中的fallback属性就可以直接关联本类中定义的失败回退处理方法，但是这样一来在控制层实现类中就需要定义大量的失败回退方法，造成代码维护困难。此时最佳的做法是创建一个专属的拦截处理类（BlockHandler），并在该类中为控制层中的方法定义匹配的拦截处理方法，如图 5-13 所示。而后就可以通过@SentinelResource 注解提供的 blockHandlerClass 和 blockHandler 属性进行关联。

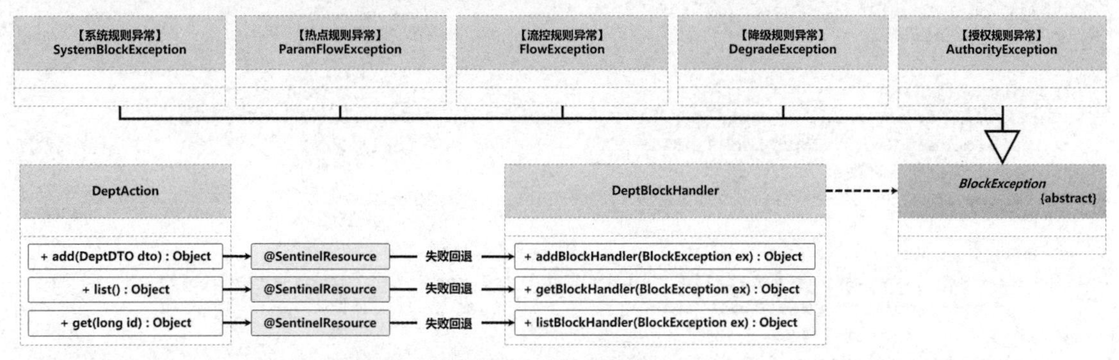

图 5-13　BlockHandler

需要注意的是，在 Sentinel 每一次进行拦截时，都会产生一个 BlockException 异常。该异常为 Exception 的子类，包含所有的拦截信息，Sentinel 中所有拦截规则的异常类全部继承此抽象类。开发者可以通过异常类的实例化对象获取所需要的异常信息。下面通过具体的操作步骤介绍如何实现这一拦截机制。

（1）【provider-dept-*子模块】创建与 DeptAction 控制器类有关的拦截处理类，并定义相关拦截处理方法。

```
package com.yootk.provider.action.block;
public class DeptBlockHandler {
    public static Object addBlockHandler(BlockException ex) {     // 部门增加拦截
        Map<String, Object> map = new HashMap<>();                 // 响应结果
        map.put("rule", ex.getRule());                             // 保存失败信息
        map.put("message", ex.getMessage());                       // 保存失败信息
        map.put("result", false);                                  // 保存失败信息
        return map;                                                // 数据响应
    }
    public static Object getBlockHandler(BlockException ex) {     // 部门查询拦截
        Map<String, Object> map = new HashMap<>();                 // 响应结果
        map.put("rule", ex.getRule());                             // 保存失败信息
        map.put("message", ex.getMessage());                       // 保存失败信息
        DeptDTO dept = new DeptDTO();                              // 实例化DTO对象
        dept.setDeptno(0L);                                        // 属性设置
        dept.setDname("【Block】部门名称");                          // 属性设置
        dept.setLoc("【Block】部门位置");                            // 属性设置
        map.put("dept", dept);                                     // 保存失败信息
        return map;                                                // 数据响应
    }
    public static Object listBlockHandler(BlockException ex) {    // 部门列表拦截
        Map<String, Object> map = new HashMap<>();                 // 响应结果
        map.put("rule", ex.getRule());                             // 保存失败信息
        map.put("message", ex.getMessage());                       // 保存失败信息
        map.put("deptList", new ArrayList<>());                    // 保存失败信息
        return map;                                                // 数据响应
    }
}
```

（2）【provider-dept-*子模块】在 DeptAction 类中关联拦截处理方法。

```
package com.yootk.provider.action;
@RestController                                                    // REST控制器
```

```
@RequestMapping("/provider/dept/*")                          // 父路径
public class DeptAction {
   @Autowired
   private IDeptService deptService;                         // 注入业务接口实例
   @GetMapping("get/{id}")                                   // 子路径
   @SentinelResource(value="/dept_get", blockHandlerClass = DeptBlockHandler.class,
         blockHandler = "getBlockHandler")                   // Sentinel保护
   public Object get(@PathVariable("id") long id) {
      return this.deptService.get(id);                       // 部门信息加载
   }
   @PostMapping("add")
   @SentinelResource(value="/dept_add", blockHandlerClass = DeptBlockHandler.class,
         blockHandler = "addBlockHandler")                   // Sentinel保护
   public Object add(@RequestBody DeptDTO deptDTO) {
      return this.deptService.add(deptDTO);                  // 部门信息添加
   }
   @GetMapping("list")                                       // 子路径
   @SentinelResource(value="/dept_list", blockHandlerClass = DeptBlockHandler.class,
         blockHandler = "listBlockHandler")                  // Sentinel保护
   public Object list() {
      return this.deptService.list();                        // 部门信息列表
   }
}
```

为控制层中的每个处理方法设置拦截处理器类以及匹配的处理方法后，当保护规则被触发时 Sentinel 就会根据配置找到对应的方法进行处理。

提示：blockHandler 优先级高于 fallback。

fallback 与 blockHandler 都是@SentinelResource 注解的属性，若开发者在定义注解时同时配置了两个属性，最终在触发保护规则时会优先通过 blockHandler 定义的方法进行处理。

5.3　Sentinel 流控规则

Sentinel 组件除了可根据访问量限流以及进行服务降级之外，实际上还提供热点、授权等流控规则。为了便于读者理解这些流控规则的使用，本节将对 Sentinel 提供的其他流控规则进行讲解。

5.3.1　热点规则

热点规则

视频名称　0509_【理解】热点规则

视频简介　热点规则是 Sentinel 独有的一种流控规则，可以直接对接口方法中指定参数的访问进行流控处理。本视频为读者讲解热点流控操作的实现以及相关配置类的定义。

热点规则会统计参数中的热点参数，并根据配置的流控阈值与模式，对包含热点参数的资源调用进行限流。热点规则可以看作一种特殊的流量控制，仅对包含热点参数的资源调用生效。如果想使用热点规则，则需要在控制层的相关方法中使用@SentinelResource 注解进行声明。

范例：修改 DeptAction 类

```
package com.yootk.provider.action;
@RestController                                              // REST控制器
@RequestMapping("/provider/dept/*")                          // 父路径
public class DeptAction {
   @Autowired
   private IDeptService deptService;                         // 注入业务接口实例
```

```
    @GetMapping("get/{id}")                              // 子路径
    @SentinelResource("/dept_get")
    public Object get(@PathVariable("id") long id) {
        return this.deptService.get(id);                 // 部门信息加载
    }
}
```

此时的 DeptAction.get()方法中并没有配置失败回退方法，一旦触发 Sentinel 热点规则会产生相关的异常对象，并通过默认方式输出异常信息。在程序启动后可以通过 Sentinel 控制台为该接口设置热点规则，如图 5-14 所示。

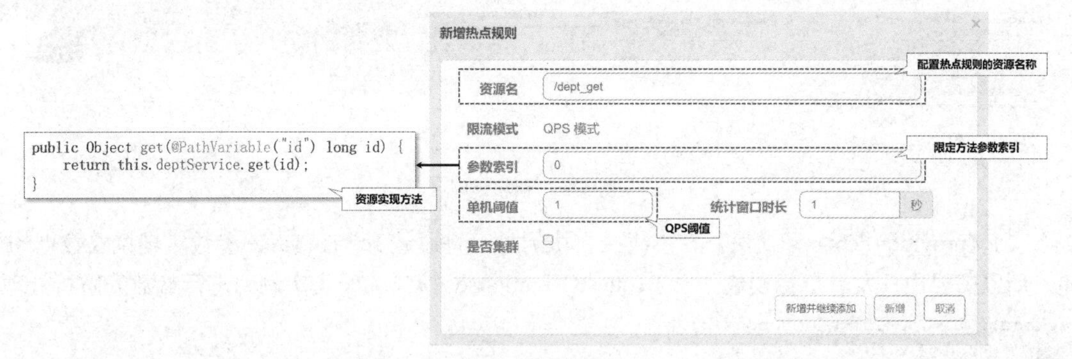

图 5-14　热点规则

在进行热点参数配置时，需要配置具体的参数索引（与方法实现中的参数索引对应），这样就可以对指定索引的参数进行 QPS 阈值设置，当参数达到阈值时就会触发流控规则。需要注意的是，此时设置的热点参数会导致热点规则对所有访问此参数的请求进行限流，如果有需要也可以针对一些特殊的值进行额外的配置，此时就需要打开“热点规则”对指定的规则进行编辑，添加“参数例外项”，如图 5-15 所示。

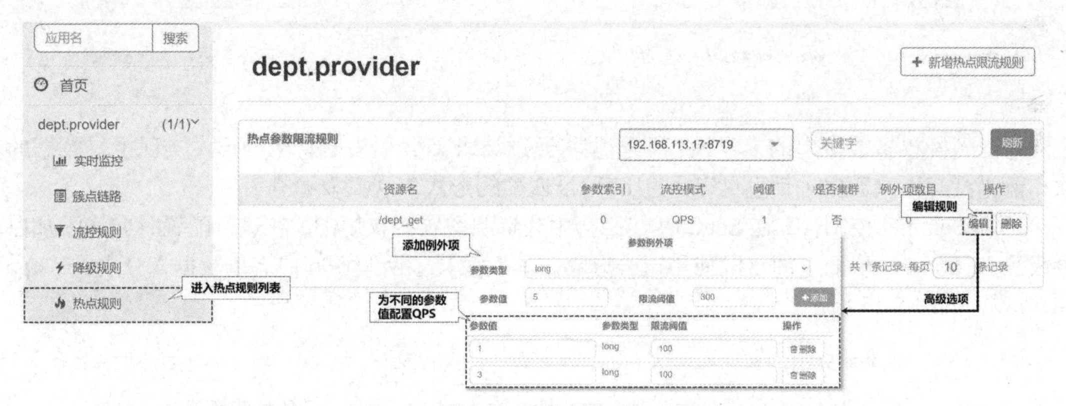

图 5-15　参数例外项

5.3.2　授权规则

视频名称　0510_【掌握】授权规则

视频简介　黑白名单是进行资源保护的重要形式，为此 Sentinel 提供了与之相关的授权规则。本视频为读者讲解授权规则的设置并分析 RequestOriginParser 接口的作用。

Sentinel 在进行资源保护时也可以根据自身的需要为资源定义白名单（符合白名单的允许访问资源）与黑名单（符合黑名单的不允许访问资源）的授权访问规则，这样就使得资源的管理更加方便。

在 Sentinel 中进行授权规则访问时需要由开发者自定义资源访问授权参数的获取规则（依靠

RequestOriginParser 接口标准实现授权数据解析)，而后就可以由 Sentinel 自行与已经保存的授权规则进行比对，如果符合授权规则则允许访问，否则不允许访问，如图 5-16 所示。下面通过具体的操作步骤介绍授权规则的使用。

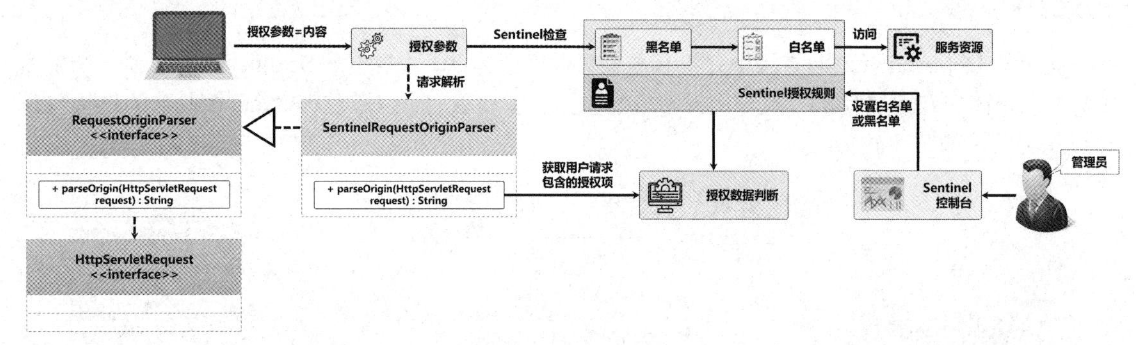

图 5-16　授权规则

(1)【provider-dept-*子模块】在本次授权访问时，将通过 serviceName 参数实现授权数据的传递，所以需要由开发者自行创建一个 RequestOriginParser 接口的子类实例，进行参数的解析处理。

```
package com.yootk.provider.config;
@Component
public class SentinelRequestOriginParser implements RequestOriginParser {
    @Override
    public String parseOrigin(HttpServletRequest request) {
        String serviceName = request.getParameter("serviceName"); // 参数接收
        if (serviceName == null || "".equals(serviceName)) {       // 参数内容为空
            serviceName = request.getHeader("serviceName");        // 头信息接收
        }
        if(!StringUtils.isEmpty(serviceName)){                     // 数据不为空
            return serviceName;                                    // 返回解析结果
        }
        return request.getRemoteAddr();                            // 没有数据根据IP地址授权
    }
}
```

考虑到授权参数传递的多样性，本程序将通过请求参数和头信息两种形式实现授权数据的获取，而当没有授权数据时，也可以通过 IP 地址匹配的形式实现授权检查。

(2)【Sentinel 控制台】在 Sentinel 控制台中为资源配置授权规则，在进行配置时允许添加白名单或黑名单，多个资源之间可以使用逗号分隔。本次设置了两个访问白名单数据，分别是“app”和“pc”，如图 5-17 所示。

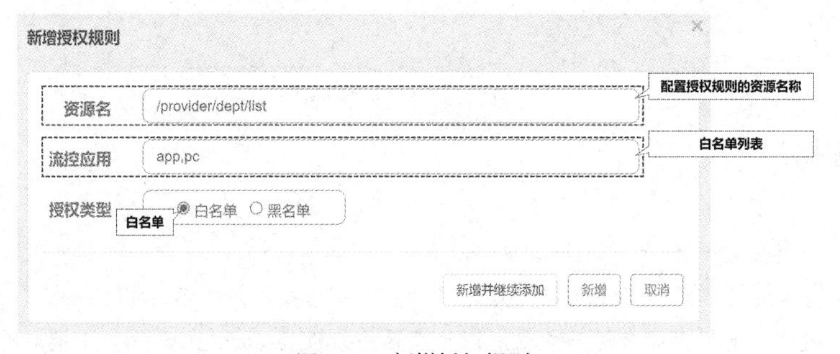

图 5-17　新增授权规则

此时客户端进行访问时就需要在访问地址后追加 serviceName 参数或者利用头信息传递 serviceName 参数。如果该参数内容符合白名单配置，则允许访问资源；如果不符合则会触发 Sentinel 保护机制，并返回错误信息。

提问：消费端如何传输授权数据？

此时的服务提供端通过 Sentinel 配置了授权规则，如果消费端在获取部门列表数据时没有 serviceName 授权数据，那么肯定无法访问，这种问题该如何解决？

回答：通过 Feign 拦截器解决。

本书第 2 章曾经为读者讲解 ClientHttpRequestInterceptor 拦截器接口的使用方法，如果没有使用 Feign 组件，则可以通过此拦截器设置头信息，而现在的项目中由于使用了 Feign 转换组件，则可以修改 Feign 配置类，并通过 RequestInterceptor 接口进行访问请求拦截。

范例：【common-api 子模块】在 Feign 配置类中追加请求拦截器

```
package com.yootk.service.config;
public class FeignConfig {
    @Bean
    public RequestInterceptor
          getFeignRequestInterceptor() {           // 请求拦截器
        return (template ->
          template.header("serviceName", "pc")); // 头信息
    }
}
```

这样在每次通过 Feign 发出请求时都会自动传递 serviceName 参数。这也是为何在本次定义 RequestOriginParser 接口实现子类时不仅定义了通过参数接收，还定义了头信息接收。

5.3.3 BlockExceptionHandler

BlockException Handler

视频名称 0511_【掌握】BlockExceptionHandler

视频简介 Sentinel 针对所有可能出现的流控处理都定义了专属的异常类型，并提供了统一的 BlockExceptionHandler 处理接口。本视频通过实例讲解异常统一处理的实现，并基于异常产生熔断机制验证此规则的使用。

每当触发 Sentinel 保护规则，实际上都会在系统内部产生 BlockException 异常实例，而后会基于默认的方式进行异常数据的响应。考虑到定制化异常信息显示的需要，Sentinel 内部提供了 BlockExceptionHandler 接口，如图 5-18 所示，开发者可以依据此接口自定义异常数据的响应。

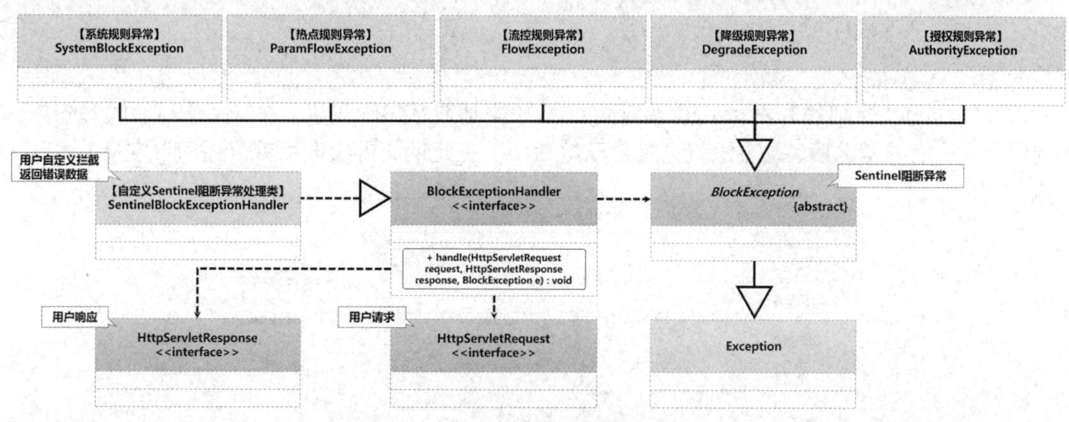

图 5-18 BlockExceptionHandler

只要在系统中配置了 BlockExceptionHandler 接口实例，那么所有产生的与 Sentinel 有关的异常都会通过 handle()方法进行处理。由于此方法提供了 HttpServletResponse 接口实例，因此可以直接在该方法中进行异常数据的响应。下面通过一个具体的实例进行演示。

（1）【provider-dept-*子模块】创建拦截异常处理类。

```
Package com.yootk.provider.action.block;
@Component                                                    // Bean注册
public class SentinelBlockExceptionHandler implements BlockExceptionHandler {
   @Override
   public void handle(HttpServletRequest request, HttpServletResponse response,
               BlockException e) throws Exception {
      Map<String, Object> errors = new HashMap<>();
      errors.put("type", e.getClass().getName());             // 异常类型
      if (e instanceof FlowException) {                       // 是否为流控异常
         errors.put("message", "服务流控");                    // 异常信息
      } else if (e instanceof DegradeException) {             // 是否为降级异常
         errors.put("message", "服务降级");                    // 异常信息
      } else if (e instanceof ParamFlowException) {           // 是否为热点参数异常
         errors.put("message", "热点参数流控");                // 异常信息
      } else if (e instanceof SystemBlockException) {         // 是否为系统拦截异常
         errors.put("message", "系统拦截");                    // 异常信息
      } else if (e instanceof AuthorityException) {           // 是否为授权异常
         errors.put("message", "授权信息错误");                // 异常信息
      } else {                                                // 未知异常
         errors.put("message", "未知错误");                    // 异常信息
      }
      errors.put("path", request.getRequestURI());            // 异常产生路径
      response.setStatus(HttpServletResponse.SC_INTERNAL_SERVER_ERROR); // 状态码
      response.setCharacterEncoding("utf-8");                 // 响应编码
      response.setHeader("Content-Type", MimeTypeUtils.APPLICATION_JSON_VALUE); // MIME
      new ObjectMapper()
            .writeValue(response.getWriter(), errors);  // Jackson数据转换
   }
}
```

（2）【provider-dept-*子模块】修改部门查询方法，使其在特定的参数下产生异常。

```
Package com.yootk.provider.action;
@RestController                                               // REST控制器
@RequestMapping("/provider/dept/*")                           // 父路径
public class DeptAction {
   @Autowired
   private IdeptService deptService;                          // 注入业务接口实例
   @GetMapping("get/{id}")                                    // 子路径
   public Object get(@PathVariable("id") long id) {
      if (id % 2 == 0) {                                      // 查询ID为偶数
         throw new RuntimeException("查询ID不能为偶数！"); // 手动抛出异常
      }
      return this.deptService.get(id);                        // 部门信息加载
   }
}
```

（3）【Sentinel 控制台】为/provider/dept/get 路径添加异常熔断规则，如图 5-19 所示。这样在方法调用产生一次异常之后，此方法就会被自动熔断 2s，在此期间再次调用就会返回自定义的 Sentinel 拦截异常信息。

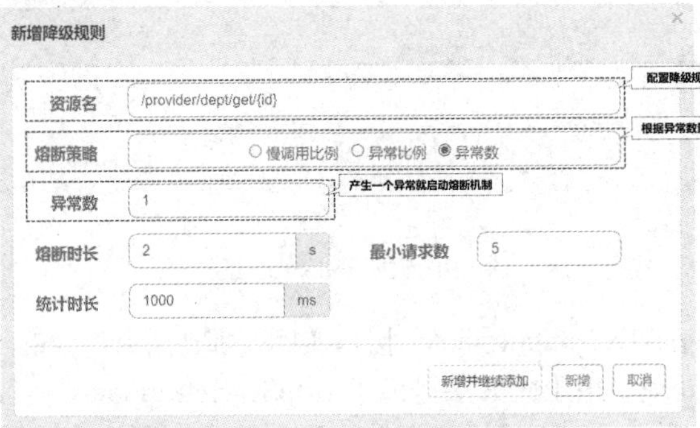

图 5-19　异常熔断

5.3.4 集群流控

视频名称 0512_【掌握】集群流控

视频简介 集群是微服务保持运行高效与稳定的重要架构，Sentinel 在设计中也充分地考虑到了集群应用环境。本视频为读者讲解集群流控的实现机制，并通过具体的 TokenServer 搭建与配置实例分析微服务集群管理的实现。

为了提高某一微服务的处理性能，开发者往往会在设计时进行微服务集群的搭建，这样同一个微服务就会有若干个不同的服务节点，如图 5-20 所示。而在进行 Sentinel 流控配置时就需要考虑到集群环境下的流量设计，例如，现在有一个“dept.provider”微服务集群，为其设置流控规则（QPS 设置为 60），这就意味着整个集群每秒可以处理的请求总量就是 60（不管集群中有多少个微服务节点）。

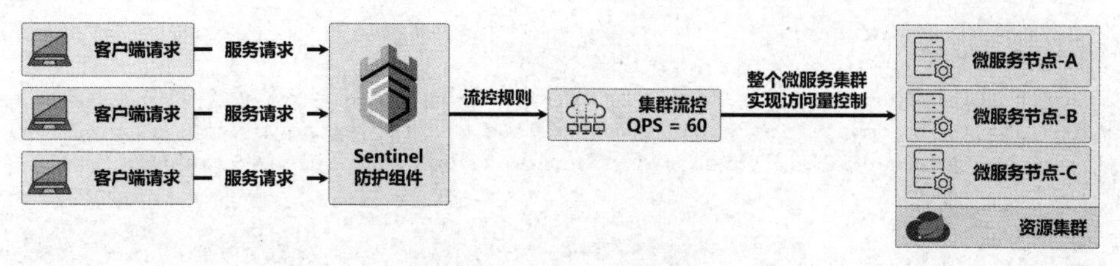

图 5-20 Sentinel 集群规则

如果想在 Sentinel 中实现集群流控，本质上需要提供一个相关访问数据的统计操作。在单一实例的情况下，这个统计操作是在每个实例中实现的。而在集群环境下，就需要提供一个专门的实例（TokenServer）进行数据统计，并且该 TokenServer 要收集所有 TokenClient 发送来的统计信息，而后根据集群流控规则来决定是否允许该请求进行资源访问，如图 5-21 所示。

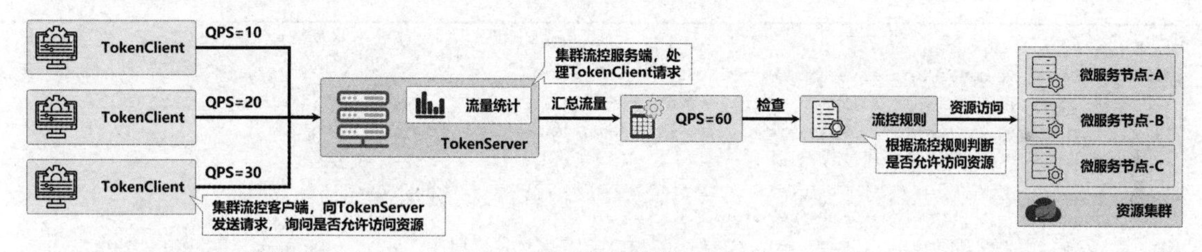

图 5-21 集群流量统计

Sentinel 从 1.4.0 版本开始引入集群流控实现模块，基于 Netty 实现了服务通信。用户在 Spring Cloud 项目模块中引入“spring-cloud-starter-alibaba-sentinel”依赖库后，就会自动引入这些相关模块。通过图 5-22 可以发现，与集群流控有关的模块一共有 3 个，这 3 个模块的具体作用如下。

- sentinel-cluster-common-default：集群流控公共模块，包含公共的接口和实体类。
- sentinel-cluster-server-default：TokenServer 实现模块，基于 Sentinel 核心逻辑进行规则扩展实现。
- sentinel-cluster-client-default：TokenClient 实现模块。

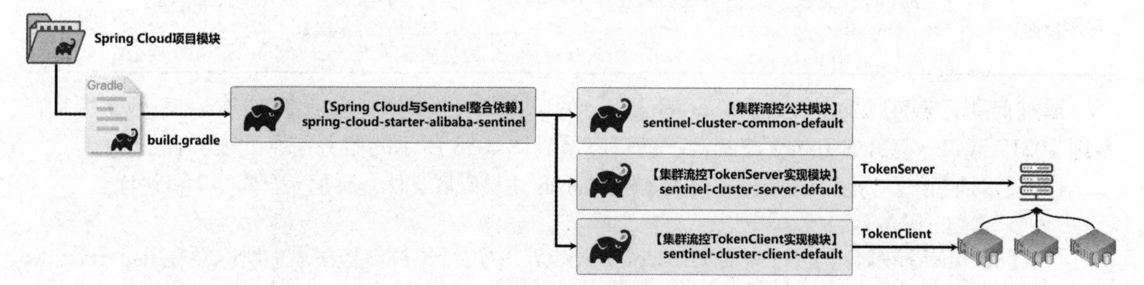

图 5-22 Sentinel 集群流控实现模块

提示：微服务实例可以实现集群流控。

微服务提供端引入“spring-cloud-starter-alibaba-sentinel”依赖库之后就可以直接进行集群流控开发的整合，其中 TokenServer 需要单独开发，而 TokenClient 会直接识别，即此时的微服务提供端会提供两项服务：资源与 TokenClient。

TokenServer 是实现集群流控的关键所在，而在实际的开发中，TokenServer 有嵌入（与一个应用实例部署在一起）与独立部署（单独启动一个 TokenServer 服务节点）两种模式。本书为了便于读者理解，将采用独立部署模式讲解 TokenServer 的开发与部署。下面通过具体的操作步骤介绍如何实现。

（1）【microcloud 项目】创建一个新的子模块，模块名称为 sentinel-token-server，随后修改 build.gradle 配置文件，配置该模块相关的依赖库。

```
project('sentinel-token-server') {
    dependencies {                    // 配置子模块依赖
        implementation('org.springframework.boot:spring-boot-starter-web')
        implementation('com.alibaba.cloud:spring-cloud-starter-alibaba-sentinel'){
            exclude group: 'com.alibaba.csp', module: 'sentinel-cluster-client-default'
        }
    }
}
```

由于“spring-cloud-starter-alibaba-sentinel”依赖库已经包含大部分 Sentinel 相关依赖，因此在本次搭建 TokenServer 时依旧采用此依赖库。为了与 TokenClient 有所区分，本次删除了 sentinel-cluster-client-default 依赖库。

（2）【sentinel-token-server 子模块】创建 TokenServer 启动类。在创建启动类时可以通过系统参数配置 Sentinel 相关连接信息（如果不配置则需要在启动时追加启动参数）。

```
package com.yootk.sentinel;
public class StartTokenServerApplication {
    // -Dcsp.sentinel.dashboard.server=sentinel-server:8888 -Dcsp.sentinel.api.port=8719
    // -Dproject.name=sentinel-token-server -Dcsp.sentinel.log.use.pid=true
    static {    // 使用系统属性代替启动参数
        System.setProperty("csp.sentinel.dashboard.server",
                "sentinel-server:8888");                                // 控制台地址
        System.setProperty("csp.sentinel.api.port", "8719");           // Sentinel端口
        System.setProperty("project.name", "token-server");            // 服务名称
        System.setProperty("csp.sentinel.log.use.pid", "true");        // 设置pid（可选）
    }
    public static void main(String[] args) throws Exception {
        ClusterTokenServer tokenServer = new SentinelDefaultTokenServer();  // 集群Token
        ClusterServerConfigManager.loadGlobalTransportConfig(new ServerTransportConfig()
                .setIdleSeconds(600).setPort(10217));                   // 传输配置
        tokenServer.start();                                            // 服务启动
    }
}
```

程序启动日志	io.netty.handler.logging.LoggingHandler - [id: 0xf989c872] REGISTERED io.netty.handler.logging.LoggingHandler - [id: 0xf989c872] BIND: 0.0.0.0/0.0.0.0:10217 io.netty.handler.logging.LoggingHandler - [id: 0xf989c872, L:/0:0:0:0:0:0:0:0:10217] ACTIVE

通过启动日志可以发现，TokenServer 基于 Netty 框架开发实现，而此时的 TokenServer 绑定在本地 10217 端口。打开 Sentinel 控制台，就可以看到图 5-23 所示的服务信息列表界面。

（3）【本地系统】为便于后续配置修改本机的 hosts 主机配置文件，添加一个新的主机名称。

```
127.0.0.1   sentinel-token-server
```

（4）【Sentinel 控制台】此时需要在“dept.provider”中进行集群流控配置，所以要在 dept.provider 中添加外部 TokenServer，具体配置如图 5-24 所示。

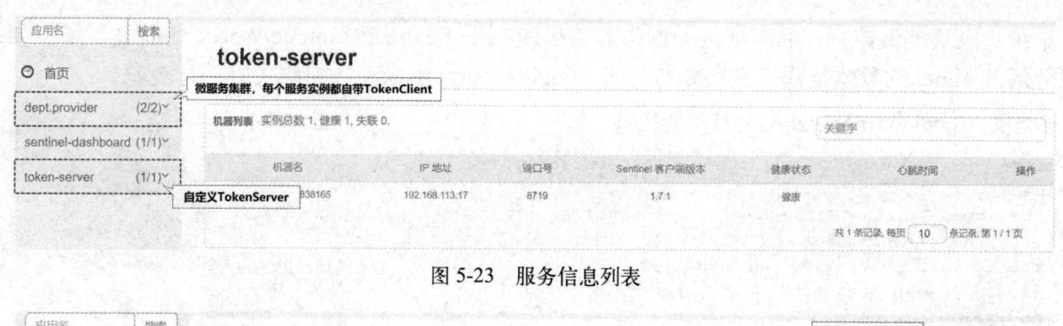

图 5-23　服务信息列表

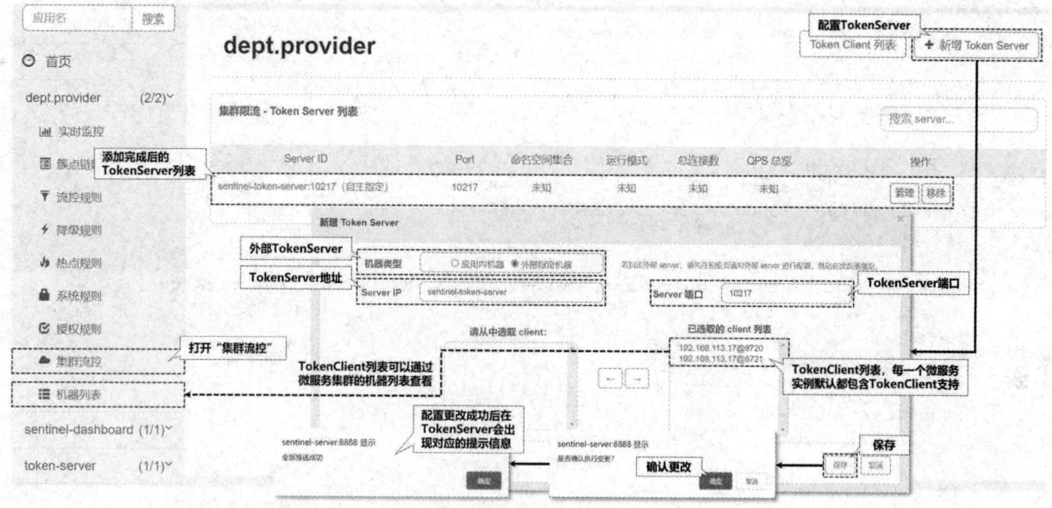

图 5-24　集群流控配置

（5）【Sentinel 控制台】为/provider/dept/list 资源新增流控规则。由于此时有若干个部门微服务的处理节点，因此需要创建集群规则，如图 5-25 所示。

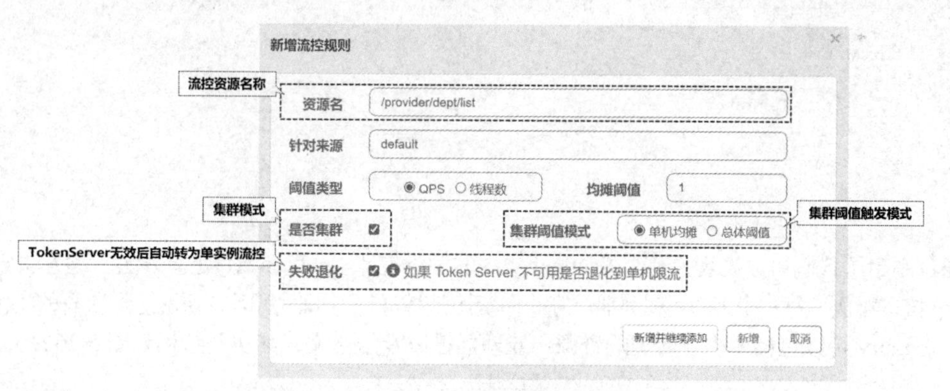

图 5-25　新增流控规则

5.4　Sentinel 实现分析

Sentinel 实现分析

视频名称　0513_【理解】Sentinel 实现分析

视频简介　Sentinel 采用注解与切面的形式实现了访问资源的流控统计，并为所有组件的方便接入提供了 Entry 类型。本视频为读者分析 Entry 类型的主要作用，并基于 Sentinel 提供的 Demo 代码讲解 Entry 与流控操作之间的关联。

微服务在与 Sentinel 整合时，只需要在 REST 接口中通过@SentinelResource 注解实现 Sentinel

流量统计以及流量控制，而实现这一操作的关键就在于 SentinelResourceAspect 切面类。所以如果想理解 Sentinel 的具体操作，就需要打开 SentinelResourceAspect 类的源代码进行观察。

范例：SentinelResourceAspect 类源代码

```
package com.alibaba.csp.sentinel.annotation.aspectj;
@Aspect                                                   // 切面实现注解
public class SentinelResourceAspect extends AbstractSentinelAspectSupport {
   @Pointcut("@annotation(com.alibaba.csp.sentinel.annotation.SentinelResource) ")
   public void sentinelResourceAnnotationPointcut() {}        // 扫描SentinelResource注解
   @Around("sentinelResourceAnnotationPointcut()")            // 环绕通知
   public Object invokeResourceWithSentinel(ProceedingJoinPoint pjp) throws Throwable {
      Method originMethod = resolveMethod(pjp);               // 解析原生方法
      SentinelResource annotation = originMethod.getAnnotation(SentinelResource.class);
      if (annotation == null) {                               // 注解不存在
         throw new IllegalStateException("Wrong state for SentinelResource annotation");
      }
      String resourceName = getResourceName(annotation.value(), originMethod);
      EntryType entryType = annotation.entryType();
      int resourceType = annotation.resourceType();
      Entry entry = null;                                     // 流控凭证
      try {
         entry = SphU.entry(resourceName, resourceType, entryType, pjp.getArgs());
         Object result = pjp.proceed();                       // 调用目标方法
         return result;
      } catch (BlockException ex) {                           // 产生阻断异常
         return handleBlockException(pjp, annotation, ex);
      } catch (Throwable ex) {
         Class<? Extends Throwable>[] exceptionsToIgnore =
                        annotation.exceptionsToIgnore();
         if (exceptionsToIgnore.length > 0 && exceptionBelongsTo(
               ex, exceptionsToIgnore)) {                     // 配置忽略异常
            throw ex;
         }
         if (exceptionBelongsTo(ex, annotation.exceptionsToTrace())) {
            traceException(ex);
            return handleFallback(pjp, annotation, ex);       // Fallback处理
         }
         throw ex;
      } finally {
         if (entry != null) {
            entry.exit(1, pjp.getArgs());
         }
      }
   }
}
```

通过观察源代码可以发现，整个 Sentinel 实现的关键在于 SphU.entry()处理方法。该方法会创建一个 Entry 接口实例，每一个 Entry 实例都对应一个资源流控凭证，同时可以方便地适配所有的主流框架。而每一个 Entry 可以包括所需要管理的资源、处理插槽以及上下文，其实现结构如图 5-26 所示。

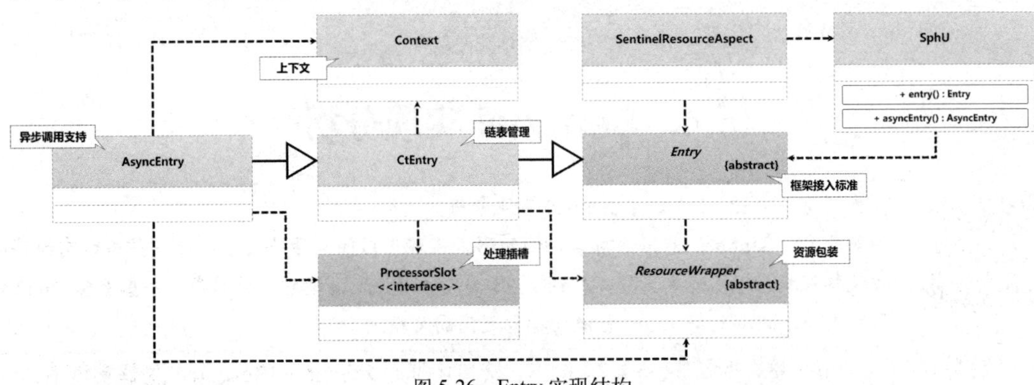

图 5-26　Entry 实现结构

所有的 Entry 可以直接通过 SphU 类所提供的 SphU.entry()或 SphU.asyncEntry()方法进行创建。下面直接通过此类实现一个自定义业务资源的流控操作管理。

范例：自定义流控管理

```
package com.yootk.test;
public class TestEntry {
    public static void main(String[] args) {
        initFlowRules();                                   // 初始化规则
        while (true) {                                     // 循环调用
            Entry entry = null;                            // 定义Entry对象
            try {
                entry = SphU.entry("YootkMessage");        // 创建Entry
                System.out.println("【业务处理】沐言科技：www.yootk.com");
            } catch (BlockException e) {                   // 流控处理
                System.out.println("【ERROR】访问量超标，流量阻断处理。");
            } finally {
                if (entry != null) {                       // Entry不为空
                    entry.exit();                          // 释放Entry
                }
            }
        }
    }
    private static void initFlowRules(){                   // 初始化流控规则
        List<FlowRule> rules = new ArrayList<>();          // 规则集合
        FlowRule rule = new FlowRule();                    // 流控规则
        rule.setResource("YootkMessage");                  // 定义资源名称
        rule.setGrade(RuleConstant.FLOW_GRADE_QPS);        // 根据QPS流控
        rule.setCount(10);                                 // 每秒允许10个请求
        rules.add(rule);                                   // 规则保存
        FlowRuleManager.loadRules(rules);                  // 规则配置
    }
}
```

此时采用手动的方式在程序中创建了一个实体，而后该实体的名称与流控规则中的流控名称相匹配，这样在进行访问时如果发现超过了配置的流控规，则会直接产生 BlockException 阻断业务处理。需要注意的是，以上操作仅仅演示了一个 Entry 基本使用形式，而要想充分理解 Entry 具体操作，还需要清楚地掌握 ResourceWrapper、ProcessorSlot、Node、Context 相关概念，下面对这些概念进行说明。

5.4.1 ResourceWrapper

ResourceWrapper

视频名称 0514_【理解】ResourceWrapper

视频简介 Sentinel 是根据资源实现保护规则配置的。本视频为读者分析 Sentinel 资源名称的获取以及 Entry 与 ResourceWrapper 之间的关联操作。

资源（Resource）是 Sentinel 的重要组成部分，Sentinel 通过资源来保护具体的业务接口或其他后台服务，管理者只需要根据资源为其配置具体的流控规则，而后所有的操作便可以由 Sentinel 负责管理。Sentinel 为了避免对原始代码的过多侵入，直接通过@SentinelResource 注解进行资源标注，而在 Sentinel 内部每一个资源都使用 ResourceWrapper 对象包装。图 5-27 展示了 ResourceWrapper 类的继承结构。

ResourceWrapper 是一个抽象类，此抽象类有两个子类：StringResourceWrapper（对字符串进行包装）、MethodResourceWrapper（对方法调用进行包装）。在 SentinelResourceAspect 类中通过父类 AbstractSentinelAspectSupport 定义的 getResourceName()方法并根据@SentinelResource 注解中的 value 属性可实现资源的配置。getResourceName()方法的源代码如下。

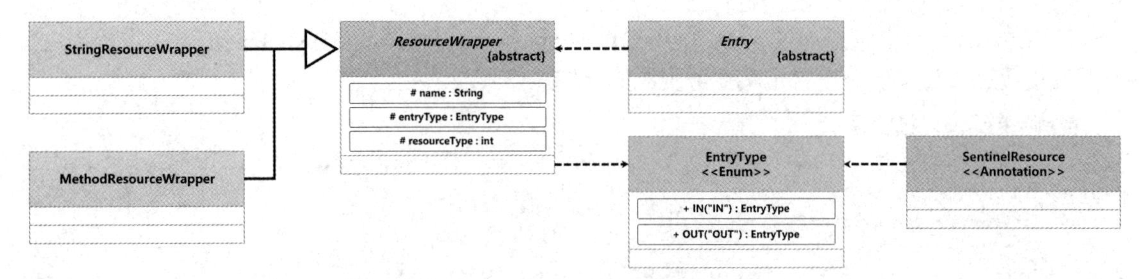

图 5-27　ResourceWrapper 类的继承结构

范例：getResourceName()源代码

```
package com.alibaba.csp.sentinel.annotation.aspectj;
public abstract class AbstractSentinelAspectSupport {
    protected String getResourceName(String resourceName, /*@NonNull*/ Method method) {
        if (StringUtil.isNotBlank(resourceName)) {       // 字符串名称不为空
            return resourceName;                         // 使用当前字符串
        }
        return MethodUtil.resolveMethodName(method);     // 解析方法名称
    }
}
```

在进行资源配置时首先会通过@SentinelResource 注解的 value 属性来进行配置。如果发现没有配置内容，则会通过 MethodUtil 类根据方法名称进行资源名称的解析，最终解析出来的资源名称会在 CtSph 子类中被封装为 ResourceWrapper 类实例，实现结构如图 5-28 所示。

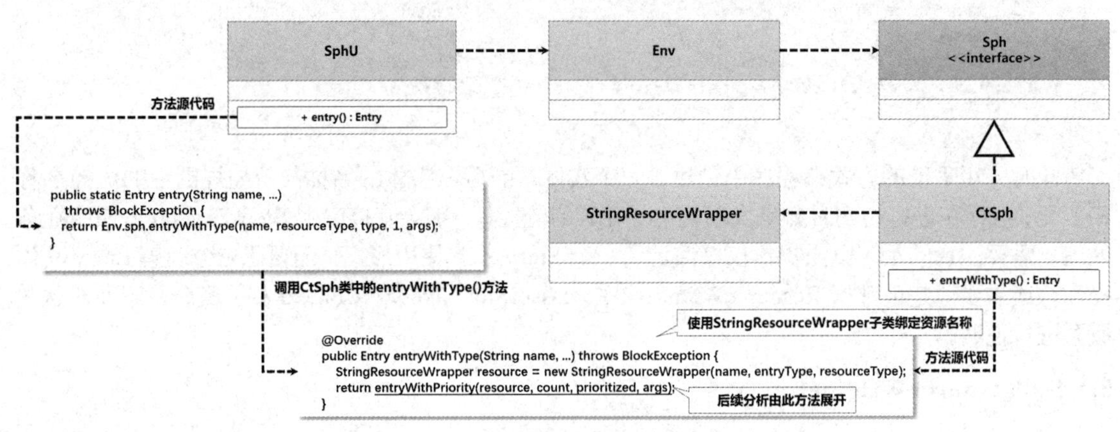

图 5-28　资源配置

5.4.2　ProcessorSlot

视频名称　0515_【理解】ProcessorSlot

视频简介　Sentinel 基于插槽与插槽链的结构实现了数据统计以及流控处理。本视频将通过 SlotChainBuilder 分析 Sentinel 的处理流程。

ProcessorSlot（插槽）是 Sentinel 实现的核心所在，在 Sentinel 进行工作处理时，所有的操作流程都是围绕着 ProcessorSlotChain（插槽链）展开的。每一个插槽都有各自的职责，它们通过一定的编排顺序来达到服务流控与降级的目的。在 Sentinel 中插槽与插槽链的实现结构如图 5-29 所示。

由于某些插槽需要依赖其他插槽才能够正常工作，因此在默认情况下插槽之间的顺序是固定的，这一点可以通过 SlotChainBuilder 接口中所提供的 DefaultSlotChainBuilder 子类观察到。

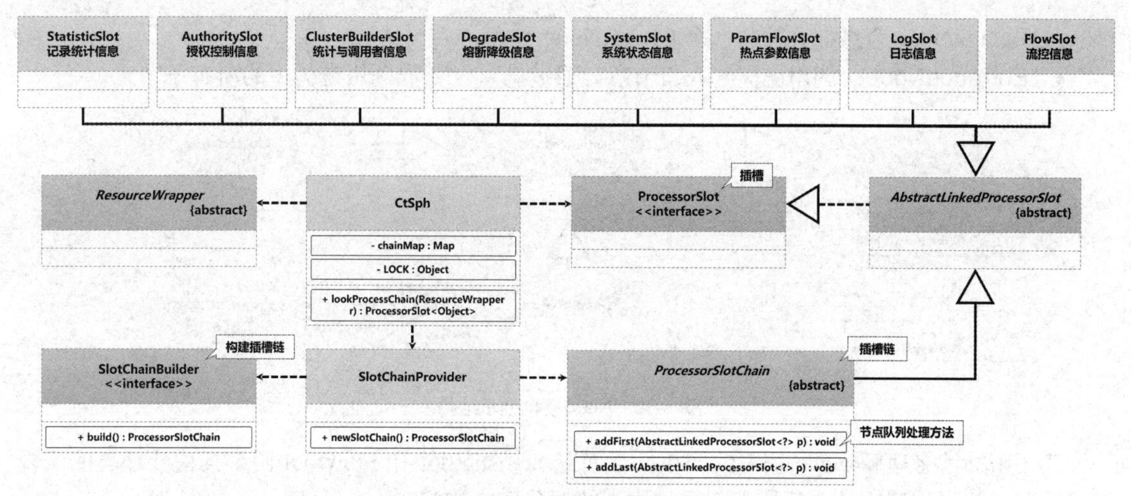

图 5-29 插槽与插槽链

范例：DefaultSlotChainBuilder 子类源代码

```
package com.alibaba.csp.sentinel.slots;
public class DefaultSlotChainBuilder implements SlotChainBuilder {
    @Override
    public ProcessorSlotChain build() {
        ProcessorSlotChain chain = new DefaultProcessorSlotChain();
        chain.addLast(new NodeSelectorSlot());       // 收集资源路径
        chain.addLast(new ClusterBuilderSlot());     // 存储资源的统计信息与调用者信息
        chain.addLast(new LogSlot());                // 日志处理
        chain.addLast(new StatisticSlot());          // 核心插槽，实现各种维度的实时监控统计
        // 以下操作为流控保护处理插槽，将依据之前的统计结果判断是否启用
        chain.addLast(new AuthoritySlot());          // 授权处理插槽
        chain.addLast(new SystemSlot());             // 系统规则插槽
        chain.addLast(new FlowSlot());               // 流控规则插槽
        chain.addLast(new DegradeSlot());            // 降级规则插槽
        return chain;
    }
}
```

通过 DefaultSlotChainBuilder.build 的源代码可以发现，一个插槽链在配置时会设置一系列的插槽，而每一个插槽都是 ProcessorSlot 接口的子类。同时在一个资源中只允许存在一个 ProcessorSlotChain 实例，所有的资源访问都必须依据插槽链中的插槽顺序进行处理。

5.4.3 Node

Node

视频名称 0516_【理解】Node

视频简介 Node 是 Sentinel 实现数据统计的标准，并且会与 Context、Entry 整合在一起应用。本视频为读者分析 Node 接口的继承结构以及数据统计子类的使用。

Sentinel 中的所有统计数据操作全部由 Node 来保存。正是因为有了这些具体的节点操作数据，在 Sentinel 中才可以实现流控、服务降级等一系列操作。Node 接口继承结构如图 5-30 所示，这些类的具体作用如下所示。

- StatisticNode：执行资源统计操作节点。
- ClusterNode：保存了某一个资源总体的运行时统计信息。不同上下文的相同资源会共享同一个 ClusterNode。
- DefaultNode：持有指定上下文中的统计信息。该节点下可能会有多个子节点，同时每个

DefaultNode 节点都会关联一个 ClusterNode 节点。
- EntranceNode：调用链树的入口节点。通过其可以获取调用链树中的所有子节点。

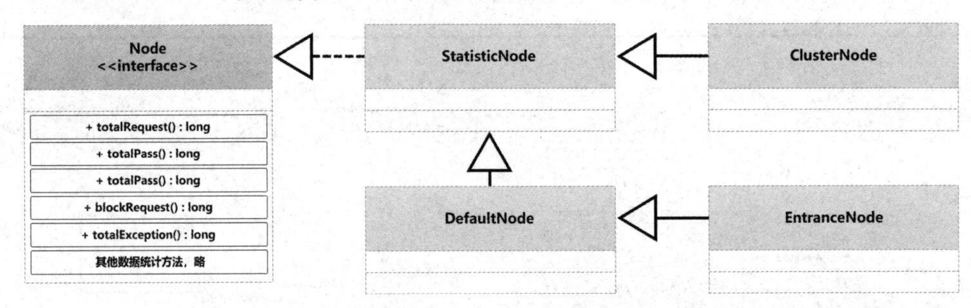

图 5-30 Node 接口继承结构

为了帮助读者理解 Node 接口的作用，下面以 StatisticSlot 和 FlowSlot 两个插槽处理类的实现进行说明。首先打开这两个插槽入口操作方法的源代码来观察。

范例：StatisticSlot.entry()源代码

```
package com.alibaba.csp.sentinel.slots.statistic;
public class StatisticSlot extends AbstractLinkedProcessorSlot<DefaultNode> {
   @Override
   public void entry(Context context, ResourceWrapper resourceWrapper,
        DefaultNode node, int count,  oolean prioritized,
        Object… args) throws Throwable {
      try {                                                       // 做一些相关的检查
         fireEntry(context, resourceWrapper, node, count, prioritized, args);
         node.increaseThreadNum();                                // 请求线程数量增长
         node.addPassRequest(count);                              // 追加通过请求数量
         if (context.getCurEntry().getOriginNode() != null) {     // 获取来源节点
            context.getCurEntry().getOriginNode().increaseThreadNum();
            context.getCurEntry().getOriginNode().addPassRequest(count);
         }
         if (resourceWrapper.getEntryType() == EntryType.IN) {    // 判断进入类型
            Constants.ENTRY_NODE.increaseThreadNum();
            Constants.ENTRY_NODE.addPassRequest(count);
         }
         for (ProcessorSlotEntryCallback<DefaultNode> handler :
               StatisticSlotCallbackRegistry.getEntryCallbacks()) {  // 回调处理
            handler.onPass(context, resourceWrapper, node, count, args);
         }
      } catch (PriorityWaitException ex) { …
      } catch (BlockException e) {                                // 阻断异常
         context.getCurEntry().setError(e);                       // 设置错误信息
         node.increaseBlockQps(count);                            // 拦截数自增
         if (context.getCurEntry().getOriginNode() != null) {
            context.getCurEntry().getOriginNode().increaseBlockQps(count);
         }
         if (resourceWrapper.getEntryType() == EntryType.IN) {
            Constants.ENTRY_NODE.increaseBlockQps(count);
         }
         for (ProcessorSlotEntryCallback<DefaultNode> handler :
               StatisticSlotCallbackRegistry.getEntryCallbacks()) {
            handler.onBlocked(e, context, resourceWrapper, node, count, args);
         }
         throw e;
      } catch (Throwable e) { … }
   }
}
```

可以发现，在插槽入口处程序会对节点的访问量进行统计，当出现异常时也会产生拦截数自增，而这些统计信息全部都记录在 DefaultNode 对象实例之中。

StatisticSlot 仅仅实现了访问量的统计，而真正的流控处理操作则肯定要通过 FlowSlot 插槽类来完成。而且在每次流控时实际上都会存在若干规则，这些规则主要通过 Rule 接口进行配置，实现结构如图 5-31 所示。

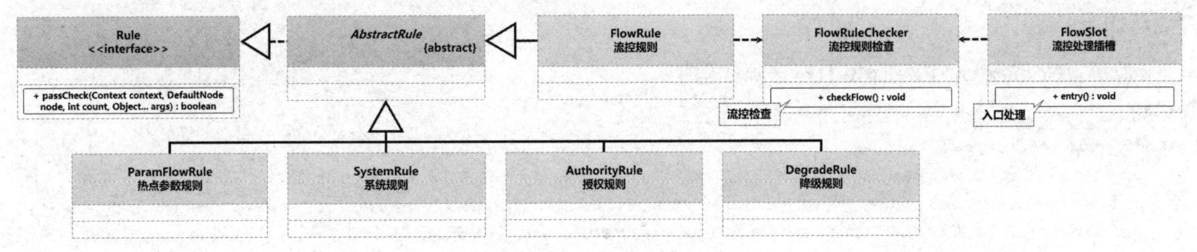

图 5-31　Sentinel 保护规则

范例：FlowSlot.entry()源代码

```
package com.alibaba.csp.sentinel.slots.block.flow;
public class FlowSlot extends AbstractLinkedProcessorSlot<DefaultNode> {
   private final FlowRuleChecker checker;
   public FlowSlot() {
      this(new FlowRuleChecker());
   }
   FlowSlot(FlowRuleChecker checker) {
      AssertUtil.notNull(checker, "flow checker should not be null");
      this.checker = checker;
   }
   @Override
   public void entry(Context context, ResourceWrapper resourceWrapper,
         DefaultNode node, int count,  oolean prioritized,
         Object… args) throws Throwable {
      checkFlow(resourceWrapper, context, node, count, prioritized);
      fireEntry(context, resourceWrapper, node, count, prioritized, args);
   }
   void checkFlow(ResourceWrapper resource, Context context, DefaultNode node,
         int count,  oolean prioritized) throws BlockException { // 流控检查
      checker.checkFlow(ruleProvider, resource, context, node, count, prioritized);
}
```

FlowSlot.entry()方法在实现时首先会调用其内部提供的 FlowSlot()方法，而后该方法会通过 FlowRuleChecker 类所提供的 checkFlow()方法对当前的资源进行流控判断，并根据判断的结果来决定是否要执行插槽入口处理。

5.4.4　Context

Context

视频名称　0517_【理解】Context

视频简介　Context 维护了整个调用链的元数据内容，是 Entry 创建与使用过程中必不可少的概念。本视频为读者讲解 Context 概念，并分析其对应的核心源代码。

每当用户发出请求进行资源访问时，都会有一个 Context 实例被创建，用于保存整个调用链的元数据内容。首先打开 Context 类中定义的成员属性来观察其组成。

范例：Context 类中成员属性

```
package com.alibaba.csp.sentinel.context;
public class Context {
   private final String name;                    // Context名称
   private DefaultNode entranceNode;             // 调用链入口节点
   private Entry curEntry;                       // 调用链的当前Entry
```

```
    private String origin = "";                     // 调用源信息
    private final boolean async;                    // 异步处理标记
}
```

通过 Context 类中的属性定义可以发现，其所保存的仅仅是当前的节点以及资源 Entry 的信息。而 Context 对象的实例则是通过 ContextUtil 类中所提供的 trueEnter()方法进行创建的。此方法的源代码如下。

范例：ContextUtil.trueEnter()方法源代码

```
package com.alibaba.csp.sentinel.context;
public class ContextUtil {
    private static ThreadLocal<Context> contextHolder = new ThreadLocal<>();
    private static volatile Map<String, DefaultNode> contextNameNodeMap = new HashMap<>();
    private static final ReentrantLock LOCK = new ReentrantLock();
    private static final Context NULL_CONTEXT = new NullContext();
    protected static Context trueEnter(String name, String origin) {
        Context context = contextHolder.get();                         // 获取当前Context
        if (context == null) {                                         // Context实例为空
            Map<String, DefaultNode> localCacheNameMap = contextNameNodeMap;
            DefaultNode node = localCacheNameMap.get(name);            // 获取节点
            if (node == null) {                                        // 操作节点为空
                if (localCacheNameMap.size() > Constants.MAX_CONTEXT_NAME_SIZE) {
                    setNullContext();
                    return NULL_CONTEXT;                               // 返回NullContext子类实例
                } else {
                    try {
                        LOCK.lock();                                   // 排他锁
                        node = contextNameNodeMap.get(name);
                        if (node == null) {                            // 节点为空
                            if (contextNameNodeMap.size() >
                                    Constants.MAX_CONTEXT_NAME_SIZE) { // 大于保存上限
                                setNullContext();
                                return NULL_CONTEXT;
                            } else {                                   // 节点不为空
                                node = new EntranceNode(new StringResourceWrapper(name,
                                        EntryType.IN), null);  v       // 创建新的节点
                                Constants.ROOT.addChild(node);         // 保存节点
                                Map<String, DefaultNode> newMap = new HashMap<>(
                                        contextNameNodeMap.size() + 1);  // 创建新的Map集合
                                newMap.putAll(contextNameNodeMap);     // 保存其他节点
                                newMap.put(name, node);                // 保存新的节点
                                contextNameNodeMap = newMap;           // Map引用变更
                            }
                        }
                    } finally {
                        LOCK.unlock();                                 // 解锁
                    }
                }
            }
            context = new Context(node, name);                         // 实例化Context
            context.setOrigin(origin);                                 // 设置来源
            contextHolder.set(context);                                // 保存Context供下次使用
        }
        return context;
    }
}
```

通过该源代码可以发现，每一个客户端的请求线程都只会有一个 Context 对象实例，Context 会完整记录每一个跨资源的链路访问统计信息。Context 类结构如图 5-32 所示。

Context 在创建 Entry 对象实例（SphU.entry()方法）的时候生成和获取，在执行 Entry.exit()方法的时候清除。由于 Context 会保存大量的链路节点数据，因此在业务处理完成并且当前操作的 Entry 实例不为空时，一定要调用 exit()方法，否则会有内存泄漏的风险。

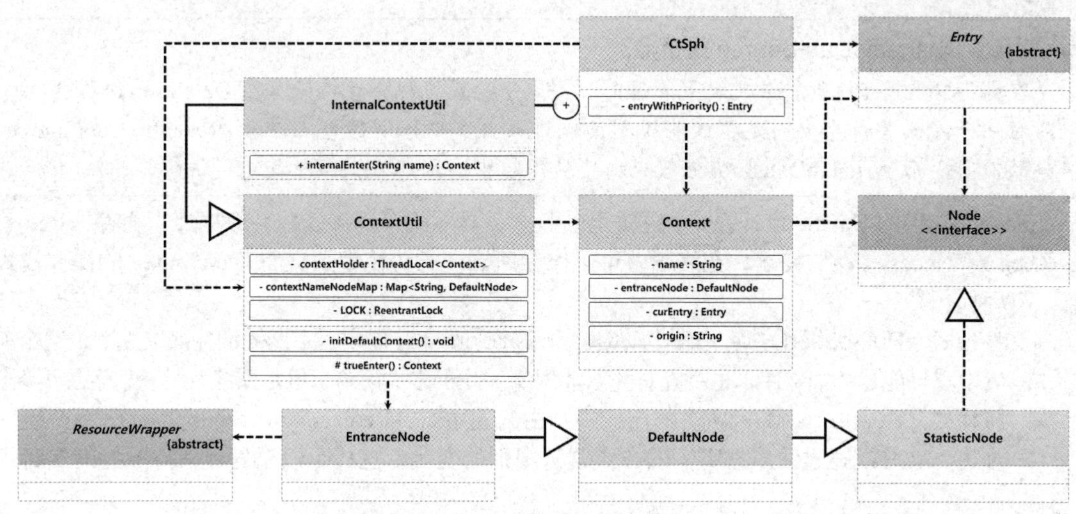

图 5-32 Context 类结构

5.5 配置规则持久化

Sentinel 规则持久化

视频名称 0518_【理解】Sentinel 规则持久化

视频简介 Sentinel 为了便于资源的维护提供了资源持久化的处理方案。本视频为读者分析默认 Sentinel 内存配置所带来的问题，并介绍 Sentinel 中的数据持久化实现方案。

使用 Sentinel 组件进行微服务流控保护时，往往先进行微服务的注册，而后管理员再通过 Sentinel 控制台进行流控规则的配置。然而在默认情况下 Sentinel 中所有的配置数据全部保存在内存之中，如图 5-33 所示。这样一旦重启相关的服务，管理员所配置的相关流控规则就会消失，对于资源防护管理会带来很大的问题。

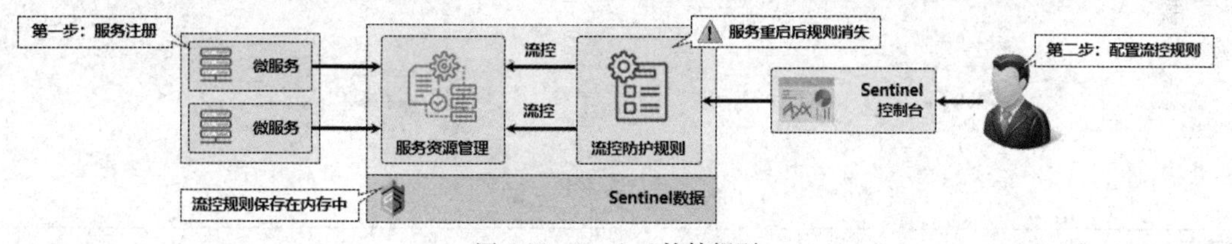

图 5-33 Sentinel 流控规则

Sentinel 为了便于资源流控规则的持久化管理，专门提供了 ReadableDataSource（配置读取）与 WritableDataSource（配置写入）两个操作接口。利用这两个接口可以向指定的存储设备实现规则的读写处理，结构如图 5-34 所示。

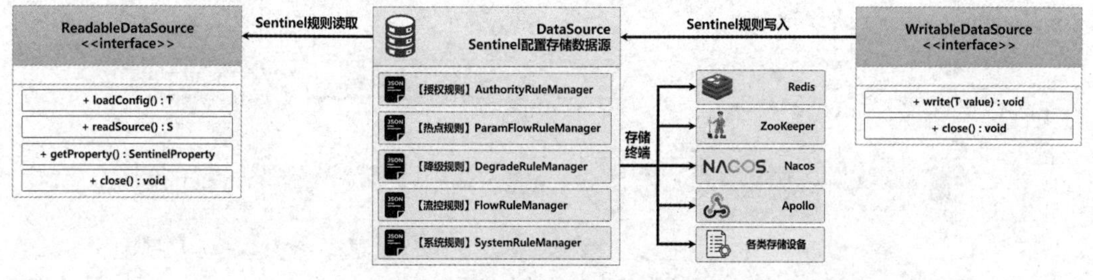

图 5-34 Sentinel 规则持久化管理接口

提示：ReadableDataSource 是重点。

在进行 Sentinel 规则持久化配置管理时，大部分配置规则会直接保存在 DataSource 之中，所以对程序开发而言最重要的是将这些规则读取过来。本书讲解的重点在于 ReadableDataSource 接口的使用，而 WritableDataSource 接口由于使用较少，因此本书不对其进行讲解。

Sentinel 提供的 DataSource 是一个逻辑上的概念，具体的存储涉及关系数据库、文件、Redis、ZooKeeper、Nacos 等存储终端，在终端中可以保存所需要的流控规则。而 DataSource 的操作模式有以下两种。

- **拉模式**（Pull-based）：客户端主动向某个 DataSource 存储中心定期轮询并读取规则，这个配置中心可能是一个文件，也可能是关系数据库。此种方式简单，但是无法及时获取配置更新。
- **推模式**（Push-based）：所有的流控规则由配置中心（Redis、ZooKeeper、Nacos 等）统一推送，客户端通过注册监听器的方式监听规则的变化，这样可以更好地保持配置的实时性和一致性。

推模式在实现数据更新时，主要依靠的是 PropertyListener 属性监听接口。该接口中定义了配置更新的处理方法，并且流控规则管理类都提供该接口的实现子类，如图 5-35 所示。这样每当指定的规则配置项发生改变时，程序都可以及时获取最新的配置项。推模式需要特定的数据终端支持，考虑到服务整合的稳定性与产品的配套性，所以在介绍规则持久化存储时本书将以 Nacos 作为存储终端。

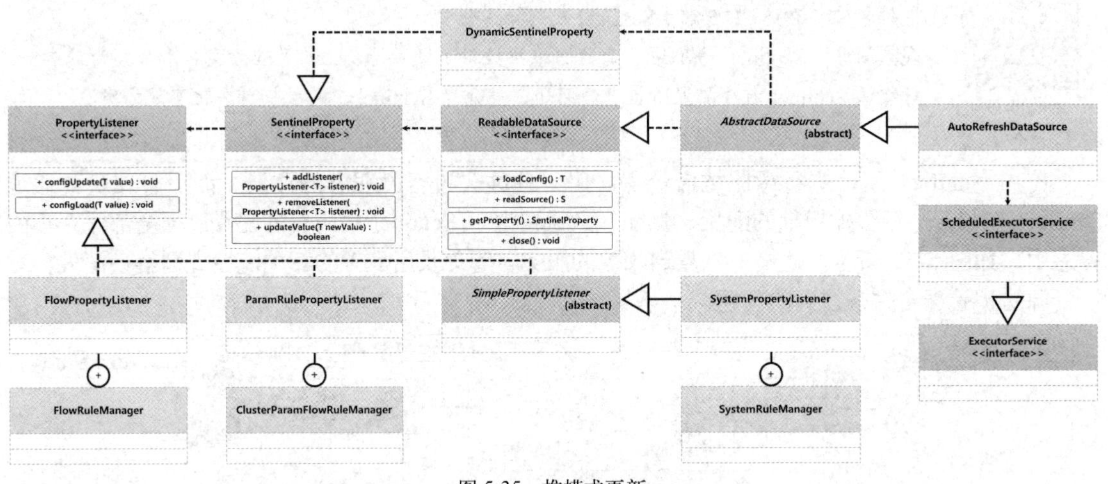

图 5-35　推模式更新

注意：Sentinel 暂不支持 Nacos 认证处理。

虽然 Sentinel 与 Nacos 都属于阿里巴巴公司提供的开源组件，但是由于 Nacos 只是工作于内网环境，因此在 Sentinel 基于 Nacos 持久化规则配置时，如果 Nacos 开启了安全认证，则程序将无法正常连接。本次的讲解中将启用一个没有安全认证的 Nacos 服务（主机名称为 sentinel-nacos-server）实现配置规则保存，如图 5-36 所示。

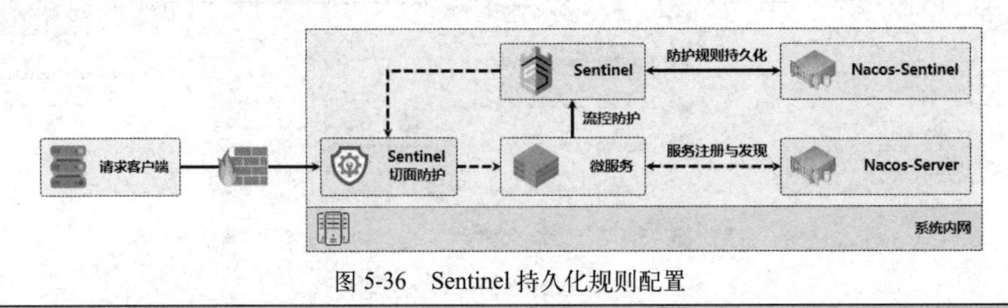

图 5-36　Sentinel 持久化规则配置

如果开发者现在所使用的 Nacos 本身没有开启安全认证，那么可以在 Nacos 中实现持久化规则的配置处理。另外要提醒读者的是，之所以首选 Nacos 作为持久化保存终端，是因为其直接提供了管理配置界面，这样就避免了服务的重复开发。

5.5.1 流控规则持久化

流控规则持久化

视频名称 0519_【掌握】流控规则持久化

视频简介 流控规则是 Sentinel 最核心也最为常用的一种保护机制，而为了服务流控管理的便利，需要对其进行持久化配置。本视频将基于一个独立的 Nacos 服务介绍如何实现流控规则的配置以及读取处理。

持久化 Sentinel 流控规则，可以直接基于 Nacos 配置项进行管理。开发者只要在微服务项目中引入相关的 DataSource 配置依赖，而后通过 application.yml 配置好所要使用的 Nacos 服务器的地址、领域模型以及数据 ID（配置项），即可实现配置的读取与更新监听，如图 5-37 所示。

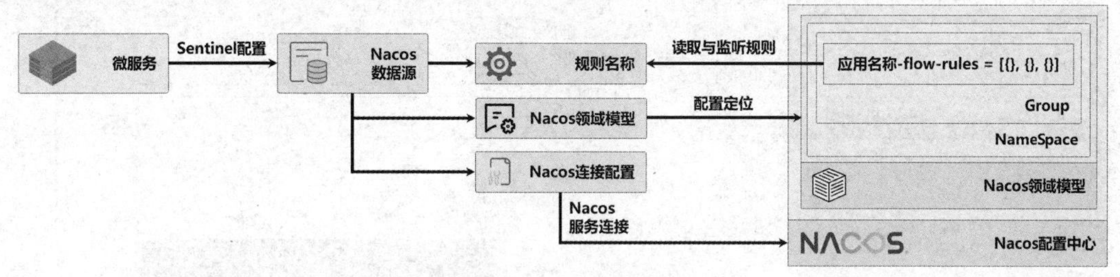

图 5-37 持久化流控规则

本次开发需要使用若干微服务、两个 Nacos 服务器以及 Sentinel。为了便于读者理解，表 5-3 给出了本次开发所使用的全部主机信息，其中 sentinel-nacos-server 为新增服务主机，在该主机中会同时运行 Nacos 和对应的 MySQL 存储进程。具体的程序开发步骤如下。

表 5-3 集群主机配置

序号	主机名称	IP 地址	服务进程	描述
1	sentinel-server	192.168.190.168	Sentinel	Sentinel 防护组件
2	sentinel-nacos-server	192.168.190.169	Nacos、MySQL	持久化流控规则存储、无安全认证
3	nacos-server	192.168.190.158	Nacos	微服务注册与发现服务
4	nacos-mysql	192.168.190.159	MySQL	提供 Nacos-Server 配置存储
5	provider-dept-8001	127.0.0.1	Spring Cloud、MySQL	微服务提供者与对应数据库
6	provider-dept-8002	127.0.0.1	Spring Cloud、MySQL	微服务提供者与对应数据库
7	provider-dept-8003	127.0.0.1	Spring Cloud、MySQL	微服务提供者与对应数据库
8	consumer-springboot-80	127.0.0.1	Spring Cloud	微服务消费者

（1）【SentinelNacos 控制台】要想实现配置项的存储，则需要在 Nacos 中创建相关的领域模型。本次将创建一个名称为“sentinel”的命名空间，用于保存全部 Sentinel 配置项，如图 5-38 所示。

（2）【SentinelNacos 控制台】在“sentinel”命名空间下创建一个新的配置项，配置项的名称为“dept.provider-flow-rules”（其中“dept.provider”是微服务应用名称，后面为规则类型），分组名称为“SENTINEL_GROUP”，如图 5-39 所示。

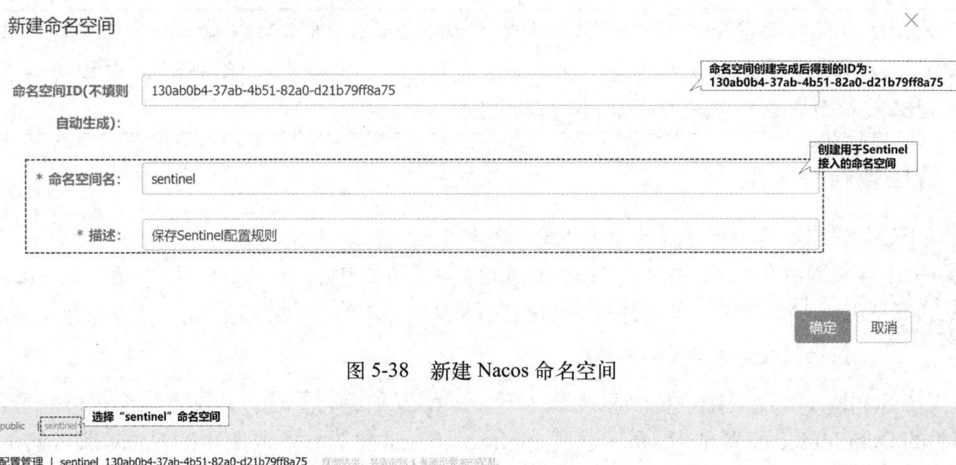

图 5-38　新建 Nacos 命名空间

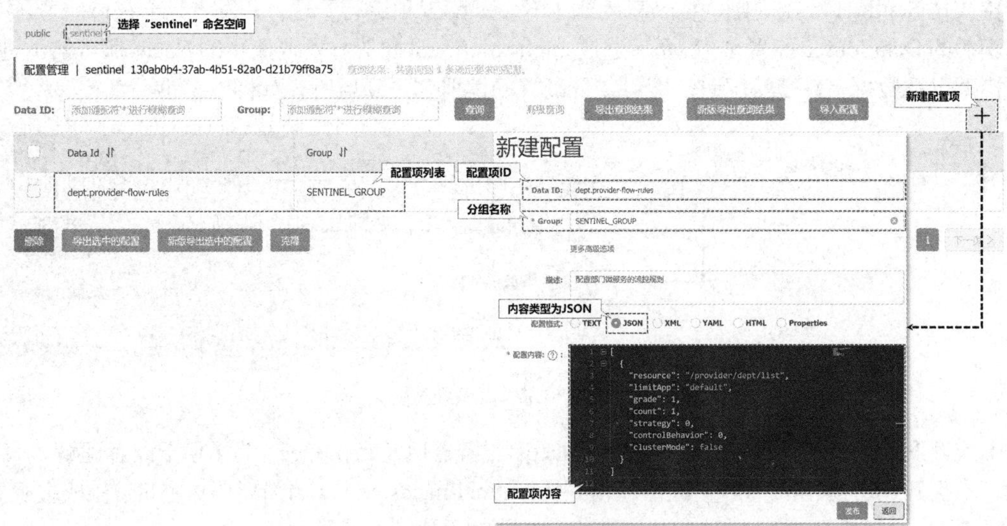

图 5-39　创建新的配置项

范例：为部门微服务定义流控规则

```
[                                             // 数组定义，此处可以配置若干个流控规则
  {                                           // 流控配置项
    "resource": "/provider/dept/list",        // 限制资源名称
    "limitApp": "default",                    // 来源应用
    "grade": 1,                               // 阈值类型（0：线程数。1：QPS）
    "count": 1,                               // 单机阈值定义
    "strategy": 0,                            // 流控模式（0：直接。1：关联。2：链路）
    "controlBehavior": 0,                     // 流控效果（0：快速失败。1：WarmUp。2：队列）
    "clusterMode": false                      // 采用非集群模式
  }]
```

（3）【microcloud 项目】修改 build.gradle 配置文件，为“provider-dept-*”添加 Nacos 数据源依赖。

```
Project('provider-dept-8003') {               // 子模块
    dependencies {                            // 重复依赖，略
        implementation('com.alibaba.csp:sentinel-datasource-nacos:1.8.1')
    }
}
```

（4）【provider-dept-*子模块】修改 application.yml 配置文件，在 spring.cloud.sentinel 配置项中添加 datasource 子项。

```
Spring:                                       # Spring配置
  application:                                # 应用配置
    name: dept.provider                       # 应用名称
  cloud:                                      # Spring Cloud配置
    sentinel:                                 # Sentinel配置
```

```
    transport:                                                # 通信配置
      port: 8719                                              # Sentinel通信端口
      dashboard: sentinel-server:8888                         # Sentinel控制面板
    eager: true
    datasource:                                               # 数据源配置
      flow-datasource:                                        # 自定义数据源名称
        nacos:                                                # Nacos存储
          server-addr: sentinel-nacos-server:8848             # 服务地址
          data-id: ${spring.application.name}-flow-rules      # 流控规则
          group-id: SENTINEL_GROUP                            # 分组ID
          namespace: 130ab0b4-37ab-4b51-82a0-d21b79ff8a75     # 命名空间ID
          data-type: json                                     # 配置类型
          rule_type: flow                                     # 规则类型
```

本程序在已有的 Sentinel 服务的配置基础之上定义了 Nacos 数据源，采用了应用名称作为配置的标记，并定义了与 Nacos 配置项相同的领域模型，这样在微服务启动时就会自动加载指定的流控规则，实现流控保护。

提示：Sentinel 数据源配置项。

在 Spring Cloud 项目之中，需要通过 spring.cloud.sentinel.datasource 配置项来定义持久化规则要使用的数据源信息，而此配置项对应的 SentinelProperties 配置类中是一个 Map 集合。该 Map 集合的 Key 为一个字符串名称，而 Value 对应的类型为 DataSourcePropertiesConfiguration。在这个类中可对 Sentinel 支持的所有数据源进行配置处理，这一点可以通过该配置类的源代码观察到。

范例：DataSourcePropertiesConfiguration 类中的成员属性

```
package com.alibaba.cloud.sentinel.datasource.config;
public class DataSourcePropertiesConfiguration {
  private FileDataSourceProperties file;          // 文件存储配置项
  private NacosDataSourceProperties nacos;        // Nacos存储配置项
  private ZookeeperDataSourceProperties zk;       // ZooKeeper存储配置项
  private ApolloDataSourceProperties  pollo;      // Apollo存储配置项
  private RedisDataSourceProperties redis;        // Redis存储配置项
  private ConsulDataSourceProperties consul;      // Consul存储配置项
}
```

此处提供了 Nacos、ZooKeeper、Apollo、Redis 等存储终端的配置项存储，而且这些数据源配置属性类全部都继承 AbstractDataSourceProperties 父类，如图 5-40 所示。

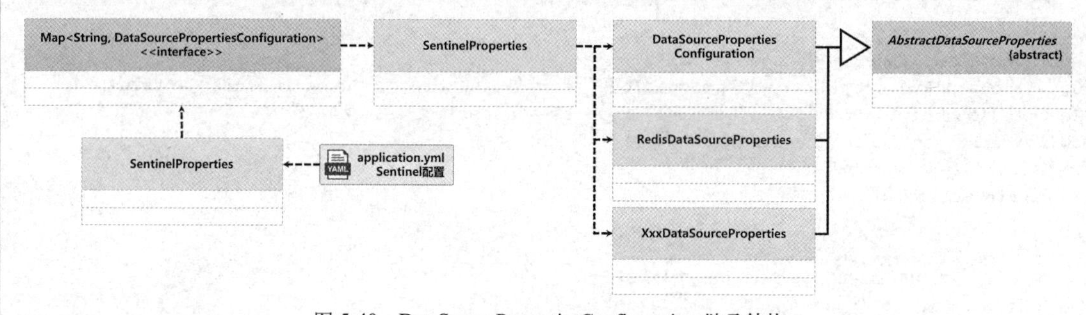

图 5-40 DataSourcePropertiesConfiguration 继承结构

读者打开 NacosDataSourceProperties 配置类可以发现，在这个类中并没有提供 Nacos 认证属性配置项，而这也是本书启用另外一个 Nacos 服务的主要原因。

（5）【Sentinel 控制台】打开 Sentinel 控制台，此时可以发现在流控规则中出现了相关的配置，如图 5-41 所示。

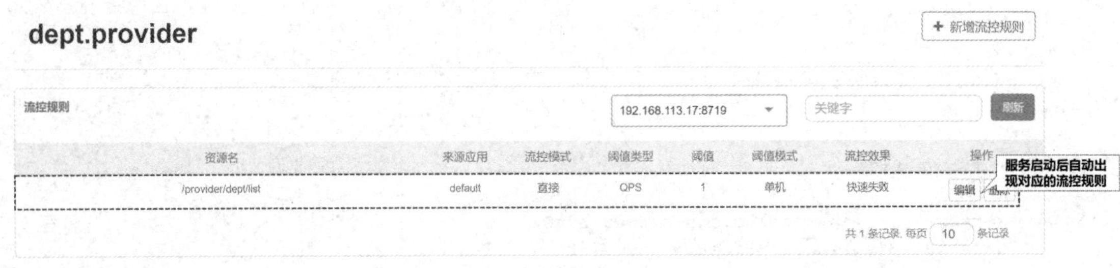

图 5-41　查看流控规则

5.5.2　流控规则解析

流控规则解析

视频名称　0520_【掌握】流控规则解析

视频简介　Sentinel 支持多种流控规则配置，如果想获取全部流控规则的 JSON 配置，则必须进行源代码的解析处理。本视频通过源代码的组成结构，为读者分析流控规则持久化配置项的定义结构。

流控规则持久化的本质在于将一个流控的配置项以 JSON 结构进行定义。这样在微服务启动时就可以根据指定的数据源配置来获取相关的解析规则，将其解析后转换为 Sentinel 可以使用的配置。转换流程如图 5-42 所示。

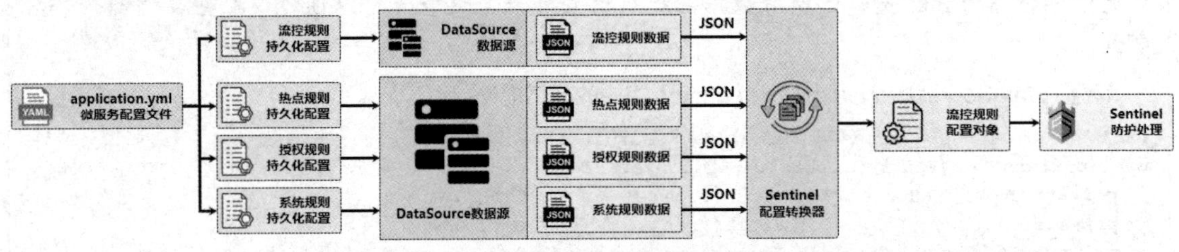

图 5-42　流控规则转换流程

Sentinel 中并不仅有一个流控规则，实际上还有系统流控、热点参数流控、授权流控等规则，那么这些规则的持久化配置 JSON 结构该如何定义呢？要想解释这些问题，首先要通过 Sentinel 配置类的结构来进行分析。

范例：SentinelAutoConfiguration 核心源代码

```java
package com.alibaba.cloud.sentinel.custom;
@Configuration(proxyBeanMethods = false)
@ConditionalOnProperty(name = "spring.cloud.sentinel.enabled", matchIfMissing = true)
@EnableConfigurationProperties(SentinelProperties.class)
public class SentinelAutoConfiguration {
   @Value("${project.name:${spring.application.name:}}")
   private String projectName;                                   // 应用名称
   @Autowired
   private SentinelProperties properties;                        // Sentinel配置属性
   @ConditionalOnClass(ObjectMapper.class)
   @Configuration(proxyBeanMethods = false)
   protected static class SentinelConverterConfiguration {     // Sentinel转换配置
      @Configuration(proxyBeanMethods = false)
      protected static class SentinelJsonConfiguration {
         private ObjectMapper objectMapper = new ObjectMapper();
         public SentinelJsonConfiguration() {
            objectMapper.configure(DeserializationFeature.FAIL_ON_UNKNOWN_PROPERTIES,
                  false);
         }
         @Bean("sentinel-json-flow-converter")
```

```
        public JsonConverter jsonFlowConverter() {          // 流控规则转换
            return new JsonConverter(objectMapper, FlowRule.class);
        }
        @Bean("sentinel-json-degrade-converter")
        public JsonConverter jsonDegradeConverter() {      // 降级规则转换
            return new JsonConverter(objectMapper, DegradeRule.class);
        }
        @Bean("sentinel-json-system-converter")
        public JsonConverter jsonSystemConverter() {       // 系统规则转换
            return new JsonConverter(objectMapper, SystemRule.class);
        }
        @Bean("sentinel-json-authority-converter")
        public JsonConverter jsonAuthorityConverter() {   // 授权规则转换
            return new JsonConverter(objectMapper, AuthorityRule.class);
        }
        @Bean("sentinel-json-param-flow-converter")
        public JsonConverter jsonParamFlowConverter() {   // 热点规则转换
            return new JsonConverter(objectMapper, ParamFlowRule.class);
        }
    }
  }
}
```

通过此配置类的源代码可以发现，所有在 application.yml 中配置的与 Sentinel 属性全部保存在 SentinelProperties 属性类中，同时在该类中还有 SentinelConverterConfiguration 内部类定义。这个类提供一系列的转换处理操作，可将读取的配置规则（JSON 结构）转为 Sentinel 内部可以使用的对象实例。这种转换操作是由专属的 JsonConverter 工具类实现的，在转换时提供目标规则的类型即可。通过此时的分析可以得出图 5-43 所示的核心关联结构。

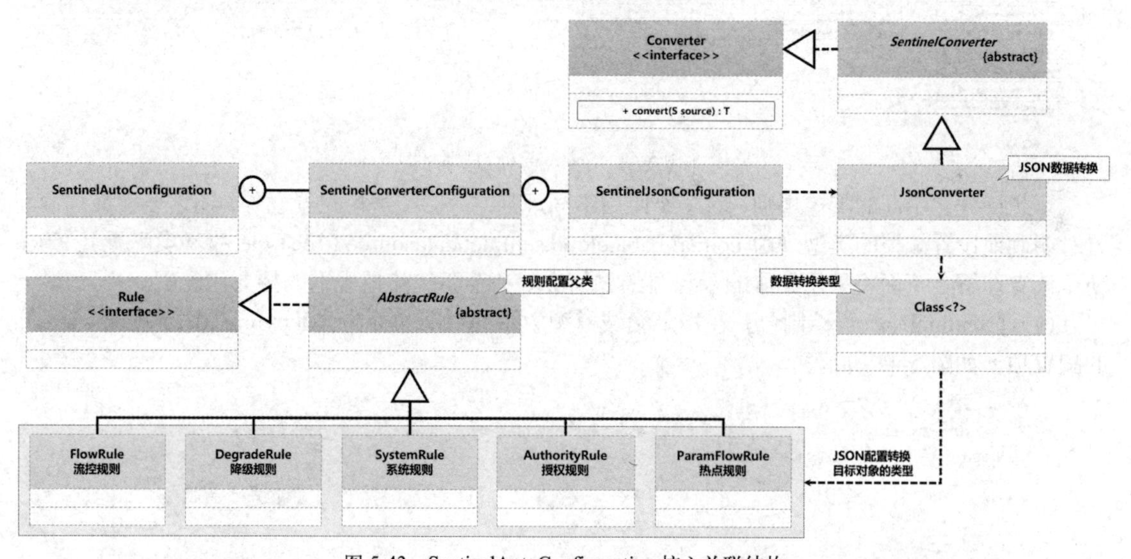

图 5-43 SentinelAutoConfiguration 核心关联结构

通过图 5-43 的结构可以清楚地发现，在使用 JsonConverter 进行转换时，需要定义转换目标的类实例。下面以 FlowRule 为例观察其源代码的组成结构。

范例：FlowRule 及其父类核心源代码

AbstractRule

```
package com.alibaba.csp.sentinel.slots.block;
public abstract class AbstractRule implements Rule {
    private String resource;                // 资源名称
    private String limitApp;                // 来源应用
}
```

FlowRule

```
package com.alibaba.csp.sentinel.slots.block.flow;
public class FlowRule extends AbstractRule {
    // 流控阈值类型，通过RuleConstant获取（0：线程数量。1：QPS）
    private int grade = RuleConstant.FLOW_GRADE_QPS;
    private double count;                          // 单机阈值定义
    // 流控模式，通过RuleConstant获取（0：直接。1：关联。2：链路）
    private int strategy = RuleConstant.STRATEGY_DIRECT;
    // 流控效果，通过RuleConstant获取（0：快速失败。1：WarmUp。2：队列）
    private int controlBehavior = RuleConstant.CONTROL_BEHAVIOR_DEFAULT;
    private int warmUpPeriodSec = 10;              // WarmUp提升时间
    private boolean clusterMode;                   // 是否采用集群模式
    private ClusterFlowConfig clusterConfig;       // 集群流控配置
}
```

通过 FlowRule 的配置类可以清楚地发现其与流控配置项 JSON 数据之间的关联，在 Sentinel 读取流控配置 JSON 后会通过 JsonConverter 工具类将字符串基于反射机制转换为 FlowRule 对象实例，这样 Sentinel 就可以基于此配置实现流控的自动配置。要想掌握其他流控模式的自动配置，就需要清楚这些规则类中属性的定义。图 5-44 为读者列出了这些规则类中的核心属性定义。

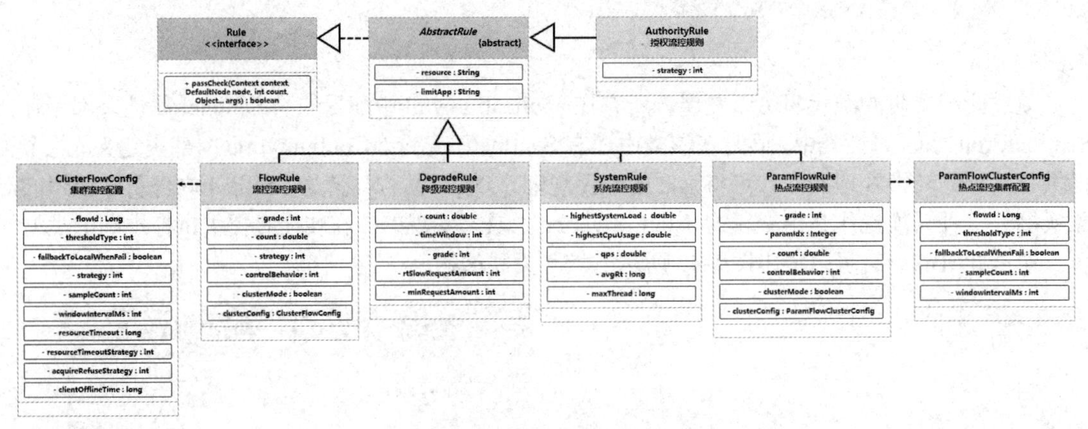

图 5-44　Sentinel 流控规则

要想持久化不同的流控规则，除了要使用不同规则的 JSON 结构配置之外，还需要在数据源配置时明确地设置流控的类型（由 com.alibaba.cloud.sentinel.datasource.RuleType 枚举类配置）。为了便于读者理解，下面将采用具体的步骤介绍如何实现一个系统流控规则的持久化配置。

（1）【SentinelNacos 控制台】在指定领域模型中添加一个名称为“dept.provider-system-rules”的配置项，如图 5-45 所示。

图 5-45　添加系统规则配置项

范例：为 dept.provider-system-rules 配置一个系统 QPS 流控规则

```
[ {  "qps": 1  } ]
```

（2）【dept-provider-*子模块】修改 application.yml 配置文件，添加要读取的系统流控规则信息。

```
spring:                                                  # Spring配置
  cloud:                                                 # Spring Cloud配置
    sentinel:                                            # Sentinel配置
      datasource:                                        # 数据源配置
        flow-datasource:                                 # 自定义数据源名称
          nacos:                                         # Nacos存储
            … : …                                        # 流控规则配置，略
        system-datasource:                               # 自定义数据源名称
          nacos:                                         # Nacos存储
            server-addr: sentinel-nacos-server:8848      # 服务地址
            data-id: ${spring.application.name}-system-rules  # 流控规则
            group-id: SENTINEL_GROUP                     # 分组ID
            namespace: 130ab0b4-37ab-4b51-82a0-d21b79ff8a75  # 命名空间ID
            data-type: json                              # 配置类型
            rule_type: system                            # 规则类型
```

配置完成后重新启动当前的微服务，此时会通过 Nacos 指定的领域模型读取 dept.provider-system-rules 配置中定义的系统规则，实现系统流控。

（3）【Sentinel 控制台】通过 Sentinel 控制台查看当前系统规则配置，如图 5-46 所示。

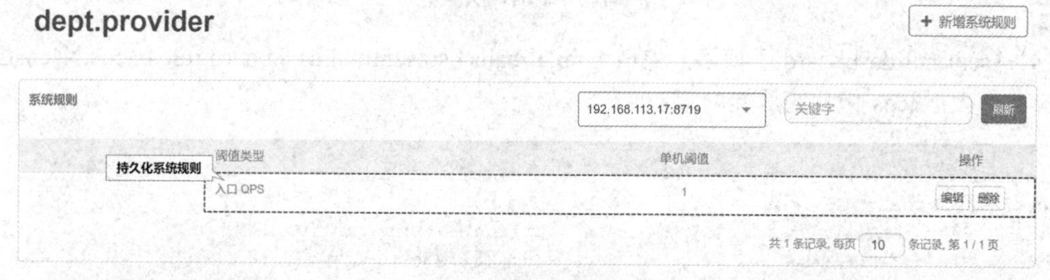

图 5-46　微服务系统规则

5.5.3　SentinelDashboard 改造

视频名称　0521_【理解】SentinelDashboard 改造

视频简介　Sentinel 源代码中提供了持久化流控规则可视化的配置支持。本视频会通过实例的形式为读者演示开启持久化流控配置的实现。

开发者除了可以通过 JSON 的形式进行流控规则的持久化配置之外，也可以通过 Sentinel 提供的控制面板进行持久化流控规则的管理。Sentinel 源代码已经为用户提供了 Nacos、ZooKeeper 等实现持久化流控管理操作的参考实现，下面将通过具体的步骤讲解如何开启 Sentinel 所提供的可视化流控规则编辑。

（1）【sentinel-dashboard 子模块】修改 src/main/resources/application.properties 配置文件，追加 Nacos 配置项。

```
nacos.address=sentinel-nacos-server:8848                  # Nacos服务地址
nacos.namespace=130ab0b4-37ab-4b51-82a0-d21b79ff8a75 # 命名空间ID
nacos.clusterName=SentinelCluster                     # 集群名称
```

（2）【sentinel-dashboard 子模块】修改 pom.xml 配置文件中关于 sentinel-datasource-nacos 依赖的使用范围。

```
<dependency>
    <groupId>com.alibaba.csp</groupId>
```

```
    <artifactId>sentinel-datasource-nacos</artifactId>    <!--Nacos数据源依赖 -->
    <scope>test</scope>                                   <!--删除此依赖范围 -->
</dependency>
```

（3）【sentinel-dashboard 子模块】该模块的测试目录下有 Nacos 持久化配置（同时还提供 Apollo、ZooKeeper 持久化存储配置），将此配置保存在 com.alibaba.csp.sentinel.dashboard.rule 包中，如图 5-47 所示。

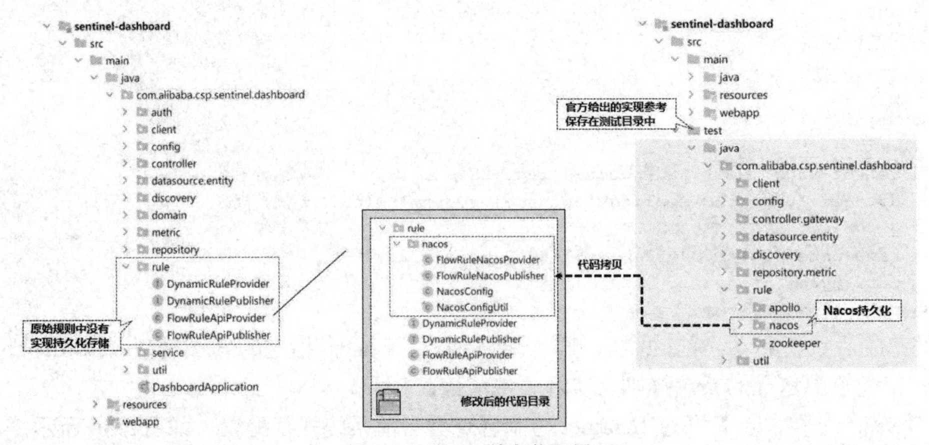

图 5-47 Nacos 持久化配置

（4）【sentinel-dashboard 子模块】修改 com.alibaba.csp.sentinel.dashboard.rule.nacos.NacosConfig 配置类，注入 Nacos 相关的属性内容。

```
package com.alibaba.csp.sentinel.dashboard.rule.nacos;
@Configuration
public class NacosConfig {
    @Value("${nacos.address}")                 // application.properties配置项
    private String address;                    // Nacos服务地址
    @Value("${nacos.namespace}")               // application.properties配置项
    private String namespace;                  // Nacos命名空间
    @Value("${nacos.clusterName}")             // application.properties配置项
    private String clusterName;                // Nacos集群名称
    @Bean
    public Converter<List<FlowRuleEntity>, String> flowRuleEntityEncoder() {
        return JSON::toJSONString;
    }
    @Bean
    public Converter<String, List<FlowRuleEntity>> flowRuleEntityDecoder() {
        return s -> JSON.parseArray(s, FlowRuleEntity.class);
    }
    @Bean
    public ConfigService nacosConfigService() throws Exception {
        Properties properties = new Properties();                           // Nacos属性
        properties.put(PropertyKeyConst.SERVER_ADDR, this.address);         // 服务地址
        properties.put(PropertyKeyConst.NAMESPACE, this.namespace);         // 命名空间
        properties.put(PropertyKeyConst.CLUSTER_NAME, this.clusterName);    // 集群名称
        return ConfigFactory.createConfigService(properties);
    }
}
```

（5）【sentinel-dashboard 子模块】修改 com.alibaba.csp.sentinel.dashboard.controller.v2.FlowControllerV2 程序类，将原始配置的流控 Bean 名称替换为 Nacos 来读取配置。

程序原始代码	package com.alibaba.csp.sentinel.dashboard.controller.v2; @RestController @RequestMapping(value = "/v2/flow")

程序原始代码

```
public class FlowControllerV2 {
    @Autowired
    @Qualifier("flowRuleDefaultProvider") // 【需修改】此为默认规则配置
    private DynamicRuleProvider<List<FlowRuleEntity>> ruleProvider;
    @Autowired
    @Qualifier("flowRuleDefaultPublisher") // 【需修改】此为默认规则配置
    private DynamicRulePublisher<List<FlowRuleEntity>> rulePublisher;
}
```

更新后的代码

```
package com.alibaba.csp.sentinel.dashboard.controller.v2;
@RestController
@RequestMapping(value = "/v2/flow")
public class FlowControllerV2 {
    @Autowired
    @Qualifier("flowRuleNacosProvider")    // 替换注入Bean实例
    private DynamicRuleProvider<List<FlowRuleEntity>> ruleProvider;
    @Autowired
    @Qualifier("flowRuleNacosPublisher")   // 替换注入Bean实例
    private DynamicRulePublisher<List<FlowRuleEntity>> rulePublisher;
}
```

（6）【sentinel-dashboard 子模块】由于此时开启了一个新的流控管理机制，因此需要开启 Sentinel 控制台中的菜单配置项，修改 src/main/webapp/resources/app/scripts/directives/sidebar/sidebar.htm 页面文件，取消相关注释。

```
<!—取消此代码之前的注释，通过此菜单配置的流控规则可以自动在Nacos中持久化保存 -->
<li ui-sref-active="active" ng-if="entry.appType==0">
  <a ui-sref="dashboard.flow({app: entry.app})">
    <I class="glyphicon glyphicon-filter"></I>  流控规则 V1</a>
</li>
```

（7）【sentinel-dashboard 子模块】对当前源代码进行打包处理。

```
mvn clean package -DskipTests;
```

（8）【sentinel-server 主机】将打包得到的 sentinel-dashboard.jar 程序文件上传到 Sentinel 服务主机，此时程序的保存路径为“/usr/local/src/sentinel-dashboard.jar”。随后通过命令启动新的 Sentinel 服务。

```
java -Dserver.port=8888 -Dproject.name=sentinel-dashboard -Dcsp.sentinel.dashboard.server=
localhost:8888 \
-Dsentinel.dashboard.auth.username=muyan -Dsentinel.dashboard.auth.password=yootk \
-jar /usr/local/src/sentinel-dashboard.jar >/usr/local/src/sentinel_new.log 2>&1 &
```

（9）【Sentinel 控制台】重新进入 Sentinel 控制台，可以发现此时增加了一个新的流控规则，如图 5-48 所示。

图 5-48 Sentinel 扩展流控规则

（10）【Sentinel 控制台】通过新增的“流控规则 V1”进行流控规则的配置，如图 5-49 所示。

（11）【SentinelNacos 控制台】此时开发者通过“流控规则 V1”所配置的全部流控规则都为持久化配置，并且会自动保存在 Sentinel-Nacos 服务器之中（默认规则名称格式为“{应用名称}-flow-rules”)，如图 5-50 所示。在微服务启动后该配置规则会自动生效。

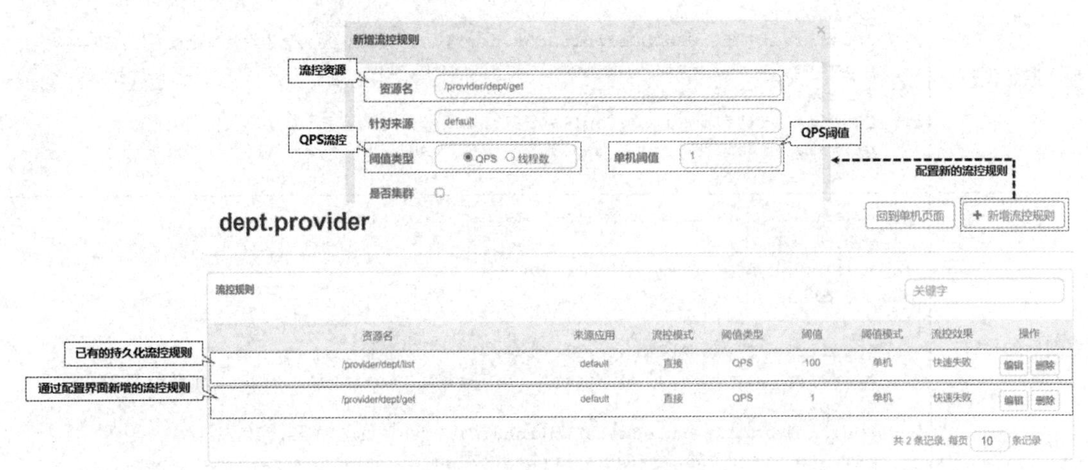

图 5-49　新增持久化流控规则

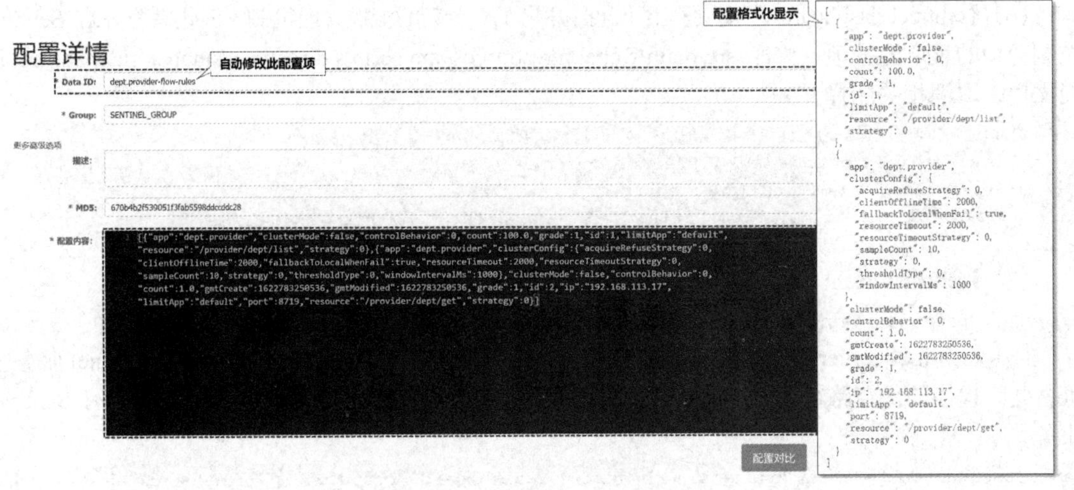

图 5-50　SentinelNacos 配置查看

5.6　本章概览

1．Sentinel 是由阿里巴巴公司提供的开源流量监控与流控保护组件，可以直接在微服务中进行整合，而后基于切面的形式实现流控管理，当触发流控规则阈值时会自动进行熔断处理。

2．Sentinel 支持多种流控规则的定义，包括流控规则、热点规则、授权规则、系统规则、降级规则。

3．Sentinel 支持集群流控，但是需要配置统一的 TokenServer 用于集群所有节点（TokenClient）访问量的统计。

4．Sentinel 内部基于 Node 接口实现所有流量数据的统计，并通过插槽实现了服务流控管理。

5．Sentinel 默认会将所有的流控规则保存在内存之中，服务重新启动后需要重新配置，这样的模式并不利于服务的管理，所以可以基于 Nacos、Redis 或 ZooKeeper 实现持久化流控规则存储。

6．Sentinel 源代码内部提供了可视化流控持久配置，开发人员可以手动开启；而并未提供其他的持久化流控规则的配置，开发者可以依据已有的代码手动实现。

第 6 章 Spring Cloud Gateway

本章学习目标

1. 掌握 Spring Cloud Gateway 组件的工作原理，并可用其实现微服务隔离；
2. 掌握 Spring Cloud Gateway 中的路由配置方法，可以实现静态路由与动态路由配置，并基于路由谓词工厂进行限制；
3. 掌握 Spring Cloud Gateway 中过滤工厂的使用方法，并可以使用过滤工厂实现拦截处理；
4. 掌握 Spring Cloud Gateway 全局过滤器的使用方法；
5. 理解 Spring Cloud 的工作原理与核心实现类的作用。

微服务的使用会导致项目中存在大量的服务节点，而为了便于节点的规范化管理，就要引入网关的支持。Spring Cloud 提供了原生的 Spring Cloud Gateway 组件，本章将为读者讲解此组件的作用以及各个配置实现。

6.1 Spring Cloud Gateway 基本使用

Spring Cloud Gateway 简介

视频名称 0601_【掌握】Spring Cloud Gateway 简介

视频简介 为了更好地进行微服务的管理并便于用户访问，Spring Cloud 提供了网关技术。本视频为读者讲解 Spring Cloud Gateway 在微服务管理中的主要作用，并分析 Zull 与 Spring Cloud Gateway 的区别。

微服务的设计都会有提供端与消费端，为了避免消费端直接与目标节点的主机进行交互，项目中引入了注册中心，这样可以根据微服务的名称实现服务实例数据的获取，从而实现最终的资源调用。但是项目业务功能的不断完善必然导致微服务节点的急剧增加，这样一来传统的依据微服务名称访问的形式就必然带来维护上的困难，此时就需要按照功能对微服务集群进行划分，如图 6-1 所示。

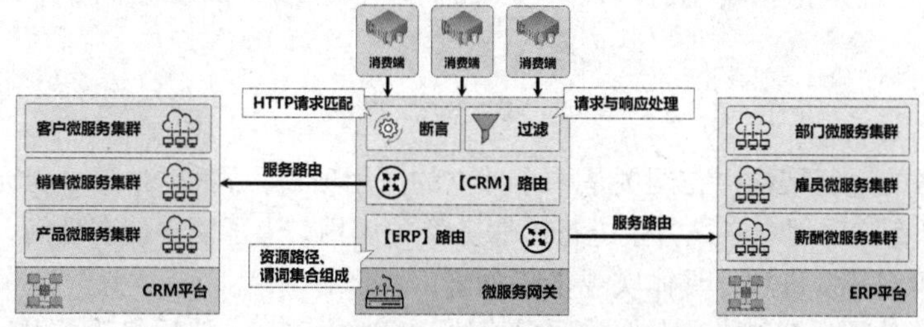

图 6-1 微服务网关

Spring Cloud Gateway 是 Spring Cloud 提供并维护的原生微服务网关项目，该项目基于 Netty、Reactor、WebFlux 构建，提供了一种简单有效的 API 路由管理方式。所有客户端请求在进行微服务资源访问时必须通过网关进行路由，这样就为整个微服务的使用提供了统一的入口。

提示：Zuul 与 Spring Cloud Gateway。

Spring Cloud 在早期开发中一直使用 Spring Cloud Netflix 套件中的 Zuul 实现微服务网关，但是由于此项目的更新停滞，因此 Spring Cloud 开发出了 Spring Cloud Gateway 网关技术。除了实现 Zuul 所有的功能之外，Spring Cloud Gateway 还提供了更强的处理性能以及强大的功能扩展。因为现在的 Spring Cloud 技术是基于 Spring Boot 2.x 技术实现的，所以本书将主要讲解 Spring Cloud Gateway。想学习 Zuul 使用的读者可以翻看笔者的《名师讲坛——Java 微服务架构实战（Spring Boot+Spring Cloud+Docker+RabbitMQ）》一书。

Spring Cloud Gateway 网关中最为重要的核心技术在于微服务的路由管理，有了该技术，才能根据用户请求的路由地址找到指定的微服务资源。在进行微服务调用时，可以通过路由执行断言（或称路由谓词工厂）进行调用的判断，并通过过滤实现请求与响应的处理，如图 6-2 所示。基于 Spring Cloud Gateway 中的过滤模式，还可以轻松地实现参数校验、认证检测、流量监控、访问流控等操作。

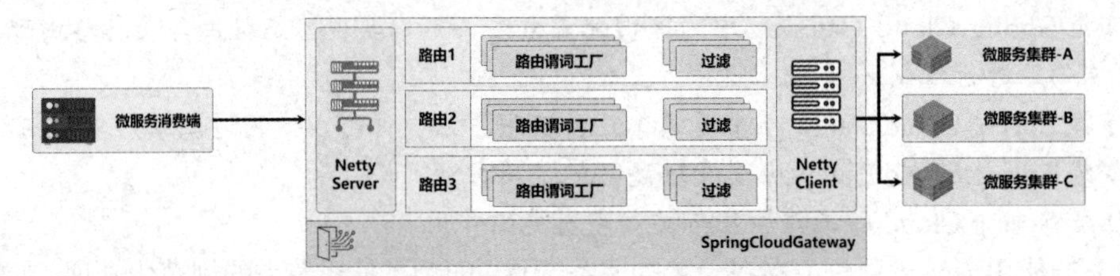

图 6-2　Spring Cloud Gateway 运行模式

6.1.1　Spring Cloud Gateway 编程起步

Spring Cloud
Gateway 编程起步

视频名称　0602_【掌握】Spring Cloud Gateway 编程起步

视频简介　Spring Cloud Gateway 基于 WebFlux 架构开发，可以提供更高效的服务处理。本视频为读者讲解网关在微服务技术中的应用架构，同时基于 Spring Cloud Gateway 的基本配置介绍如何实现部门微服务的网关代理访问操作。

Spring Cloud Gateway 主要通过注册中心代理所有相关的微服务资源。因为所有的服务资源都需要通过 Nacos 注册中心获取，所以需要将网关注册到 Nacos 服务列表之中，随后就可以通过网关地址来进行微服务访问，如图 6-3 所示。

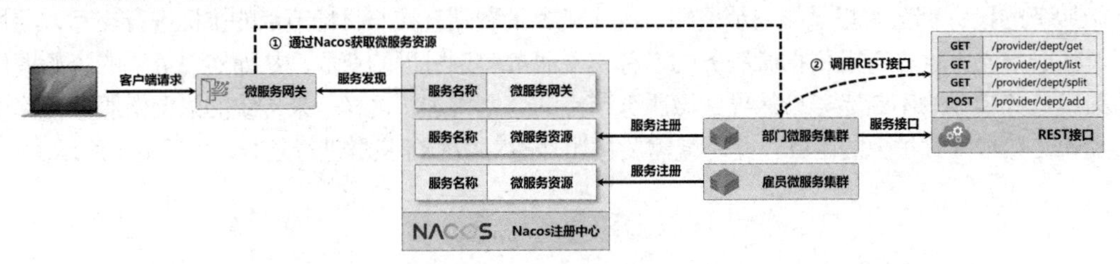

图 6-3　网关与微服务结构

在整个微服务的运行架构中，网关是一个独立的应用项目，同时 Spring Cloud Gateway 基于 WebFlux 开发，可直接采用响应式方式进行微服务资源调用。为了便于读者理解，下面将通过一个具体的程序进行实例开发讲解，具体实现步骤如下。

（1）【microcloud 项目】创建一个新的项目模块 gateway-9501，随后修改 build.gradle 配置文件，追加相关依赖，此时不需要添加 spring-cloud-starter-web 依赖。

```
project('gateway-9501') {                              // 子模块
    dependencies {                                     // 配置子模块依赖
        implementation('org.springframework.cloud:spring-cloud-starter-gateway')
```

```
        implementation('com.alibaba.cloud:spring-cloud-starter-alibaba-nacos-discovery')
    }
}
```

（2）【gateway-9501 子模块】创建 src/main/resources/application.yml 配置文件，进行网关应用的端口、Nacos 发现服务以及网关资源查找配置。

```
server:                                                    # 服务端配置
  port: 9501                                               # 监听端口
spring:                                                    # Spring配置
  application:                                             # 应用配置
    name: microcloud.gateway                               # 应用名称
  cloud:                                                   # Spring Cloud配置
    nacos:                                                 # Nacos注册中心
      discovery:                                           # 发现服务
        username: nacos                                    # 用户名
        password: nacos                                    # 密码
        service: ${spring.application.name}                # 服务名称
        server-addr: nacos-server:8848                     # 服务地址
        namespace: 650fab32-c7dc-4ae1-8ac4-2dbdefd7e617    # 命名空间
        group: MICROCLOUD_GROUP                            # 组名称
        cluster-name: MuyanGateway                         # 匹配集群名称
    gateway:                                               # 网关配置
      discovery:                                           # 服务发现
        locator:                                           # 资源定位
          enabled: true                                    # 通过服务发现查找其他微服务
```

在 application.yml 中配置 spring.cloud.gateway.discovery.locator.enabled = true 时（该配置项默认为 false）才可以根据微服务的名称进行服务调用。

（3）【gateway-9501 子模块】定义程序启动类。由于此时需要通过 Nacos 查询服务资源，因此要使用发现服务注解。

```
package com.yootk.gateway;
@SpringBootApplication
@EnableDiscoveryClient                                          // 启用Nacos发现服务注解
public class StartGatewayApplication9501 {
    public static void main(String[] args) {
        SpringApplication.run(StartGatewayApplication9501.class, args); // 服务启动
    }
}
```

（4）【Nacos 控制台】Spring Cloud Gateway 项目启动之后需要在 Nacos 中注册，并通过 Nacos 提供的发现服务实现所有微服务资源的调用。此时打开 Nacos 控制台，可以看见图 6-4 所示的服务列表。

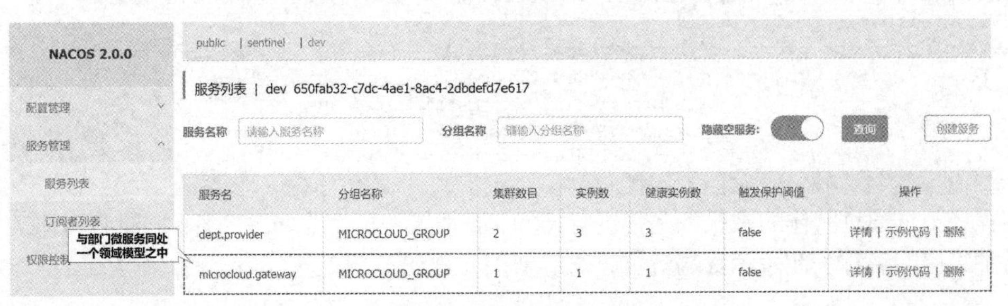

图 6-4 Nacos 服务列表

（5）【本地系统】为便于网关应用的访问，修改 hosts 配置文件，追加新的主机名称。

```
127.0.0.1   gateway-9501
```

（6）【Postman】网关与微服务启动之后，就可以通过网关实现微服务的调用，调用的格式为 http://{网关地址}/{微服务名}/{请求路径}。以调用部门微服务为例，可以使用表 6-1 给出的地址。

表 6-1　网关调用地址

序号	功能	模式	网关调用地址
01	部门列表	GET	gateway-9501:9501/dept.provider/provider/dept/list
02	部门查找	GET	gateway-9501:9501/dept.provider/provider/dept/get/{id}
03	部门分页	GET	gateway-9501:9501/dept.provider/provider/dept/split?cp={cp}&ls={ls}&col={col}&kw={kw}
04	部门增加	POST	gateway-9501:9501/dept.provider/provider/dept/add

6.1.2　消费端整合 Spring Cloud Gateway

消费端整合 Spring Cloud Gateway

视频名称　0603_【理解】消费端整合 Spring Cloud Gateway

视频简介　微服务架构需要提供 Spring Boot 消费端处理。本视频介绍如何基于已有的 Feign 组件进行修改，实现网关调用与服务映射路径的配置。

微服务架构群中引入网关服务之后，消费端就应该根据网关服务名称实现相关微服务的调用。在实际的项目中消费端往往会基于 Feign 接口映射方式进行微服务调用，此时就需要通过@FeignClient 注解配置网关名称，而映射接口中每一个方法也需要追加微服务的名称后才可以正常访问，如图 6-5 所示。

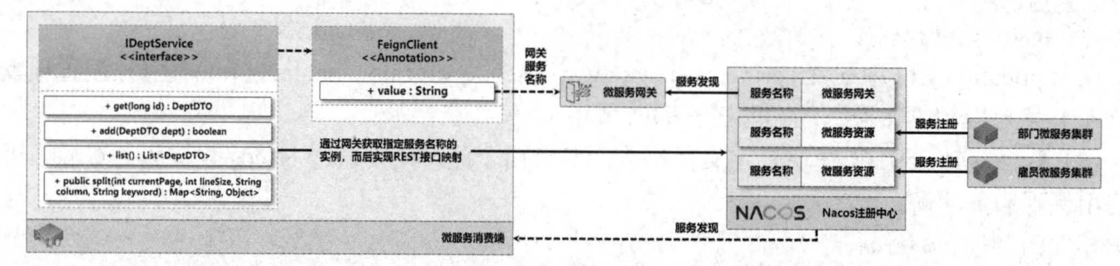

图 6-5　消费端处理

范例：【common-api 子模块】修改 IDeptService 业务接口，直接通过网关服务名称实现接口映射

```
package com.yootk.service;
@FeignClient(value="microcloud.gateway", configuration = FeignConfig.class,
    fallbackFactory = DeptServiceFallbackFactory.class)                // 服务降级配置
public interface IDeptService {                                        // 部门业务接口
    @GetMapping("/dept.provider/provider/dept/get/{deptno}")           // 接口映射
    public DeptDTO get(@PathVariable("deptno")long id);
    @PostMapping("/dept.provider/provider/dept/add")                   // 接口映射
    public boolean add(DeptDTO dept);
    @GetMapping("/dept.provider/provider/dept/list")                   // 接口映射
    public List<DeptDTO> list();
    @GetMapping("/dept.provider/provider/dept/split")                  // 接口映射
    public Map<String, Object> split(@RequestParam("cp") int currentPage,
                            @RequestParam("ls") int lineSize,
                            @RequestParam("col") String column,
                            @RequestParam("kw") String keyword);
}
```

本程序不再通过具体的微服务名称进行访问，而是统一更换为网关名称，同时在微服务访问地址前追加相应的名称，此时消费端就可以通过网关实现服务调用。

6.1.3　静态路由配置

静态路由配置

视频名称　0604_【掌握】静态路由配置

视频简介　Spring Cloud Gateway 中的路由除了可自动根据微服务的名称进行匹配之外，还可以在 application.yml 中进行静态配置。本视频为读者讲解单一资源的路由以及集群资源的路由访问配置。

在默认情况下，开发者只要配置了网关微服务，就会自动通过 Nacos 发现服务获取相关的微服务信息，以实现微服务的代理调用。在 Spring Cloud Gateway 的处理中也可以对访问的微服务进行各种配置，如图 6-6 所示。

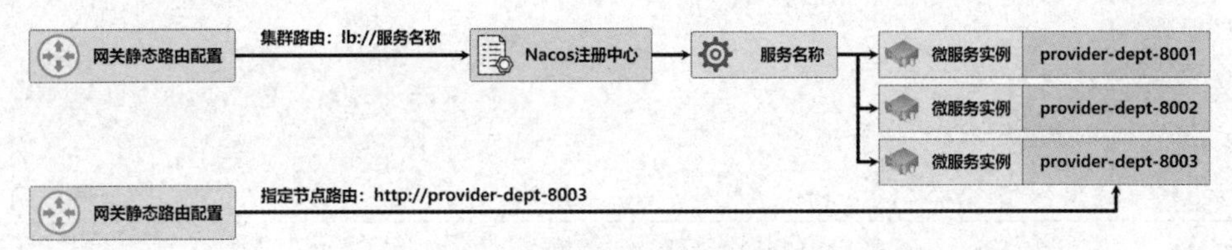

图 6-6 微服务网关绑定

所有的静态路由配置都可以直接在 application.yml 中进行定义，开发者可以根据需要绑定单一微服务路径或者绑定 Nacos 中注册的服务名称（一个服务名称对应多个微服务实例），并根据断言配置实现微服务调用。为便于读者理解路由的配置操作，下面将通过几个具体的配置形式进行演示。

（1）【gateway-9501 子模块】绑定单一 URI 路径资源。

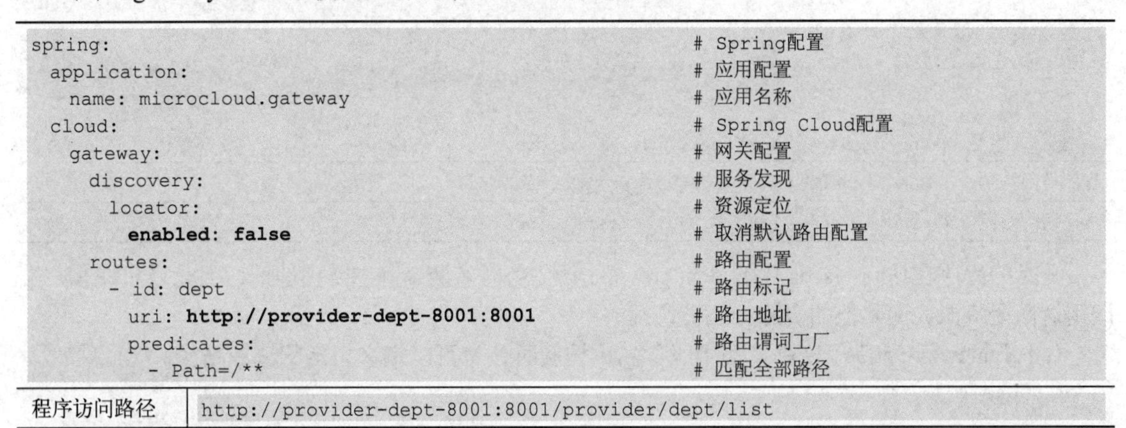

```
spring:                                          # Spring配置
  application:                                   # 应用配置
    name: microcloud.gateway                     # 应用名称
  cloud:                                         # Spring Cloud配置
    gateway:                                     # 网关配置
      discovery:                                 # 服务发现
        locator:                                 # 资源定位
          enabled: false                         # 取消默认路由配置
      routes:                                    # 路由配置
        - id: dept                               # 路由标记
          uri: http://provider-dept-8001:8001    # 路由地址
          predicates:                            # 路由谓词工厂
            - Path=/**                           # 匹配全部路径
```

程序访问路径	http://provider-dept-8001:8001/provider/dept/list

此时的配置项直接将路由访问映射到了“provider-dept-8001:8001”实例之中，同时在 predicates 配置路由访问断言时实现了全部资源路径的匹配。

（2）【gateway-9501 子模块】如果此时觉得直接进行全部路径的映射不安全，也可以直接进行指定访问路径的匹配。

```
spring:                                          # Spring配置
  application:                                   # 应用配置
    name: microcloud.gateway                     # 应用名称
  cloud:                                         # Spring Cloud配置
    gateway:                                     # 网关配置
      routes:                                    # 路由配置
        - id: dept                               # 路由标记
           uri: http://provider-dept-8001:8001/provider/dept/get/{id} # 路由地址
           predicates:                           # 路由谓词工厂
            - Path=/provider/dept/get/{id}       # 匹配指定路径
```

程序访问路径	http://gateway-9501:9501/provider/dept/get/1

本程序实现了在“provider-dept-8001:8001”实例中指定资源的路由配置，这样只允许访问 Path 定义的路径，而其他未配置的路径将不允许进行访问。

（3）【gateway-9501 子模块】虽然此时已经实现了微服务的访问处理，但是仅仅依靠一个 URI 的配置只能够实现单一微服务的访问。现实的项目开发中是需要考虑集群访问的，可以通过“lb://微服务名称”的形式进行配置。

```
spring:                                              # Spring配置
  application:                                       # 应用配置
    name: microcloud.gateway                         # 应用名称
  cloud:                                             # Spring Cloud配置
    gateway:                                         # 网关配置
      routes:                                        # 路由配置
        - id: dept                                   # 路由标记
          uri: lb://dept.provider                    # 服务资源匹配
          predicates:                                # 路由谓词工厂
            - Path=/**                               # 匹配全部路径
```

程序访问路径	http://gateway-9501:9501/provider/dept/list

（4）【gateway-9501 子模块】在网关中配置微服务集群时，如果设置匹配路径为“/**”，那么会匹配所有的路径。当然根据先前的配置也可以匹配单一路径，直接修改 Path 属性即可。

```
spring:                                              # Spring配置
  application:                                       # 应用配置
    name: microcloud.gateway                         # 应用名称
  cloud:                                             # Spring Cloud配置
    gateway:                                         # 网关配置
      routes:                                        # 路由配置
        - id: dept                                   # 路由标记
          uri: lb://dept.provider                    # 服务资源匹配
          predicates:                                # 路由谓词工厂
            - Path=/provider/dept/get/{id},/provider/dept/list
```

程序访问路径	http://gateway-9501:9501/provider/dept/get/1
程序访问路径	http://gateway-9501:9501/provider/dept/list

此时的程序在访问 Nacos 中的 dept.provider 微服务时有两个静态路由地址，因此只允许客户端调用这两个资源，而其他的资源无法被访问。

（5）【gateway-9501 子模块】基于服务名称访问时，也可以定义指定的资源路径。

```
spring:                                                  # Spring配置
  application:                                           # 应用配置
    name: microcloud.gateway                             # 应用名称
  cloud:                                                 # Spring Cloud配置
    gateway:                                             # 网关配置
      routes:                                            # 路由配置
        - id: dept                                       # 路由标记
          uri: lb://dept.provider/provider/dept/get/{id} # 服务资源匹配
          predicates:                                    # 路由谓词工厂
            - Path=/provider/dept/get/{id}
```

程序访问路径	http://gateway-9501:9501/provider/dept/get/1

由于程序在当前路由配置时将路由地址与/provider/dept/get/{id}访问路径进行了关联，因此最终只能使用部门查找。

6.2　RoutePredicateFactory

RoutePredicate Factory 简介

视频名称　0605_【掌握】RoutePredicateFactory 简介

视频简介　RoutePredicateFactory（又被称为路由谓词工厂）是在网关中实现服务代理转发的重要前提条件。本视频通过宏观的概念为读者分析内置断言提供的配置模式，并且基于类结构的形式总结路由谓词工厂的功能。

Spring Cloud Gateway 主要依据路由实现请求的转发处理，在进行请求转发时往往需要进行路由断言处理，符合配置条件时才允许执行后续操作；如果配置条件不符合将无法实现转发，从而导致调用失败，如图 6-7 所示。

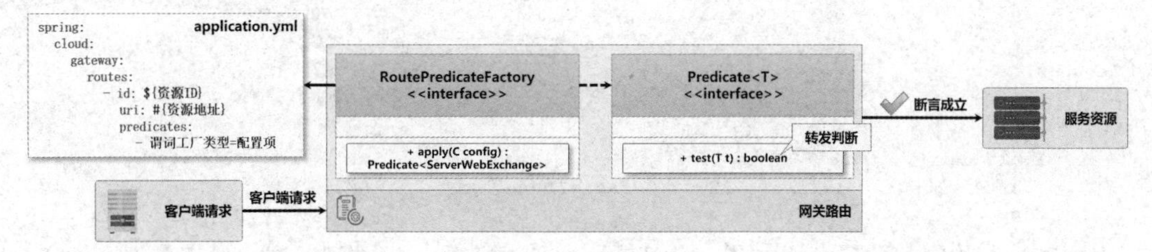

图 6-7 路由断言处理

所有的网关 Predicate 配置实际上都是通过 RoutePredicateFactory 工厂类型定义的，每当启动网关应用时程序都会自动地提供相应的工厂实例（对应子类保存在 org.springframework.cloud.gateway.handler.predicate 包中），这些信息可以在项目启动日志中进行观察。

```
[ restartedMain] o.s.c.g.r.RouteDefinitionRouteLocator    : Loaded RoutePredicateFactory [After]
[ restartedMain] o.s.c.g.r.RouteDefinitionRouteLocator    : Loaded RoutePredicateFactory [Before]
[ restartedMain] o.s.c.g.r.RouteDefinitionRouteLocator    : Loaded RoutePredicateFactory [Between]
[ restartedMain] o.s.c.g.r.RouteDefinitionRouteLocator    : Loaded RoutePredicateFactory [Cookie]
[ restartedMain] o.s.c.g.r.RouteDefinitionRouteLocator    : Loaded RoutePredicateFactory [Header]
[ restartedMain] o.s.c.g.r.RouteDefinitionRouteLocator    : Loaded RoutePredicateFactory [Host]
[ restartedMain] o.s.c.g.r.RouteDefinitionRouteLocator    : Loaded RoutePredicateFactory [Method]
[ restartedMain] o.s.c.g.r.RouteDefinitionRouteLocator    : Loaded RoutePredicateFactory [Path]
[ restartedMain] o.s.c.g.r.RouteDefinitionRouteLocator    : Loaded RoutePredicateFactory [Query]
[ restartedMain] o.s.c.g.r.RouteDefinitionRouteLocator    : Loaded RoutePredicateFactory
[ReadBodyPredicateFactory]
[ restartedMain] o.s.c.g.r.RouteDefinitionRouteLocator    : Loaded RoutePredicateFactory [RemoteAddr]
[ restartedMain] o.s.c.g.r.RouteDefinitionRouteLocator    : Loaded RoutePredicateFactory [Weight]
[ restartedMain] o.s.c.g.r.RouteDefinitionRouteLocator    : Loaded RoutePredicateFactory
[CloudFoundryRouteService]
```

Spring Cloud Gateway 中支持的断言类型包括日期时间、Cookie、请求头信息、主机信息、路径（Path）等，只有符合指定的断言条件才允许进行请求转发。RoutePredicateFactory 类的继承结构如图 6-8 所示。

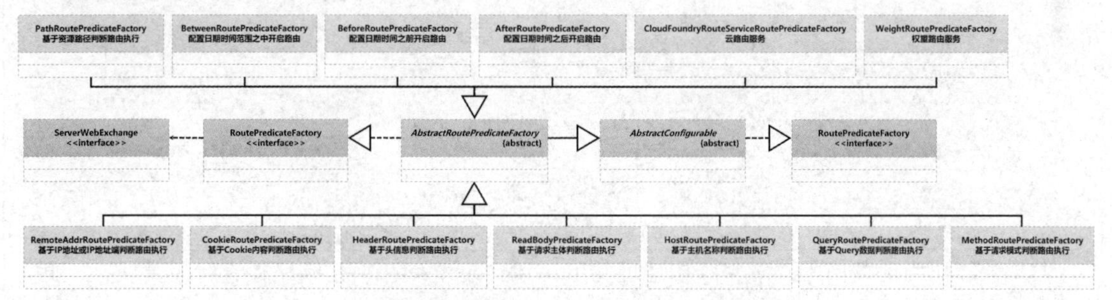

图 6-8 RoutePredicateFactory 类的继承结构

6.2.1 内置 RoutePredicateFactory 子类

视频名称 0606_【掌握】内置 RoutePredicateFactory 子类

视频简介 在 Spring Cloud Gateway 项目中可以直接使用已有的内置路由谓词工厂类实现网关访问的执行判断。本视频通过实例为读者演示内置路由谓词工厂类的使用。

Spring Cloud Gateway 内置的路由谓词工厂类主要提供请求日期时间以及 HTTP 请求内容的访问限制。本节将介绍通过 application.yml 实现常用的路由判断操作。

（1）【gateway-9501 子模块】定义网关在指定日期之后允许访问。

```
spring:                                                    # Spring配置
  cloud:                                                   # Spring Cloud配置
    gateway:                                               # 网关配置
      routes:                                              # 路由配置
        - id: dept                                         # 路由标记
          uri: lb://dept.provider                          # 服务资源匹配
          predicates:                                      # 路由谓词工厂
            - After=2050-02-17T21:15:32+08:00[Asia/Shanghai]  # 指定日期时间之后访问
            - Path=/**                                     # 匹配全部资源路径
```

本次在已有的网关配置中添加了一个“After”路由谓词工厂类型，这样只有在到达指定时间之后网关才允许客户端访问请求资源，而在此之前进行访问会显示错误信息。与此配置类似的还有 Before 和 Between（两个日期时间之间使用“,”分隔）两个路由谓词工厂类。

提问：日期时间如何配置？

在使用 After 配置日期时间限制时，采用的并不是“年-月-日 时:分:秒”的字符串。如果想将指定的日期时间变为此格式的字符串，该如何实现呢？

回答：通过 ZonedDateTime 实现。

本系列图书中的《Java 进阶开发实战（视频讲解版）》曾经为读者解释过 LocalDateTime、Instant、ZonedDateTime 类。实际上如果想使用特定的日期时间，建议采用如下代码生成。

范例：生成日期时间数据

```java
package com.yootk.test;
import java.time.Instant;
import java.time.LocalDateTime;
import java.time.ZoneId;
import java.time.ZonedDateTime;
import java.time.format.DateTimeFormatter;
public class CreateGatewayDatetime {
    public static void main(String[] args) {
        DateTimeFormatter formatter = DateTimeFormatter
                .ofPattern("yyyy-MM-dd HH:mm:ss");
        ZoneId zoneId = ZoneId.systemDefault();
        LocalDateTime localDateTime = LocalDateTime
                .parse("2017-02-17 21:15:32", formatter);
        ZonedDateTime now = ZonedDateTime
                .of(localDateTime, zoneId);
        System.out.println(now);
    }
}
```

程序执行结果	2017-02-17T21:15:32+08:00[Asia/Shanghai]

这个数据包含日期时间、时区等，而在 AfterRoutePredicateFactory 类处理时，实际上也是基于当前的日期时间与配置的日期时间进行比对的，如果符合访问规则则允许用户进行资源调用。

（2）【gateway-9501 子模块】客户端请求时必须带有指定的 Cookie 数据才允许实现资源调用。

```
spring:                                      # Spring配置
  cloud:                                     # Spring Cloud配置
    gateway:                                 # 网关配置
      routes:                                # 路由配置
        - id: dept                           # 路由标记
          uri: lb://dept.provider            # 服务资源匹配
          predicates:                        # 路由谓词工厂
```

```
        - Cookie=muyan-yootk-key, muyan\-\w+        # Cookie匹配
        - Path=/**                                  # 全部资源
```

此时客户端在访问时必须通过头信息传递一个名称为“muyan-yootk-key”的 Cookie，同时对应的数据格式必须为“muyan-xxx”。如果没有此 Cookie 或者数据格式不正确，则无法进行访问。如果使用 Postman 工具进行接口测试，则可以采用图 6-9 所示的方式进行配置。

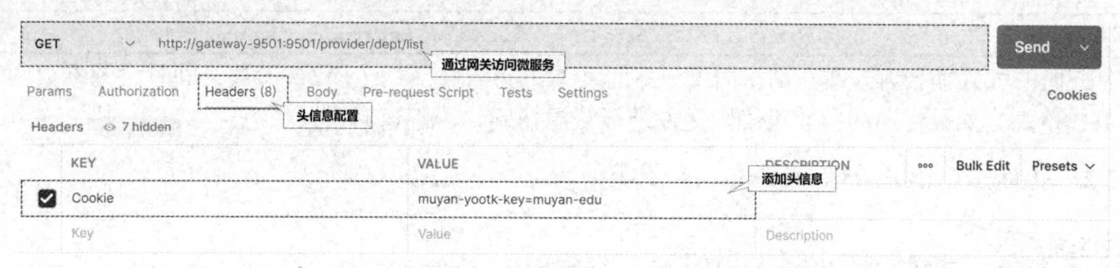

图 6-9　Postman 网关请求配置

（3）【gateway-9501 子模块】客户端请求由于是基于 HTTP 提交的，因此也可以通过头信息传递数据。现在假设在每次访问时都需要传递一个名称为“X-Muyan-Request-Id”的头信息，内容为数字，则可以按照如下方式配置。

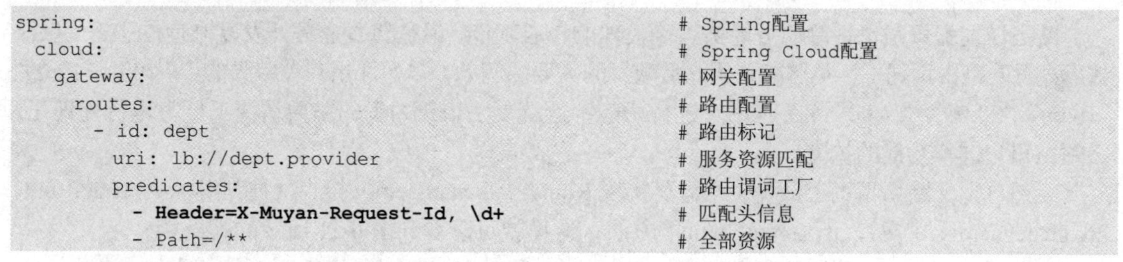

```
spring:                                             # Spring配置
  cloud:                                            # Spring Cloud配置
    gateway:                                        # 网关配置
      routes:                                       # 路由配置
        - id: dept                                  # 路由标记
          uri: lb://dept.provider                   # 服务资源匹配
          predicates:                               # 路由谓词工厂
            - Header=X-Muyan-Request-Id, \d+        # 匹配头信息
            - Path=/**                              # 全部资源
```

本次配置自定义了一个请求头信息的名称，只要在请求时传入指定的数据，就可以通过网关实现微服务资源的访问，配置信息如图 6-10 所示。

图 6-10　配置信息

（4）【gateway-9501 子模块】配置只允许特定主机进行网关服务调用。

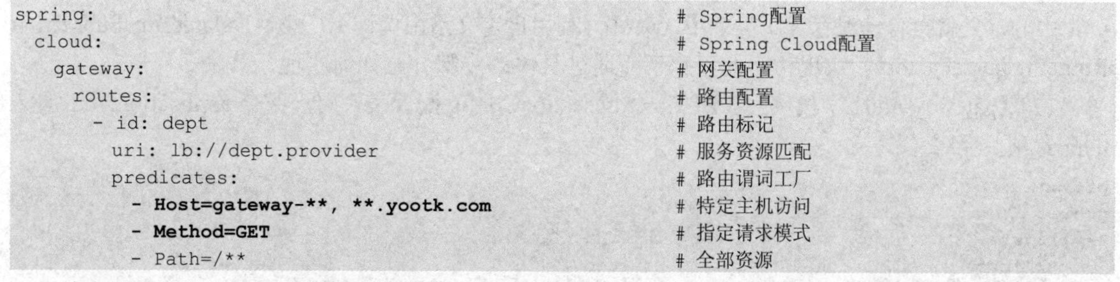

```
spring:                                             # Spring配置
  cloud:                                            # Spring Cloud配置
    gateway:                                        # 网关配置
      routes:                                       # 路由配置
        - id: dept                                  # 路由标记
          uri: lb://dept.provider                   # 服务资源匹配
          predicates:                               # 路由谓词工厂
            - Host=gateway-**, **.yootk.com         # 特定主机访问
            - Method=GET                            # 指定请求模式
            - Path=/**                              # 全部资源
```

（5）【gateway-9501 子模块】在进行 HTTP 请求时可以将请求参数通过 URL 重写的方式进行传递，而在网关中则可以根据请求地址的附加参数的内容来进行判断。

```
spring:                                             # Spring配置
  cloud:                                            # Spring Cloud配置
```

```
    gateway:                                                # 网关配置
      routes:                                               # 路由配置
        - id: dept                                          # 路由标记
          uri: lb://dept.provider                           # 服务资源匹配
          predicates:                                       # 路由谓词工厂
            - Query=msg, yootk-\w+                          # 地址重写参数判断
            - Path=/**                                      # 全部资源
```

服务调用路径	http://gateway-9501:9501/provider/dept/list?**msg=yootk-java**

此时的配置需要客户端在请求时通过地址重写的方式传递一个 msg 参数，同时设置参数内容的组成格式。如果传递的参数正确则允许进行资源访问，否则返回错误信息。

6.2.2　扩展 RoutePredicateFactory 子类

扩展 RoutePredicate Factory 子类

视频名称　0607_【理解】扩展 RoutePredicateFactory 子类

视频简介　项目开发中会有各类特殊需要，所以对于网关的访问也有特殊的要求。为了便于用户功能的扩展，可以自定义 RoutePredicateFactory 子类。本视频通过一个时间点访问应用讲解自定义路由谓词工厂类的程序实现。

路由谓词配置是保证微服务能被正确代理的核心功能，但是随着业务开发复杂度的上升，仅仅依靠内置的路由谓词工厂类可能无法满足项目的需要。例如，只允许用户在每天的“08:00”“12:00”“16:00”“20:00”“24:00”5 个时间点进行访问，这就无法依靠内置的路由谓词工厂类进行处理了，必须由用户进行功能的扩充。

要想自定义路由谓词工厂类，只需要实现 RoutePredicateFactory 接口（推荐继承 AbstractRoutePredicateFactory 子类），而后覆写 apply()方法，这样就可以得到图 6-11 所示的类结构。

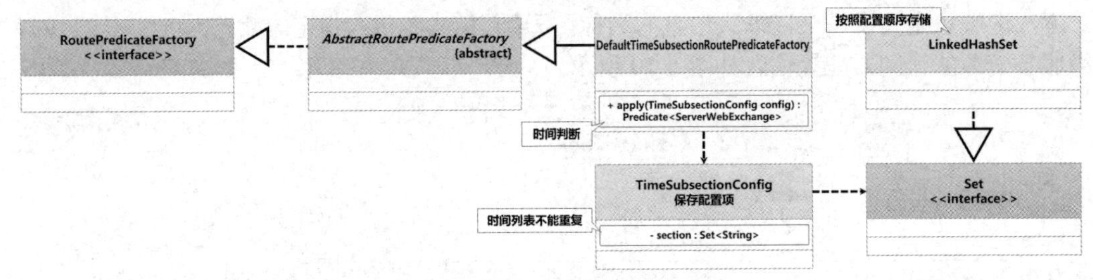

图 6-11　自定义路由谓词工厂类结构

在实现时间点访问的网关配置中，所有的时间点都需要用户根据需要自己进行配置，而为了避免配置的重复，可以通过 Set 集合进行数据的保存（同时考虑到配置顺序可以使用 LinkedList 子类）。为了便于所有配置项的管理，应该创建一个 TimeSubsectionConfig 配置类，这样用户在 application.yml 中的定义就会自动保存在此类之中，最终再利用自定义路由谓词工厂类（DefaultTimeSubsectionRoutePredicateFactory）实现访问处理。下面通过具体的步骤介绍如何实现。

（1）【gateway-9501 子模块】创建 TimeSubsectionConfig 配置类，用于保存 application.yml 定义的配置项。

```
package com.yootk.gateway.predicate.config;
public class TimeSubsectionConfig {
   private Set<String> section = new LinkedHashSet<>();        // 保存所有配置时间段
   public void setSection(List<String> section) {              // 不是标准的Setter方法
      this.section.addAll(section);
   }
   public Set<String> getSection() {
      return section;
   }
}
```

（2）【gateway-9501 子模块】创建自定义路由谓词工厂类。

```
package com.yootk.gateway.predicate.factory;
@Component
public class DefaultTimeSubsectionRoutePredicateFactory extends
        AbstractRoutePredicateFactory<TimeSubsectionConfig> {
    // 定义日期时间格式化模板，此时只获取当前日期时间中的“小时:分钟”部分
    private static final DateTimeFormatter FORMATTER =
                DateTimeFormatter.ofPattern("HH:mm");
    public DefaultTimeSubsectionRoutePredicateFactory() {
        super(TimeSubsectionConfig.class);                          // 传递配置类
    }
    @Override
    public Predicate<ServerWebExchange> apply(TimeSubsectionConfig config) { // 执行判断
        return serverWebExchange -> {
            String now = LocalTime.now().format(FORMATTER);     // 获取当前时间
            return config.getSection().contains(now);           // 时间判断
        };
    }
    @Override
    public List<String> shortcutFieldOrder() {
        // 定义属性名称，如果有多个属性则可以根据配置时出现的“,”进行前后顺序的区分
        return Collections.singletonList("section");
    }
    @Override
    public ShortcutType shortcutType() {                            // 设置内容使用“,”分隔
        return ShortcutType.GATHER_LIST;                            // 传递List集合
    }
}
```

为了方便配置的处理，本程序直接继承 AbstractRoutePredicateFactory 父类，这样可以方便地定义配置属性，以及配置项之间的分隔符（默认为“,”）。在进行访问验证时，如果当前的日期时间正好处于配置的时间点，则允许请求通过（apply()方法返回 true），否则不允许请求通过。

（3）【gateway-9501 子模块】修改 application.yml 配置访问时间项列表。

```
spring:                                                     # Spring配置
  cloud:                                                    # Spring Cloud配置
    gateway:                                                # 网关配置
      routes:                                               # 路由配置
        - id: dept                                          # 路由标记
          uri: lb://dept.provider                           # 服务资源匹配
          predicates:                                       # 路由谓词工厂
            - DefaultTimeSubsection=08:00,16:00,20:00,24:00 # 列表项
            - Path=/**                                      # 全部资源
```

配置完成后重新启动当前的网关应用，可以在后台发现已经自动新增了一个 DefaultTimeSubsection 路由谓词工厂类的配置项，而后就可以根据当前的时间来判断请求是否可以正常响应。

6.3 GatewayFilterFactory

网关过滤简介

视频名称 0608_【掌握】网关过滤简介

视频简介 为了便于网关进行自动的请求或响应处理，网关提供了过滤工厂的概念。本视频通过实例为读者简要说明过滤工厂的作用，并通过简单的配置讲解其具体使用方法。

微服务项目中引入网关之后，可以直接通过网关实现微服务的转发代理，而在进行微服务转发时，可以通过网关过滤器实现统一的请求或响应处理操作，如图 6-12 所示。例如，可以在请求或响应时添加固定的头信息数据，或者设置请求的状态等。Spring Cloud Gateway 中的网关有 pre（网

关转发请求之前）与 post（网关转发请求之后）两种状态。

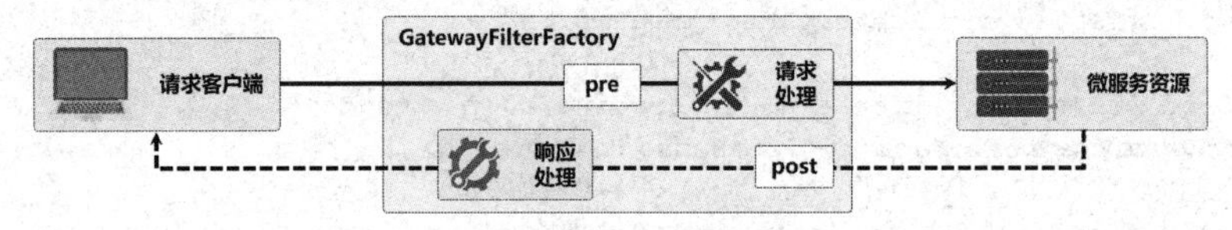

图 6-12　网关过滤器

范例：【gateway-9501 子模块】修改 application.yml 配置文件，追加过滤器配置

```
spring:                                                          # Spring配置
  cloud:                                                         # Spring Cloud配置
    gateway:                                                     # 网关配置
      routes:                                                    # 路由配置
        - id: dept                                               # 路由标记
          uri: lb://dept.provider                                # 服务资源匹配
          predicates:                                            # 路由谓词工厂
            - Path=/**                                           # 全部资源
          filters:                                               # 网关过滤工厂
            - AddRequestHeader=Request-Token-Muyan, www.yootk.com   # 添加头信息
```

此后每次进行网关代理时，在所有访问 dept.provider 微服务的请求上，都会通过过滤器自动添加一个名称为“Request-Token-Muyan”的头信息。

提示：通过微服务消费端观察头信息。

此时通过网关访问的微服务都会自动根据过滤器追加上一个名称为“Request-Token-Muyan”头信息数据，而这个数据可以在目标资源上直接获取。

范例：获取请求头信息

```
package com.yootk.provider.action;
@RestController                                          // REST控制器
@RequestMapping("/provider/dept/*")                      // 父路径
@Slf4j                                                   // 日志输出
public class DeptAction {
    @GetMapping("list")                                  // 子路径
    public Object list() {
        HttpServletRequest request = ((ServletRequestAttributes)
                RequestContextHolder.getRequestAttributes()).getRequest();
        log.info("【Dept微服务】Request-Token-Muyan = {}",
                    request.getHeader("Request-Token-Muyan"));
        return this.deptService.list();                  // 部门信息列表
    }
}
```

程序执行结果	【Dept微服务】Request-Token-Muyan = www.yootk.com

通过部门微服务调用方法的输出可以发现，通过网关设置的请求头信息已经发送给目标微服务，这样就可以利用过滤器实现请求或响应的修改。

另外需要提醒读者的是，如果此时用户的请求中有相同的头信息，则在经过网关后（设置的头信息名称相同）会出现覆盖，即最终传递到目标服务资源的头信息数据为网关过滤器配置的数据。

6.3.1　内置网关过滤工厂类

内置网关过滤工厂类

视频名称　0609_【掌握】内置网关过滤工厂类

视频简介　Spring Cloud Gateway 内置了许多过滤工厂类。本视频为读者分析过滤工厂类的继承结构，并介绍常用的内置过滤工厂类及其使用实例。

AddRequestHeader 过滤工厂类（对应 AddRequestHeaderGatewayFilterFactory 子类）实际上是一个系统内置的网关过滤工厂类，其基本继承结构如图 6-13 所示。

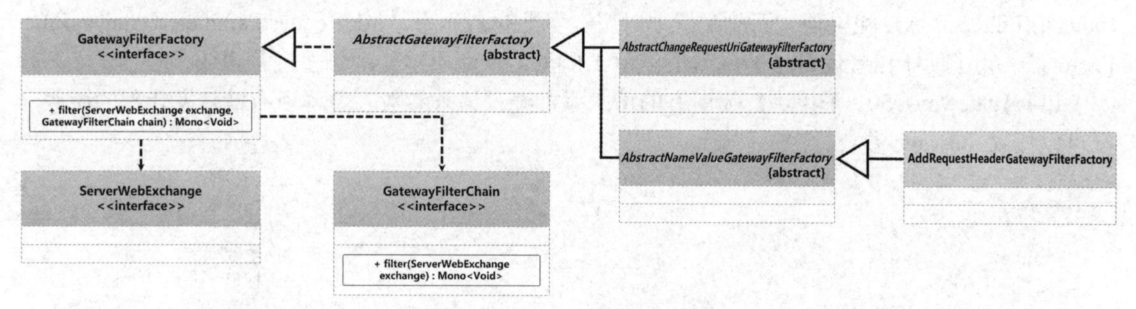

图 6-13　AddRequestHeader 过滤工厂类

可以发现，在 Spring Cloud Gateway 中所有的过滤工厂类全部实现 GatewayFilterFactory 父接口，同时在该类提供的 filter()方法中定义了 GatewayFilterChain 的过滤链处理类型，这样可以直接将用户请求交由下一个过滤器进行处理。表 6-2 为读者列出了内置的过滤工厂类。

表 6-2　Spring Cloud Gateway 内置的过滤工厂类

序号	网关过滤工厂名称	过滤工厂类	描述
01	AddRequestHeader	AddRequestHeaderGatewayFilterFactory	增加请求头信息
02	AddRequestParameter	AddRequestParameterGatewayFilterFactory	增加请求参数
03	AddResponseHeader	AddResponseHeaderGatewayFilterFactory	增加响应头信息
04	DedupeResponseHeader	DedupeResponseHeaderGatewayFilterFactory	剔除重复响应头信息
05	Hystrix	HystrixGatewayFilterFactory	为路由引入 Hystrix 断路器
06	FallbackHeaders	FallbackHeadersGatewayFilterFactory	添加失败回退头信息
07	PrefixPath	PrefixPathGatewayFilterFactory	为原始请求添加前缀
08	PreserveHostHeader	PreserveHostHeaderGatewayFilterFactory	是否要发送原始主机信息
09	RequestRateLimiter	RequestRateLimiterGatewayFilterFactory	实现令牌桶流控
10	RedirectTo	RedirectToGatewayFilterFactory	请求重定向 URL
11	RemoveRequestHeader	RemoveRequestHeaderGatewayFilterFactory	删除请求头信息
12	RemoveRequestParameter	RemoveRequestParameterGatewayFilterFactory	删除请求参数
13	RemoveResponseHeader	RemoveResponseHeaderGatewayFilterFactory	删除响应头信息
14	RewritePath	RewritePathGatewayFilterFactory	重写原始请求路径
15	RewriteResponseHeader	RewriteResponseHeaderGatewayFilterFactory	重写原始响应头信息
16	RequestHeaderToRequestUri	RequestHeaderToRequestUriGatewayFilterFactory	请求头信息转为请求 URI 地址
17	SaveSession	SaveSessionGatewayFilterFactory	转发前强制保存会话
18	SecureHeaders	SecureHeadersGatewayFilterFactory	添加与安全有关的请求头信息
19	SetPath	SetPathGatewayFilterFactory	修改原始请求路径
20	StripPrefix	StripPrefixGatewayFilterFactory	截断原始请求路径前缀
21	SetRequestHeader	SetRequestHeaderGatewayFilterFactory	修改原始请求头信息
22	SetResponseHeader	SetResponseHeaderGatewayFilterFactory	修改原始响应头信息
23	SetStatus	SetStatusGatewayFilterFactory	设置状态码
24	Retry	RetryGatewayFilterFactory	对指定的响应进行重试
25	RequestSize	RequestSizeGatewayFilterFactory	设置请求接收的最大数据包
26	ModifyRequestBody	ModifyRequestBodyGatewayFilterFactory	修改原始请求主体内容
27	ModifyResponseBody	ModifyResponseBodyGatewayFilterFactory	修改原始响应主体内容
28	MapRequestHeader	MapRequestHeaderGatewayFilterFactory	请求头转换

所有的内置过滤工厂类名称都应该使用 GatewayFilterFactory 进行标注，而后在 application.yml 中写上类型名称即可。例如，要添加请求参数，则使用 AddRequestParameter 作为“spring.cloud.gateway.routers[x].filters.Xxx”配置项，而此配置项对应实现类的名称为“AddRequestParameterGatewayFilterFactory”。为了便于读者理解这些内置的过滤工厂类，下面通过几个例子进行说明。

（1）【gateway-9501 子模块】为所有的请求添加统一请求参数，这样所有被转发代理的微服务都可以接收 message 参数。

```
spring:                                                     # Spring配置
  cloud:                                                    # Spring Cloud配置
    gateway:                                                # 网关配置
      routes:                                               # 路由配置
        - id: dept                                          # 路由标记
          uri: lb://dept.provider                           # 服务资源匹配
          predicates:                                       # 路由谓词工厂
            - Path=/**                                      # 全部资源
          filters:                                          # 网关过滤工厂
            - AddRequestHeader=Request-Token-Muyan, www.yootk.com   # 添加头信息
            - AddRequestParameter=message, edu.yootk.com    # 添加请求参数
```

（2）【gateway-9501 子模块】网关中提供了 MapRequestHeader 请求头转换过滤器，利用此过滤器可以将一个请求头名称转换为另一个请求头的名称，如果用户所发送的请求包含转换目标请求头，则不进行转换处理。

```
spring:                                                     # Spring配置
  cloud:                                                    # Spring Cloud配置
    gateway:                                                # 网关配置
      routes:                                               # 路由配置
        - id: dept                                          # 路由标记
          uri: lb://dept.provider                           # 服务资源匹配
          predicates:                                       # 路由谓词工厂
            - Path=/**                                      # 全部资源
          filters:                                          # 网关过滤工厂
            - MapRequestHeader=Request-Token-Muyan, Muyan-Yootk-Key  # 请求头转换
```

用户在发送请求时如果只传递了“Request-Token-Muyan”头信息，则此头信息会被映射为“Muyan-Yootk-Key”再发送给目标资源，这样两个头信息的内容是相同的；而如果此时的客户端传递了“Muyan-Yootk-Key”头信息，则目标资源接收时会接收用户传递的头信息内容。

（3）【gateway-9501 子模块】请求重定向到指定路径。

```
spring:                                                     # Spring配置
  cloud:                                                    # Spring Cloud配置
    gateway:                                                # 网关配置
      routes:                                               # 路由配置
        - id: dept                                          # 路由标记
          uri: lb://dept.provider                           # 服务资源匹配
          predicates:                                       # 路由谓词工厂
            - Path=/**                                      # 全部资源
          filters:                                          # 网关过滤工厂
            - RemoveRequestHeader=Request-Token-Muyan       # 删除头信息
            - RedirectTo=302, https://www.yootk.com         # 必须设置3xx状态码
```

网关在进行请求转发时会首先通过 RemoveRequestHeader 过滤工厂类删除“Request-Token-Muyan”的请求头信息内容，而后利用“RedirectTo”跳转到指定的路径进行响应。

6.3.2　自定义过滤工厂类

自定义过滤工厂类

视频名称　0610_【掌握】自定义过滤工厂类

视频简介　Spring Cloud Gateway 支持用户自定义过滤工厂类的配置实现。本视频为读者讲解自定义过滤工厂类实现的基本结构，并通过实例介绍如何开发用于日志记录的网关过滤工厂类。

不同的项目对于网关的过滤有不同的处理要求，而这些要求在内置过滤工厂类中不一定都有实现，所以开发者需要根据自身的需要来创建属于自己的过滤工厂类。可以直接实现 GatewayFilterFactory 或继承相关抽象子类来进行定义。本节将介绍如何实现一个用于访问日志记录的过滤工厂类，如图 6-14 所示。

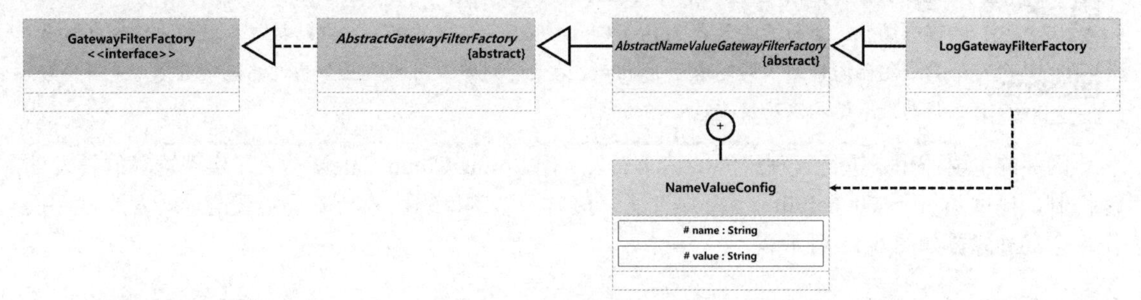

图 6-14　自定义过滤工厂类

为了便于开发，本次将直接继承 AbstractNameValueGatewayFilterFactory 抽象类。由于在自定义过滤工厂类时可以配置参数名称以及定义内容，因此可以直接在 LogGatewayFilterFactory 子类中利用 NameValueConfig 内部类来进行配置的接收。下面通过具体的步骤进行讲解。

（1）【gateway-9501 子模块】创建日志网关过滤工厂类。

```
package com.yootk.gateway.filter;
@Slf4j
@Component
public class LogGatewayFilterFactory extends AbstractNameValueGatewayFilterFactory {
    @Override
    public GatewayFilter apply(NameValueConfig config) {
        return (exchange, chain) -> {
            ServerHttpRequest request = exchange.getRequest().mutate().build();
            ServerWebExchange webExchange = exchange.mutate().request(request).build();
            log.info("配置参数：{}, {}", config.getName(), config.getValue());
            log.info("请求路径：{}、请求模式：{}", request.getPath(), request.getMethod());
            return chain.filter(webExchange);
        };
    }
}
```

后台日志输出	c.y.g.filter.LogGatewayFilterFactory　: 配置参数：muyan, yootk c.y.g.filter.LogGatewayFilterFactory　: 请求路径：/provider/dept/list、请求模式：GET

本程序通过 NameValueConfig 对象实例获取 application.yml 中配置的参数以及内容。为了便于读者观察，直接将配置项以日志的形式输出，同时输出当前请求的模式以及访问路径。

（2）【gateway-9501 子模块】在 application.yml 中配置日志过滤工厂类，并定义参数。

```
spring:                                              # Spring配置
  cloud:                                             # Spring Cloud配置
    gateway:                                         # 网关配置
      routes:                                        # 路由配置
        - id: dept                                   # 路由标记
          uri: lb://dept.provider                    # 服务资源匹配
          predicates:                                # 路由谓词工厂
            - Path=/**                               # 全部资源
          filters:                                   # 网关过滤工厂
            - Log=muyan, yootk                       # 网关日志
            - RemoveRequestHeader=Request-Token-Muyan  # 删除头信息
```

配置完成后，每当用户发出请求都会通过 LogGatewayFilterFactory 类进行访问日志的记录。由于此时配置有多个过滤器，因此所有的过滤器会按照顺序依次执行。

6.4 全局过滤器

全局过滤器简介

视频名称 0611_【掌握】全局过滤器简介

视频简介 全局过滤器是 Spring Cloud Gateway 中的重要组成结构。本视频为读者分析全局过滤器的作用，同时基于 Actuator 为读者列举常用的内置全局过滤器的名称，并简要分析其继承结构以及核心过滤器的作用。

网关的主要作用是进行代理资源的请求转发，而 Spring Cloud Gateway 为了实现这样的转发处理机制，提供了一个 GlobalFilter（全局过滤）接口，该接口的核心功能是实现请求转发的处理操作。全局过滤器如图 6-15 所示。

图 6-15 全局过滤器

通过图 6-15 可以发现，网关的请求最终都是由 GlobalFilter 实现转发处理的，所以开发者可以通过全局过滤器来实现统一的业务处理功能，如认证与授权检测、服务流控等。而要想实现网关的正常运转，实际上可在应用中进行许多配置来实现全局过滤器。可以采用如下步骤通过 Actuator 对之进行加载。

（1）【microcloud 项目】修改 build.gradle 配置文件，在 gateway-9501 子模块中添加 Actuator 相关依赖。

```
implementation('org.springframework.boot:spring-boot-starter-actuator')
```

（2）【gateway-9501 子模块】修改 application.yml 并开启全部 Actuator 访问终端。

```
management:                                  # Actuator监控配置
  server:                                    # 服务配置
    port: 9090                               # 监听接口
  endpoints:                                 # 监控端点
    web:                                     # Web接口
      exposure:                              # 访问终端配置
        include: "*"                         # 访问全部端点
      base-path: /actuator                   # 监控访问路径
```

（3）【Postman】配置完成后重新启动网关应用，此时可以通过/actuator/gateway/globalfilters 路径获取全部内置的全局过滤器信息，如图 6-16 所示。

图 6-16 内置的全局过滤器

通过此时的 Actuator 监控信息可以发现，系统内部提供了很多内置的全局过滤器。正是这些过滤器的存在，让用户可以方便地通过 application.yml 实现网关的配置。表 6-3 为读者列出了几个常用的全局过滤器。

表 6-3　常用全局过滤器

序号	全局过滤器	描述
01	自定义全局过滤器	通过 GlobalFilter 创建若干全局过滤器，并通过 Order 注解配置顺序
02	ForwardRoutingFilter	实现本地网关内的 forward 转发，不再转发到下游微服务
03	NettyRoutingFilter	基于 Netty 实现下游 HTTP 或 HTTPS 资源转发
04	NettyWriteResponseFilter	将代理响应写回网关客户端
05	ReactiveLoadBalancerClientFilter	根据微服务名称（负载均衡）实现下游资源转发
06	RouteToRequestUrlFilter	将请求包含的原始 URL 转换成网关进行请求转发所使用的 URL
07	GatewayMetricsFilter	监控整合过滤
08	WebsocketRoutingFilter	转发 WebSocket 请求

打开这些全局过滤器可以发现，所有的过滤器都实现了 org.springframework.cloud.gateway. filter. GlobalFilter 接口，该接口只提供一个过滤的处理方法。

范例：GlobalFilter 接口定义

```
package org.springframework.cloud.gateway.filter;
public interface GlobalFilter {
  Mono<Void> filter(ServerWebExchange exchange, GatewayFilterChain chain);
}
```

在该接口中可以通过 filter()方法来实现具体的过滤操作，同时在该方法中可以通过 ServerWebExchange 接口获取请求与响应的有关信息，而后通过 GatewayFilterChain 接口实现对其他过滤器的调用。图 6-17 为读者展示了 GlobalFilter 继承结构。

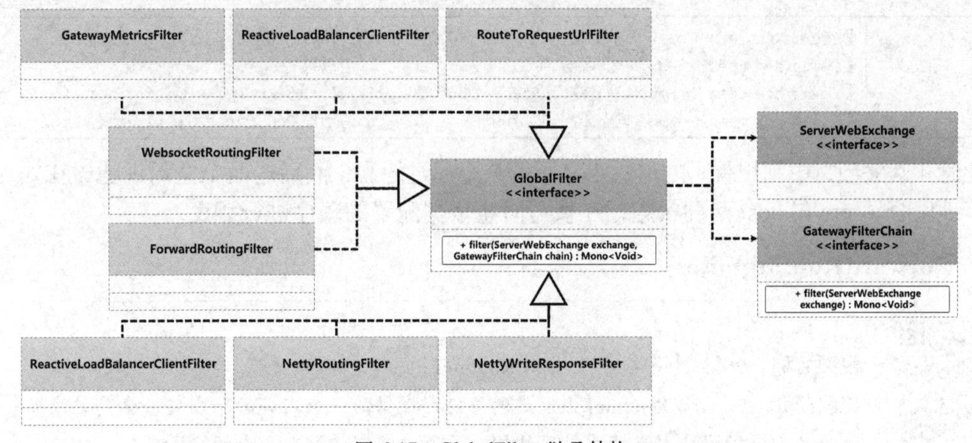

图 6-17　GlobalFilter 继承结构

6.4.1　自定义全局过滤器

视频名称　0612_【掌握】自定义全局过滤器

视频简介　Spring Cloud Gateway 根据全局过滤器实现请求转发的相关处理操作。本视频通过实例介绍如何自定义全局过滤器，并通过 Order 注解介绍如何实现过滤器执行顺序的配置。

虽然 Spring Cloud Gateway 提供许多内置的全局过滤器，但是开发者依然可以通过 GlobalFilter

接口进行自定义全局过滤器的配置，同时也可以利用@Order 注解来配置执行顺序。下面通过具体的实例进行介绍。

范例：【gateway-9501 子模块】创建全局过滤器，同时在该过滤器中提供两个过滤处理操作

```
package com.yootk.gateway.filter.global;
@Configuration
@Slf4j
public class CombinedGlobalFilter {                    // 全局过滤器配置
    @Bean
    @Order(-2)                                         // 数字越小优先级越高
    public GlobalFilter getFirstFilter() {             // 定义全局过滤器
        return (exchange, chain) -> {
            log.info("【FirstFilter - pre状态】请求ID：{}。请求路径：{}",
                    exchange.getRequest().getId(), exchange.getRequest().getPath());
            return chain.filter(exchange).then(Mono.fromRunnable(() -> {
                log.info("【FirstFilter - post状态】请求ID：{}。请求路径：{}",
                        exchange.getRequest().getId(), exchange.getRequest().getPath());
            }));
        };
    }
    @Bean
    @Order(-1)                                         // 数字越小优先级越高
    public GlobalFilter getSecondFilter() {            // 定义全局过滤器
        return (exchange, chain) -> {
            log.info("【SecondFilter - pre状态】请求ID：{}。请求路径：{}",
                    exchange.getRequest().getId(), exchange.getRequest().getPath());
            return chain.filter(exchange).then(Mono.fromRunnable(() -> {
                log.info("【SecondFilter - post状态】请求ID：{}。请求路径：{}",
                        exchange.getRequest().getId(), exchange.getRequest().getPath());
            }));
        };
    }
}
```

程序执行结果	【FirstFilter - pre状态】请求ID：866d4b14-1。请求路径：/provider/dept/list 【SecondFilter - pre状态】请求ID：866d4b14-1。请求路径：/provider/dept/list 【SecondFilter - post状态】请求ID：866d4b14-1。请求路径：/provider/dept/list 【FirstFilter - post状态】请求ID：866d4b14-1。请求路径：/provider/dept/list

此时在网关项目中利用编码的形式定义了两个全局过滤器，而后这两个全局过滤器在处理时已经明确地区分了 pre 与 post 两种状态，在执行时也依据配置的顺序进行触发。

6.4.2 ForwardRoutingFilter

ForwardRouting Filter

视频名称 0613_【掌握】ForwardRoutingFilter

视频简介 ForwardRoutingFitler 是网关内部提供的一个内部的路由转发全局过滤器。本视频为读者分析该过滤器的源代码结构，并通过实例讲解网关内的路由转发。

Spring Cloud Gateway 网关是基于 WebFlux 技术开发出来的，这样除了可以实现其下游资源的转发之外，也可以在网关的内部定义所需的服务接口，采用“forward://路径”的形式实现，如图 6-18 所示。

通过图 6-18 可以清楚地发现，在用户发送“/globalforward”访问路径时，程序会自动匹配路由 ID 为“forward_example”的配置项，而后根据路由网关过滤工厂配置的前缀与访问路径实现网关内部资源的调用。下面就依据当前的配置介绍具体的操作实现，开发步骤如下。

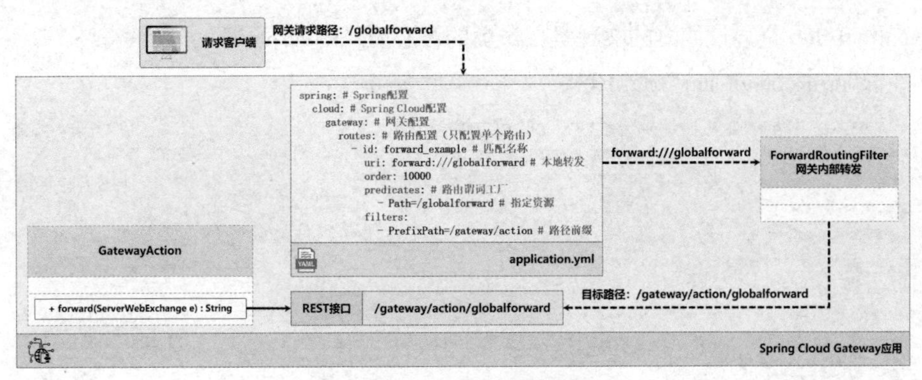

图 6-18　网关内部转发

（1）【gateway-9501 子模块】在网关内部定义一个访问资源，用于实现内部转发操作。

```
package com.yootk.gateway.action;
@RestController
@RequestMapping("/gateway/action/*")
public class GatewayAction {
    @RequestMapping("globalforward")
    public Map<String, String> forward(ServerWebExchange exchange) { // 跳转资源
        Map<String, String> result = new HashMap<>();              // 保存结果
        result.put("message", "forward");                           // 提示信息
        result.put("requestId", exchange.getRequest().getId());     // 请求ID
        result.put("requestPath", exchange.getRequest().getPath().toString()); // 请求路径
        return result;
    }
}
```

（2）【gateway-9501 子模块】修改 application.yml，配置一个本地转发路由。

```
spring:                                          # Spring配置
  cloud:                                         # Spring Cloud配置
    gateway:                                     # 网关配置
      routes:                                    # 路由配置（只配置单个路由）
        - id: forward_example                    # 匹配名称
          uri: forward:///globalforward          # 本地转发
          predicates:                            # 路由谓词工厂
            - Path=/globalforward                # 指定资源
          filters:                               # 网关过滤工厂
            - PrefixPath=/gateway/action         # 路径前缀
```

（3）【Postman】利用测试工具访问 forward 配置的访问路径。通过最终的执行结果可以发现，此时的访问已经由网关进行了内部资源的转发，如图 6-19 所示。

图 6-19　网关内部资源转发

通过以上执行过程可以清楚地看到“forward://”跳转的作用，而此跳转操作对应的是

ForwardRoutingFilter 全局过滤器的实现类，该类的核心源代码如下。

范例：ForwardRoutingFilter 核心源代码

```
package org.springframework.cloud.gateway.filter;
public class ForwardRoutingFilter implements GlobalFilter, Ordered {// 全局过滤器
    @Override                                                      // 覆写GlobalFilter接口方法
    public Mono<Void> filter(ServerWebExchange exchange, GatewayFilterChain chain) {
        URI requestUrl = exchange.getRequiredAttribute(GATEWAY_REQUEST_URL_ATTR);
        String scheme = requestUrl.getScheme();                    // 获取请求模式
        if (isAlreadyRouted(exchange) || !"forward".equals(scheme)) {  // 为forward模式
            return chain.filter(exchange);                         // 执行下一个过滤处理
        }
        if (log.isTraceEnabled()) {
            log.trace("Forwarding to URI: " + requestUrl);
        }
        return this.getDispatcherHandler().handle(exchange);       // DispatcherHandler处理
    }
}
```

通过 ForwardRoutingFilter 核心源代码可以发现，在每次执行"forward://"处理时，如果没有执行过，则交由 DispatcherHandler 类进行转发处理；而如果执行过，则继续执行下一个过滤。此类的继承结构如图 6-20 所示。

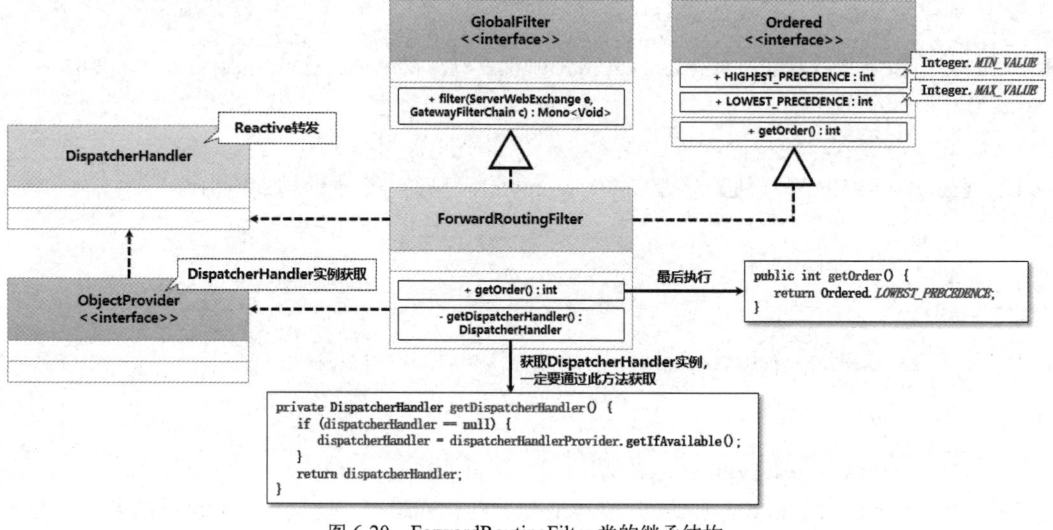

图 6-20　ForwardRoutingFilter 类的继承结构

6.4.3　Netty 全局路由

视频名称　0614_【掌握】Netty 全局路由

视频简介　Spring Cloud Gateway 可以直接实现指定资源的路由转发处理，而这一转发的操作是基于 Netty 实现的（NettyRoutingFilter 与 NettyWriteResponseFilter）。本视频为读者分析这两个内置全局过滤器的使用。

NettyRoutingFilter 是由 Netty 实现的网关全局过滤器，其可以根据"http://"或"https://"前缀实现过滤处理，使用基于 Netty 实现的 HttpClient 向后端微服务发出资源请求；而微服务资源进行响应时，可以通过 NettyWriteResponseFilter 将响应数据发送给请求客户端，如图 6-21 所示。

NettyRoutingFilter 与 NettyWriteResponseFilter 两个类都实现了 GlobalFilter 父接口，其中 NettyRoutingFilter 类最终要通过 HttpClient 实现类进行资源调用，所以在该类对象实例化时需要传入 HttpClient、HttpClientProperties 对象实例；而 NettyWriteResponseFilter 实现类由于需要进行数据

响应处理，因此在使用构造方法实例化时需要传入 MediaType 对象实例。Netty 全局路由实现结构如图 6-22 所示。

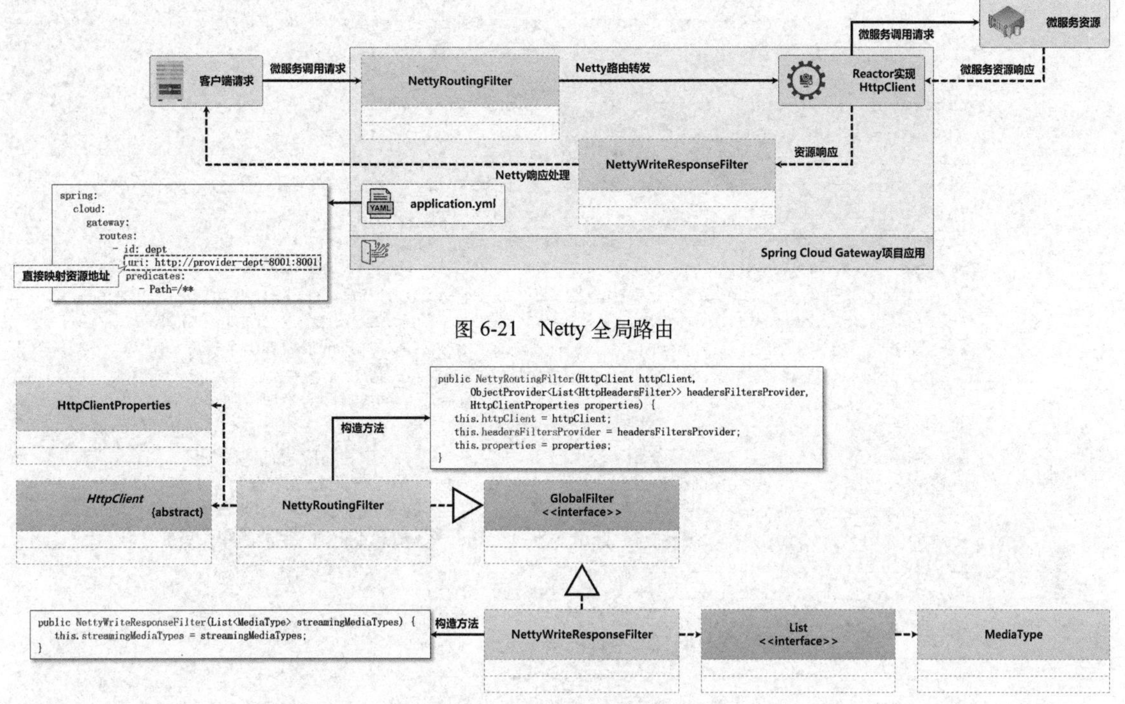

图 6-21 Netty 全局路由

```
public NettyRoutingFilter(HttpClient httpClient,
        ObjectProvider<List<HttpHeadersFilter>> headersFiltersProvider,
        HttpClientProperties properties) {
    this.httpClient = httpClient;
    this.headersFiltersProvider = headersFiltersProvider;
    this.properties = properties;
}
```

```
public NettyWriteResponseFilter(List<MediaType> streamingMediaTypes) {
    this.streamingMediaTypes = streamingMediaTypes;
}
```

图 6-22 Netty 全局路由实现结构

范例：NettyRoutingFilter 核心源代码

```
package org.springframework.cloud.gateway.filter;
public class NettyRoutingFilter implements GlobalFilter, Ordered {
   @Override
   @SuppressWarnings("Duplicates")
   public Mono<Void> filter(ServerWebExchange exchange, GatewayFilterChain chain) {
      URI requestUrl = exchange.getRequiredAttribute(GATEWAY_REQUEST_URL_ATTR);
      String scheme = requestUrl.getScheme();        // 获取当前请求模式
      if (isAlreadyRouted(exchange)                  // 已经处理过路由
            || (!"http".equals(scheme) && !"https".equals(scheme))) {// 不是http或https
         return chain.filter(exchange);              // 交给下一个过滤器处理
      }
      setAlreadyRouted(exchange);                    // 设置已路由标记
      ServerHttpRequest request = exchange.getRequest();
      final HttpMethod method = HttpMethod.valueOf(request.getMethodValue());
      final String url = requestUrl.toASCIIString();
      HttpHeaders filtered = filterRequest(getHeadersFilters(), exchange);
      final DefaultHttpHeaders httpHeaders = new DefaultHttpHeaders();
      filtered.forEach(httpHeaders::set);            // 设置头信息
      boolean preserveHost = exchange
            .getAttributeOrDefault(PRESERVE_HOST_HEADER_ATTRIBUTE, false);
      Route route = exchange.getAttribute(GATEWAY_ROUTE_ATTR);     // 获取路由配置
      Flux<HttpClientResponse> responseFlux = getHttpClient(route, exchange)
            .headers(headers -> {                    // 配置请求头信息
               headers.add(httpHeaders);             // 设置头信息内容
               headers.remove(HttpHeaders.HOST);     // 删除主机信息
               if (preserveHost) {                   // 主机保护配置
                  String host = request.getHeaders().getFirst(HttpHeaders.HOST);
                  headers.add(HttpHeaders.HOST, host);
               }
```

```
        }).request(method).uri(url).send((req, nettyOutbound) -> { // 发送请求
            if (log.isTraceEnabled()) {
                nettyOutbound
                        .withConnection(connection -> log.trace("outbound route: "
                                + connection.channel().id().asShortText()
                                + ", inbound: " + exchange.getLogPrefix()));
            }
            return nettyOutbound.send(request.getBody().map(this::getByteBuf));
        }).responseConnection((res, connection) -> {        // 接收响应
            exchange.getAttributes().put(CLIENT_RESPONSE_ATTR, res);
            exchange.getAttributes().put(CLIENT_RESPONSE_CONN_ATTR, connection);
            ServerHttpResponse response = exchange.getResponse();
            HttpHeaders headers = new HttpHeaders();
            res.responseHeaders().forEach(                      // 配置响应头信息
                    entry -> headers.add(entry.getKey(), entry.getValue()));
            String contentTypeValue = headers.getFirst(HttpHeaders.CONTENT_TYPE);
            if (StringUtils.hasLength(contentTypeValue)) {  // 判断请求内容类型长度
                exchange.getAttributes().put(ORIGINAL_RESPONSE_CONTENT_TYPE_ATTR,
                        contentTypeValue);                      // 保存响应类型
            }
            setResponseStatus(res, response);                   // 设置响应状态
            HttpHeaders filteredResponseHeaders = HttpHeadersFilter.filter(
                    getHeadersFilters(), headers, exchange, Type.RESPONSE);
             // 如果“transfer-encoding”与“content-length”头信息不正确，则删除相关头信息
            if (!filteredResponseHeaders
                    .containsKey(HttpHeaders.TRANSFER_ENCODING)
                    && filteredResponseHeaders
                        .containsKey(HttpHeaders.CONTENT_LENGTH)) {
                response.getHeaders().remove(HttpHeaders.TRANSFER_ENCODING);
            }
            exchange.getAttributes().put(CLIENT_RESPONSE_HEADER_NAMES,
                filteredResponseHeaders.keySet());              // 保存响应头信息属性
            response.getHeaders().putAll(filteredResponseHeaders);
            return Mono.just(res);
        });
    Duration responseTimeout = getResponseTimeout(route);   // 获取响应超时时间
    if (responseTimeout != null) {                          // 超时配置不为空
        responseFlux = responseFlux
            .timeout(responseTimeout, Mono.error(new TimeoutException(
                "Response took longer than timeout: " + responseTimeout)))
            .onErrorMap(TimeoutException.class,
                th -> new ResponseStatusException(HttpStatus.GATEWAY_TIMEOUT,
                    th.getMessage(), th));                      // 超时处理
    }
    return responseFlux.then(chain.filter(exchange));       // 交给其他过滤器进行异步响应
  }
}
```

通过 NettyRoutingFilter.filter()方法可以发现，此过滤器会向访问终端发出请求，同时会根据用户的配置进行相关头信息的传递。访问终端在接收请求时并没有进行响应的具体处理，而是将其交给了其他的过滤器，即 NettyWriteResponseFilter 过滤器类来进行处理。下面来观察此类的核心源代码。

范例：NettyWriteResponseFilter 类核心源代码

```
package org.springframework.cloud.gateway.filter;
public class NettyWriteResponseFilter implements GlobalFilter, Ordered {
  @Override
  public Mono<Void> filter(ServerWebExchange exchange, GatewayFilterChain chain) {
    return chain.filter(exchange)                               // 请求响应时处理
        .doOnError(throwable -> cleanup(exchange))              // 错误处理
        .then(Mono.defer(() -> {
          Connection connection = exchange.getAttribute(CLIENT_RESPONSE_CONN_ATTR);
```

```
            if (connection == null) {                                       // 没有连接实例
                return Mono.empty();                                        // 响应空信息
            }
            if (log.isTraceEnabled()) {
                log.trace("NettyWriteResponseFilter start inbound...");
            }
            ServerHttpResponse response = exchange.getResponse();
            final Flux<DataBuffer> body = connection.inbound().receive().retain()
                    .map(byteBuf -> wrap(byteBuf, response));               // 数据转为DataBuffer实例
            MediaType contentType = null;                                   // 响应类型
            try {
                contentType = response.getHeaders().getContentType();
            } catch (Exception e) { }
            return (isStreamingMediaType(contentType)
                    ? response.writeAndFlushWith(body.map(Flux::just)) // 数据响应
                    : response.writeWith(body));
        })).doOnCancel(() -> cleanup(exchange));
    }
}
```

可以发现 NettyWriteResponseFilter 并没有对 pre 状态做任何处理，只是对 post 状态进行了处理，并将获取的响应内容以 DataBuffer 的形式进行了异步响应。

6.4.4 ReactiveLoadBalancerClientFilter

ReactiveLoad BalancerClient Filter

视频名称 0615_【掌握】ReactiveLoadBalancerClientFilter

视频简介 Spring Cloud Gateway 为了提高服务的处理性能，提供了 Reactive 响应式编程处理支持。本视频为读者分析默认负载均衡转发的实现子类，并基于依赖库以及配置文件开启 ReactiveLoadBalancerClientFilter 全局过滤器。

网关可以采用“lb://服务名称”的形式实现某一个微服务集群的转发代理操作，同时在进行调用时还提供负载均衡的支持。而 Spring Cloud Gateway 中针对负载均衡代理有两个 GlobalFilter 实现类：默认的负载均衡代理类（LoadBalancerClientFilter）、响应式负载均衡代理类（ReactiveLoadBalancerClientFilter），如图 6-23 所示。

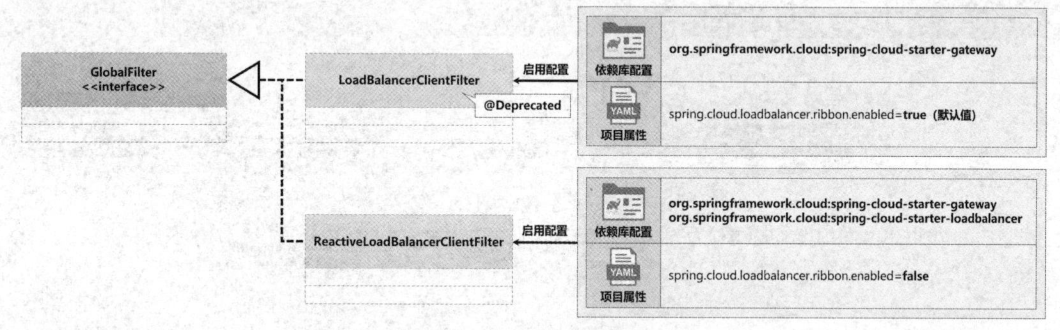

图 6-23 负载均衡全局过滤器

在默认情况下，只要在网关项目中引入了“spring-cloud-starter-gateway”依赖库就可以通过 LoadBalancerClientFilter 类实现负载均衡代理操作，但是这个实现子类已经使用了@Deprecated 注解，故不再推荐使用，所以开发者就需要通过依赖库配置与 application.yml 配置的形式来启用 ReactiveLoadBalancerClientFilter 子类，这样才可以获得更好的处理性能。下面通过具体的步骤讲解此全局过滤器的启用。

（1）【microcloud 项目】修改 build.gradle 配置文件，为 gateway-9501 子模块添加所需要的依赖。

```
implementation('org.springframework.cloud:spring-cloud-starter-loadbalancer')
```

（2）【gateway-9501 子模块】修改 application.yml 配置文件，关闭默认的负载均衡配置。

```
spring:                                      # Spring配置
  cloud:                                     # Spring Cloud配置
    loadbalancer:                            # 负载均衡配置
      ribbon:                                # Ribbon组件
        enabled: false                       # 关闭默认配置
```

（3）【Postman】配置完成后如果想验证当前的配置是否正确，则可以通过 Actuator 接口查看当前全局过滤器的配置。

```
http://gateway-9501:9090/actuator/gateway/globalfilters
```

程序执行结果	"org.springframework.cloud.gateway.filter.ReactiveLoadBalancerClientFilter@79d0b688": 10150

通过执行结果可以发现，网关中所使用的负载均衡加载过滤器已经更换为 ReactiveLoadBalancerClientFilter 实现类，这样在网关处理时就可以基于响应式编程的模式为系统提供更好的处理性能。

提示：Loadbalancer 模块与 Caffeine 缓存。

在项目引入 spring-cloud-starter-loadbalancer 依赖后，每当进行网关应用启动时，会在控制台输出如下警告信息:

```
iguration$LoadBalancerCaffeineWarnLogger : Spring Cloud LoadBalancer is currently working with the default cache. You can switch to using Caffeine cache, by adding it to the classpath.
```

该警告信息主要是告诉开发者现在项目中缺少 Caffeine 依赖库，因为 Spring Boot 2.0 之后的开发版本中，默认使用的本地缓存是 Caffeine 组件。为了解决此警告问题，可以在项目中添加如下依赖库。

范例：添加依赖库

```
implementation('com.github.ben-manes.caffeine:caffeine:3.0.2')
```

Caffeine 是一款高性能的内存缓存组件，是 Java 8 对 Guava 缓存的重写版本，可以提供接近最佳的命中率，基于 LRU 算法实现，同时支持多种缓存过期策略。

ReactiveLoadBalancerClientFilter 过滤器是基于 Reactive 响应式开发框架实现的，其可以基于注册中心的定义来获取所需要的服务实例对象。此类的继承结构如图 6-24 所示。

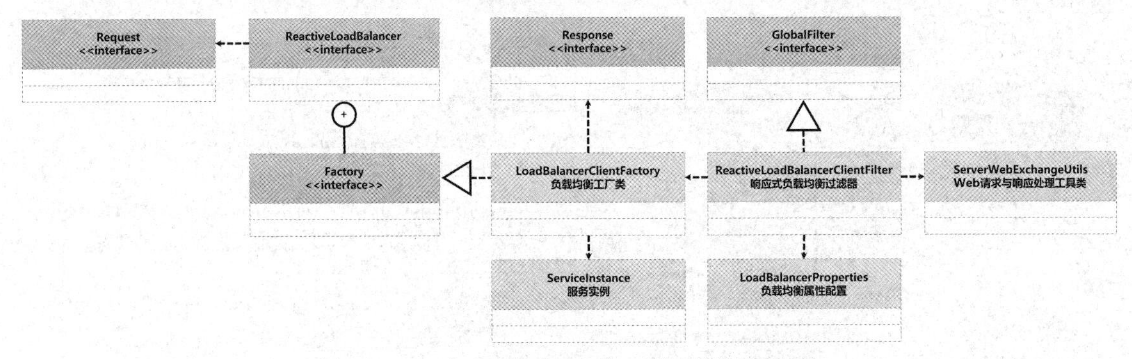

图 6-24　ReactiveLoadBalancerClientFilter 类的继承结构

范例：ReactiveLoadBalancerClientFilter 类核心源代码

```
package org.springframework.cloud.gateway.filter;
public class ReactiveLoadBalancerClientFilter implements GlobalFilter, Ordered {
   @Override
   @SuppressWarnings("Duplicates")
   public Mono<Void> filter(ServerWebExchange exchange, GatewayFilterChain chain) {
      URI url = exchange.getAttribute(GATEWAY_REQUEST_URL_ATTR);
      String schemePrefix = exchange.getAttribute(GATEWAY_SCHEME_PREFIX_ATTR);
      if (url == null
```

```
        || (!"lb".equals(url.getScheme())
        && !"lb".equals(schemePrefix))) {                // 判断是否为lb开头
      return chain.filter(exchange);                     // 不是lb模式，交给下一个过滤器
    }
    addOriginalRequestUrl(exchange, url);                // 保存原始请求路径
    if (log.isTraceEnabled()) {
      log.trace(ReactiveLoadBalancerClientFilter.class.getSimpleName() + " … ");
    }
    return choose(exchange).doOnNext(response -> {       // 获取服务实例并进行响应处理
      if (!response.hasServer()) {
        throw NotFoundException.create(properties.isUse404(),…);
      }
      URI uri = exchange.getRequest().getURI();          // 获取资源路径
      String overrideScheme = null;
      if (schemePrefix != null) {                        // 模式前缀不为空
        overrideScheme = url.getScheme();                // 获取前缀
      }
      DelegatingServiceInstance serviceInstance = new DelegatingServiceInstance(
            response.getServer(), overrideScheme);       // 获取服务实例对象
      URI requestUrl = reconstructURI(serviceInstance, uri);
      if (log.isTraceEnabled()) {
        log.trace("LoadBalancerClientFilter url chosen: " + requestUrl);
      }
      exchange.getAttributes().put(GATEWAY_REQUEST_URL_ATTR, requestUrl);
    }).then(chain.filter(exchange));                     // 处理完成后交给下个过滤器处理
  }
  protected URI reconstructURI(ServiceInstance serviceInstance, URI original) {
    return LoadBalancerUriTools.reconstructURI(serviceInstance, original);
  }
  private Mono<Response<ServiceInstance>> choose(ServerWebExchange exchange) {
    URI uri = exchange.getAttribute(GATEWAY_REQUEST_URL_ATTR);
    ReactorLoadBalancer<ServiceInstance> loadBalancer = this.clientFactory
        .getInstance(uri.getHost(), ReactorLoadBalancer.class,
            ServiceInstance.class);                      // 获取负载均衡选择器实例
    if (loadBalancer == null) {
      throw new NotFoundException("No loadbalancer available for " + uri.getHost());
    }
    return loadBalancer.choose(createRequest());         // 随机选择一个服务实例
  }
  private Request createRequest() {                      // 创建请求
    return ReactiveLoadBalancer.REQUEST;
  }
}
```

从此时的实现子类中可以清楚地发现，整体的处理操作基于响应式编程模式完成。随后在每次请求时，如果发现采用的是 lb 模式，则会通过 ReactorLoadBalancer 接口实例来获取一个 ServiceInstance（集群中任意一个节点实例）以实现请求资源的转发。

6.4.5 GatewayMetricsFilter

GatewayMetrics Filter

视频名称 0616_【掌握】GatewayMetricsFilter

视频简介 Spring Cloud Gateway 内部提供了网关服务度量监控数据。本视频为读者展示监控接口的开启，并介绍如何基于 Prometheus 与 Grafana 实现数据采集与可视化监控显示。

网关中除了可实现代理转发的过滤器之外，也提供对服务监控度量过滤器的支持。开发者可以利用此过滤器并结合 Actuator 服务监控网关服务数据，同时这些数据还可以方便地通过 Prometheus 采集与 Grafana 进行可视化处理，如图 6-25 所示。

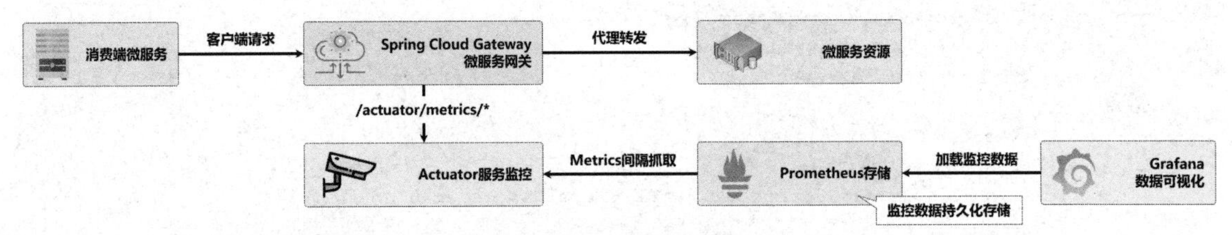

图 6-25 网关监控数据持久化

提示：关于 Prometheus 与 Grafana 服务安装与配置。

本套丛书中关于 Spring Boot 的书已经为读者详细讲解了 Actuator 监控服务处理，并且详细演示了 Prometheus 与 Grafana 服务实现。读者如果不熟悉可以翻阅相关图书。本次演示将以最终的网关数据采集介绍为主，不介绍具体的服务搭建。

（1）【microcloud 项目】修改 build.gradle 配置文件，为 gateway-9501 子模块添加所需要的依赖库。

```
implementation('io.micrometer:micrometer-registry-prometheus:1.7.0')
implementation('io.micrometer:micrometer-core:1.7.0')
```

（2）【gateway-9501 子模块】修改 application.yml 配置文件，启用 metrics 监控。

```
spring:                                    # Spring配置
  cloud:                                   # Spring Cloud配置
    gateway:                               # 网关配置
      metrics:                             # metrics配置
        enabled: true                      # 获取监控数据
```

（3）【Postman】网关监控数据配置完成就可以利用/actuator/metrics/gateway.requests 路径获取网关监控数据。图 6-26 展示了通过 Postman 测试工具实现的监控数据调用界面。

```
{
  "name": "gateway.requests",
  "description": null,
  "baseUnit": "seconds",
  "measurements": [
    { "statistic": "COUNT", "value": 6.0 },
    { "statistic": "TOTAL_TIME", "value": 0.1519575 },
    { "statistic": "MAX", "value": 0.0 }
  ],
  "availableTags": [
    { "tag": "routeUri", "values": [ "lb://dept.provider" ] },
    { "tag": "routeId", "values": [ "dept" ] },
    { "tag": "httpMethod", "values": [ "GET" ] },
    { "tag": "outcome", "values": [ "SUCCESSFUL" ] },
    { "tag": "status", "values": [ "OK" ] },
    { "tag": "httpStatusCode", "values": [ "200" ] }
  ]
}
```

图 6-26 调用网关监控数据

此时所返回的监控数据实际上主要有 3 项。

（4）【prometheus-server】修改 Prometheus 配置文件并配置网关服务的监控路径。

```
vi /usr/local/prometheus/prometheus.yml
```

`global:`	数据抓取全局参数
`  scrape_interval: 15s`	数据抓取间隔为 15 秒（默认为 1 分钟）
`  evaluation_interval: 15s`	规则评估间隔为 15 秒（默认为 1 分钟）

`scrape_configs:`	数据抓取配置
`  - job_name: 'microcloud-gateway'`	定义数据抓取作业并设置抓取目标（target）名称
`    scrape_interval: 10s`	每隔 10 秒抓取一次（局部配置生效）
`    scrape_timeout: 5s`	每次抓取超时时间
`    metrics_path: '/actuator/prometheus'`	数据抓取路径
`    static_configs:`	配置抓取主机列表
`      - targets: ['gateway-server:9090']`	抓取 Actuator 监控数据

（5）【prometheus-server】检查当前 Prometheus 配置是否正确。

```
/usr/local/prometheus/promtool check config /usr/local/prometheus/prometheus.yml
```

（6）【prometheus-server】启动 Prometheus 服务。

```
systemctl start prometheus;
```

（7）【Prometheus 控制台】查看 Prometheus 控制台（控制台地址：prometheus-server:9999），此时用户可以通过 3 个监控项来实现网关服务的监控，这 3 项分别为每秒请求数量（gateway_requests_seconds_count）、每秒最大请求量（gateway_requests_seconds_max）、每秒请求量总和（gateway_requests_seconds_sum）。图 6-27 展示了每秒请求量总和的统计。

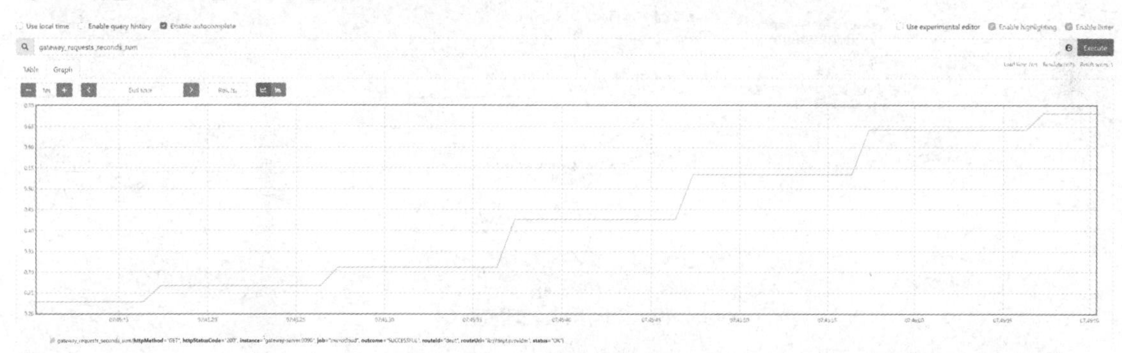

图 6-27　网关监控数据

（8）【grafana-server】为了实现更加丰富的可视化展示，可以启动 Grafana 服务。

```
systemctl start grafana;
```

（9）【Grafana 控制台】首先通过 Grafana 控制台（控制台地址：grafana-server:3000/），随后通过 Prometheus 添加所需要的监控项，并选择展示的图形效果，就可以得到图 6-28 所示的界面。

图 6-28　Grafana 可视化展示

6.5　Spring Cloud Gateway 工作原理

Spring Cloud Gateway 自动配置类

视频名称　0617_【理解】Spring Cloud Gateway 自动配置类

视频简介　Spring Cloud Gateway 基于 Starter 自动配置模式进行服务处理。本视频为读者分析网关请求代理的完整流程，并基于依赖库中的 spring.factories 介绍配置类的作用。

Spring Cloud Gateway 是基于 Reactor 技术实现的响应式服务。每当有用户向网关发出微服务调用请求时，HttpWebHandlerAdapter 类会接收所有的请求并创建一个网关上下文对象，随后在 DispatcherHandler 中通过 Predicate 对该请求上下文进行分发处理。如果路由匹配成功，则会执行后续的过滤链来访问指定微服务资源。其处理流程如图 6-29 所示。

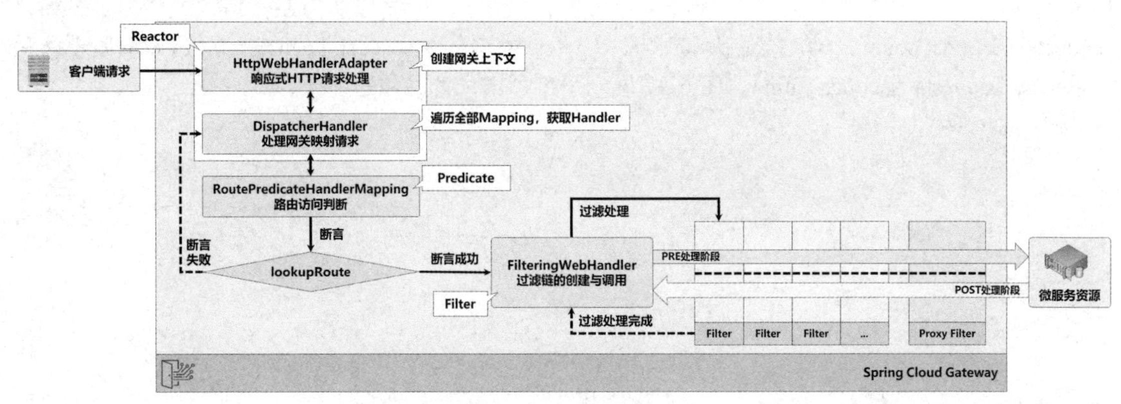

图 6-29　网关处理流程

在 Spring Cloud 中引入网关处理只需要在项目中配置"spring-cloud-starter-gateway"依赖库，而后就可以进行自动配置并启用网关代理支持。打开该模块内部的 pom.xml 配置文件可以清楚地发现，所有的网关操作的核心处理全部是由 spring-cloud-gateway-core 模块完成的，而所有的自动配置类也都在此模块中定义，如图 6-30 所示。

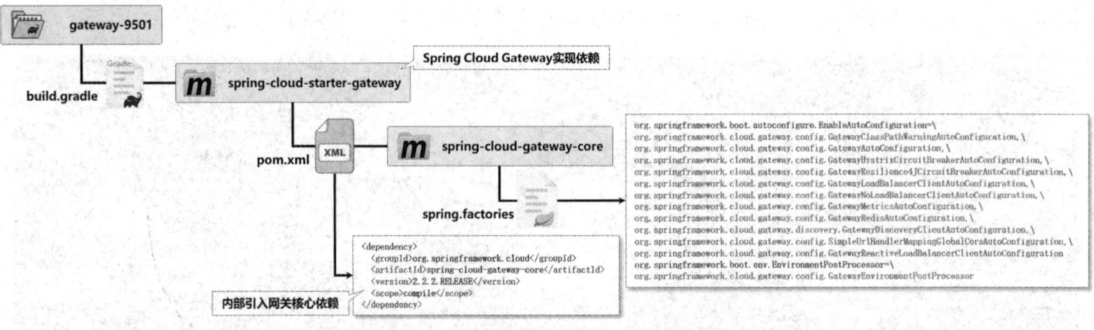

图 6-30　Spring Cloud Gateway 自动配置处理

打开 spring-cloud-gateway-core/META-INF/spring.factories 配置文件后可以发现，此文件定义了一系列的自动配置类。要想理解整个网关处理流程，就必须首先清楚这些自动配置类的作用，如表 6-4 所示。

实际上以上所列出的自动配置类很多都与前面讲解过的概念有关，例如，在使用网关时如果导入了 Spring MVC 的相关依赖，则会出现警告，这一警告信息是由 GatewayClassPathWarningAutoConfiguration 类负责的。我们学习过的 GatewayFilterFactory 与 GlobalFilter 实际上也都有对应的自动配置类，如图

6-31 所示。在整个 Spring Cloud Gateway 中最需要进行结构分析的就是 GatewayAutoConfiguration，下面对此类的结构进行分析。

表 6-4 Spring Cloud Gateway 自动配置类

序号	自动配置类	描述
01	o.s.c.g.c.GatewayClassPathWarningAutoConfiguration	类库检测警告处理类，网关不能使用 MVC 依赖
02	o.s.c.g.c.GatewayAutoConfiguration	网关自动配置类
03	o.s.c.g.c.GatewayHystrixCircuitBreakerAutoConfiguration	Hystrix 熔断自动配置类
04	o.s.c.g.c.GatewayResilience4JCircuitBreakerAutoConfiguration	滑动窗口统计自动配置类
05	o.s.c.g.c.GatewayLoadBalancerClientAutoConfiguration	默认负载均衡加载自动配置类
06	o.s.c.g.c.GatewayNoLoadBalancerClientAutoConfiguration	非负载均衡加载自动配置类
07	o.s.c.g.c.GatewayMetricsAutoConfiguration	网关度量统计自动配置类
08	o.s.c.g.c.GatewayRedisAutoConfiguration	基于 Redis 实现的流控处理自动配置类
09	o.s.c.g.d.GatewayDiscoveryClientAutoConfiguration	网关发现服务自动配置类
10	o.s.c.g.c.SimpleUrlHandlerMappingGlobalCorsAutoConfiguration	URL 全局处理映射自动配置类
11	o.s.c.g.c.GatewayReactiveLoadBalancerClientAutoConfiguration	响应式负载均衡加载自动配置类

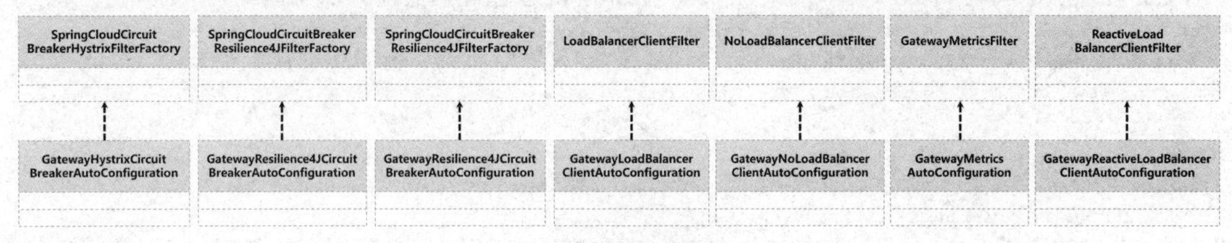

图 6-31 启动类关联

6.5.1 GatewayAutoConfiguration

GatewayAuto
Configuration

视频名称 0618_【理解】GatewayAutoConfiguration

视频简介 GatewayAutoConfiguration 是网关服务配置的核心实现类。本视频通过源代码介绍该类的关联结构，并讲解其与路由谓词工厂、网关过滤工厂之间的关联。

GatewayAutoConfiguration 是实现网关相关定义的自动配置类。由于 Spring Cloud Gateway 是基于 Reactor 结构开发的，因此在该类声明处会使用一系列的注解与响应式编程类结构产生关联。同时该类内部也会为读者实例化必要的对象，如路由定位器（RouteLocator）、过滤链（FilterWebHandler）、路由访问处理（HandlerMapping）、过滤器等，这样就可以得到图 6-32 所示的 Bean 关联结构。

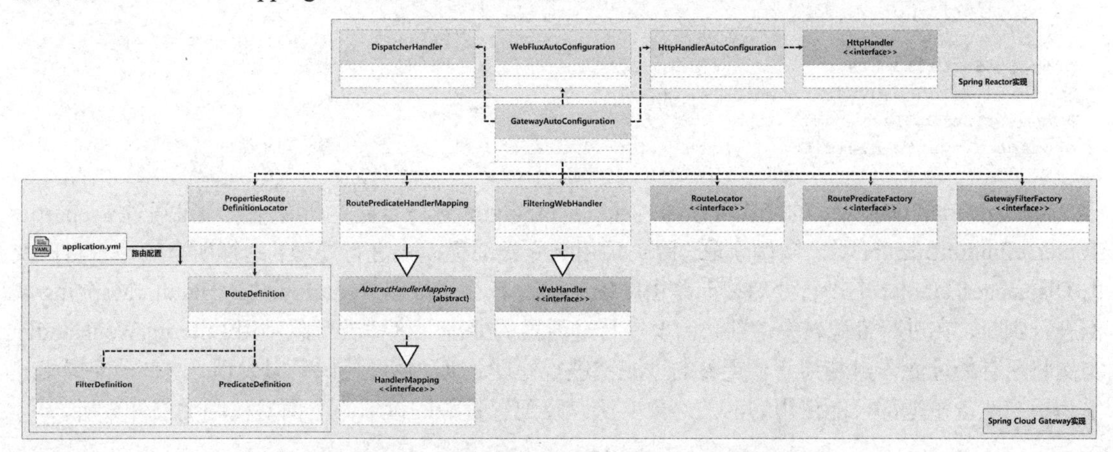

图 6-32 GatewayAutoConfiguration 自动配置类

范例：GatewayAutoConfiguration 核心源代码

```
package org.springframework.cloud.gateway.config;
@Configuration(proxyBeanMethods = false)
@ConditionalOnProperty(name = "spring.cloud.gateway.enabled", matchIfMissing = true)
@EnableConfigurationProperties
@AutoConfigureBefore({ HttpHandlerAutoConfiguration.class,
    WebFluxAutoConfiguration.class })
@AutoConfigureAfter({ GatewayLoadBalancerClientAutoConfiguration.class,
    GatewayClassPathWarningAutoConfiguration.class })
@ConditionalOnClass(DispatcherHandler.class)                    // 响应式请求分发处理
public class GatewayAutoConfiguration {
  @Bean
  @ConditionalOnMissingBean
  public PropertiesRouteDefinitionLocator propertiesRouteDefinitionLocator(
      GatewayProperties properties) {                          // 网关配置属性
    return new PropertiesRouteDefinitionLocator(properties);
  }
  @Bean                                                        // 路由定位器
  @Primary
  @ConditionalOnMissingBean(name = "cachedCompositeRouteLocator")
  public RouteLocator cachedCompositeRouteLocator(List<RouteLocator> routeLocators) {
    return new CachingRouteLocator(                            // 缓存路由定位器
        new CompositeRouteLocator(Flux.fromIterable(routeLocators)));
  }
@Bean
  public RouteLocator routeDefinitionRouteLocator(GatewayProperties properties,
      List<GatewayFilterFactory> gatewayFilters,
      List<RoutePredicateFactory> predicates,
      RouteDefinitionLocator routeDefinitionLocator,
      ConfigurationService configurationService) {             // 获取路由配置
    return new RouteDefinitionRouteLocator(routeDefinitionLocator, predicates,
        gatewayFilters, properties, configurationService);
  }
@Bean
  public FilteringWebHandler filteringWebHandler(List<GlobalFilter> globalFilters) {
    return new FilteringWebHandler(globalFilters);             // 过滤链处理
  }
  @Bean
  public RoutePredicateHandlerMapping routePredicateHandlerMapping(
      FilteringWebHandler webHandler, RouteLocator routeLocator,
      GlobalCorsProperties globalCorsProperties, Environment environment) {
    return new RoutePredicateHandlerMapping(webHandler, routeLocator,
        globalCorsProperties, environment);                    // 路由转发映射
  }
  @Configuration(proxyBeanMethods = false)
  @ConditionalOnClass({ HystrixObservableCommand.class, RxReactiveStreams.class })
  protected static class HystrixConfiguration {}               // Hystrix配置
  @Configuration(proxyBeanMethods = false)
  @ConditionalOnClass(Health.class)
  protected static class GatewayActuatorConfiguration {}       // Acutator配置
}
```

Spring Cloud Gateway 的配置可以通过 application.yml 进行定义，所有配置项可通过 PropertiesRouteDefinitionLocator 对象实例获取。网关采用响应式编程模式进行处理，这样所有的请求可以交由 DispatcherHandler 来进行处理，而路由转发的关键就在于依靠 RoutePredicateHandlerMapping 实例进行判断，在进行资源转发之前，需要进行过滤器的处理，这样就可以利用 FilteringWebHandler 实例将所有的过滤器封装为一个完整的过滤链进行处理。通过该源代码可以发现，该自动配置类几乎为用户准备了全部可能使用的网关对象实例以及配置资源的读取。下面对核心类展开分析。

6.5.2 RouteLocator

视频名称 0619_【理解】RouteLocator

视频简介 网关中需要保存大量的路由信息地址，而所有路由信息的获取可以通过RouteLocator接口完成。本视频介绍RouteLocator接口的继承结构，并通过启动类的源代码介绍3个子类内部的关联。

为了实现所有路由信息的加载处理，Spring Cloud Gateway 提供了 RouteLocator 接口，同时该接口内部有 3 个实现子类：RouteDefinitionRouteLocator（路由定义）、CompositeRouteLocator（组合方式）、CachingRouteLocator（缓存实现），如图 6-33 所示。

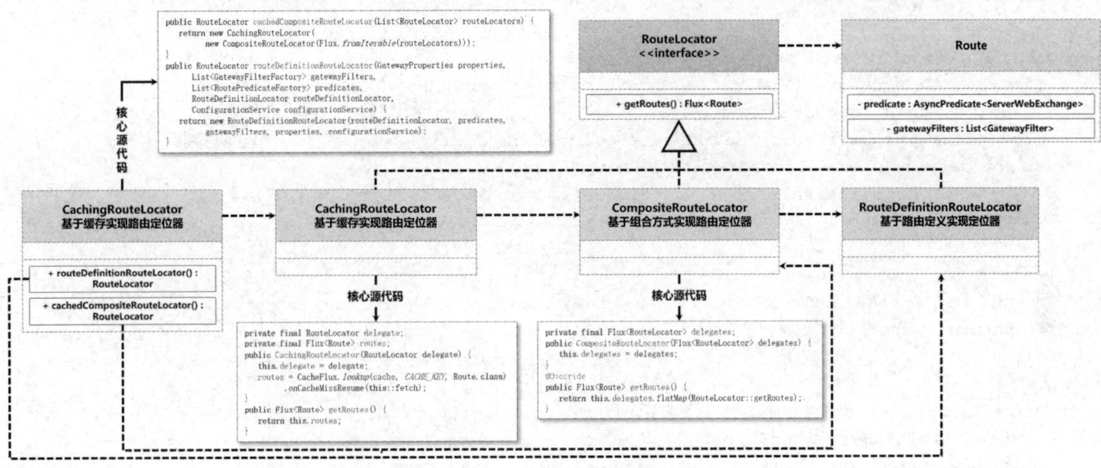

图 6-33 RouteLocator 接口

GatewayAutoConfiguration 类的 routeDefinitionRouteLocator()方法会提供 RouteDefinitionRouteLocator 子类的实例化处理操作，而 cachedCompositeRouteLocator()方法会将所有的 RouteLocator 接口实例包装在 CompositeRouteLocator 子类实例之中，最终用户可以使用的路由信息都会通过 CachingRouteLocator（基于缓存的路由定位器）实例完成获取处理，所以 RouteLocator 接口的实现关键在于 RouteDefinitionRouteLocator 子类。下面打开此子类的源代码进行观察。

范例：RouteDefinitionRouteLocator 子类源代码

```
package org.springframework.cloud.gateway.route;
public class RouteDefinitionRouteLocator
    implements RouteLocator, BeanFactoryAware, ApplicationEventPublisherAware {
  public static final String DEFAULT_FILTERS = "defaultFilters";
  protected final Log logger = LogFactory.getLog(getClass());
  private final RouteDefinitionLocator routeDefinitionLocator;      // 路由定位配置
  private final ConfigurationService configurationService;          // 配置路由服务
  // 保存全部配置的路由谓词工厂与网关过滤工厂实例
  private final Map<String, RoutePredicateFactory> predicates = new LinkedHashMap<>();
  private final Map<String, GatewayFilterFactory> gatewayFilterFactories =
            new HashMap<>();
  private final GatewayProperties gatewayProperties;                 // 网关属性
  public RouteDefinitionRouteLocator(RouteDefinitionLocator routeDefinitionLocator,
      List<RoutePredicateFactory> predicates,
      List<GatewayFilterFactory> gatewayFilterFactories,
      GatewayProperties gatewayProperties,
      ConfigurationService configurationService) {
    this.routeDefinitionLocator = routeDefinitionLocator;            // 属性保存
    this.configurationService = configurationService;                // 属性保存
    initFactories(predicates);                                       // 初始化路由谓词工厂
```

```
    gatewayFilterFactories.forEach(
            factory -> this.gatewayFilterFactories.put(factory.name(), factory));
    this.gatewayProperties = gatewayProperties;
}
private void initFactories(List<RoutePredicateFactory> predicates) {
    predicates.forEach(factory -> {                                    // 集合迭代
        String key = factory.name();                                   // 获取配置名称
        if (this.predicates.containsKey(key)) {
            this.logger.warn("A RoutePredicateFactory named ...");
        }
        this.predicates.put(key, factory);                             // 保存路由断言
        if (logger.isInfoEnabled()) { logger.info("Loaded ..."); }
    });
}
@Override
public Flux<Route> getRoutes() {                                       // 获取全部路由
    Flux<Route> routes = this.routeDefinitionLocator.getRouteDefinitions()
            .map(this::convertToRoute);                                // 获取路由
    if (!gatewayProperties.isFailOnRouteDefinitionError()) {           // 配置属性判断
        routes = routes.onErrorContinue((error, obj) -> {
            if (logger.isWarnEnabled()) { logger.warn("RouteDefinition id ... "); }
        });
    }
    return routes.map(route -> {                                       // 路由处理
        if (logger.isDebugEnabled()) { logger.debug("RouteDefinition matched ..."); }
        return route;
    });
}
private Route convertToRoute(RouteDefinition routeDefinition) {        // 路由解析
    AsyncPredicate<ServerWebExchange> predicate = combinePredicates(routeDefinition);
    List<GatewayFilter> gatewayFilters = getFilters(routeDefinition);
    return Route.async(routeDefinition).asyncPredicate(predicate)
            .replaceFilters(gatewayFilters).build();                   // 构建路由对象
}
@SuppressWarnings("unchecked")
List<GatewayFilter> loadGatewayFilters(String id,
        List<FilterDefinition> filterDefinitions) {                    // 获取网关过滤集合
    ArrayList<GatewayFilter> ordered = new ArrayList<>(filterDefinitions.size());
    for (int i = 0; i < filterDefinitions.size(); i++) {               // 过滤配置
        FilterDefinition definition = filterDefinitions.get(i);
        GatewayFilterFactory factory = this.gatewayFilterFactories
                .get(definition.getName());                            // 过滤工厂
        if (factory == null) {                                         // 过滤工厂为空
            throw new IllegalArgumentException("Unable ...");
        }
        if (logger.isDebugEnabled()) { logger.debug("RouteDefinition …"); }
        Object configuration = this.configurationService.with(factory)
                .name(definition.getName())
                .properties(definition.getArgs())
                .eventFunction((bound, properties) -> new FilterArgsEvent(
                    RouteDefinitionRouteLocator.this, id, (Map<String, Object>) properties))
                .bind();                                               // 获取路由配置
        if (configuration instanceof HasRouteId) {                     // 配置为路由ID判断
            HasRouteId hasRouteId = (HasRouteId) configuration;        // 实例转型
            hasRouteId.setRouteId(id);                                 // 保存路由ID
        }
        GatewayFilter gatewayFilter = factory.apply(configuration);   // 获取网关过滤接口
        if (gatewayFilter instanceof Ordered) {                        // 可以排列顺序
            ordered.add(gatewayFilter);                                // 配置过滤执行顺序
        } else {                                                       // 按照解析顺序执行
            ordered.add(new OrderedGatewayFilter(gatewayFilter, i + 1));
        }
```

```
        }
        return ordered;
    }
    private List<GatewayFilter> getFilters(RouteDefinition routeDefinition) {
        List<GatewayFilter> filters = new ArrayList<>();              // 保存网关过滤
        if (!this.gatewayProperties.getDefaultFilters().isEmpty()) {  // 网关属性判断
            filters.addAll(loadGatewayFilters(DEFAULT_FILTERS,
                    this.gatewayProperties.getDefaultFilters()));     // 获取网关过滤器
        }
        if (!routeDefinition.getFilters().isEmpty()) {                // 路由属性存在
            filters.addAll(loadGatewayFilters(routeDefinition.getId(),
                    routeDefinition.getFilters()));                   // 获取网关过滤器
        }
        AnnotationAwareOrderComparator.sort(filters);                 // 注解顺序排序
        return filters;
    }
    private AsyncPredicate<ServerWebExchange> combinePredicates(      // 断言整合
            RouteDefinition routeDefinition) {
        List<PredicateDefinition> predicates = routeDefinition.getPredicates();
        AsyncPredicate<ServerWebExchange> predicate = lookup(routeDefinition,
                predicates.get(0));                                   // 路由查找
        for (PredicateDefinition andPredicate : predicates.subList(1,
                predicates.size())) {                                 // 迭代断言定义
            AsyncPredicate<ServerWebExchange> found = lookup(routeDefinition,
                    andPredicate);                                    // 查找路由断言子项
            predicate = predicate.and(found);
        }
        return predicate;
    }
    @SuppressWarnings("unchecked")
    private AsyncPredicate<ServerWebExchange> lookup(RouteDefinition route,
            PredicateDefinition predicate) {                          // 路由查找
        RoutePredicateFactory<Object> factory = this.predicates.get(predicate.getName());
        if (factory == null) {
            throw new IllegalArgumentException("Unable to find RoutePredicateFactory…");
        }
        if (logger.isDebugEnabled()) { logger.debug("RouteDefinition …"); }
        return factory.applyAsync(config);                            // 获取Predicate实例
    }
}
```

通过该源代码可以清楚地发现，RouteDefinitionRouteLocator 类在实例化时需要明确通过构造方法接收所有的路由配置信息（RouteDefinitionLocator 接口实例）、全部的路由谓词工厂（RoutePredicateFactory 接口集合）、网关过滤工厂（GatewayFilterFactory 接口实例）、网关配置属性（GatewayProperties 类实例）、配置服务（ConfigurationService）等实例对象，而后在构造方法内部将获取的 RoutePredicateFactory 与 GatewayFilterFactory 接口集合转换为 Map 集合进行存储。这样可以去除重复的配置项，同时又可以方便地进行路由断言与过滤操作。

由于 RouteLocator 3 个子类定义的关系，RouteDefinitionRouteLocator 类中所覆写的 getRoutes()方法会由组合路由定位子类（CompositeRouteLocator）进行调用，这样就可以根据已有的 RouteDefinitionLocator 接口实例并结合 convertToRoute()方法进行 Route 类对象实例的拼凑，从而获取一个完整的实例。

需要注意的是，最终在网关中所有的路由数据全部要在 CachingRouteLocator 子类中保存，而该类的构造方法中明确调用了 RouteDefinitionRouteLocator.lookup()方法。该方法可以根据路由配置和断言配置来获取一个 AsyncPredicate 接口实例，从而根据返回结果判断当前的路由是否可以执行。

6.5.3 FilteringWebHandler

视频名称 0620_【理解】FilteringWebHandler

视频简介 过滤器是网关进行请求和响应处理的关键一步。本视频通过网关自动配置类分析 FilteringWebHandler 类的作用与过滤链的形成逻辑。

过滤是网关在进行最终服务资源代理前的最后一步，同时在网关中需要将所有配置的过滤器转为过滤链的形式，这样就可以依据过滤链的配置进行请求与响应过滤操作。在 GatewayAutoConfiguration 类中可通过 FilteringWebHandler 类实现过滤链的处理操作，该类的关联结构如图 6-34 所示。

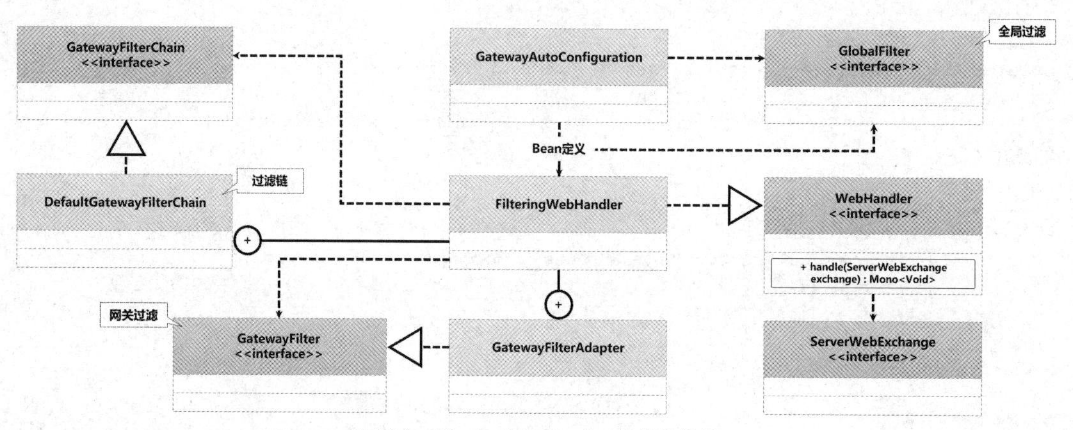

图 6-34 FilteringWebHandler 类关联结构

范例：FilteringWebHandler 源代码

```
package org.springframework.cloud.gateway.handler;
public class FilteringWebHandler implements WebHandler {
  protected static final Log logger = LogFactory.getLog(FilteringWebHandler.class);
  private final List<GatewayFilter> globalFilters;                    // 网关过滤器集合
  public FilteringWebHandler(List<GlobalFilter> globalFilters) {
    this.globalFilters = loadFilters(globalFilters);                  // 全局过滤器转为网关过滤器
  }
  private static List<GatewayFilter> loadFilters(List<GlobalFilter> filters) {
    return filters.stream().map(filter -> {                           // 集合处理
      GatewayFilterAdapter gatewayFilter = new GatewayFilterAdapter(filter);
      if (filter instanceof Ordered) {                                // 是否为Ordered接口实例
        int order = ((Ordered) filter).getOrder();
        return new OrderedGatewayFilter(gatewayFilter, order);
      }
      return gatewayFilter;
    }).collect(Collectors.toList());                                  // 转为List集合
  }
  @Override
  public Mono<Void> handle(ServerWebExchange exchange) {             // 过滤链处理
    Route route = exchange.getRequiredAttribute(GATEWAY_ROUTE_ATTR);
    List<GatewayFilter> gatewayFilters = route.getFilters();          // 获取路由对象
    List<GatewayFilter> combined = new ArrayList<>(this.globalFilters);
    combined.addAll(gatewayFilters);                                  // 保存过滤集合
    AnnotationAwareOrderComparator.sort(combined);                    // 过滤器排序
    if (logger.isDebugEnabled()) {
      logger.debug("Sorted gatewayFilterFactories: " + combined);
    }
    return new DefaultGatewayFilterChain(combined).filter(exchange); // 过滤链
  }
```

```
  private static class DefaultGatewayFilterChain implements GatewayFilterChain {
     private final int index;                                      // 过滤链索引
     private final List<GatewayFilter> filters;                    // 网关过滤器集合
     DefaultGatewayFilterChain(List<GatewayFilter> filters) {
        this.filters = filters;                                    // 保存过滤器集合
        this.index = 0;                                            // 设置默认索引
     }
     private DefaultGatewayFilterChain(DefaultGatewayFilterChain parent, int index) {
        this.filters = parent.getFilters();                        // 父过滤链
        this.index = index;
     }
     public List<GatewayFilter> getFilters() {                     // 返回网关过滤器集合
        return filters;
     }
     @Override
     public Mono<Void> filter(ServerWebExchange exchange) {        // 过滤处理
        return Mono.defer(() -> {
           if (this.index < filters.size()) {                      // 过滤链触发条件
              GatewayFilter filter = filters.get(this.index);      // 获取过滤器
              DefaultGatewayFilterChain chain = new DefaultGatewayFilterChain(this,
                    this.index + 1);                               // 定义过滤链
              return filter.filter(exchange, chain);               // 配置过滤链顺序
           } else {
              return Mono.empty();                                 // 处理完成
           }
        });
     }
  }
  private static class GatewayFilterAdapter implements GatewayFilter {
     private final GlobalFilter delegate;                          // 全局过滤
     GatewayFilterAdapter(GlobalFilter delegate) {                 // 构造方法
        this.delegate = delegate;                                  // 过滤器保存
     }
     @Override
     public Mono<Void> filter(ServerWebExchange exchange, GatewayFilterChain chain) {
        return this.delegate.filter(exchange, chain);              // 异步过滤处理
     }
     @Override
     public String toString() { … }
  }
}
```

此类的源代码为了便于过滤链的管理，将全局过滤器与网关过滤器合并，在合并处理时使用了一个 GatewayFilterAdapter 类实现包装。同时该类最重要的一点就是过滤链（GatewayFilterChain 接口）的实现，所以定义了一个 DefaultGatewayFilterChain 内部类，并在此类中实现了每一个过滤器顺序的定义，这样在过滤时就会按照当前的处理结果执行。

6.5.4 RoutePredicateHandlerMapping

RoutePredicate HandlerMapping

视频名称 0621_【理解】RoutePredicateHandlerMapping

视频简介 网关可以通过 application.yml 进行路由访问配置，而路由访问的断言匹配处理操作是由 RoutePredicateHandlerMapping 类实现的。本视频分析该类的源代码。

RoutePredicateHandlerMapping（路由断言处理映射）实现了用户请求路由的访问，即开发者在 application.yml 中配置路由访问路径项（Path 配置项），路由匹配成功后会返回一个 FilteringWebHandler 对象实例以实现后续的过滤处理。该类的 Bean 定义是由 GatewayAutoConfiguration 配置类实现的，其关联结构如图 6-35 所示。

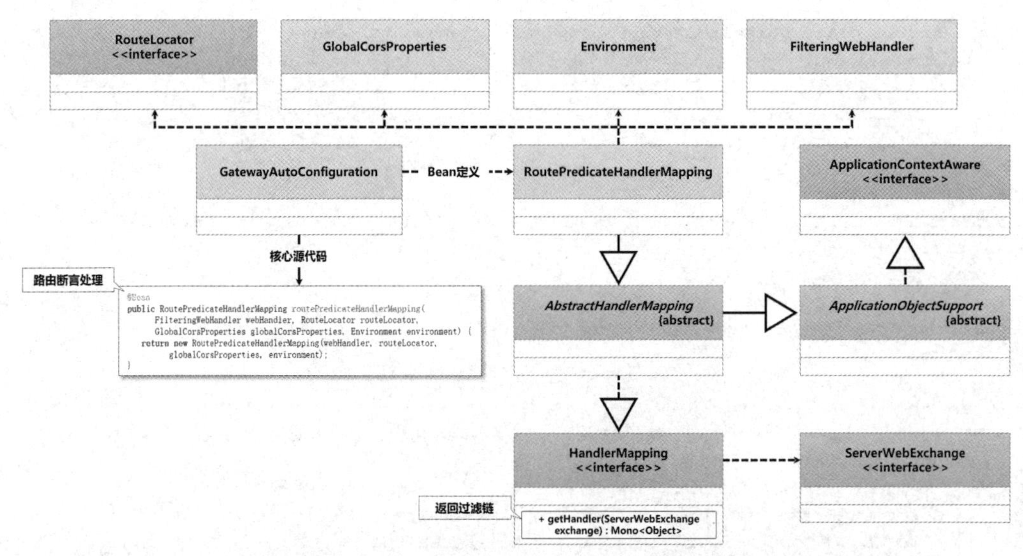

图 6-35 RoutePredicateHandlerMapping 类关联结构

范例：RoutePredicateHandlerMapping 源代码

```
package org.springframework.cloud.gateway.handler;
public class RoutePredicateHandlerMapping extends AbstractHandlerMapping {
  private final FilteringWebHandler webHandler;                         // 过滤处理链
  private final RouteLocator routeLocator;                              // 路由定位器
  private final Integer managementPort;                                 // 服务端口
  private final ManagementPortType managementPortType;                  // 端口状态
  public RoutePredicateHandlerMapping(FilteringWebHandler webHandler,
      RouteLocator routeLocator, GlobalCorsProperties globalCorsProperties,
      Environment environment) {
    this.webHandler = webHandler;                                       // 保存过滤链
    this.routeLocator = routeLocator;                                   // 保存路由定位器实例
    // 获取当前项目配置的management.server.port端口信息
    this.managementPort = getPortProperty(environment, "management.server.");
    this.managementPortType = getManagementPortType(environment);      // 端口类型
    setOrder(1);                                                        // 设置执行顺序
    setCorsConfigurations(globalCorsProperties.getCorsConfigurations()); // 跨域处理
  }
  private ManagementPortType getManagementPortType(Environment environment) {
    // 根据当前配置的端口判断得到ManagementPortType 对象实例
  }
  private static Integer getPortProperty(Environment environment, String prefix) {
    return environment.getProperty(prefix + "port", Integer.class);    // 端口配置
  }
  @Override
  protected Mono<?> getHandlerInternal(ServerWebExchange exchange) {
    // 如果设置的管理端口与服务器端口不同，则不处理管理端口上的请求
    if (this.managementPortType == DIFFERENT && this.managementPort != null
        && exchange.getRequest().getURI().getPort() == this.managementPort) {
      return Mono.empty();                                              // 返回空数据
    }
    exchange.getAttributes().put(GATEWAY_HANDLER_MAPPER_ATTR, getSimpleName());
    // 进行GATEWAY_PREDICATE_ROUTE_ATTR属性的相关操作，以确定断言状态
    return lookupRoute(exchange)                                        // 路由查找
        .flatMap((Function<Route, Mono<?>>) r -> {
          exchange.getAttributes().remove(GATEWAY_PREDICATE_ROUTE_ATTR);
          if (logger.isDebugEnabled()) {
            logger.debug("Mapping [...]");
          }
          exchange.getAttributes().put(GATEWAY_ROUTE_ATTR, r);
```

```
            return Mono.just(webHandler);                          // 返回FilteringWebHandler
        }).switchIfEmpty(Mono.empty().then(Mono.fromRunnable(() -> {
            exchange.getAttributes().remove(GATEWAY_PREDICATE_ROUTE_ATTR);
            if (logger.isTraceEnabled()) {
                logger.trace("No RouteDefinition found for […]");
            } })));
    }
    @Override
    protected CorsConfiguration getCorsConfiguration(Object handler,
        ServerWebExchange exchange) {}                             // 获取跨域配置
    private String getExchangeDesc(ServerWebExchange exchange) {}  // 获取转发描述
    protected Mono<Route> lookupRoute(ServerWebExchange exchange) {  // 路由查找
        return this.routeLocator.getRoutes()                       // 获取路由数据
            .concatMap(route -> Mono.just(route).filterWhen(r -> {  // 路由筛选
                exchange.getAttributes().put(GATEWAY_PREDICATE_ROUTE_ATTR, r.getId());
                return r.getPredicate().apply(exchange); })        // 返回断言结果
                    .doOnError(e -> logger.error("Error …"))       // 错误处理
                    .onErrorResume(e -> Mono.empty()))             // 错误信息
            .next().map(route -> {                                 // 路由处理
                if (logger.isDebugEnabled()) {
                    logger.debug("Route matched: " + route.getId());
                }
                validateRoute(route, exchange);                    // 路由验证（未实现）
                return route; });
    }
    public enum ManagementPortType {                               // 端口类型
        DISABLED,                                                  // 服务端口关闭
        SAME,                                                      // 与服务端口相同
        DIFFERENT;                                                 // 与服务端口不同
    }
}
```

本程序的构造方法会接收 FilteringWebHandler 核心对象实例，而所有的路由匹配处理操作全部由 lookupRoute()方法来实现。该方法会通过 RouteLocator 类的实例来实现路由数据的查询，而后会与当前的请求资源进行匹配。如果匹配成功则 Predicate 结果为 true，否则为 false。在断言成功后会返回 FilteringWebHandler 对象实例，从而实现后续的过滤处理。

6.6 动态路由

动态路由简介

视频名称 0622_【掌握】动态路由简介

视频简介 网关是微服务项目架构实现的核心技术，而网关的稳定运行将影响到整个微服务架构的状态。本视频为读者分析静态网关实现的问题，并分析动态网关配置的实现。

微服务架构引入网关技术之后，就可以得到图 6-36 所示的微服务核心架构。在整个架构之中，网关将成为某一个资源能否被正常调用的关键。

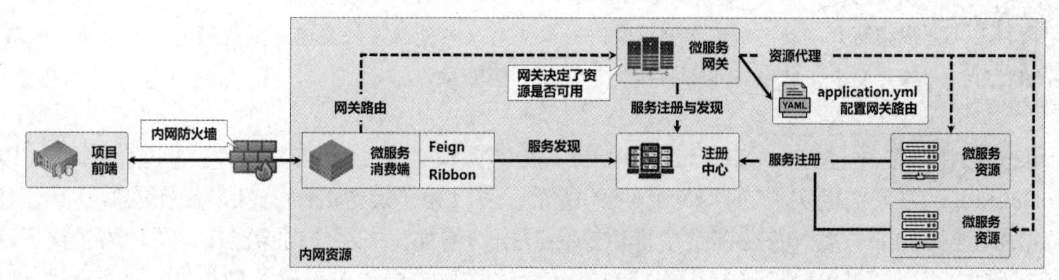

图 6-36　微服务核心架构

在传统的网关开发过程中，所有需要网关代理的资源都必须通过 application.yml 文件进行定义，而这些配置都是通过静态方式实现的。当有某一个新的微服务资源上线时，除了要考虑到新资源的服务部署之外，还需要进行微服务网关的重新启动，否则将无法加载新的资源，而网关的重启过程势必影响其他微服务的正常运行。要想解决此类问题就必须引入动态网关技术进行资源管理，如图 6-37 所示。

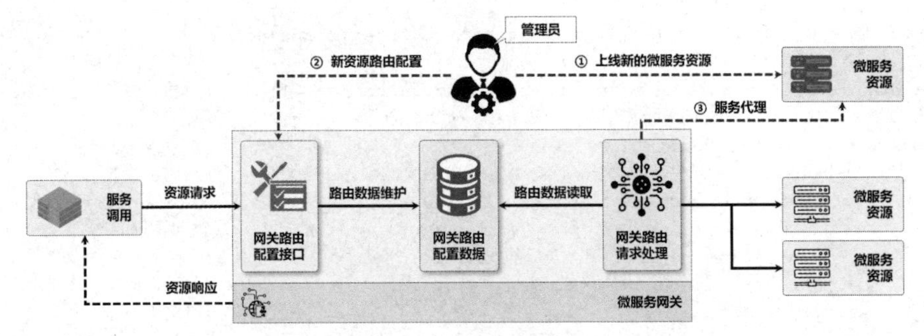

图 6-37　动态网关技术

在进行路由设置时需要明确传递路由的 ID、Predicate、Filter 等配置项，所以一般可以通过一个 JSON 结构来进行此配置项的定义。而在 Spring Cloud Gateway 中路由的信息是由 RouteDefinition 类对象定义的，这样就可以将 JSON 数据转为 RouteDefinition 实例，随后利用 RouteDefinitionWriter 接口进行配置写入，如图 6-38 所示。要想让动态路由生效，还必须通过 Spring 事件处理机制发送一个 RefreshRoutesEvent 事件，这样才可以对写入的路由信息进行保存。

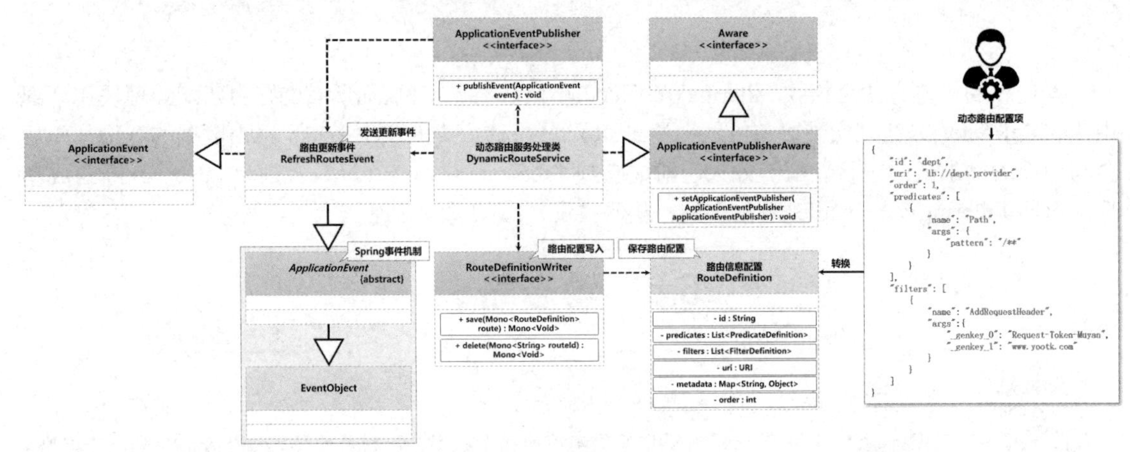

图 6-38　动态路由配置结构

6.6.1　动态路由模型

视频名称　0623_【掌握】动态路由模型

视频简介　动态路由模型需要进行配置写入与更新事件发布。本视频通过一个统一的路由操作类来介绍动态路由的增加、修改、删除操作。

动态路由需要通过 RouteDefinitionWriter 接口完成，而该接口主要是通过路由 ID 以及 RouteDefinition 对象实例实现的，如图 6-39 所示。所以为了操作的统一性，最佳的做法是创建一个专属的动态路由服务类，并在该类中提供路由数据的增加、修改与删除操作。同时为了便于网关数据的管理，可以通过 REST 进行操作接口发布，这样只需要传入正确的数据即可进行网关维护。

下面通过具体的步骤介绍如何实现。

图 6-39　动态路由

（1）【gateway-9501 子模块】创建路由数据操作类。

```
package com.yootk.gateway.service;
@Service
@Slf4j
public class DynamicRouteService implements ApplicationEventPublisherAware {
    @Autowired
    private RouteDefinitionWriter routeDefinitionWriter;                    // 路由配置写入
    private ApplicationEventPublisher publisher;                            // 事件发布器
    @Override
    public void setApplicationEventPublisher(
        ApplicationEventPublisher applicationEventPublisher) {
        this.publisher = applicationEventPublisher;
    }
    public boolean add(RouteDefinition definition) {                        // 增加路由
        log.info("增加路由配置项，新的路由ID为：{}", definition.getId());      // 日志输出
        try {
            routeDefinitionWriter.save(Mono.just(definition)).subscribe(); // 配置写入
            this.publisher.publishEvent(new RefreshRoutesEvent(this));     // 路由更新事件
        } catch (Exception e) {
            log.error("路由增加失败，增加的路由ID为：{}", definition.getId()); // 日志输出
            return false;                                                  // 操作失败
        }
        return true;                                                       // 操作成功
    }
    public boolean update(RouteDefinition definition) {                     // 更新路由
        log.info("修改路由配置项，修改的路由ID为：{}", definition.getId());    // 日志输出
        try {
            this.delete(definition.getId());                               // 删除已有配置
            routeDefinitionWriter.save(Mono.just(definition)).subscribe(); // 写入新配置
            this.publisher.publishEvent(new RefreshRoutesEvent(this));     // 路由更新事件
        } catch (Exception e) {
            log.error("路由更新失败，更新的路由ID为：{}", definition.getId()); // 日志输出
            return false;                                                  // 操作失败
        }
        return true;                                                       // 操作成功
    }
    public Mono<ResponseEntity<Object>> delete(String id) {                 // 删除路由
        log.info("删除路由配置项，删除的路由ID为：{}", id);                   // 日志输出
        return this.routeDefinitionWriter.delete(Mono.just(id))
                .then(Mono.defer(() -> {
            return Mono.just(ResponseEntity.ok().build());
        })).onErrorResume((t) -> {
            return t instanceof NotFoundException;
        }, (t) -> {
            return Mono.just(ResponseEntity.notFound().build());
        });
    }
}
```

（2）【gateway-9501 子模块】创建 REST 接口，用于实现路由信息的接收与业务方法调用。

```
package com.yootk.gateway.action;
@RestController
@RequestMapping("/routes/*")
public class DynamicRouteAction {
    @Autowired
    private DynamicRouteService dynamicRouteService;                 // 注入路由操作服务对象
    @PostMapping("add")
    public Boolean add(@RequestBody RouteDefinition definition) {     // 增加路由
        return this.dynamicRouteService.add(definition);              // 增加路由项
    }
    @DeleteMapping("delete/{id}")
    public Mono<ResponseEntity<Object>> delete(@PathVariable String id) { // 删除路由
        return this.dynamicRouteService.delete(id);                   // 删除路由项
    }
    @PostMapping("update")
    public Boolean update(@RequestBody RouteDefinition definition) {  // 更新路由
        return this.dynamicRouteService.update(definition);           // 更新路由项
    }
}
```

（3）【Postman】由于在配置时需要传递 JSON 数据结构，为简单起见，本次将通过 Postman 工具进行动态路由的操作。以增加路由信息的操作为例，传递如下 JSON 数据即可，配置如图 6-40 所示。

```
{   "id": "dept",                                   // 配置路由ID
    "uri": "lb://dept.provider",                    // 配置路由访问地址
    "order": 1,                                     // 定义路由执行顺序
    "predicates": [                                 // 配置路由断言
        {   "name": "Path",                         // 定义断言路径
            "args": { "pattern": "/**"    }         // 断言匹配模式
        }
    ],
    "filters": [                                    // 配置过滤器
        {   "name": "AddRequestHeader",             // 追加请求头信息
            "args":{                                // 配置参数内容
                "_genkey_0": "Request-Token-Muyan", // 头信息名称
                "_genkey_1": "www.yootk.com"        // 头信息内容
            }
        }
    ]}
```

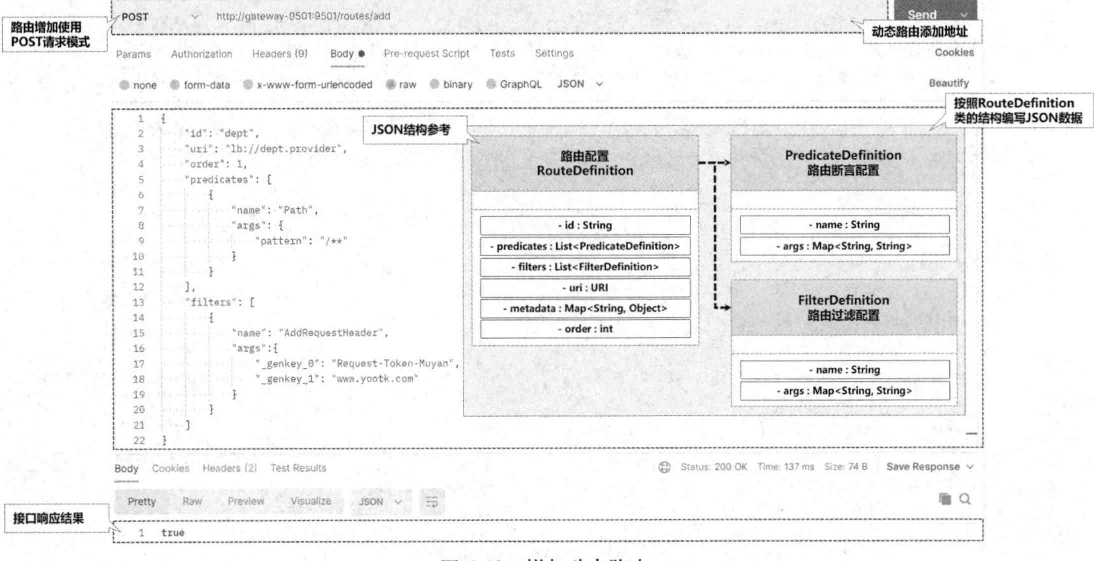

图 6-40　增加动态路由

（4）【Postman】所有配置的路由信息可以通过 Actuator 接口（/actuator/gateway/routes）查看，

如图 6-41 所示。

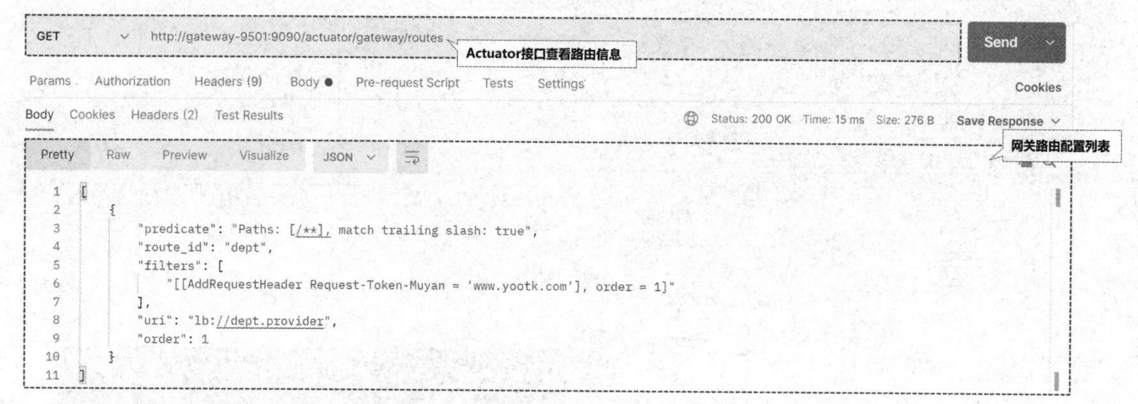

图 6-41　查看路由信息

6.6.2　动态路由配置持久化

视频名称　0624_【掌握】动态路由配置持久化

视频简介　Nacos 是 Spring Cloud Alibaba 实现的核心，除了可以承担起服务注册与发现支持之外，还提供了配置数据的管理功能。本视频将介绍使用 Nacos 实现网关配置数据持久化管理操作，以实现配置的初始化加载与更新加载。

除了动态配置网关数据之外，还需要进行有效的持久化管理，这样才可以保证在每次服务启动之后网关配置数据不会丢失。在进行数据持久化配置时，还需要充分考虑持久化配置数据发生改变时，网关也需要及时进行更新处理，所以最佳的做法就是通过 Nacos 进行数据存储，如图 6-42 所示。

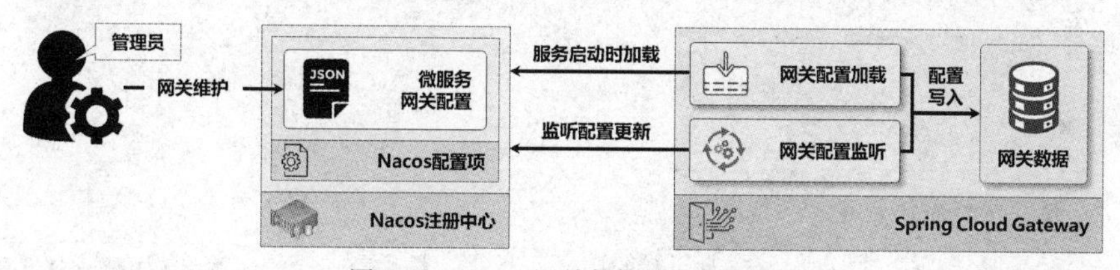

图 6-42　Nacos 网关数据持久化配置

由于网关启动时就需要进行配置数据的加载，因此本次开发将通过 CommandLineRunner 接口实现类容器启动后的数据加载操作，也将启动一个 Nacos 服务监听。一旦监听到指定的路由配置项发生更改，就可以获取新的配置项并实现路由数据的更新操作。为便于理解，下面通过具体的操作步骤介绍这一功能的实现。

（1）【Nacos 控制台】在 Nacos 中创建一个名称为“gateway.config”的配置项，并保存网关配置，如图 6-43 所示。

（2）【microcloud 项目】修改 gateway-9501 子模块中关于 Nacos 的依赖，更换为与之匹配的版本。

```
project('gateway-9501') {                                            // 子模块
    dependencies {                                                   // 配置子模块依赖
        implementation('com.alibaba.cloud:' +
                'spring-cloud-starter-alibaba-nacos-config') {       // Nacos配置
            exclude group: 'com.alibaba.nacos', module: 'nacos-client'   // 删除nacos-client
        }
        implementation('com.alibaba.nacos:nacos-client:2.0.0')       // NacosClient依赖
    }
}
```

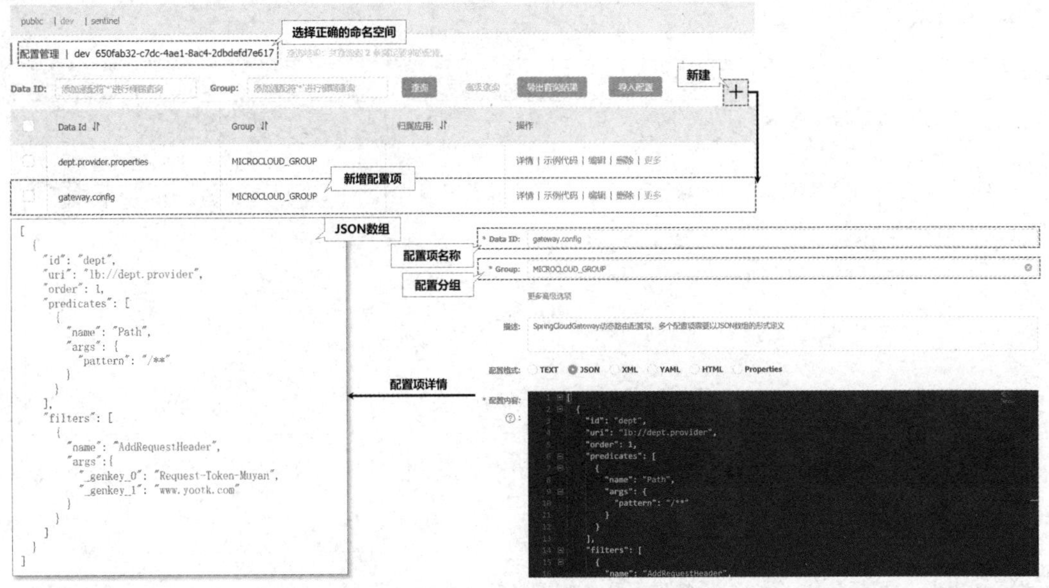

图 6-43 添加网关配置项

（3）【gateway-9501 子模块】创建 Nacos 配置类，该配置类主要保存 Nacos 连接配置项。

```
package com.yootk.gateway.config;
@Component
@Data
// 为便于理解，本次将读取application.yml中已经存在的Nacos配置项
// 在实际开发中可以根据项目环境的需要连接其他的Nacos服务器
@ConfigurationProperties(prefix = "spring.cloud.nacos.discovery")        // 属性配置前缀
public class GatewayNacosConfig {
    private String serverAddr;                                           // 服务地址
    private String namespace;                                            // 命名空间
    private String group;                                                // 配置分组
    private String username;                                             // 用户名
    private String password;                                             // 密码
    private String dataId = "gateway.config";                            // 数据ID
    private long timeout = 2000;                                         // 读取超时
    public Properties getNacosProperties() {                             // 返回Nacos属性
        Properties properties = new Properties();                        // Nacos属性配置
        properties.put(PropertyKeyConst.SERVER_ADDR, this.serverAddr);   // 服务地址
        properties.put(PropertyKeyConst.NAMESPACE, this.namespace);      // 命名空间
        properties.put(PropertyKeyConst.USERNAME, this.username);        // 用户名
        properties.put(PropertyKeyConst.PASSWORD, this.password);        // 密码
        return properties;
    }
}
```

为了减少 Nacos 服务节点的数量，本次开发利用了已有的 Nacos 注册中心保存网关路由配置项。同时为了减少 application.yml 的配置项，本次开发读取了 spring.cloud.nacos.discovery 中的相关配置项。

（4）【gateway-9501 子模块】网关配置数据需要在微服务启动后立即获取，所以可以创建 CommandLineRunner 接口子类。这样在容器启动完成后就可以通过 run()方法实现在 Nacos 配置加载的同时启动配置监听，以及时获取最新配置项。

```
package com.yootk.gateway.listener;
@Component
@Slf4j
public class GatewayNacosRouteListener implements CommandLineRunner {   // 容器启动后执行
    @Autowired
    private DynamicRouteService dynamicRouteService;                     // 动态路由操作
    @Autowired
    private GatewayNacosConfig nacosConfig;                              // Nacos配置
```

```
    private ObjectMapper mapper = new ObjectMapper();           // Jackson解析
    @Override
    public void run(String... args) throws Exception {          // 启动后执行
        nacosDynamicRouteListener();                            // 动态路由监听
    }
    public void nacosDynamicRouteListener() {                   // 动态路由监听
        try {
            ConfigService configService = NacosFactory.createConfigService(
                    this.nacosConfig.getNacosProperties());     // 创建ConfigService实例
            String content = configService.getConfig(this.nacosConfig.getDataId(),
                    this.nacosConfig.getGroup(), 5000);         // 获取配置内容
            log.info("【网关启动】读取Nacos网关配置项：{}", content);  // 日志输出
            GatewayNacosRouteListener.this.setRoute(content);   // 路由配置
            configService.addListener(nacosConfig.getDataId(),
                    nacosConfig.getGroup(), new Listener() {    // 配置监听
                @Override
                public void receiveConfigInfo(String configInfo) {
                    log.info("【网关更新】读取Nacos网关配置项：{}", configInfo); // 日志输出
                    GatewayNacosRouteListener.this.setRoute(configInfo); // 路由配置
                }
                @Override
                public Executor getExecutor() { return null; }
            });
        } catch (NacosException e) { e.printStackTrace(); }
    }
    private void setRoute(String configInfo) {                  // 路由配置
        try {
            RouteDefinition[] routes = mapper.readValue(configInfo,
                    RouteDefinition[].class);                   // 将JSON转为RouteDefinition数组
            for (RouteDefinition definition : routes) {         // 数组循环
                this.dynamicRouteService.update(definition);    // 路由更新
            }
        } catch (JsonProcessingException e) { e.printStackTrace(); }
    }
}
```

配置完成后可以重新启动网关应用。通过后台日志可以清楚地发现通过 Nacos 抓取的配置项数据，而管理员通过 Nacos 控制台修改配置后，也可以触发 Nacos 监听配置实现路由更新。

6.7 本章概览

1．Spring Cloud 服务架构可以利用网关进行服务资源的隔离。Spring Cloud 中的网关技术有两种，一种是由 Netflix 提供的 Zuul，另外一种是由 Spring 自行研发的 Spring Cloud Gateway 组件，其中后者是基于 Netty 开发的，处理性能更高效。

2．Spring Cloud Gateway 需要通过 Nacos 注册中心抓取微服务的配置信息，随后就可以进行代理转发，而为了便于服务的管理，可以基于路由（静态路由或动态路由）的形式进行设置。

3．微服务转发时需要根据 RoutePredicateFactory（路由谓词工厂）的配置实现服务代理转发。除了内置的路由谓词工厂（如 After、Before、Cookie、Header 等）之外，也可以根据项目需要进行扩充。

4．用户在进行网关服务请求时，除了经过路由谓词工厂的处理之外，还需要进行网关过滤处理，可以基于过滤器的形式传递统一的数据到目标资源。

5．Spring Cloud Gateway 提供了全局过滤器的支持，利用全局过滤器可以实现最终的路由转发、Netty 服务处理以及服务监控等操作，同时还可以结合 Prometheus 与 Grafana 进行监控。

6．项目中引入 Spring Cloud Gateway 相关依赖后，所有服务的初始化全部是由 GatewayAuto Configuration 配置类实现的，包括路由定位器、过滤器、断言器等。

7．为了便于网关的管理，可以通过动态路由进行配置，动态路由可以基于项目的内存存储，也可以基于持久化方式存储。考虑到服务数据的动态更新，本章介绍了使用 Nacos 作为网关数据存储终端。

第 7 章 微服务安全与监控管理

本章学习目标

1. 掌握 JWT 技术与 Spring Cloud 实现的关联，并可以基于 JWT 技术进行微服务安全改造；
2. 掌握 JWT 在微服务集群架构中的作用，同时可以实现认证与授权管理操作；
3. 掌握 Spring Boot Admin 工具的使用方法，并可以使用此工具实现微服务运行状态的监控；
4. 理解 Spring Boot Admin 与 E-mail 的整合处理，并可以实现服务的上线与离线监控。

微服务除了要提供稳定的服务之外，最重要的就是保证资源的安全，同时可以对微服务的每个节点的运行状态进行有效的监控数据采集与实时化的展现。本章将综合地为读者讲解如何基于 JWT 技术实现微服务认证管理，以及微服务监控工具 Spring Boot Admin 的实际应用。

7.1 Spring Cloud 认证管理

Spring Cloud
认证管理简介

视频名称 0701_【了解】Spring Cloud 认证管理简介

视频简介 为了有效地进行服务资源的保护，需要进行统一的认证管理。本视频为读者分析 Spring Cloud 中的认证管理模式以及各自的特点。

本书所讲解的微服务技术仅仅实现了微服务的核心功能。虽然微服务工作在内网，但是也需要进行有效的微服务安全保护，这样可以防范可能出现的安全漏洞，如图 7-1 所示。

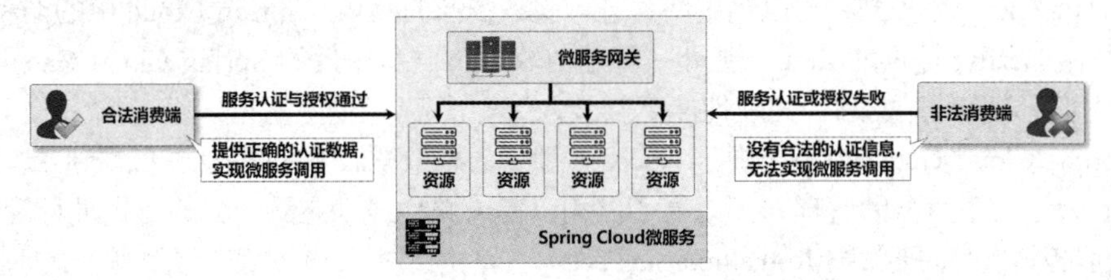

图 7-1 微服务安全访问

因为微服务集群架构设计之中会有大量的服务节点，所以需要一种统一的认证服务来对其进行管理。可能很多读者首先想到的就是 OAuth2 认证服务，如图 7-2 所示。

在 Spring Boot 中可以基于 Spring Security 来实现 OAuth2 的相关认证，这样就可以轻松地实现 OAuth2 服务端的搭建，结合 Spring Security 提供的@EnableGlobalMethodSecurity 注解，还可以轻松地实现授权检查的处理。但是基于 OAuth2 实现的统一认证会有如下几个问题。

（1）OAuth2 是为了 Web 应用而设计的

用 OAuth2 进行平台接入管理会比较方便，它也是一个实现的标准，但是其在最初设计时并未考虑到与 Spring Cloud 的技术整合，所以有“先天性”的不足。

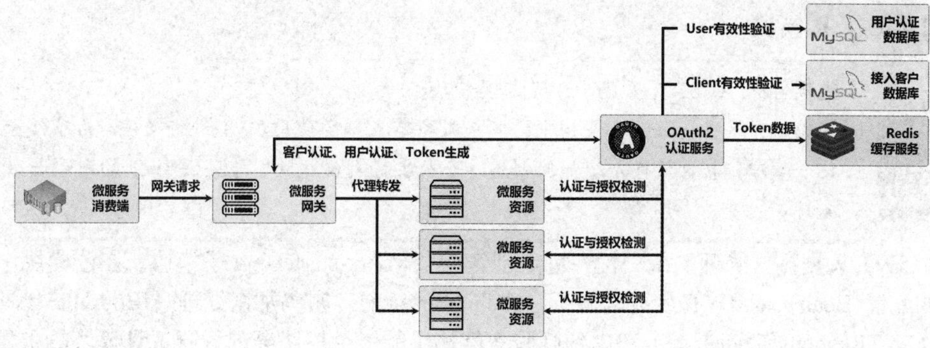

图 7-2 OAuth2 认证服务

（2）处理逻辑烦琐

OAuth2 认证需要首先进行客户端接入，而后生成授权码，最后通过客户端信息、授权码生成访问 Token。这样烦琐的逻辑必然带来较低的处理性能。

（3）Spring Security 支持不友好

在 Spring Security 之中已经默认取消了 OAuth2 支持维护，所以在整合中有可能出现非正常的因素，导致开发成本上升。

提示：OAuth2 属于架构必备技术。

技术的发展具有时效性。虽然现阶段的架构设计已经不提倡使用 OAuth2 了，但是现存的互联网平台中依然存在大量的 OAuth2 技术应用，所以开发者依然需要掌握该技术。本套丛书用 Shiro 与 Spring Security 两种方式讲解过 OAuth2，未掌握此技术的读者可以自行翻阅相关图书。

关于如何使用 Spring Cloud Netflix 套件与 OAuth2 整合开发，读者可以参考笔者的《名师讲坛——Java 微服务架构实战（Spring Boot+Spring Cloud+Docker+RabbitMQ）》一书自行学习。

在新版本的 Spring Cloud 项目开发中，考虑到服务处理性能和代码的可维护性，本书并不推荐使用 OAuth2 方式进行统一的认证管理。Spring Cloud 官方推荐的技术是基于 JWT（JSON Web Token）的方式进行认证，如图 7-3 所示。这样只需要在每次请求前获取 JWT 数据，并在每次请求时传递 JWT 数据，即可轻松地实现认证与授权检查。

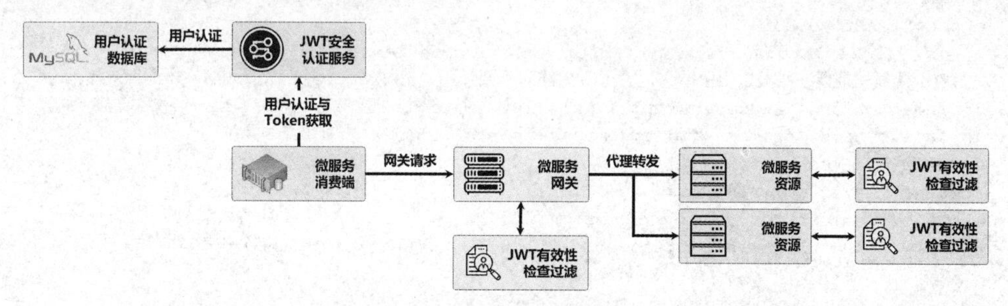

图 7-3 JWT 安全认证

由于 JWT 数据自身组成结构的特点，开发者并不需要为 JWT 数据提供额外的存储。消费端在调用微服务之前首先要获取相应的 JWT 数据，同时微服务网关和微服务资源也可以通过过滤器对 JWT 有效性进行检查，这种方式更适合现在的微服务架构设计。

提示：JWT 在 Spring Boot 中有详细讲解。

本套丛书中的《Spring Boot 开发实战（视频讲解版）》一书已经为读者详细讲解了 JWT 组成结构以及有效性检测处理形式，所以本书不再对这一基础知识进行讲解，而是直接进行实战应用。JWT 属于现代微服务设计的重要组成技术，不熟悉 JWT 的读者请自行翻阅本套丛书进行知识补充。

7.1.1　JWT 工具模块

视频名称　0702_【掌握】JWT 工具模块

视频简介　整个微服务架构中的各个服务节点都会使用 JWT 进行处理。为了便于代码的统一管理，本视频将介绍如何创建一个公共的开发模块，并提供 JWT 的配置与数据处理支持。

为了 JWT 数据操作管理的统一性，可以创建一个 yootk-starter-jwt 子模块。在该模块中可以提供相关的配置 Bean，如 JWT 生成与解析（ITokenService）、密码加密处理（IEncryptService）、响应状态（JWTResponseCode）等。考虑到微服务架构中的多个模块会进行此工具模块的引用，可以将其定义为一个自动配置模块，如图 7-4 所示。这样在进行该模块引用时，就可以自动提示 application.yml 的配置项，为开发者带来方便。由于此模块所涉及的代码较多，下面将通过具体的步骤进行讲解。

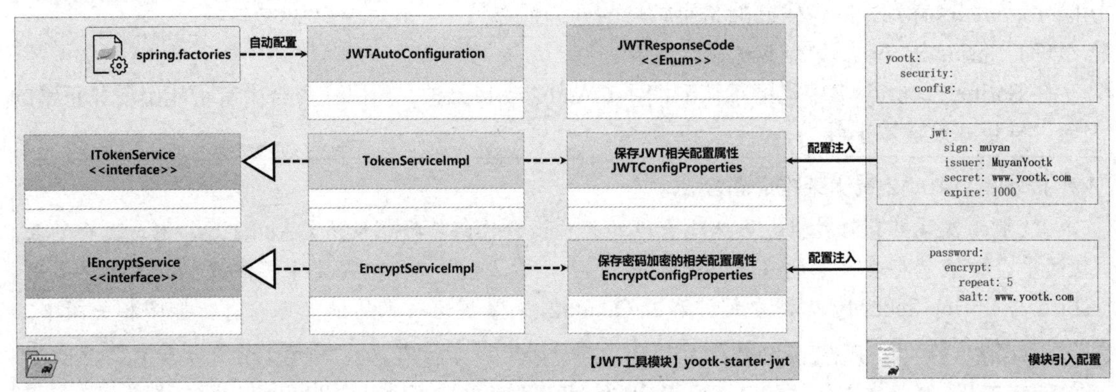

图 7-4　JWT 工具模块

（1）【microcloud 项目】创建一个 yootk-starter-jwt 子模块，随后修改 build.gradle 配置文件，添加所需要的依赖库。

```
project('yootk-starter-jwt') {                                    // 子模块
    dependencies {                                                // 配置子模块依赖
        annotationProcessor(
                'org.springframework.boot:spring-boot-configuration-processor')
        // 由于此模块需要在网关应用中导入，为了避免冲突不要使用“spring-boot-starter-web”依赖
        implementation('javax.servlet:javax.servlet-api:4.0.1')
        implementation('commons-codec:commons-codec:1.15')      // Base64编码
        compile('io.jsonwebtoken:jjwt:0.9.1')                    // JWT工具组件
        compile('javax.xml.bind:jaxb-api:2.3.1')                 // JWT相关依赖
        compile('com.sun.xml.bind:jaxb-impl:2.3.0')              // JWT相关依赖
        compile('com.sun.xml.bind:jaxb-core:2.3.0')              // JWT相关依赖
    }
}
```

（2）【yootk-starter-jwt 子模块】由于此项目模块最终需要进行打包关联，因此修改 build.gradle 进行相关任务的关闭。

```
jar { enabled = true }                                            // 保留jar任务
javadocTask { enabled = false }                                   // 关闭JavaDoc任务
javadocJar { enabled = false }                                    // 关闭打包JavaDoc任务
bootJar { enabled = false }                                       // 关闭Spring Boot任务
```

（3）【yootk-starter-jwt 子模块】为了便于用户响应信息，创建一个 JWT 响应代码的枚举类。

```
package com.yootk.jwt.code;
import javax.servlet.http.HttpServletResponse;
public enum JWTResponseCode {                                     // JWT响应代码
```

```
    SUCCESS_CODE(HttpServletResponse.SC_OK, "Token正确，服务正常访问！"), // 成功响应
    TOKEN_TIMEOUT_CODE(HttpServletResponse.SC_BAD_REQUEST,
                "Token信息已经失效，需要重新申请！"),          // Token失效
    NO_AUTH_CODE(HttpServletResponse.SC_NOT_FOUND,
                "没有找到匹配的Token信息，无法进行服务访问！"); // 没有Token
    private int code;                                        // 响应代码
    private String message;                                  // 提示信息
    private JWTResponseCode(int code, String message) {      // 双参构造
        this.code = code;
        this.message = message;
    }
    @Override
    public String toString() {
        return "{\"code\":" + this.code + ",\"message\":\"" + this.message + "\"}";
    }
}
```

（4）【yootk-starter-jwt 子模块】创建 JWTConfig Properties 配置类，用于保存 application.yml 定义的 JWT 配置项。

```
package com.yootk.jwt.config;
@Data                                                          // 自动生成类结构
@ConfigurationProperties(prefix = "yootk.security.config.jwt")  // 属性配置
public class JWTConfig Properties {                            // 保存JWT配置项
    private String sign;                                       // 证书签名信息
    private String issuer;                                     // 证书签发人
    private String secret;                                     // 加密密钥
    private long expire;                                       // 失效时间（单位：秒）
}
```

（5）【yootk-starter-jwt 子模块】创建 JWTSecurityService 来定义 JWT 相关的数据处理操作。

```
package com.yootk.jwt.service;
public interface ITokenService {
    /**
     * 获取当前JWT数据的加密Key
     * @return SecretKey接口实例
     */
    public SecretKey generalKey();
    /**
     * 根据指定的加密算法以及加密Key创建一个Token数据
     * @param id TokenID数据
     * @param subject JWT数据需要携带的用户认证或授权数据
     * @return 生成的Token信息
     */
    public String createToken(String id, Map<String, Object> subject);
    /**
     * 根据已有的Token解析出所包含的数据信息
     * @param token 要解析的Token数据
     * @return Jws接口实例（指定Token数据所包含的数据）
     * @throws JwtException JWT数据解析出现的异常
     */
    public Jws<Claims> parseToken(String token) throws JwtException;
    /**
     * 验证Token是否有效
     * @param token 要验证的Token数据
     * @return Token有效返回true，否则返回false
     */
    public boolean verifyToken(String token);
    /**
     * Token刷新（延缓失效时间）
     * @param token 要延缓失效时间的Token数据
     * @return 新的Token数据，如果该Token已经失效，则返回null
     */
    public String refreshToken(String token);
}
```

（6）【yootk-starter-jwt 子模块】创建 TokenServiceImpl 子类来实现 Token 的相关操作。需要注意的是，在进行 Token 处理时需要通过 JWTConfigProperties 配置 Bean 来获取相关信息。

```
package com.yootk.jwt.service.impl;
public class TokenServiceImpl implements ITokenService {              // Token操作实现
    @Autowired                                                        // 依赖注入
    private ObjectMapper objectMapper;                                // JSON数据操作
    @Autowired                                                        // 依赖注入
    private JWTConfigProperties jwtConfigProperties;                  // 注入JWT配置属性
    @Value("${spring.application.name}")                              // SpEL注入
    private String applicationName;                                   // 应用名称
    private SignatureAlgorithm signatureAlgorithm = SignatureAlgorithm.HS256; // 签名算法
    @Override
    public SecretKey generalKey() {                                   // 生成密钥
        byte[] encodedKey = Base64.decodeBase64(Base64.encodeBase64(
                this.jwtConfigProperties.getSecret().getBytes()));
        SecretKeySpec key = new SecretKeySpec(encodedKey, 0, encodedKey.length, "AES");
        return key;
    }
    @Override
    public String createToken(String id, Map<String, Object> subject) {
        Date nowDate = new Date();                                    // 签发时间
        Date expireDate = new Date(nowDate.getTime() +
                this.jwtConfigProperties.getExpire() * 1000);         // 证书过期时间
        Map<String, Object> claims = new HashMap<>();                 // claims信息
        claims.put("site", "www.yootk.com");                          // 附加信息
        Map<String, Object> headers = new HashMap<>();                // 头部信息
        headers.put("author", "李兴华");                               // 头部信息
        headers.put("module", this.applicationName);                  // 头部信息
        JwtBuilder builder = null;                                    // 失效时间
        try {
            builder = Jwts.builder()                                  // JwtBuilder实例
                    .setClaims(claims)                                // 设置附加数据
                    .setHeader(headers)                               // 设置头信息
                    .setId(id)                                        // JWT唯一标记
                    .setIssuedAt(nowDate)                             // 签发时间
                    .setIssuer(this.jwtConfigProperties.getIssuer())  // 签发人
                    .setSubject(this.objectMapper.writeValueAsString(subject)) // 用户数据
                    .signWith(signatureAlgorithm, this.generalKey())  // 签名算法
                    .setExpiration(expireDate);
        } catch (JsonProcessingException e) { e.printStackTrace(); }
        return builder.compact();                                     // 创建Token
    }
    @Override
    public boolean verifyToken(String token) {
        try {                                                         // Token数据解析
            Jwts.parser().setSigningKey(this.generalKey())
                    .parseClaimsJws(token).getBody();                 // 获取数据
            return true;                                              // Token正确
        } catch (Exception e) { return false; }                       // Token错误
    }
    @Override
    public Jws<Claims> parseToken(String token) throws JwtException {
        if (this.verifyToken(token)) {                                // Token校验
            Jws<Claims> claims = Jwts.parser()                        // 获取接口实例
                    .setSigningKey(this.generalKey())                 // 签名密钥
                    .parseClaimsJws(token);                           // 数据解析
            return claims;
        }
        return null;
    }
    @Override
```

```
    public String refreshToken(String token) {
        if (this.verifyToken(token)) {                          // Token校验
            Jws<Claims> jws = this.parseToken(token);           // Token解析
            return this.createToken(jws.getBody().getId(),
                    this.objectMapper.readValue(jws.getBody().getSubject(),Map.class));
        }
        return null;
    }
}
```

（7）【yootk-starter-jwt 子模块】创建密码加密配置类。

```
package com.yootk.jwt.config;
@Data                                                                   // 自动生成类结构
@ConfigurationProperties(prefix = "yootk.security.config.password.encrypt") // 属性配置
public class EncryptConfigProperties {                                  // 密码加密配置
    private Integer repeat;                                             // 获取加密次数
    private String salt;                                                // 获取加密盐值
}
```

（8）【yootk-serter-jwt 子模块】创建密码加密业务操作接口。

```
package com.yootk.jwt.service;
public interface IEncryptService {
    /**
     * 密码加密处理
     * @param password 要加密的明文数据
     * @return 加密后的密文数据（不可逆）
     */
    public String getEncryptPassword(String password);
}
```

（9）【yootk-serter-jwt 子模块】创建 EncryptServiceImpl 子类，使用 MD5 与 Base64 两种方式进行加密。

```
package com.yootk.jwt.service.impl;
public class EncryptServiceImpl implements IEncryptService {
    @Autowired
    private EncryptConfigProperties encryptConfig;          // 获取加密属性配置
    public static MessageDigest MD5_DIGEST;                  // MD5加密算法
    private static final Base64.Encoder BASE64_ENCODER = Base64.getEncoder(); // 编码器
    static {
        try {
            MD5_DIGEST = MessageDigest.getInstance("MD5"); // 获取MD5密码加密器
        } catch (NoSuchAlgorithmException e) { e.printStackTrace();}
    }
    @Override
    public String getEncryptPassword(String password) {
        String saltPassword = "{" + this.encryptConfig.getSalt() +
                "}" + password;                              // 设置盐值密码
        for (int x = 0; x < this.encryptConfig.getRepeat(); x++) {
            saltPassword = BASE64_ENCODER.encodeToString(MD5_DIGEST.digest(
                    saltPassword.getBytes()));               // 密码加密
        }
        return saltPassword;
    }
}
```

（10）【yootk-serter-jwt 子模块】创建 JWT 启动配置类，并添加要注册的 Bean 对象。

```
package com.yootk.jwt.autoconfig;
@Configuration                                               // 启动配置类
@EnableConfigurationProperties({JWTConfigProperties.class,
        EncryptConfigProperties.class})                      // 启用属性配置
public class JWTAutoConfiguration {
    @Bean("tokenService")
```

```
    public ITokenService getTokenServiceBean() {           // 创建ITokenService接口实例
        return new TokenServiceImpl();                     // 实例化接口子类
    }
    @Bean("encryptService")
    public IEncryptService getEncryptServiceBean() {       // 创建IEncryptService接口实例
        return new EncryptServiceImpl();                   // 实例化接口子类
    }
}
```

（11）【yootk-serter-jwt 子模块】在 src/main/resources 源代码目录中创建 META-INF/spring.factories 配置文件，并在此文件中定义自动配置类名称。

```
org.springframework.boot.autoconfigure.EnableAutoConfiguration=\
  com.yootk.jwt.autoconfig.JWTAutoConfiguration
```

（12）【yootk-serter-jwt 子模块】模块开发完成后进行编译。

```
gradle build
```

（13）【yootk-serter-jwt 子模块】为了便于功能测试，创建“src/main/resources/application.yml”配置文件，进行 JWT 以及密码加密的相关配置。需要注意的是，后续引入此模块的应用也都需要编写相同的配置项（内容自定义）。

```
yootk:                                          # 自定义配置项
  security:                                     # 安全配置相关
    config:                                     # 配置分类
      jwt:                                      # JWT配置
        sign: muyan                             # JWT证书签名
        issuer: MuyanYootk                      # 证书签发人
        secret: www.yootk.com                   # 加密密钥（公共密钥）
        expire: 100                             # 有效时长（单位：秒）
      password:                                 # 密码配置
        encrypt:                                # 加密配置
          repeat: 5                             # 重复加密的次数
          salt: www.yootk.com                   # 加密盐值
```

（14）【yootk-serter-jwt 子模块】创建 TestTokenService 类对 Token 的操作进行测试。

```
package com.yootk.test;
@ExtendWith(SpringExtension.class)                                      // JUnit 5测试工具
@WebAppConfiguration                                                    // 启动Web配置
@SpringBootTest(classes = StartJWTApplication.class)                    // 启动类
public class TestTokenService {
    @Autowired
    private ITokenService tokenService;                                 // 注入业务接口
    private String jwt = "eyJhdXRob...";                                // Token数据
    @Test
    public void testCreateJWT() {
        Map<String, Object> map = new HashMap<>();                      // 保存subject数据
        map.put("mid", "muyan");                                        // 主体数据
        map.put("name", "沐言科技");                                     // 主体数据
        map.put("rids", "USER;ADMIN;DEPT;EMP;ROLE");                    // 主体数据
        String id = "yootk-" + UUID.randomUUID();                       // 随机生成一个ID
        System.out.println(this.tokenService.createToken(id, map));     // 创建Token
    }
    @Test
    public void testParseJWT() {
        Jws<Claims> jws = this.tokenService.parseToken(jwt);            // Token解析
        System.out.println(jws.getSignature());                         // 签名数据
        JwsHeader headers = jws.getHeader();                            // 头信息
        headers.forEach((headerName, headerValue) -> {
            System.out.println("【JWT头信息】" + headerName + " = " + headerValue);
        });
        Claims claims = jws.getBody();                                  // 主体数据
        claims.forEach((bodyName, bodyValue) -> {
            System.out.println("【JWT数据】" + bodyName + " = " + bodyValue);
```

```
        });
    }
    @Test
    public void testVerifyJWT() {
        System.out.println(this.tokenService.verifyToken(jwt));       // Token校验
    }
    @Test
    public void testRefreshJWT() {
        System.out.println(this.tokenService.refreshToken(jwt));      // 获取新Token
    }
}
```

在创建 Token 的时候通过附加信息的形式传递用户的相关数据（ID、角色信息等），这样就可以利用这些数据信息实现用户授权的检测处理。

（15）【yootk-serter-jwt 子模块】创建一个密码测试类，验证 IEncryptService 接口是否正常工作。

```
package com.yootk.test;
@ExtendWith(SpringExtension.class)                          // JUnit 5测试工具
@WebAppConfiguration                                         // 启动Web配置
@SpringBootTest(classes = StartJWTApplication.class)       // 启动类
public class TestEncryptService {
    @Autowired
    private IEncryptService encryptService;                 // 注入业务接口
    @Test
    public void testCreatePassword() {
        System.out.println(this.encryptService.getEncryptPassword("hello"));
    }
}
```

程序执行结果	Wx7vJ71XD3TgJg5uiETnKA==

7.1.2 Token 认证服务

Token 认证服务

视频名称 0703_【掌握】Token 认证服务

视频简介 JWT 是整个微服务架构安全的核心，所以需要搭建统一的 Token 服务器。本视频为读者分析 Token 服务器的作用，并通过具体的实例讲解 Token 的创建。

在微服务架构之中，要想获取有意义的 JWT 数据信息，则需要凭借有效的身份信息（用户名和密码）来获取用户数据，并在认证成功之后将此数据保存在 JWT 数据结构中。所以为了便于 Token 的操作，可以创建一个 TokenServer 应用来进行此功能的统一管理，如图 7-5 所示。

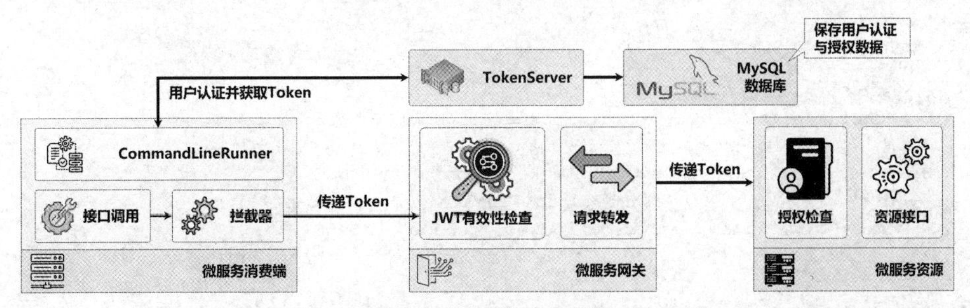

图 7-5 TokenServer 应用

消费端每次启动时，都可以根据获取的用户名和密码向 Token 服务器进行 Token 申请；而在 Token 申请的时候则可以通过 JWT 的结构特点对用户的授权数据进行保存，随后就可以通过网关来进行 JWT 有效性的检查；在进行微服务资源调用的时候，再通过传递的 JWT 数据进行授权信息检测，从而实现授权管理。考虑到设计的需要，本次开发所使用的数据表结构如图 7-6 所示。具体的开发步骤如下所示。

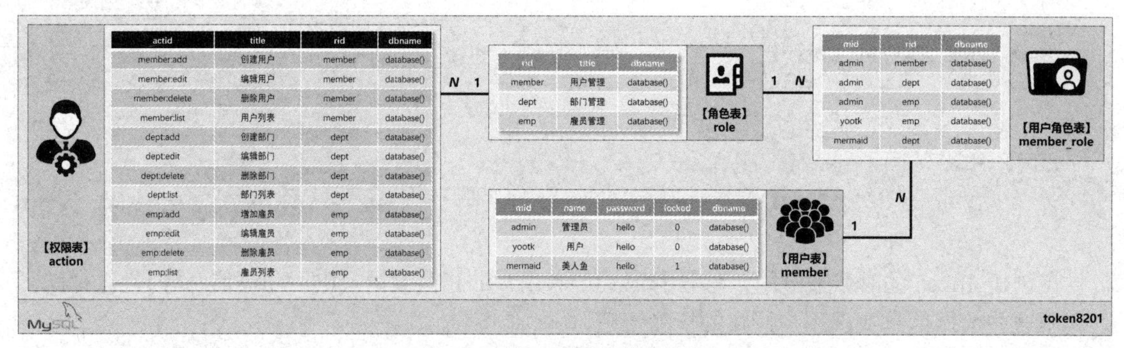

actid | title | rid | dbname
member:add | 创建用户 | member | database()
member:edit | 编辑用户 | member | database()
member:delete | 删除用户 | member | database()
member:list | 用户列表 | member | database()
dept:add | 创建部门 | dept | database()
dept:edit | 编辑部门 | dept | database()
dept:delete | 删除部门 | dept | database()
dept:list | 部门列表 | dept | database()
emp:add | 增加雇员 | emp | database()
emp:edit | 编辑雇员 | emp | database()
emp:delete | 删除雇员 | emp | database()
emp:list | 雇员列表 | emp | database()
【权限表】
action
N
1
rid | title | dbname
member | 用户管理 | database()
dept | 部门管理 | database()
emp | 雇员管理 | database()
【角色表】
role
1
N
mid | rid | dbname
admin | member | database()
admin | dept | database()
admin | emp | database()
yootk | emp | database()
mermaid | dept | database()
【用户角色表】
member_role
N
1
mid | name | password | locked | dbname
admin | 管理员 | hello | 0 | database()
yootk | 用户 | hello | 0 | database()
mermaid | 美人鱼 | hello | 1 | database()
【用户表】
member
MySQL
token8201

图 7-6　认证与授权数据表结构

（1）【数据库】编写数据库脚本，创建用户、角色、权限等表。

```
DROP DATABASE IF EXISTS token8201;
CREATE DATABASE token8201 CHARACTER SET UTF8 ;
USE token8201 ;
CREATE TABLE member(
   mid                         VARCHAR(50) NOT NULL,
   name                        VARCHAR(30),
   password                    VARCHAR(32),
   locked                      INT,
   dbname                      VARCHAR(50),
   CONSTRAINT pk_mid PRIMARY KEY (mid)
) engine='innodb';
CREATE TABLE role(
   rid                         VARCHAR(50) ,
   title                       VARCHAR(200) ,
   dbname                      VARCHAR(50),
   CONSTRAINT pk_rid PRIMARY KEY(rid)
) engine='innodb' ;
CREATE TABLE action(
   actid                       VARCHAR(50) ,
   title                       VARCHAR(200) ,
   rid                         VARCHAR(50) ,
   dbname                      VARCHAR(50),
   CONSTRAINT pk_actid PRIMARY KEY(actid)
) engine='innodb' ;
CREATE TABLE member_role(
   mid                         VARCHAR(50) ,
   rid                         VARCHAR(50) ,
   dbname                      VARCHAR(50)
) engine='innodb' ;
// 1表示活跃、0表示锁定，用户密码铭文：hello
INSERT INTO member(mid, name, password, locked, dbname) VALUES
          ('admin', '管理员', 'Wx7vJ71XD3TgJg5uiETnKA==', 0, database()) ;
INSERT INTO member(mid, name, password, locked, dbname) VALUES
          ('yootk', '用户', 'Wx7vJ71XD3TgJg5uiETnKA==', 0, database()) ;
INSERT INTO member(mid, name, password, locked, dbname) VALUES
          ('mermaid', '美人鱼', 'Wx7vJ71XD3TgJg5uiETnKA==', 1, database()) ;
// 定义角色信息
INSERT INTO role(rid, title, dbname) VALUES ('member', '用户管理', database()) ;
INSERT INTO role(rid, title, dbname) VALUES ('dept', '部门管理', database()) ;
INSERT INTO role(rid, title, dbname) VALUES ('emp', '雇员管理', database()) ;
// 定义权限信息
INSERT INTO action(actid, title, rid, dbname) VALUES
          ('member:add', '创建用户', 'member', database()) ;
INSERT INTO action(actid, title, rid, dbname) VALUES
          ('member:edit', '编辑用户', 'member', database()) ;
```

```
INSERT INTO action(actid, title, rid, dbname) VALUES
        ('member:delete', '删除用户', 'member', database()) ;
INSERT INTO action(actid, title, rid, dbname) VALUES
        ('member:list', '用户列表', 'member', database()) ;
INSERT INTO action(actid, title, rid, dbname) VALUES
        ('dept:add', '创建部门', 'dept', database()) ;
INSERT INTO action(actid, title, rid, dbname) VALUES
        ('dept:edit', '编辑部门', 'dept', database()) ;
INSERT INTO action(actid, title, rid, dbname) VALUES
        ('dept:delete', '删除部门', 'dept', database()) ;
INSERT INTO action(actid, title, rid, dbname) VALUES
        ('dept:list', '部门列表', 'dept', database()) ;
INSERT INTO action(actid, title, rid, dbname) VALUES
        ('emp:add', '增加雇员', 'emp', database()) ;
INSERT INTO action(actid, title, rid, dbname) VALUES
        ('emp:edit', '编辑雇员', 'emp', database()) ;
INSERT INTO action(actid, title, rid, dbname) VALUES
        ('emp:delete', '删除雇员', 'emp', database()) ;
INSERT INTO action(actid, title, rid, dbname) VALUES
        ('emp:list', '雇员列表', 'emp', database()) ;
// 定义用户与角色的关系
INSERT INTO member_role(mid, rid, dbname) VALUES ('admin', 'member', database()) ;
INSERT INTO member_role(mid, rid, dbname) VALUES ('admin', 'dept', database()) ;
INSERT INTO member_role(mid, rid, dbname) VALUES ('admin', 'emp', database()) ;
INSERT INTO member_role(mid, rid, dbname) VALUES ('yootk', 'emp', database()) ;
INSERT INTO member_role(mid, rid, dbname) VALUES ('mermaid', 'dept', database()) ;
COMMIT ;
```

（2）【microcloud 项目】创建 token-server-8201 子模块，该模块主要实现用户的认证管理以及 Token 生成。随后修改 build.gradle 配置文件，为该模块配置所需要的依赖库。

```
project('token-server-8201') {                                          // 子模块
    dependencies {
        implementation('com.alibaba.cloud:' +
                'spring-cloud-starter-alibaba-sentinel')                 // Sentinel保护
        implementation('com.alibaba.cloud:' +
                'spring-cloud-starter-alibaba-nacos-discovery') {        // Nacos依赖
            exclude group: 'com.alibaba.nacos', module: 'nacos-client'  // 删除nacos-client
        }
        implementation('com.alibaba.cloud:' +
                'spring-cloud-starter-alibaba-nacos-config') {           // Nacos配置
            exclude group: 'com.alibaba.nacos', module: 'nacos-client'  // 删除nacos-client
        }
        implementation('com.alibaba.nacos:nacos-client:2.0.0')          // NacosClient依赖
        implementation(project(':common-api'))                           // 引入公共子模块
        implementation(project(':yootk-starter-jwt'))                    // 引入JWT子模块
        implementation('com.baomidou:mybatis-plus-boot-starter:3.4.2')  // MyBatisPlus依赖
        implementation('mysql:mysql-connector-java:8.0.23')             // MySQL依赖
        implementation('com.alibaba:druid:1.2.5')                       // Druid依赖
    }
}
```

（3）【token-server-8201 子模块】在 src/main/resources 源代码目录中创建 application.yml 配置文件，定义 JWT 操作所需要的各个配置项。

```
spring:                                          # Spring配置
  application:                                   # 应用配置
    name: token.provider                         # 应用名称
server:                                          # 服务端配置
  port: 8201                                     # 8201端口监听
yootk:                                           # 自定义配置项
  security:                                      # 安全配置相关
    config:                                      # 配置分类
      jwt:                                       # JWT配置
```

```
    sign: muyan                                        # JWT证书签名
    issuer: MuyanYootk                                 # 证书签发人
    secret: www.yootk.com                              # 加密密钥（公共密钥）
    expire: 100                                        # 有效时长（单位：秒）
  password:                                            # 密码配置
    encrypt:                                           # 加密配置
      repeat: 5                                        # 重复加密的次数
      salt: www.yootk.com                              # 加密盐值
```

以上所定义的配置项可以根据不同的项目进行调整，而密码的配置项必须要与创建密码的相关配置项相同，否则将无法实现正确的密码加密处理。需要注意的是，由于 JWT 生成时会将应用信息保存在 JWT 头信息之中，因此一定要在该应用中定义 spring.application.name 配置项。

提示：application.yml 重复配置。

本次的程序开发除了以上 JWT 相关配置外，还应该包含 Druid、MyBatis、Sentinel、Nacos 配置项。由于配置项内容较多而且配置重复，读者可以参考前面的微服务定义进行复制与修改。本操作重复性较强，读者可以通过视频学习完整的操作。

（4）【token-server-8201 子模块】根据当前给出的数据表结构创建实体映射类。

Member

```java
package com.yootk.provider.vo;
@Data
@TableName("member")                          // 映射表名称
public class Member {
    @TableId                                  // 主键字段
    private String mid;
    private String name;
    private String password;
    private Integer locked;
    private String dbname;
}
```

Role

```java
package com.yootk.provider.vo;
@Data
@TableName("role")                            // 映射表名称
public class Role {
    @TableId                                  // 主键字段
    private String rid ;
    private String title ;
    private String dbname ;
}
```

Action

```java
package com.yootk.provider.vo;
@Data
@TableName("action")                          // 映射表名称
public class Action {
    @TableId                                  // 主键字段
    private String actid;
    private String title;
    private String rid;
    private String dbname;
}
```

（5）【token-server-8201 子模块】创建 DAO 接口。

IMemberDAO

```java
package com.yootk.provider.dao;
@Mapper
public interface IMemberDAO extends BaseMapper<Member> { }
```

IRoleDAO

```java
package com.yootk.provider.dao;
@Mapper
public interface IRoleDAO extends BaseMapper<Member> {
    public Set<String> findAllByMember(String mid); // 获取用户角色
}
```

<table>
<tr><td>IActionDAO</td><td><pre>package com.yootk.provider.dao;
@Mapper
public interface IActionDAO extends BaseMapper<Member> {
    public Set<String> findAllByMember(String mid); // 获取用户权限
}</pre></td></tr>
</table>

（6）【token-server-8201 子模块】由于此时需要进行数据查询处理，为了方便将创建 3 个 Mapper 配置文件。

<table>
<tr><td>MemberMapper.xml</td><td><pre><mapper namespace="com.yootk.provider.dao.IMemberDAO"></mapper></pre></td></tr>
<tr><td>RoleMapper.xml</td><td><pre><mapper namespace="com.yootk.provider.dao.IRoleDAO">
    <select id="findAllByMember" parameterType="string"
          resultType="string">
      SELECT rid FROM member_role WHERE mid=#{mid}
  </select>
</mapper></pre></td></tr>
<tr><td>ActionMapper.xml</td><td><pre><mapper namespace="com.yootk.provider.dao.IActionDAO">
    <select id="findAllByMember" parameterType="string"
          resultType="string">
        SELECT actid FROM action WHERE rid IN(
          SELECT rid FROM member_role WHERE mid=#{mid})
    </select>
</mapper></pre></td></tr>
</table>

（7）【common-api 子模块】为便于数据传输，将创建 MemberDTO 数据传输类。

```
package com.yootk.common.dto;
@Data
public class MemberDTO {
    private String mid;
    private String password;
}
```

（8）【common-api 子模块】创建 IMemberService 业务接口实现认证与授权信息获取。

```
package com.yootk.service;
public interface IMemberService {
    /**
     * 实现用户登录认证业务处理
     * @param memberDTO 通过MemberDTO包裹要处理的mid、password
     * @return 此时需要返回的内容通过Map集合包装，具体内容包括
     * 1.key = status、value = 登录状态（true、false）
     * 2.key = mid、value = 用户名
     * 3.key = name、value = 真实姓名
     * 4.key = resource、value = 用户授权信息（内嵌Map）
     * 4-1.key = roles、value = 用户拥有的全部角色
     * 4-2.key = actions、value = 用户拥有的全部权限
     */
    public Map<String, Object> login(MemberDTO memberDTO);
}
```

（9）【token-server-8201 子模块】创建 IMemberService 业务接口实现子类，并使用 DAO 接口进行认证处理。

```
package com.yootk.provider.service.impl;
@Service
public class MemberServiceImpl implements IMemberService {
    @Autowired
    private IMemberDAO memberDAO;                                    // 注入DAO接口实例
    @Autowired
    private IRoleDAO roleDAO;                                        // 注入DAO接口实例
    @Autowired
```

```
    private IActionDAO actionDAO;                                          // 注入DAO接口实例
    @Override
    public Map<String, Object> login(MemberDTO memberDTO) {                // 用户登录
        Map<String, Object> result = new HashMap<>();                      // 保存登录结果
        Member member = this.memberDAO.selectById(memberDTO.getMid());    // 根据ID查询
        if (member == null || !member.getPassword().equals(
                memberDTO.getPassword())|| member.getLocked().equals(1)) { // 密码验证
            result.put("status", false);                                   // 登录失败标记
        } else {
            result.put("status", true);                                    // 登录成功标记
            result.put("mid", memberDTO.getMid());                         // 保存用户ID
            result.put("name", member.getName());                          // 保存用户姓名
            Map<String, Object> resource = new HashMap<>();                // 用户资源
            resource.put("roles", this.roleDAO.findAllByMember(memberDTO.getMid()));
            resource.put("actions", this.actionDAO.findAllByMember(memberDTO.getMid()));
            result.put("resource", resource);                              // 保存资源
        }
        return result;
    }
}
```

（10）【token-server-8201 子模块】编写 TestMemberService 测试类，测试 login()方法能否正常工作。

```
package com.yootk.test;
@ExtendWith(SpringExtension.class)                                 // JUnit 5测试工具
@WebAppConfiguration                                               // 启动Web配置
@SpringBootTest(classes = StartTokenApplication8201.class)         // 启动类
public class TestMemberService {
    @Autowired
    private IMemberService memberService;                          // 注入业务接口实例
    @Autowired
    private IEncryptService encryptService;                        // 注入加密接口实例
    @Test
    public void testLogin() {
        MemberDTO member = new MemberDTO();                        // 实例化DTO对象
        member.setMid("admin");                                    // 用户名
        member.setPassword(this.encryptService.getEncryptPassword("hello")); // 密码
        System.out.println(this.memberService.login(member));      // 登录业务
    }
}
```

程序执行结果	{resource={roles=[member, emp, dept], actions=[emp:list, dept:edit, dept:list, emp:edit, member:add, dept:add, emp:add, member:edit, dept:delete, member:delete, member:list, emp:delete]}, name=管理员, mid=admin, status=true}

（11）【token-server-8201 子模块】由于本应用提供了一个 Token 管理的资源生产端功能，因此需要对消费端提供 REST 接口，创建 TokenAction 控制器类，并进行相关业务接口调用。

```
package com.yootk.provider.action;
@RestController                                                        // REST响应
@RequestMapping("/token/*")                                            // 父路径
public class TokenAction {
    @Autowired
    private IMemberService memberService;                              // 注入业务接口实例
    @Autowired
    private IEncryptService encryptService;                            // 注入加密接口实例
    @Autowired
    private ITokenService tokenService;                                // 注入Token接口实例
    @RequestMapping("create")
    public Object login(MemberDTO memberDTO) {
        memberDTO.setPassword(this.encryptService.getEncryptPassword(
                memberDTO.getPassword()));                             // 登录密码加密
        Map<String, Object> result = this.memberService.login(memberDTO);
        if (((Boolean) result.get("status"))) {                        // 用户登录成功
            return this.tokenService.createToken(result.get("mid").toString(),
```

```
                (Map<String, Object>) result.get("resource"));     // 创建Token
        }
        return null;
    }
    @RequestMapping("parse")
    public Object parseToken(String token) {
        return this.tokenService.parseToken(token);                // Token解析
    }
}
```

（12）【本地系统】为便于服务调用，可修改本地 hosts 文件，添加 Token 服务主机地址。

```
127.0.0.1   token-server-8201
```

（13）【Postman】通过 Postman 测试工具进行服务调用，如图 7-7 所示。

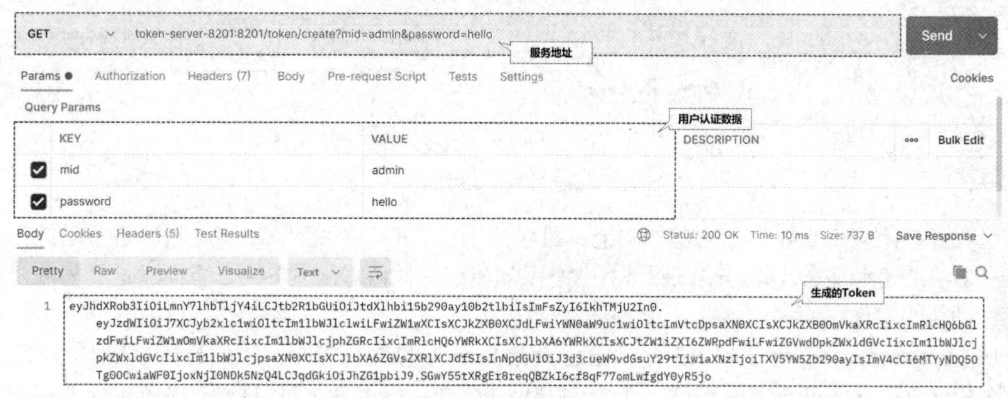

图 7-7　Token 生成测试

（14）【Postman】本应用除了提供 Token 生成支持之外，还提供 Token 解析支持。将已有的 Token 数据传递到指定的接口就可以得到图 7-8 所示的结果。

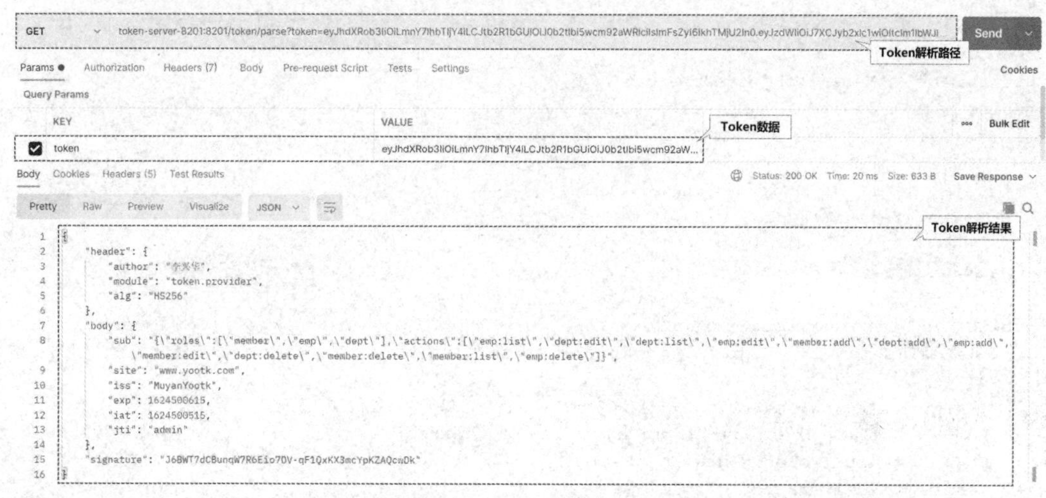

图 7-8　Token 解析测试

7.1.3　JWT 授权检测

视频名称　0704_【掌握】JWT 授权检测

视频简介　考虑到安全性问题，微服务的资源提供者需要进行有效的授权检测，这样就可以有效地进行微服务管理。本视频将介绍如何对已有的微服务进行改造，并对 JWT 公共模块进行修改，基于注解与过滤器的形式实现微服务资源的授权保护。

微服务中的所有资源需要有效的授权管理，而每一个生成的 JWT 数据都可以包含完整的授权信息，这样就可以基于注解的方式对指定的 REST 资源进行保护。保护的形式可以分为 3 种：JWT 认证保护（有正确的 Token 信息）、授权保护、权限保护。所有的 JWT 校验操作都应该在资源访问前进行，所以本次演示将以拦截器的模式进行 JWT 的相关操作，实现结构如图 7-9 所示。

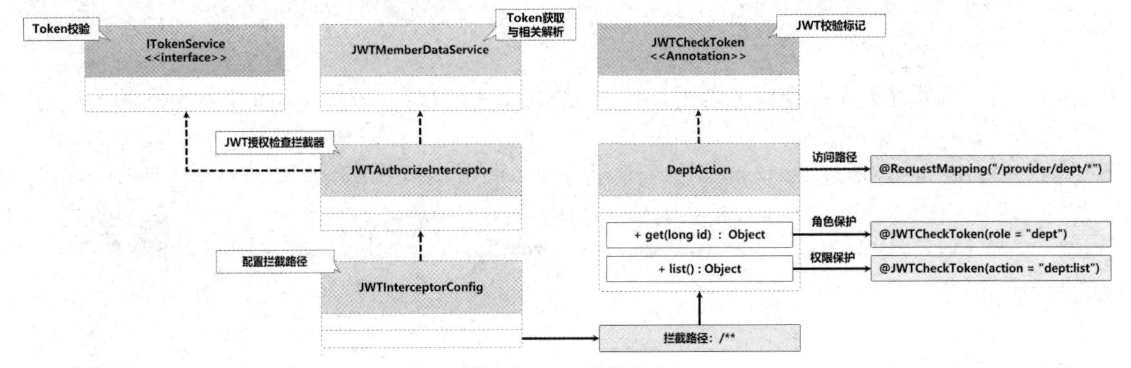

图 7-9　JWT 数据拦截

考虑到多个应用需要相同的操作功能，可以基于已有的 yootk-starter-jwt 模块进行功能扩充，而后在对应的资源服务（如部门微服务）之中添加相应的拦截器配置即可。下面通过具体的步骤介绍这一功能的实现。

（1）【yootk-starter-jwt 子模块】此时已经确定了 JWT 的数据结构，为了便于其他模块进行 JWT 的相关操作，建议创建一个用于 JWT 数据剥离的工具类，该工具类可以获取 JWT 里面保存的角色与权限数据。

```
package com.yootk.jwt.util;
public class JWTMemberDataService {
   @Autowired
   private ITokenService tokenService;                               // 注入Token业务接口实例
   @Autowired
   private ObjectMapper objectMapper;
   public Map<String, String> headers(String token) {                // 获取JWT中的头信息
      Jws<Claims> claims = this.tokenService.parseToken(token);      // Token解析操作
      Map<String, String> headers = new HashMap<>();                 // 保存全部头信息
      claims.getHeader().forEach((key, value) -> {                   // 头信息迭代
         headers.put(key.toString(), value.toString());              // 保存数据
      });
      return headers;                                                // 返回头信息
   }
   public Set<String> roles(String token) {                          // 获取全部的角色信息
      Jws<Claims> claims = this.tokenService.parseToken(token);      // Token解析操作
      try {
         Map<String, List<String>> map = this.objectMapper.readValue(
               claims.getBody().getSubject(), Map.class);            // 将主体数据转为Map集合
         Set<String> roles = new HashSet<>();                        // 保存角色集合
         roles.addAll(map.get("roles"));                             // 保存角色数据
         return roles;                                               // 返回角色集合
      } catch (JsonProcessingException e) { return null; }
   }
   public Set<String> actions(String token) {                        // 获取全部的权限信息
      Jws<Claims> claims = this.tokenService.parseToken(token);      // Token解析操作
      try {
         Map<String, List<String>> map = this.objectMapper.readValue(
               claims.getBody().getSubject(), Map.class);            // 将主体数据转为Map集合
         Set<String> actions = new HashSet<>();                      // 保存权限集合
         actions.addAll(map.get("actions"));                         // 保存权限数据
         return actions;                                             // 返回权限集合
```

```
        } catch (JsonProcessingException e) { return null; }
    }
    public String id(String token) {                                // 获取ID
        Jws<Claims> claims = this.tokenService.parseToken(token);   // Token解析操作
        return claims.getBody().getId();                            // 返回保存ID
    }
    public String getToken(HttpServletRequest request, String name) {  // 获取Token数据
        String token = request.getParameter(name);                  // 通过参数获取Token
        if (token == null || "".equals(token)) {                    // 没有接收Token
            token = request.getHeader(name);                        // 头信息接收Token
        }
        return token;
    }
}
```

(2)【yootk-starter-jwt 子模块】为了便于其他模块使用、修改 JWTAutoConfiguration 配置类，追加新的注册 Bean。

```
package com.yootk.jwt.autoconfig;
@Configuration                                              // 启动配置类
@EnableConfigurationProperties({JWTConfigProperties.class,
        EncryptConfigProperties.class})                     // 启用属性配置
public class JWTAutoConfiguration {                         // 该类中的其他重复代码，略
    @Bean("memberDataService")                              // 用户数据操作Bean
    public JWTMemberDataService getMemberDataService() {
        return new JWTMemberDataService();
    }
}
```

(3)【yootk-starter-jwt 子模块】考虑到微服务授权检查的便捷性，可以创建一个@JWTCheckToken 注解，该注解可以根据需要启用，进行指定的授权检查。为了简化开发，本次只通过角色和权限之一进行授权检查。

```
package com.yootk.jwt.annotation;
@Target({ElementType.METHOD})                               // 类和方法上使用
@Retention(RetentionPolicy.RUNTIME)                         // 运行时生效
public @interface JWTCheckToken {
    boolean required() default true;                        // 启用配置
    String role() default "";                               // 角色检查（二选一）
    String action() default "";                             // 权限检查（二选一）
}
```

(4)【yootk-starter-jwt 子模块】模块修改完成后对该模块重新编译。

```
gradle build
```

(5)【microcloud 项目】为部门微服务模块添加 yootk-starter-jwt 依赖。

```
implementation(project(':yootk-starter-jwt'))       // 引入JWT子模块
```

(6)【provider-dept-*】修改 DeptAction 程序类，对指定的方法添加 JWT 验证注解。

```
package com.yootk.provider.action;
public class DeptAction {
    // 本次只修改两个操作方法的注解定义，其他重复代码略
    @JWTCheckToken(role = "dept")                           // 角色检查
    public Object get(@PathVariable("id") long id) { … }
    @JWTCheckToken(action = "dept:list")                    // 权限检查
    public Object list() { … }
}
```

(7)【provider-dept-*】创建拦截器，在该拦截器中可以对有@JWTCheckToken 注解的方法进行授权检查。

```
package com.yootk.provider.interceptor;
public class JWTAuthorizeInterceptor implements HandlerInterceptor {
    private static final String TOKEN_NAME = "yootk-token";            // Token参数名称
    @Autowired
```

```
    private JWTMemberDataService memberDataService;                    // 用户数据操作
    @Autowired
    private ITokenService tokenService;                                // Token数据操作
    @Override
    public boolean preHandle(HttpServletRequest request, HttpServletResponse response,
                Object handler) throws Exception {
        boolean flag = true;                                           // 拦截标记
        if (!(handler instanceof HandlerMethod)) {                     // 类型不匹配
            return flag;                                               // 不拦截
        }
        HandlerMethod handlerMethod = (HandlerMethod) handler;        // 获取操作方法对象
        Method method = handlerMethod.getMethod();                     // 获取操作方法
        if (method.isAnnotationPresent(JWTCheckToken.class)) {         // 是否存在指定注解
            response.setCharacterEncoding("UTF-8");                    // 响应编码
            response.setContentType("application/json; charset=utf-8"); // 响应类型
            JWTCheckToken checkToken = method.getAnnotation(
                        JWTCheckToken.class);                          // 获取方法定义的指定注解
            if (checkToken.required()) {                               // 检测状态，默认为true
                String token = this.memberDataService.getToken(request, TOKEN_NAME);
                if (!StringUtils.hasLength(token)) {                   // 没有Token数据
                    flag = false;                                      // 不允许访问
                    response.getWriter().println(JWTResponseCode.NO_AUTH_CODE); // 错误响应
                } else {                                               // Token存在
                    if (!this.tokenService.verifyToken(token)) {       // Token校验失败
                        flag = false;                                  // 不允许访问
                        response.getWriter().println(
                            JWTResponseCode.TOKEN_TIMEOUT_CODE);       // 错误响应
                    } else {                                           // Token校验成功
                        if (!(checkToken.role() == null ||
                            "".equals(checkToken.role()))) {           // 需要进行角色校验
                            if (this.memberDataService.roles(token)
                                .contains(checkToken.role())) {        // 拥有此角色
                                flag = true;                           // 允许访问
                            } else {
                                response.getWriter().println(
                                        JWTResponseCode.NO_AUTH_CODE); // 错误响应
                                flag = false;                          // 不允许访问
                            }
                        } else if (!(checkToken.action() == null ||
                            "".equals(checkToken.action()))) {         // 需要进行权限校验
                            if (this.memberDataService.actions(token)
                                .contains(checkToken.action())) {      // 拥有此权限
                                flag = true;                           // 允许访问
                            } else {
                                response.getWriter().println(
                                        JWTResponseCode.NO_AUTH_CODE); // 错误响应
                                flag = false;                          // 不允许访问
                            }
                        } else {
                            flag = true;                               // 允许访问
                        }
                    }
                }
            }
        }
        return flag;
    }
}
```

（8）【provider-dept-*】创建拦截器配置类并定义访问路径。

```
package com.yootk.provider.config;
@Configuration
public class JWTInterceptorConfig implements WebMvcConfigurer {
```

```
   @Override
   public void addInterceptors(InterceptorRegistry registry) {   // 拦截器注册
      registry.addInterceptor(this.getDefaultHandlerInterceptor())
            .addPathPatterns("/**");                             // 追加拦截器
   }
   @Bean
   public HandlerInterceptor getDefaultHandlerInterceptor() {    // 获取拦截器实例
      return new JWTAuthorizeInterceptor();                      // 获取拦截器实例
   }
}
```

（9）【provider-dept-*】修改 application.yml 配置文件，添加 yootk-starter-jwt 的相关配置项（代码略）。

（10）【Postman】部门微服务修改完成后，可以利用 Postman 工具向部门微服务发出服务调用请求，同时可以通过头信息或参数传递 Token 信息，配置如图 7-10 所示。

图 7-10 服务调用并传递 Token

7.1.4 网关认证过滤

网关认证过滤

视频名称 0705_【掌握】网关认证过滤

视频简介 网关是微服务资源整合的关键，所以在资源代理转发前都必须进行有效的 JWT 认证。本视频将介绍通过具体的实例对已有的网关应用进行修改，并利用全局过滤器的形式实现 JWT 数据结构的校验处理。

网关作为服务资源调用的核心组件，在消费端每一次进行资源访问时也应该进行有效的 Token 验证（只验证 Token 的有效性，不验证 Token 中的具体内容）。如果消费端发送的 Token 有效则进行代理转发，否则显示错误信息。可以通过 GlobalFilter 方式来实现这样的验证操作，如图 7-11 所示。具体的开发步骤如下。

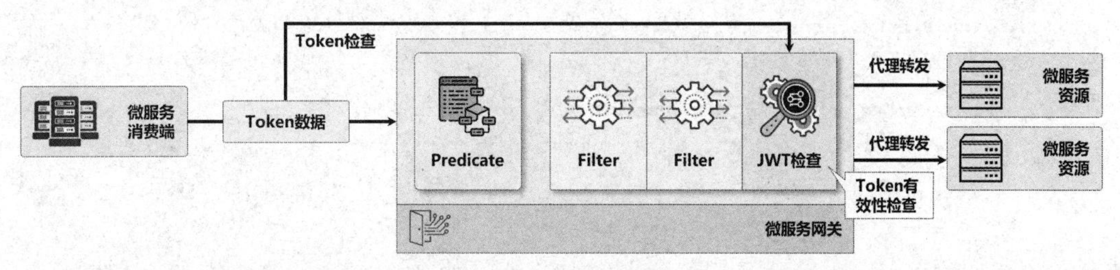

图 7-11 网关认证过滤

（1）【microcloud 项目】修改 build.gradle 配置文件，为 gateway-9501 子模块添加 yootk-starter-jwt 依赖。

```
implementation(project(':yootk-starter-jwt'))          // 引入JWT子模块
```

（2）【gateway-9501 子模块】由于所有的微服务最终都需要通过网关进行调用，因此对 Token 的获取路径就应该跳过检查。可以通过 application.yml 配置文件进行相关定义。

```
# 有关“yootk-starter-jwt”的自动配置项代码略
gateway:                                          # 自定义网关配置
  config:                                         # 配置项
    jwt:                                          # JWT相关配置
      header-name: yootk-token                    # 头信息参数名称
      skip-auth-urls:                             # 非检查路径
        - /token/create
```

（3）【gateway-9501 子模块】为了便于属性的注入，创建一个 GatewayJWTConfigProperties 配置类。

```
package com.yootk.gateway.config;
@Component
@Data
@ConfigurationProperties("gateway.config.jwt")          // 属性配置前缀
public class GatewayJWTConfigProperties {
    private List<String> skipAuthUrls;                  // 跳过路径集合
    private String headerName;                          // 头信息名称
}
```

（4）【gateway-9501 子模块】创建一个全局过滤器。

```
package com.yootk.gateway.filter.global;
@Component
@Slf4j
public class JwtTokenCheckFilter implements GlobalFilter {
    @Autowired
    private GatewayJWTConfigProperties jwtConfig;                  // JWT配置项
    @Autowired
    private ITokenService tokenService;                            // Token业务操作
    @Override
    public Mono<Void> filter(ServerWebExchange exchange, GatewayFilterChain chain) {
        String url = exchange.getRequest().getURI().getPath();     // 获取当前访问路径
        if (this.jwtConfig.getSkipAuthUrls() != null &&
          this.jwtConfig.getSkipAuthUrls().contains(url)) {        // 跳过不需要验证的路径
            return chain.filter(exchange);                         // 非拦截路径
        }
        String token = exchange.getRequest().getHeaders().get(
                this.jwtConfig.getHeaderName()).get(0);            // 获取Token
        log.info("网关Token检测，Token = {}", token);                // 日志输出
        ServerHttpResponse response = exchange.getResponse();      // 获取响应实例
        if (StringUtils.isBlank(token)) {                          // Token数据为空
            DataBuffer buffer = response.bufferFactory().wrap(
                    JWTResponseCode.NO_AUTH_CODE.toString()
                     .getBytes(StandardCharsets.UTF_8));           // 错误响应
            return response.writeWith(Flux.just(buffer));          // 错误响应
        } else {                                                   // Token存在
            if (this.tokenService.verifyToken(token)) {            // 校验成功
                return chain.filter(exchange);                     // 执行后续处理
            } else {
                DataBuffer buffer = response.bufferFactory().wrap(
                        JWTResponseCode.TOKEN_TIMEOUT_CODE.toString()
                              .getBytes(StandardCharsets.UTF_8));
                return response.writeWith(Flux.just(buffer));      // 错误响应
            }
        }
    }
}
```

由于网关主要起到资源访问的代理作用，因此消费端发送请求到网关时，可以将获取的 Token 数据以头信息的方式进行传输，随后网关会依据 Token 数据的有效性进行处理。

（5）【Nacos 控制台】由于当前的微服务网关采用的是动态路由管理模式，因此需要开发者通过 Nacos 控制台修改网关配置项，添加 Token 应用的访问地址，如图 7-12 所示。

（6）【Postman】动态路由修改完成后，就可以通过网关进行 Token 的创建，如图 7-13 所示。

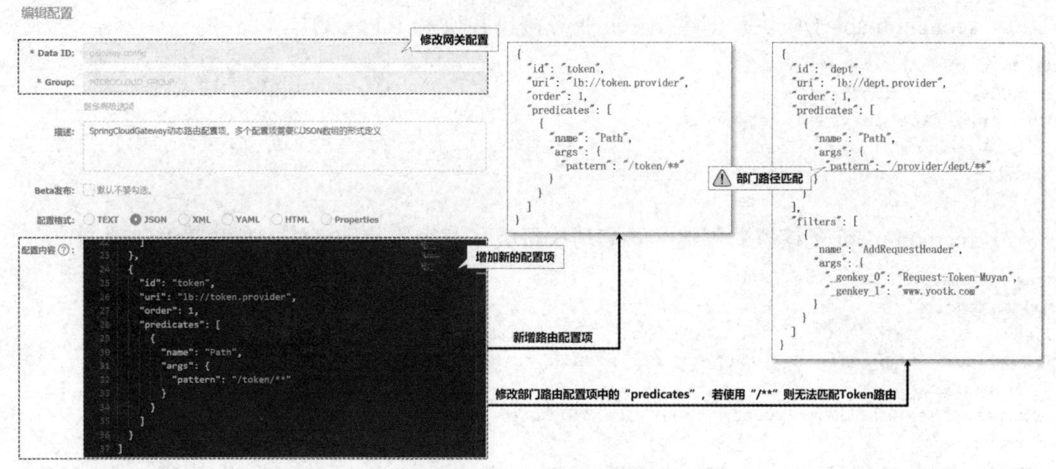

图 7-12　动态路由配置

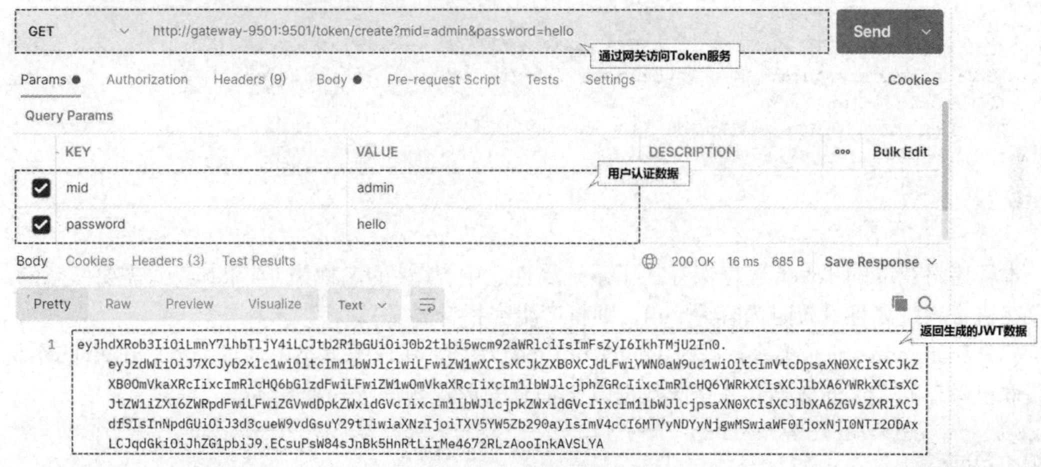

图 7-13　通过网关创建 Token

7.1.5　消费端获取 JWT

消费端获取 JWT

视频名称　0706_【掌握】消费端获取 JWT

视频简介　微服务资源最终是通过消费端应用进行调用的，这样在进行资源调度时就必须进行 Token 数据的获取与传递。本视频通过实例讲解如何通过 CommandLineRunner 进行 Token 获取，并利用 Feign 拦截器实现 Token 数据的头信息传递操作。

此时的网关需要进行 Token 有效性检查，而资源微服务需要进行 Token 授权检查，所以在消费端进行服务资源调用时就必须通过头信息的形式进行 Token 传递。考虑到代码整体的结构性需要，可以利用 CommandLineRunner 接口在容器启动时进行 Token 加载，并将其保存在系统属性之中。在每次请求时基于拦截器的方式通过系统属性加载 Token 数据，并以头信息的形式发送 Token 数据，就可以实现服务的正确调用，操作流程如图 7-14 所示。具体实现步骤如下。

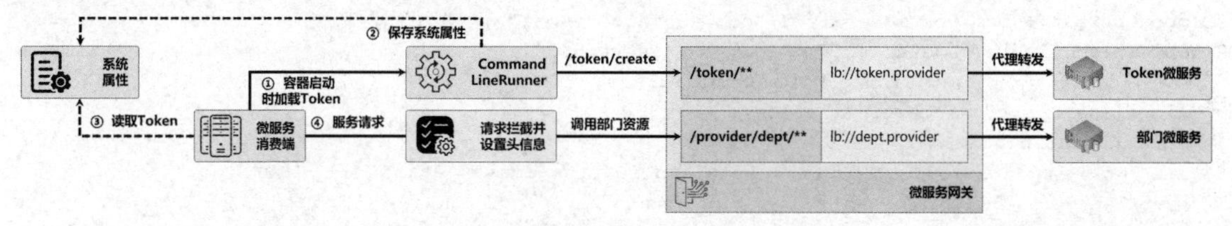

图 7-14　消费端获取 JWT 操作流程

（1）【common-api 子模块】创建 Token 业务接口并映射 Token 的操作路径。

```
package com.yootk.service;
@FeignClient(value="microcloud.gateway")                              // 网关调用
public interface IMemberTokenService {
   @GetMapping("/token/create")                                       // 获取Token路径
   public String login(MemberDTO memberDTO);
}
```

（2）【common-api 子模块】创建容器启动执行类，在容器启动完成后加载 Token 数据。

```
package com.yootk.service.load;
@Component
@Slf4j
public class FeignTokenLoaderRunner implements CommandLineRunner {   // 启动后运行
   @Autowired
   private IMemberTokenService memberTokenService;                    // 业务接口实例
   @Override
   public void run(String... args) throws Exception {
      MemberDTO dto = new MemberDTO() ;                               // 实例化DTO对象
      dto.setMid("admin");                                            // 用户名
      dto.setPassword("hello");                                       // 密码
      String token = this.memberTokenService.login(dto);              // 获取Token
      if (token != null) {
         log.info("获取Token数据成功：{}", token);                      // 日志输出
         System.setProperty("yootk.token", token);                    // 保存属性
      }
   }
}
```

本程序将获取的 Token 数据保存在了系统属性之中，属性的名称为 yootk.token。需要注意的是，如果在进行系统属性设置时内容为 null，则将产生空指针异常。

（3）【common-api 子模块】修改 FeignConfig 配置类，在该配置类中通过 yootk.token 的系统属性获取已保存的 Token 数据，并将此数据以头信息的形式与请求一起发送。

```
package com.yootk.service.config;
public class FeignConfig {
   @Bean
   public Logger.Level level() {
      return Logger.Level.FULL;                                       // 输出所有请求的细节
   }
   @Bean
   public RequestInterceptor getFeignRequestInterceptor() {  // 请求拦截器
      return (template -> {
         template.header("serviceName", "pc");                       // Sentinel授权规则
         template.header("yootk-token",
               System.getProperty("yootk.token"));                   // Token传递
      });                                                             // 设置头信息
   }
}
```

（4）【consumer-springboot-80 子模块】修改程序启动类，并定义 FeignConfig 配置类。

```
package com.yootk.consumer;
@SpringBootApplication
@EnableDiscoveryClient                                                // 启用发现服务
@ComponentScan({"com.yootk.service","com.yootk.consumer"})           // 配置扫描包
@EnableFeignClients(basePackages={"com.yootk.service"},
      defaultConfiguration = FeignConfig.class)
public class StartConsumerApplication {
   public static void main(String[] args) {
      SpringApplication.run(StartConsumerApplication.class, args); // 服务启动
   }
}
```

（5）【Postman】配置完成后可以启动消费端，这样在启动时就会通过 FeignTokenLoaderRunner 类加载 Token 数据，随后在每次请求时通过头信息发送 Token，消费端按照以往的方式调用微服务资源即可，如图 7-15 所示。

图 7-15　消费端服务调用

注意：JWT 数据有效时间。

消费端在每次调用时都必须获得有效的 JWT 数据，在不引入 JWT 刷新机制的前提下，可以尝试延长 JWT 有效时间。如果觉得此种方式依然不方便，也可以对整体代码进行改造，例如，以追加过滤器或拦截器的形式在每次请求前进行 Token 的刷新操作。

7.2　Spring Boot Admin

Spring Boot Admin 服务端

视频名称　0707_【掌握】Spring Boot Admin 服务端

视频简介　良好的微服务需要有效的监控与管理，Spring 官方提供了 Spring Boot Admin 管理工具，即通过 Actuator 实现服务监控。本视频通过实例讲解服务监控功能的启用。

Spring Boot 的应用程序可以通过 Actuator 提供服务监控数据，因其每次访问时都只能获取当前的状态数据，所以为了便于观察需要将这些数据汇总在一起进行统一监控（如 Prometheus 等）。而 Spring Cloud 在服务管理阶段就必须考虑到全部节点的监控（如资源、网关、消费端等），因此推荐使用 Spring Boot Admin 开源项目进行管理。

该组件是以依赖库的形式出现的，开发者只需要建立 Spring Boot 项目，在相应的微服务中引入依赖后就可以通过注册中心实现监控数据的抓取。该组件内置了完整的前端监控界面，以便管理员监控，其结构如图 7-16 所示。

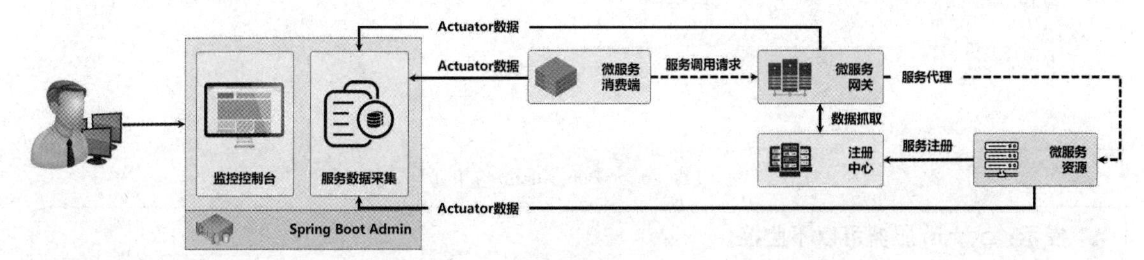

图 7-16　Spring Boot Admin 结构

要想实现 Spring Boot Admin 服务监控，就需要对各个微服务进行相应的改造，使其可以向 Spring Boot Admin 服务端开放监控数据。服务端的具体搭建步骤如下。

（1）【microcloud 项目】创建一个新的服务监控子模块 microcloud-admin-8000，随后修改 build.gradle 配置文件，添加所需要的服务依赖。

```
project('microcloud-admin-8000') {  // 子模块
    dependencies {
        implementation('org.springframework.boot:spring-boot-starter-web')
        implementation('de.codecentric:spring-boot-admin-starter-server:2.3.0')
        implementation('com.alibaba.cloud:' +
                'spring-cloud-starter-alibaba-nacos-discovery') {          // Nacos依赖
            exclude group: 'com.alibaba.nacos', module: 'nacos-client'     // 删除nacos-client
        }
        implementation('com.alibaba.cloud:' +
                'spring-cloud-starter-alibaba-nacos-config') {             // Nacos配置
            exclude group: 'com.alibaba.nacos', module: 'nacos-client'     // 删除nacos-client
        }
        implementation('com.alibaba.nacos:nacos-client:2.0.0')             // NacosClient依赖
    }
}
```

（2）【microcloud-admin-8000 子模块】在该项目中配置 bootstrap.yml 与 application.yml 文件并进行 Nacos 的相关配置，参考本书第 3 章。

（3）【microcloud-admin-8000 子模块】创建应用程序启动类并添加启动注解。

```
package com.yootk.admin;
@EnableAdminServer                                                     // 启用Spring Boot Admin
@EnableDiscoveryClient                                                 // Nacos发现服务
@SpringBootApplication                                                 // Spring Boot启动注解
public class StartSpringBootAdmin8000 {
    public static void main(String[] args) {
        SpringApplication.run(StartSpringBootAdmin8000.class, args);   // 服务启动
    }
}
```

（4）【本地系统】为了便于服务访问，修改 hosts 配置文件，添加新的主机映射项。

```
127.0.0.1   microcloud-admin
```

（5）【浏览器】打开浏览器输入 Spring Boot Admin 服务访问地址，可以得到图 7-17 所示的界面。

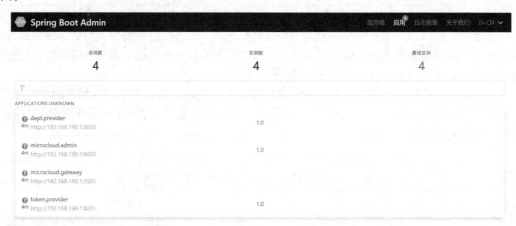

图 7-17　Spring Boot Admin 应用列表

通过此时的运行结果可以发现，当前的 microcloud.admin 项目模块也被 Spring Boot Admin 监控了（因为此时的微服务数据是通过 Nacos 发现服务抓取而来的）。如果不想看到此微服务的信息，则可以修改项目中的 application.yml，取消 Nacos 注册支持。

7.2.1 Spring Boot Admin 安全配置

Spring Boot Admin 安全配置

视频名称 0708_【掌握】Spring Boot Admin 安全配置

视频简介 监控服务可以获取全部微服务节点数据，这样就必须进行有效的安全保护。本视频介绍将 Spring Security 与 Spring Boot Admin 结合，实现用户认证管理。

通过 Nacos 注册中心成功获取全部微服务信息后，Spring Boot Admin 没有安全保护，任何用户都可以直接访问。为了安全考虑，应该为项目追加有效的认证管理。在 Spring Boot Admin 中默认使用 Spring Security 进行安全保护。下面将通过具体的操作步骤介绍如何进行安全配置。

（1）【microcloud 项目】修改 build.gradle 配置文件，为 microcloud-admin-8000 子模块添加 Spring Security 依赖库。

```
implementation('org.springframework.boot:spring-boot-starter-security')
```

（2）【microcloud-admin-8000 子模块】为项目添加一个 AdminSecurityConfig 类，进行 Spring Security 相关配置。

```
package com.yootk.admin.config;
@Configuration
public class AdminSecurityConfig extends WebSecurityConfigurerAdapter {
   private final String adminContextPath;
   public AdminSecurityConfig(AdminServerProperties adminServerProperties) {
      this.adminContextPath = adminServerProperties.getContextPath();
   }
   @Override
   protected void configure(HttpSecurity http) throws Exception {
      SavedRequestAwareAuthenticationSuccessHandler successHandler = new
               SavedRequestAwareAuthenticationSuccessHandler();
      successHandler.setTargetUrlParameter("redirectTo");
      http.authorizeRequests()
              .antMatchers(adminContextPath + "/assets/**").permitAll()  // 配置访问路径
              .antMatchers(adminContextPath + "/login").permitAll()      // 配置访问路径
              .anyRequest().authenticated()                             // 认证访问
              .and().formLogin().loginPage(adminContextPath + "/login")  // 登录处理
              .successHandler(successHandler)                           // 登录成功
              .and().logout().logoutUrl(adminContextPath + "/logout")    // 注销配置
              .and().httpBasic().and().csrf().disable();                // 关闭CSRF
   }
}
```

（3）【microcloud-admin-8000 子模块】在 src/main/resources 源代码目录中创建 application.yml 配置文件，追加当前系统的用户名与密码（固定信息）。

```
spring:                                                          # Spring配置
  security:                                                      # 安全配置
    user:                                                        # 固定账户
      name: muyan                                                # 用户名
      password: yootk                                            # 密码
```

重新启动 microcloud-admin-8000 子模块的应用之后，再次访问“http://microcloud-admin:8000”路径将会自动出现用户登录界面，登录成功后即可使用 Spring Boot Admin 控制台。

7.2.2 Spring Boot Admin 客户端接入

Spring Boot Admin 客户端接入

视频名称 0709_【掌握】Spring Boot Admin 客户端接入

视频简介 Spring Boot Admin 中的监控数据由各个微服务通过 Actuator 进行发送，这样就需要进行监控节点的整合。本视频介绍对已有的微服务架构进行修改，实现控制台的服务状态监控以及统计信息的展示。

Spring Boot Admin 服务搭建完成之后需要接收所有微服务节点发送来的监控数据信息，利用这些信息可以判断当前微服务的实例状态（在线或离线）、JVM 监控信息，还可以动态修改日志状态。而要想实现这样的控制，就需要对已有的微服务进行配置。下面展示具体的实现步骤。

（1）【microcloud 项目】修改 build.gradle 配置文件，为部门微服务、网关微服务、消费端微服务、Token 微服务这几个模块添加如下依赖。

```
implementation('de.codecentric:spring-boot-admin-starter-client:2.3.0')
implementation('org.springframework.boot:spring-boot-starter-actuator')
```

（2）【微服务模块】修改需要接入 Spring Boot Admin 服务的微服务中的 application.yml 配置文件，打开监控数据。

```
management:                                  # 管理配置
 health:                                     # 健康监控配置
  redis:                                     # Redis健康启用配置
   enabled: false                            # 关闭
  sentinel:                                  # Sentinel监控启用配置
   enabled: false                            # 关闭
  nacosDiscovery:                            # Nacos健康启用配置
   enabled: false                            # 关闭
 endpoints:                                  # 终端配置
  web:                                       # Web终端
   exposure:                                 # 配置暴露终端
    include: "*"                             # 暴露所有节点
  health:                                    # 健康配置
   sensitive: false                          # 关闭过滤敏感信息
 endpoint:                                   # 终端配置
  health:                                    # 健康监控
   show-details: ALWAYS                      # 显示详细信息
```

由于在进行健康监控的时候，Spring Boot 会自动开启一些健康启用监控项，有可能不正确地获取微服务的健康状态，因此将 Redis、Sentinel、nacosDiscover 的健康状态监控关闭。

（3）【微服务模块】修改需要接入 Spring Boot Admin 服务的微服务中的 application.yml 配置文件，添加 Spring Boot Admin 服务地址以及认证信息。

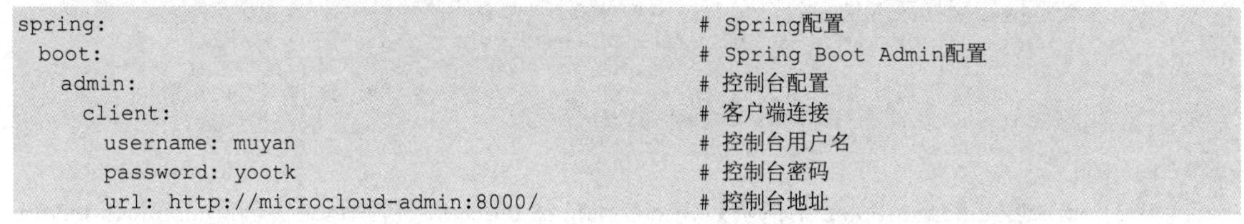

```
spring:                                      # Spring配置
 boot:                                       # Spring Boot Admin配置
  admin:                                     # 控制台配置
   client:                                   # 客户端连接
    username: muyan                          # 控制台用户名
    password: yootk                          # 控制台密码
    url: http://microcloud-admin:8000/       # 控制台地址
```

（4）【Spring Boot Admin 控制台】配置完成后重新启动相应的微服务，随后再次登录 Spring Boot Admin 控制台，就可以看见全部的微服务信息。单击微服务可以查看完整的 Actuator 监控数据，如图 7-18 所示。

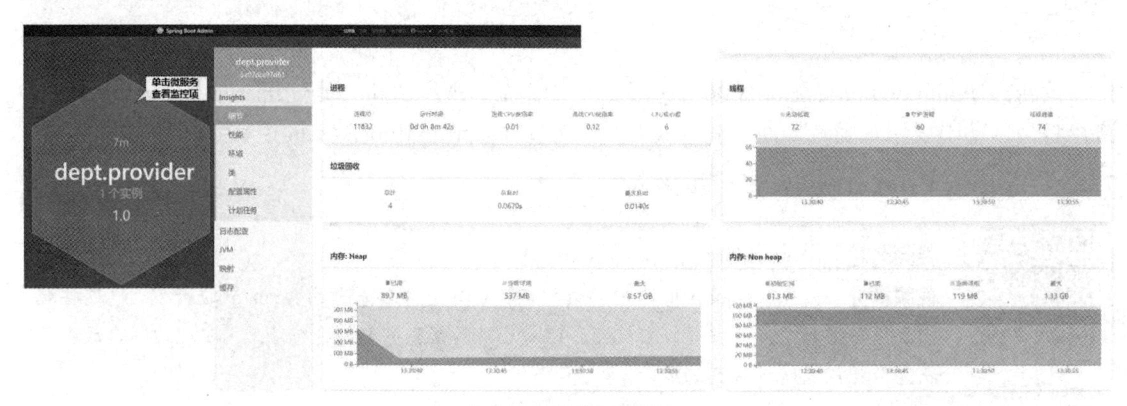

图 7-18　查看监控数据

7.2.3 微服务离线警报

微服务离线警报

视频名称 0710_【掌握】微服务离线警报

视频简介 完善的服务运行离不开警报信息的配置。Spring Cloud 集群架构中有大量的微服务节点，如果某一个服务节点出现问题则应该立即通知管理员，这样就可以通过 Spring Boot Admin 提供的邮件整合服务进行警报信息的发送。

Spring Boot Admin 可以进行微服务状态的监控，一旦微服务状态发生了更改（如实例上线或下线），则应该立即将此状态告诉管理员，如图 7-19 所示。这样管理员才可以随时掌握微服务实例的运行状态。

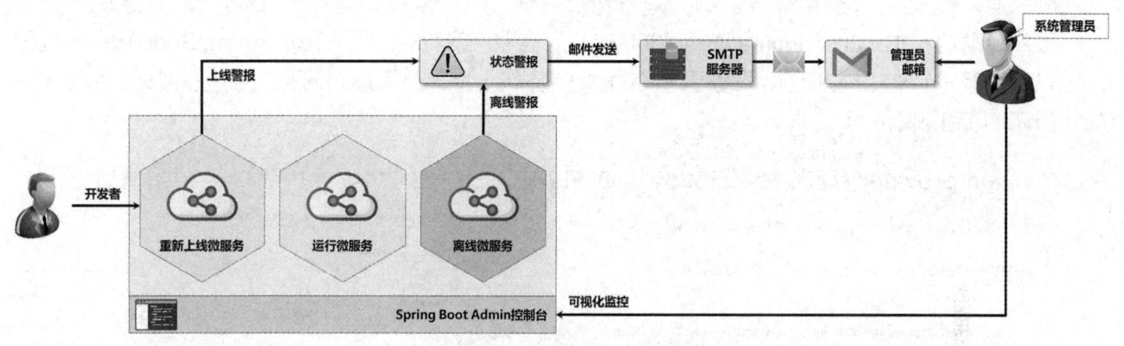

图 7-19 Spring Boot Admin 警报

要想实现这样的邮件警报发送机制，可以通过 Spring Boot Admin 整合 Java mail 相关依赖，同时利用已经存在的 SMTP 进行邮件发送处理。下面通过具体的操作步骤进行这一功能的实现。

（1）【microcloud 项目】修改 build.gradle 配置文件，为 microcloud-admin-8000 子模块添加 mail 依赖库。

```
implementation('org.springframework.boot:spring-boot-starter-mail')
```

（2）【microcloud-admin-8000 子模块】修改 application.yml 配置文件，添加 mail 的相关配置项。

```
spring:                                                      # Spring配置
  mail:                                                      # 邮箱配置
    host: smtp.qq.com                                        # 邮箱地址
    username: 784420216@qq.com                               # 用户名
    password: ehxdklqXsjwubeaf                               # 授权码
  boot:                                                      # Spring Boot Admin配置
    admin:                                                   # 控制台配置
      notify:                                                # 警报配置
        mail:                                                # 邮箱配置
          to: 2273627816@qq.com                              # 收件人
          from: 784420216@qq.com                             # 寄件人
```

（3）【microcloud-admin-8000 子模块】为了便于管理警报，可以创建一个警报通知配置类。

```
package com.yootk.admin.config;
@Configuration
public class MicroServiceNotificationConfiguration {
    private final InstanceRepository instanceRepository;           // 存储实例
    private final ObjectProvider<List<Notifier>> providers;        // 警报策略
    public MicroServiceNotificationConfiguration(InstanceRepository instanceRepository,
          ObjectProvider<List<Notifier>> providers) {
        this.instanceRepository = instanceRepository;
        this.providers = providers;
    }
    @Bean
    public FilteringNotifier getFilteringNotifier() {              // 定义过滤通知策略
```

```
        CompositeNotifier delegate = new CompositeNotifier(this.providers
                .getIfAvailable(Collections::emptyList));          // 混合策略
        return new FilteringNotifier(delegate, this.instanceRepository);
    }
    @Primary
    @Bean
    public RemindingNotifier remindingNotifier() {                  // 服务上线与下线警报
        RemindingNotifier notifier = new RemindingNotifier(getFilteringNotifier(),
                this.instanceRepository);                           // 通知提醒
        notifier.setReminderPeriod(Duration.ofMinutes(5));          // 每10分钟发送一次提醒
        notifier.setCheckReminderInverval(Duration.ofSeconds(10));  // 检测周期
        return notifier;
    }
}
```

重新启动当前的 microcloud-admin-8000 应用，随后关闭或启动一个被 Spring Boot Admin 监控的微服务。若一个检测周期之后服务已不存在或已重新上线，则管理员可以在对应的邮箱中收到图 7-20 所示的警报邮件。

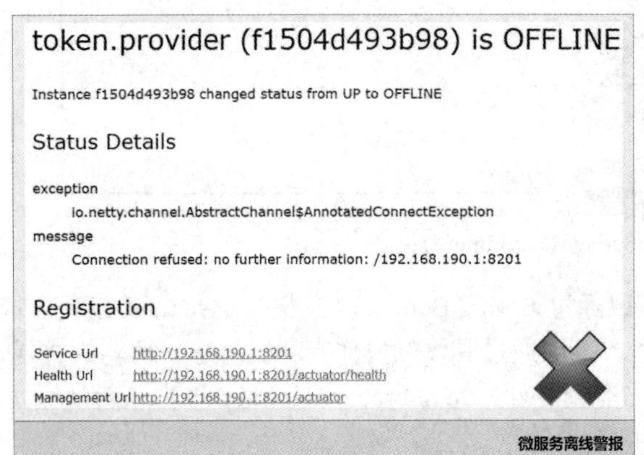

图 7-20 Spring Boot Admin 警报邮件

7.3 本章概览

1．Spring Cloud 的安全认证可以基于 OAuth2 完成，但是 OAuth2 并非为 Spring Cloud 设计，所以在处理性能上有所欠缺，也存在开发不稳定的缺点。

2．JWT 拥有严格的组成结构，可以根据需要传入附加信息，开发中可以基于这些附加信息实现服务认证与授权管理。

3．引入 JWT 认证之后，微服务消费端每次进行资源调用时都需要传递正确的 JWT 数据，否则无法实现访问。

4．JWT 数据有时间效应，开发者可以根据自己的需要修改有效期的配置，或利用其他方式实现 Token 刷新。

5．Spring Boot Admin 是一款为 Spring Boot 设计的监控组件，其可以结合 Actuator 进行微服务的状态监控。

6．Spring Boot Admin 可以与 mail 组件整合，实现实例上线与下线的警报信息发送。

第 8 章 RocketMQ

本章学习目标

1. 掌握 RocketMQ 消息组件的主要作用、集群服务搭建以及程序开发方法；
2. 掌握 RocketMQ 的工作原理、实现架构，及其与主流消息组件的区别；
3. 掌握 RocketMQ 的消息发送与处理逻辑中的标签、识别码以及 NameSpace 的使用方法；
4. 掌握事务消息处理的原理与具体实现方法。

RocketMQ 是由阿里巴巴公司推出的一款基于 Java 开发的高性能消息组件，适用于有高并发与高可靠要求的应用环境。本章将为读者全面讲解该消息组件的主要作用、集群服务搭建以及程序开发。

8.1 RocketMQ 安装与配置

RocketMQ 简介

视频名称 0801_【掌握】RocketMQ 简介

视频简介 消息组件是在系统开发中较为常见的服务应用。本视频为读者宏观地讲解消息组件的主要作用，同时分析常见的几款消息组件的特点与实际应用场景。

在标准的项目开发与设计之中，最常见的处理形式就是客户端向服务端发出操作请求，而后等待服务端业务处理完成与数据响应。但是这种操作所带来的最严重的后果就是当服务端处理性能较差，或者业务处理逻辑耗时较多，该客户端会持续占用服务端的线程资源，导致后续其他请求无法正常处理。

以一个购物商城的系统为例，如图 8-1 所示。商城用户向交易系统发出了一个交易请求，但是一个完整的交易涉及订单系统、购物车系统、库房系统、物流系统等多个关联系统的操作，整个业务处理有可能需要几分钟甚至数十个小时，而在处理期间用户不可能一直处于等待响应的状态，尤其是在高并发访问期间，这样的处理逻辑会造成服务瘫痪。

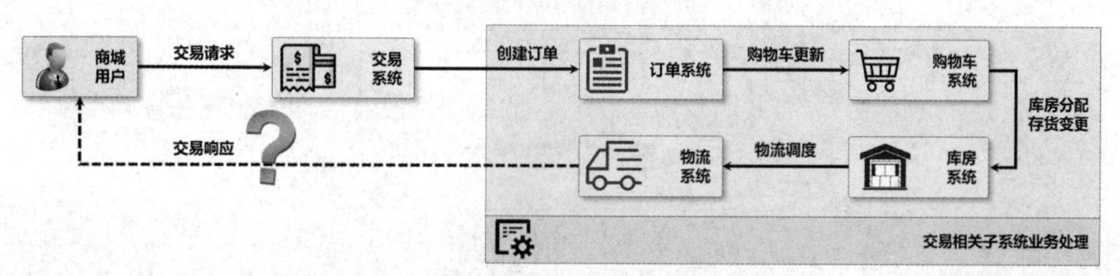

图 8-1 购物商城请求与响应简单架构设计

为了解决此类问题，最佳做法是在项目之中引入消息系统，即用户通过交易系统发出一个交易数据后会立即得到响应，但是此交易数据并非立即被处理，而是进入一个消息系统等待其他相关子系统的处理，如图 8-2 所示。这样就可以极大地减少服务端处理时间，也可以提高项目的吞吐量，起到业务操作缓冲的作用，而这就是消息组件的主要作用。

图 8-2 消息组件处理

消息组件是一种基于队列处理模式的服务端应用，在消息组件中有消息生产者与消息消费者两个处理终端，如图 8-3 所示。不管生产者生产了多少消息，这些消息都会保存在消息组件之中，可以很好地实现削峰填谷的处理效果，避免大规模用户请求所带来的服务瘫痪等问题。所有保存在消息组件中的消息数据可以依据自身的业务逻辑需要由消费端按顺序进行处理，这样就可以保证整个业务处理逻辑的可靠性以及服务运行的稳定性，同时，基于消息处理机制的方式还可以使得各种第三方平台的接入更加方便。

图 8-3 消息组件的作用

在 Java 技术发展的早期，Apache 公司制定了 JMS 服务标准（实现组件为 ActiveMQ），但是 JMS 并不是协议本身，所以性能较差，已经逐步退出主流技术的行列。为了提高消息处理的性能以及稳定性，Apache 公司后续又推出了 AMQP（Advanced Message Queuing Protocol，高级消息队列协议），并基于此协议开发了 RabbitMQ 消息组件。该组件可以提供稳定的消息服务，多用于银行系统之间的服务整合，也是现阶段进行服务整合时使用较多的消息组件。后来由于大数据技术的发展，Apache 公司开发了基于数据流方式的 Kafka 消息组件，它可以承受百万级的日志采集量，也是现在使用最广泛的消息组件之一。

提示：刷盘与落盘。

由于消息中间件中需要进行大量消息数据的存储，因此对数据的存储就有了“刷盘”与“落盘”两种实现机制，如图 8-4 所示。

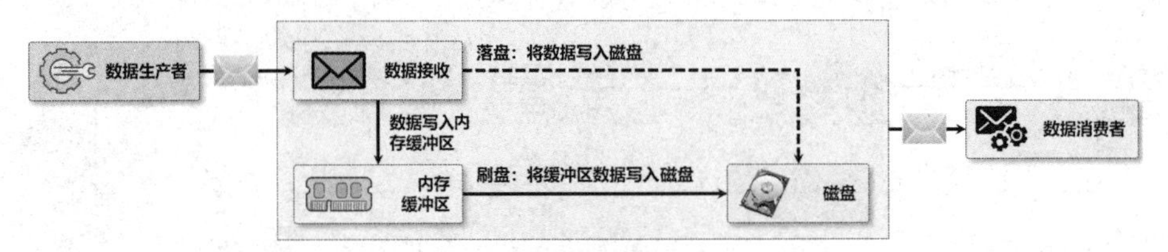

图 8-4 刷盘与落盘

通过图 8-4 可以清楚地发现两种实现机制的操作特点。落盘主要是为了保证数据持久化，这样即使服务器死机也不会导致数据丢失，但是由于 I/O 的限制，其处理性能较差。而刷盘会将数据先写入内存缓冲区，而后由操作系统决定何时写入磁盘进行持久化存储，因此，一旦出现服务器死机，那些保存在内存缓冲区未写入磁盘的数据就有可能丢失。由于此种实现机制主要对内存进行读写，因此其处理性能较好。

而对刷盘的处理一般有同步刷盘与异步刷盘两种方式，这两种方式的特点如下。

① **同步刷盘**（SYNC_FLUSH）：数据写入内存缓冲区后立刻刷盘，在保证刷盘成功的前提下进行客户端响应，此种操作拥有较高的数据可靠性。

② **异步刷盘**（ASYNC_FLUSH）：写入处理速度快，吞吐量大，内存缓冲区数据积累到一定量时快速写入，但是不能够保证数据的可靠性。

通过如上分析可以发现，对于当前的项目开发来讲，最为常用的 3 种消息组件分别是 RabbitMQ、RocketMQ、Kafka，如表 8-1 所示。可以发现，RocketMQ 是一种设计性能比 RabbitMQ 更高的消息组件，但其并不适合于实现日志采集的操作。

表 8-1 常用消息组件

序号	比较点	RabbitMQ	RocketMQ	Kafka
01	开发语言	ERLang	Java	Scala
02	吞吐量	万级	十万级	百万级
03	可用性	高，主从架构，基于镜像模式实现，数据量大时可能有性能问题	非常高，采用分布式主从架构	非常高，采用分布式主从架构
04	可靠性	同步刷盘	同步刷盘、异步刷盘	异步刷盘，容易丢数据
05	堆积能力	一般	非常好	非常好
06	顺序消费	支持。一个消息失败，此消息的顺序会被打乱	支持。顺序消息场景下，消费失败时消息队列会暂停	支持。一台 Broker 死机后，消息会乱序
07	定时消息	支持	支持	不支持
08	事务消息	不支持	支持	不支持
09	消息重试	支持	支持	不支持
10	死信队列	支持	支持	不支持
11	访问权限	认证与授权	无（4.4.0 版本后支持）	无
12	产品定位	非日志采集的可靠消息传输	非日志采集的可靠消息传输	日志采集消息

RocketMQ 是由阿里巴巴公司推出的一款分布式、队列模型的开源消息中间件，图 8-5 所示为 RocketMQ 发展史。它经历了 10 次“双十一”大促活动的洗礼，依然保证较高的处理性能与较好的稳定性。RocketMQ 现在已经交由 Apache 负责维护，并于 2017 年 9 月 25 日成为 Apache 的顶级项目。

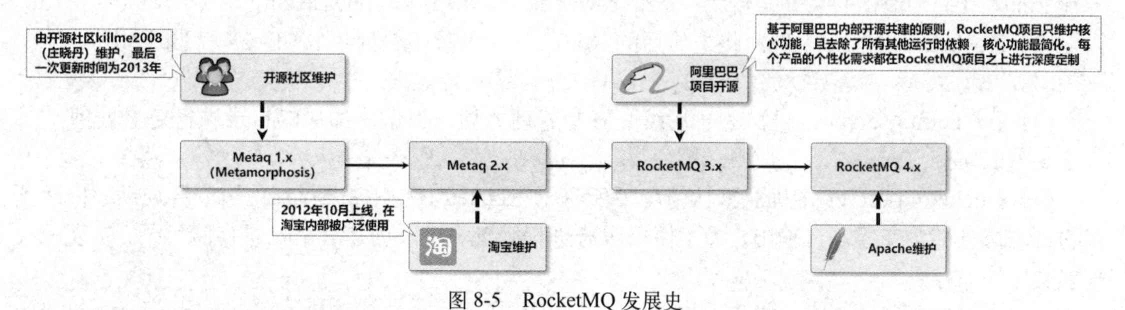

图 8-5 RocketMQ 发展史

8.1.1 RocketMQ 服务搭建

RocketMQ 服务搭建

视频名称 0802_【掌握】RocketMQ 服务搭建

视频简介 官方推荐在 Linux 系统中进行 RocketMQ 服务搭建。本视频为读者讲解 RocketMQ 的组件获取、服务安装与启动的相关实例。

RocketMQ 组件是基于 Java 开发的，开发者只需要在配置了 JDK 8 的主机上进行应用部署。要想获得 RocketMQ 部署包，可以访问 Apache 官网进行下载，如图 8-6 所示，本次所使用的程序版本号为 4.9.0。

图 8-6　下载 RocketMQ 部署包

一个完整的 RocketMQ 服务至少需要启动两个进程（NameServer、Broker）。为了便于读者理解 RocketMQ 的使用，本次将采用图 8-7 所示的方式在一台主机中进行服务搭建，并基于 RocketMQ 提供的测试工具进行消息的发送与消费。具体操作步骤如下。

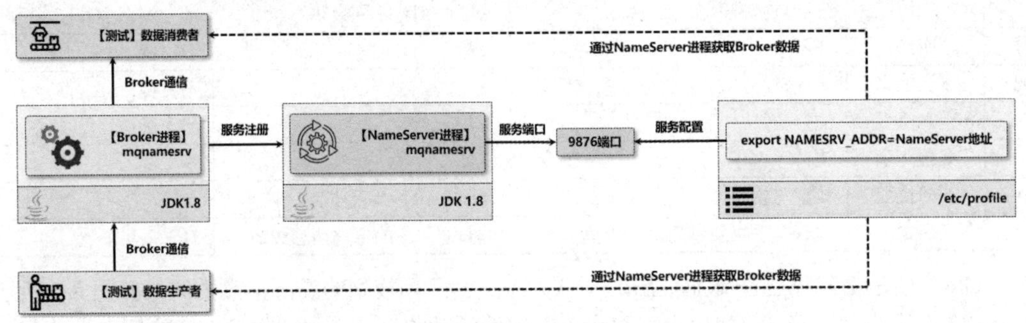

图 8-7　RocketMQ 服务

（1）【rocketmq-server 主机】将下载得到的 RocketMQ 包（rocketmq-all-4.9.0-bin-release.zip）上传到 Linux 系统，保存的完整路径为/var/ftp/rocketmq-all-4.9.0-bin-release.zip。

（2）【rocketmq-server 主机】将上传后的 RocketMQ 包解压缩到“/usr/local”目录之中。

```
unzip /var/ftp/rocketmq-all-4.9.0-bin-release.zip -d /usr/local/
```

（3）【rocketmq-server 主机】为了后续配置与管理方便，对解压缩后的目录进行更名处理。

```
mv /usr/local/rocketmq-all-4.9.0-bin-release /usr/local/rocketmq
```

（4）【rocketmq-server 主机】默认情况下所有的 RocketMQ 相关进程都采用前台启动，同时所有的日志信息都会在控制台输出。为了便于服务进程管理，可以创建一个日志存储目录，实现相关日志文件的保存。

```
mkdir /usr/local/rocketmq/logs
```

（5）【rocketmq-server 主机】采用后台模式启动 NameServer 进程（默认端口为 9876），并设置日志保存路径。

```
nohup /usr/local/rocketmq/bin/mqnamesrv > /usr/local/rocketmq/logs/rocketmq-namesrv.log 2>&1 &
```

（6）【rocketmq-server 主机】在默认配置下，RocketMQ 中的数据内容会保存在“用户目录/store”路径之中，而为了便于这些数据的统一管理，建议创建一个 RocketMQ 数据存储目录，以保存 BrokerService 的数据。

```
mkdir -p /usr/data/rocketmq/store/commitlog
```

（7）【rocketmq-server 主机】修改 broker.conf 配置文件并通过“storePathRootDir”配置项定义存储目录。

打开配置文件	`vi /usr/local/rocketmq/conf/broker.conf`	
增加配置项	NameServer 地址	`namesrvAddr=rocketmq-server:9876`
	数据存储路径	`storePathRootDir=/usr/data/rocketmq/store`
	CommitLog 存储路径	`storePathCommitLog=/usr/data/rocketmq/store/commitlog`

（8）【rocketmq-server 主机】采用后台模式启动 BrokerServer 进程。

```
nohup /usr/local/rocketmq/bin/mqbroker -c /usr/local/rocketmq/conf/broker.conf > /usr/local/
rocketmq/logs/rocketmq-broker.log 2>&1 &
```

程序执行结果	The broker[rocketmq-server, 192.168.190.170:10911] boot success. serializeType=JSON and name server is rocketmq-server:9876

由于此时已经配置了 broker.conf 文件，因此在启动 BrokerServer 服务时，必须通过“-c”参数设置配置文件路径，才可以加载指定的“NameServer”进程服务地址，以便发送 Broker 服务数据，这样消息的生产者与消费者才可以通过 NameServer 找到匹配的 Broker 实现消息收发。而如果在启动时未设置此配置文件，则会以默认配置的方式启动 Broker。

提示：关于 JVM 内存分配。

在默认情况下用户启动 NameServer 会占用约 4GB 的内存，而启动 BrokerServer 会占用约 8GB 的内存。如果开发者的服务主机内存不足，则有可能出现服务进程启动失败。在测试环境下，可以根据需要进行 JVM 参数调整，具体步骤如下。

① 调整 NameServer 的内存配置。

打开配置文件	`vi /usr/local/rocketmq/bin/runserver.sh`
修改 JVM 参数	`JAVA_OPT="${JAVA_OPT} -server -Xms1g -Xmx1g -Xmn1g -XX:MetaspaceSize=128m -XX:MaxMetaspaceSize=320m"`

② 调整 BrokerServer 的内存配置。

打开配置文件	`vi /usr/local/rocketmq/bin/runbroker.sh`
修改 JVM 参数	`JAVA_OPT="${JAVA_OPT} -server -Xms1g -Xmx1g -Xmn1g"`

以上 JVM 内存参数可以根据不同的主机环境进行调整。由于当前是在测试环境下，因此只为每个进程分配了 1GB 的内存。

在默认情况下 Broker 启动时会占用 3 个接口：10911（默认的服务监听端口，生产者与消费者基于此端口通信）、10912（“默认端口号 +1”，实现 HA 机制监听端口）、10909（“默认端口号-2”，实现主从同步处理）。如果开发者需要修改此端口，在“conf/broker.conf”配置文件中通过“listenPort=默认端口”配置项修改即可。关于 Broker 的相关配置项的定义，可以参考附录。

（9）【rocketmq-server 主机】为便于 RocketMQ 服务测试，可以创建一个“NAMESRV_ADDR”属性项，该属性项的内容为 NameServer 的服务地址，配置步骤如下。

打开配置文件	`vi /etc/profile`
增加配置项	`export NAMESRV_ADDR=rocketmq-server:9876`
配置项生效	`source /etc/profile`

（10）【rocketmq-server 主机】修改防火墙配置，添加服务端口访问权限。

添加访问端口	`firewall-cmd --zone=public --add-port=9876/tcp --permanent` `firewall-cmd --zone=public --add-port=10911/tcp --permanent`

添加访问端口	firewall-cmd --zone=public --add-port=10909/tcp --permanent firewall-cmd --zone=public --add-port=10912/tcp --permanent
重新加载配置	firewall-cmd --reload

（11）【rocketmq-server 主机】服务进程启动完成后，使用 RocketMQ 内置的生产者进行服务测试。

```
/usr/local/rocketmq/bin/tools.sh org.apache.rocketmq.example.quickstart.Producer
```

程序执行结果	SendResult [sendStatus=SEND_OK, msgId=7F00000108491B6D358698E768F503E5, offsetMsgId=C0A8BEAA00002A9F0000000000031629, messageQueue=MessageQueue [topic=TopicTest, brokerName=rocketmq-server, queueId=3], queueOffset=249]

生产者测试端启动时会自动创建一个 TopicTest 消息主题，同时会根据 NameServer 进程找到要使用的 Broker 进程。如果此时消费端没有启动，则程序会将发送的信息存储在 RocketMQ 之中，等待消费端启动时再进行消费处理。

（12）【rocketmq-server 主机】生产者数据发送完成后所有的数据会保存在 Broker 之中。启动消费端进程接收数据。

```
/usr/local/rocketmq/bin/tools.sh org.apache.rocketmq.example.quickstart.Consumer
```

程序执行结果	ConsumeMessageThread_20 Receive New Messages: [MessageExt [brokerName=rocketmq-server, queueId=0, storeSize=203, queueOffset=158, sysFlag=0, bornTimestamp=1625042101906, bornHost=/192.168.190.170:59798, storeTimestamp=1625042101906, storeHost=/192.168.190.170:10911, msgId=C0A8BEAA00002A9F000000000001F650, commitLogOffset=128592, bodyCRC=688614417, reconsumeTimes=0, preparedTransactionOffset=0, toString()=Message{topic='TopicTest', flag=0, properties={MIN_OFFSET=0, MAX_OFFSET=250, CONSUME_START_TIME=1625042124585, UNIQ_KEY=7F00000108491B6D358698E76692027A, CLUSTER=DefaultCluster, WAIT=true, TAGS=TagA}, body=[72, 101, 108, 108, 111, 32, 82, 111, 99, 107, 101, 116, 77, 81, 32, 54, 51, 52], transactionId='null'}]]

消费端测试启动后，会通过默认的 TopicTest 主题获取保存在 RocketMQ 中的主体（body）消息内容，还会返回一些消息的附加信息（如 tag、storeHost、queueId、queueOffset 等）。

8.1.2 访问控制列表

访问控制列表

视频名称 0803_【掌握】访问控制列表

视频简介 消息组件在实际项目中需要实现对外的服务整合，这样一来为了保证服务的安全就必须进行认证与授权管理。本视频通过实例为读者讲解访问控制列表的开启以及配置。

此时一个基础的 RocketMQ 应用服务已经搭建完成，同时通过测试环境实现了测试消息的发送与接收。但是在实际的项目开发中，由于消息组件往往应用在许多重要的领域之中（如金融、电信、电商、医疗等），同时又有项目平台的对接需求，因此需要将消息服务暴露在公网之中。在 RocketMQ 4.4.0 后 Apache 提供了 ACL（Access Control List，访问控制列表）安全机制，这样就可以通过认证与授权的方式保证整个系统的安全与运行稳定，如图 8-8 所示。

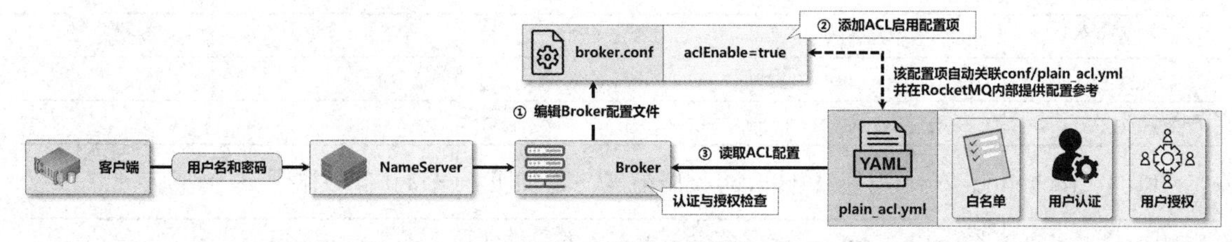

图 8-8 ACL 安全机制

RocketMQ 提供的 ACL 安全机制包含认证信息（用户名和密码）、资源（消息主题、消费组）、权限（资源允许的操作形式）、角色（管理员与普通用户）4 个核心概念的实现。这些概念可以通过表 8-2 所示的配置项在“${ROCKETMQ_HOME}/conf/plain_acl.yml”配置文件中进行定义。

表 8-2 ACL 配置项

序号	配置项	描述	配置数据
01	globalWhiteRemoteAddresses	全局访问 IP 地址白名单	192.168.*.*、192.168.190.*、127.0.0.1
02	accounts	配置访问账户	认证与授权配置项集合
03	accessKey	用户名	字符串（建议长度超过 7 位）
04	secretKey	密码	字符串（建议长度超过 7 位）
05	whiteRemoteAddress	用户访问 IP 地址白名单	192.168.*.*、192.168.190.*、127.0.0.1
06	admin	是否为管理员账户	true、false
07	defaultTopicPerm	默认的主题权限	DENY、PUB、SUB、PUB|SUB
08	defaultGroupPerm	默认的订阅组（消费组）权限	DENY、PUB、SUB、PUB|SUB
09	topicPerms	配置每个主题的操作权限	DENY、PUB、SUB、PUB|SUB
10	groupPerms	配置每个订阅组（消费组）权限	DENY、PUB、SUB、PUB|SUB

plain_acl.yml 在进行配置时采用的是 YAML 定义格式，所有要配置的账户信息全部在 accounts 中进行定义，而在 globalWhiteRemoteAddresses、topicPerms、groupPerms 中可以采用列表的方式进行定义。为便于读者理解，下面通过具体的步骤介绍如何配置。

（1）【rocketmq-server 主机】修改 broker.conf 配置文件，启用 ACL 支持。

打开配置文件	`vi /usr/local/rocketmq/conf/broker.conf`
添加配置项	`aclEnable=true`

（2）【rocketmq-server 主机】编辑 plain_acl.yml 配置文件。

```
vi /usr/local/rocketmq/conf/plain_acl.yml
globalWhiteRemoteAddresses:               // 定义全局访问IP地址白名单
- 10.10.103.*                             // 白名单IP地址范围
- 192.168.190.*                           // 白名单IP地址范围
accounts:                                 // 配置账户信息
- accessKey: RocketMQMuyan                // 认证用户名
  secretKey: helloyootk                   // 认证密码
  whiteRemoteAddress:                     // 用户访问IP地址白名单
  admin: false                            // 当前配置的账户为普通用户
  defaultTopicPerm: PUB|SUB               // 默认的主题权限为可读写
  defaultGroupPerm: PUB|SUB               // 默认的订阅组权限为可读写
  topicPerms:                             // 用户主题权限配置
  - TopicMuyan=DENY                       // 配置指定主题的操作权限
  - TopicYootk=PUB|SUB                    // 配置指定主题的操作权限
  - TopicBenchmark=PUB|SUB                // 配置指定主题的操作权限
  groupPerms:                             // 订阅组权限
  - yootk-group=DENY                      // 配置指定订阅组的操作权限
  - muyan-group=PUB|SUB                   // 配置指定订阅组的操作权限
  - happy-group=SUB                       // 配置指定订阅组的操作权限
- accessKey: RocketMQAdmin                // 认证用户名
  secretKey: hello123456                  // 认证密码
  whiteRemoteAddress: 192.168.1.*         // 用户访问IP地址白名单
  admin: true                             // 当前配置的账户为管理员
```

（3）【rocketmq-server 主机】重新启动 Broker 服务进程。

```
nohup /usr/local/rocketmq/bin/mqbroker -c /usr/local/rocketmq/conf/broker.conf > /usr/local/
rocketmq/logs/rocketmq-broker.log 2>&1 &
```

由于当时配置的 IP 地址白名单提供本机的访问支持网段（192.168.190.*），因此此时可以启用

RocketMQ 提供的生产者与消费者应用进行测试。

8.1.3　RocketMQ 控制台

视频名称　0804_【理解】RocketMQ 控制台

视频简介　有服务必有监控已经成为当代软件项目设计的宗旨。虽然 RocketMQ 本身提供了强大的消息服务支持，但是为了保证其运行的稳定，需要通过 RocketMQ-Externals 开源项目实现监控。本视频讲解 GitHub 的扩展程序的下载以及服务配置。

RocketMQ 本身只提供了消息服务的功能，为了便于消息服务的管理以及访问流量监控，还需要通过 Apache 给出的 RocketMQ-Externals 扩展项目实现 RocketMQ 控制台功能。该项目基于 Maven 开发，开发者可以通过 GitHub 的服务地址下载源代码，并根据自身的需要对源代码进行修改（配置 RocketMQ 的 NameServer 地址），以实现最终的 RocketMQ 连接，如图 8-9 所示。下面通过具体的操作介绍这一服务的部署。

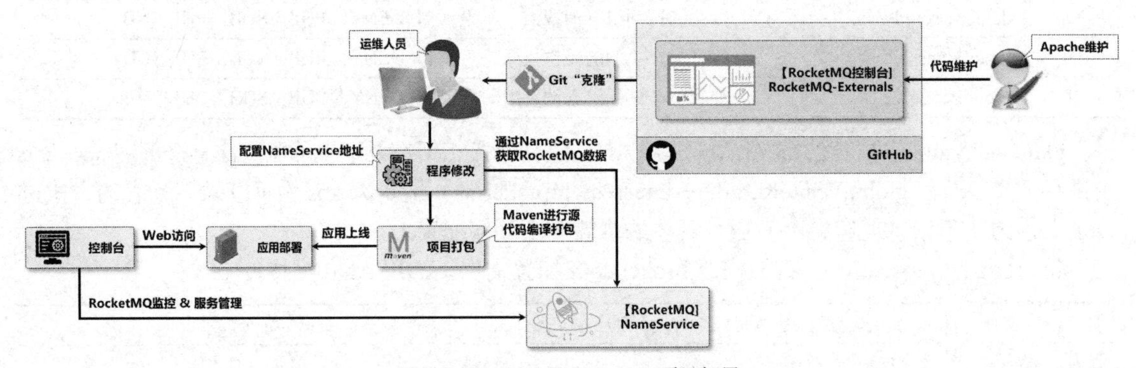

图 8-9　RocketMQ-Externals 项目部署

（1）【GitHub 下载】RocketMQ 控制台是以源代码的形式给出的，开发者可以通过 GitHub 项目地址进行项目克隆。

```
git clone git@github.com:muyan-yootk/rocketmq-externals.git
```

（2）【rocketmq-console 子模块】修改模块之中的 application.yml 配置，明确设置 NameServer 服务地址。

添加 NameServer 地址	`rocketmq.config.namesrvAddr=rocketmq-server:9876`
配置认证用户名	`rocketmq.config.accessKey=RocketMQMuyan`
配置认证密码	`rocketmq.config.secretKey=helloyootk`
开启控制台安全认证	`rocketmq.config.loginRequired=true`

（3）【rocketmq-console 子模块】由于此时已经开启了控制台安全认证支持，因此需要修改 users.properties 配置文件，添加控制台账户，配置格式为"用户名=密码,类型"，而账户类型有两种：管理员（1）、用户（0）。

添加管理员账户	`admin=hello,1`
添加用户账户	`muyan=yootk`

（4）【Maven 编译】RocketMQ-Externals 中有众多子模块，其中 rocketmq-console 子模块为 RocketMQ 控制台，需要开发者通过 Maven 工具进行项目打包。

```
mvn clean package -Dmaven.test.skip=true
```

（5）【rocketmq-server 主机】将打包生成的 rocketmq-console-ng-2.0.0.jar 包上传到服务器。本次

程序包保存的完整路径为“/usr/local/src/rocketmq-console-ng-2.0.0.jar”。

（6）【rocketmq-server 主机】启动 rocketmq-console 应用，并设置日志保存路径。

```
java -jar /usr/local/src/rocketmq-console-ng-2.0.0.jar > /usr/local/rocketmq/logs/rocketmq-console.log 2>&1 &
```

（7）【rocket-server 主机】配置防火墙规则，追加 8080 端口访问规则。

添加访问端口	`firewall-cmd --zone=public --add-port=8080/tcp --permanent`
重新加载配置	`firewall-cmd --reload`

（8）【浏览器】通过浏览器访问 RocketMQ 控制台，访问地址为“rocketmq-server:8080”，如图 8-10 所示。

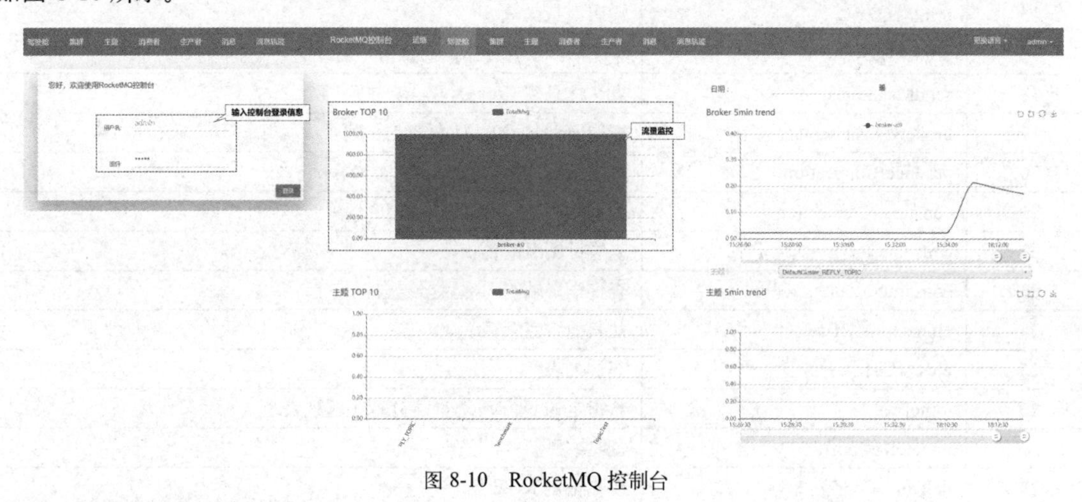

图 8-10　RocketMQ 控制台

8.1.4　RocketMQ 管理命令

RocketMQ 管理命令

视频名称　0805_【理解】RocketMQ 管理命令

视频简介　RocketMQ 提供了 mqadmin 命令，利用该命令可以管理当前消息系统之中的主题、生产者与消费者。本视频通过实例为读者讲解这一命令。

为了便于管理员进行 RocketMQ 的维护，RocketMQ 组件提供了 mqadmin 命令。该命令由一系列的子指令所组成，使用者输入 mqadmin 命令即可得到所有的子指令，这些子指令的作用如表 8-3 所示。通过这些子指令可实现 RocketMQ 的配置与监控处理操作。

表 8-3　mqadmin 子指令的作用

序号	指令名称	描述
01	updateTopic	创建或更新主题
02	deleteTopic	从指定的 NameServer 或 Broker 中删除主题
03	updateSubGroup	创建或更新订阅组
04	deleteSubGroup	从指定的 Broker 中删除订阅组
05	updateBrokerConfig	更新 Broker 配置项
06	updateTopicPerm	更新主题读写模式（6：支持读写。2：禁读。4：禁写）
07	topicRoute	获取主题路由信息
08	topicStatus	获取主题状态信息

续表

序号	指令名称	描述
09	topicClusterList	获取集群主题列表
10	brokerStatus	获取 Broker 的运行时状态数据
11	queryMsgById	根据消息 ID 查询消息数据
12	queryMsgByKey	根据消息 Key 查询消息数据
13	queryMsgByUniqueKey	根据唯一 Key 查询消息
14	queryMsgByOffset	根据 Offset 查询消息
15	QueryMsgTraceById	查询消息详情
16	printMsg	输出消息详情
17	printMsgByQueue	采用队列形式输出消息详情
18	sendMsgStatus	发送信息到 Broker
19	brokerConsumeStats	抓取 Broker 中的消费端状态数据
20	producerConnection	查询所有连接的生产者信息
21	consumerConnection	查询所有连接的消费端信息
22	consumerProgress	查看订阅组消费状态
23	consumerStatus	查询消费端的内部数据结构
24	cloneGroupOffset	从其他组克隆 Offset
25	clusterList	集群列表
26	topicList	从指定的 NameServer 中获取主题列表
27	updateKvConfig	创建或更新 Kv 数据
28	deleteKvConfig	删除指定 Kv 数据
29	wipeWritePerm	擦除所有写入的权限
30	resetOffsetByTime	根据时间戳重置 Offset
31	updateOrderConf	创建、更新或删除 Order 配置
32	cleanExpiredCQ	从 Broker 中清除失效的消费队列
33	cleanUnusedTopic	从 Broker 中清除未使用的主题
34	startMonitoring	开启监控
35	statsAll	主题和消费端程序统计
36	allocateMQ	分配消息队列
37	checkMsgSendRT	检查消息与发送响应时间
38	clusterRT	集群消息列表与响应时间
39	getNamesrvConfig	获取 NameServer 配置数据
40	updateNamesrvConfig	更新 NameServer 配置项
41	getBrokerConfig	从指定的集群或指定的 Broker 获取配置项
42	queryCq	查询消费组指令
43	sendMessage	发送消息
44	consumeMessage	消息消费
45	updateAclConfig	更新 ACL 配置
46	deleteAccessConfig	删除 ACL 配置
47	clusterAclConfigVersion	查看集群中的 ACL 配置列表
48	updateGlobalWhiteAddr	更新全局白名单配置
49	getAccessConfigSubCommand	列出全部 ACL 配置信息

mqadmin 提供的子指令较多，一般可以采用“mqadmin 子指令 -h”的形式查看对应的执行参数。下面通过介绍几个子指令进行讲解。

提示：子指令程序实现类。

mqadmin 命令提供的所有子指令都有其对应的程序实现类，开发者可以导入 RocketMQ 的项目源代码，所有的程序实现类都在“tools”子模块中定义，如图 8-11 所示。

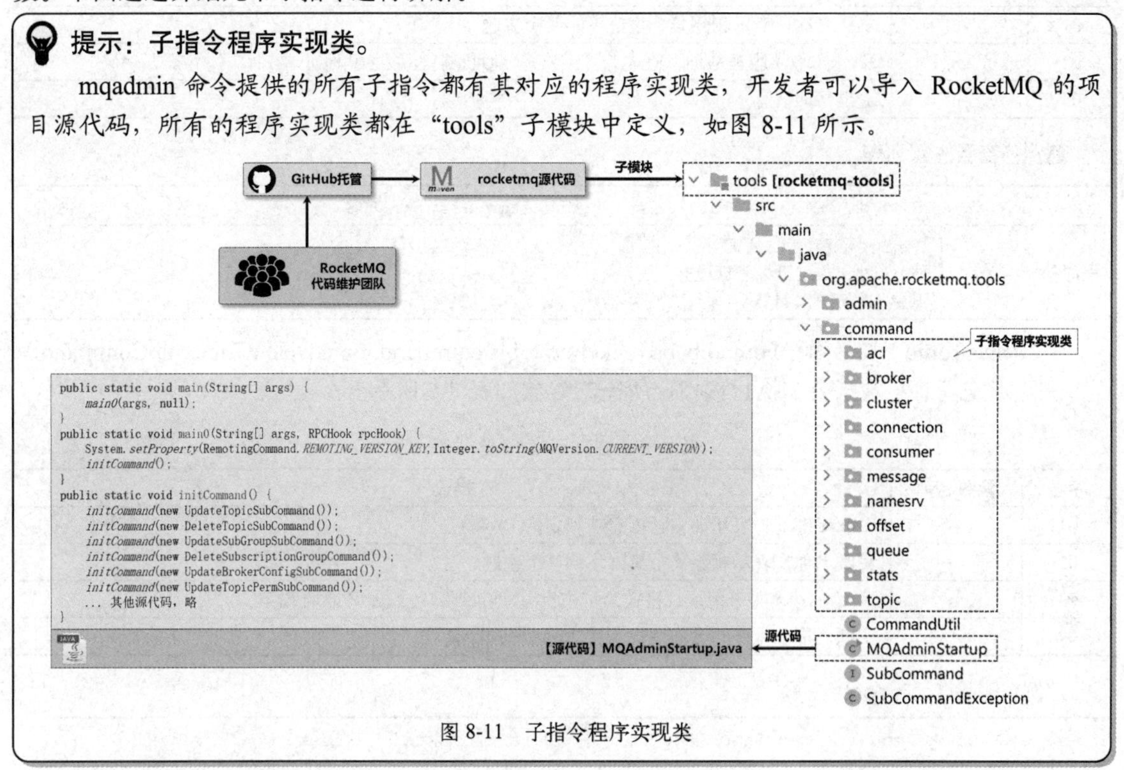

图 8-11　子指令程序实现类

1. updateTopic（类名称：com.alibaba.rocketmq.tools.command.topic.UpdateTopicSubCommand）

创建消息主题指令，执行该指令所需的参数如表 8-4 所示。

表 8-4　创建主题指令参数

序号	参数名称	描述
01	-b	Broker 地址，表示该主题创建在指定的 Broker 中
02	-c	集群名称，表示该主题创建在指定的集群中
03	-h	输出帮助信息
04	-n	NameServer 服务地址，格式为“主机 IP 地址:端口;主机 IP 地址:端口;…”
05	-p	指定新的主题操作权限（W、R、WR）
06	-r	读取队列数量（默认为 8）
07	-w	可写队列数量（默认为 8）
08	-t	主题名称，允许的名称格式为“^[a-zA-Z0-9_-]+$”

范例：创建一个名称为 TopicYootk 的消息主题

```
/usr/local/rocketmq/bin/mqadmin updateTopic -b rocketmq-server:10911 -t TopicYootk
```

程序执行结果

```
create topic to rocketmq-server:10911 success.
TopicConfig [topicName=yootkTopic, readQueueNums=8, writeQueueNums=8, perm=RW-,
topicFilterType=SINGLE_TAG, topicSysFlag=0, order=false]
```

2. topicList（类名称：com.alibaba.rocketmq.tools.command.broker.UpdateBrokerConfigSubCommand）

查看全部的主题列表指令，执行该指令所需的参数如表 8-5 所示。

表 8-5 主题列表指令参数

序号	参数名称	描述
01	-c	集群名称，表示删除某个集群下的某个主题
02	-n	NameServer 服务地址，格式为“主机 IP 地址:端口;主机 IP 地址:端口;…”
03	-h	输出帮助信息

范例：查看主题列表

```
/usr/local/rocketmq/bin/mqadmin topicList -n rocketmq-server:9876
```

程序执行结果	TopicTest（消息测试创建的主题） TopicYootk（开发者创建的主题） 其他主题名称，略

3. deleteTopic（类名称：com.alibaba.rocketmq.tools.command.topic.DeleteTopicSubCommand）

删除指定消息主题指令，执行该指令所需的参数如表 8-6 所示。

表 8-6 删除主题指令参数

序号	参数名称	描述
01	-b	Broker 地址，表示该主题创建在指定的 Broker 之中
02	-c	集群名称，表示删除某个集群下的某个主题
03	-n	NameServer 服务地址，格式为“主机 IP 地址:端口;主机 IP 地址:端口;…”
04	-t	主题名称，允许的名称格式为“^[a-zA-Z0-9_-]+$”

范例：删除 TopicYootk 主题

```
/usr/local/rocketmq/bin/mqadmin deleteTopic -c DefaultCluster -n rocketmq-server:9876 -t
TopicYootk
```

程序执行结果	delete topic [TopicYootk] from cluster [DefaultCluster] success. delete topic [TopicYootk] from NameServer success.

4. updateSubGroup（类名称：com.alibaba.rocketmq.tools.command.consumer.UpdateSubGroupSubCommand）

创建或修改订阅组指令，执行该指令所需的参数如表 8-7 所示。

表 8-7 创建或修改订阅组指令参数

序号	参数名称	描述
01	-b	Broker 地址，表示该主题创建在指定的 Broker 之中
02	-c	集群名称，表示删除某个集群下的某个主题
03	-d	是否允许广播方式消费
04	-g	订阅组名称
05	-i	从指定的 Broker 开始消费
06	-m	是否允许从队列最小位置开始消费（默认为 false）
07	-n	NameServer 服务地址，格式为“主机 IP 地址:端口;主机 IP 地址:端口;…”
08	-q	消费失败的消息保存到一个重试队列，每个订阅组配置的重试队列个数
09	-r	重试消费的最大次数，超过则投递到死信队列，不再投递并报警
10	-s	消费功能是否开启
11	-w	发现消息堆积后，将 Consumer 的消费请求重定向到其他 Slave 主机

范例：创建 muyan-group 订阅组

```
/usr/local/rocketmq/bin/mqadmin updateSubGroup -c DefaultCluster -n rocketmq-server:9876 -g
muyan-group
```

程序执行结果	create subscription group to 192.168.190.170:10911 success. SubscriptionGroupConfig [groupName=muyan-group, consumeEnable=true, consumeFromMinEnable=false, consumeBroadcastEnable=false, retryQueueNums=1, retryMaxTimes=16, brokerId=0, whichBrokerWhenConsumeSlowly=1, notifyConsumerIdsChangedEnable=true]

5. deleteSubGroup（类名称：com.alibaba.rocketmq.tools.command.consumer.DeleteSubscriptionGroupCommand）

删除订阅组配置指令，执行该指令所需的参数如表 8-8 所示。

表 8-8　删除订阅组指令参数

序号	参数名称	描述
01	-b	Broker 地址，表示该主题创建在指定的 Broker 之中
02	-c	集群名称，表示删除某个集群下的某个主题
03	-g	订阅组名称
04	-n	NameServer 服务地址，格式为“主机 IP 地址:端口;主机 IP 地址:端口;…”

范例：删除订阅组

```
/usr/local/rocketmq/bin/mqadmin deleteSubGroup -c DefaultCluster -n rocketmq-server:9876 -g
muyan-group
```

程序执行结果	delete subscription group [muyan-group] from broker [192.168.190.170:10911] in cluster [DefaultCluster] success. delete topic [%RETRY%muyan-group] from cluster [DefaultCluster] success. delete topic [%RETRY%muyan-group] from NameServer success. delete topic [%DLQ%muyan-group] from cluster [DefaultCluster] success. delete topic [%DLQ%muyan-group] from NameServer success.

6. updateBrokerConfig（类名称：com.alibaba.rocketmq.tools.command.broker.UpdateBrokerConfigSubCommand）

更新 Broker 配置文件指令，执行该指令所需的参数如表 8-9 所示。

表 8-9　更新 Broker 配置文件指令参数

序号	参数名称	描述
01	-b	Broker 地址，表示该主题创建在指定的 Broker 之中
02	-c	集群名称，表示删除某个集群下的某个主题
03	-k	修改或增加的 Key
04	-v	修改或增加的 Value
05	-n	NameServer 服务地址，格式为“主机 IP 地址:端口;主机 IP 地址:端口;…”

范例：更新 Broker 刷盘方式为同步

```
/usr/local/rocketmq/bin/mqadmin updateBrokerConfig -c DefaultCluster -n rocketmq-server:9876 -k
flushDiskType -v SYNC_FLUSH
```

程序执行结果	update broker config success, 192.168.190.170:10911

此时更新完成后会自动修改 broker.conf 配置文件中的内容，同时会自动向 broker.conf 中写入所有 Broker 配置项。打开此文件后查询“flushDiskType”配置项时可以发现内容已经更新。

7. topicRoute（类名称：com.alibaba.rocketmq.tools.command.topic.TopicRouteSubCommand）

查看主题路由信息指令，执行该指令所需的参数如表 8-10 所示。

表 8-10　主题路由信息指令参数

序号	参数名称	描述
01	-n	NameServer 服务地址，格式为“主机 IP 地址:端口;主机 IP 地址:端口;...”
02	-t	主题名称，允许的名称格式为“^[a-zA-Z0-9_-]+$”
03	-l	使用列表格式输出路由详情

范例：查看主题路由信息

```
/usr/local/rocketmq/bin/mqadmin topicRoute -n rocketmq-server:9876 -t TopicTest
```

程序执行结果

```
{
  "brokerDatas": [                      // Broker数据
    {
      "brokerAddrs": {                  // Broker地址列表
        0: "192.168.190.170:10911"  // Broker地址配置项
      },
      "brokerName": "broker-a",         // Broker名称（broker.conf配置）
      "cluster": "DefaultCluster"   // 集群名称（broker.conf配置）
    }
  ],
  "filterServerTable": {},              // 过滤数据表
  "queueDatas": [                       // 队列数据
    {
      "brokerName": "broker-a",         // Broker名称（broker.conf配置）
      "perm": 6,                        // 读写模式（6：支持读写。2：禁读。4：禁写）
      "readQueueNums": 4,               // 队列读取数量
      "topicSysFlag": 0,                // 主题标记
      "writeQueueNums": 4               // 写入队列数量
    }
  ]
}
```

8.1.5　Benchmark 压力测试

Benchmark 压力测试

视频名称　0806_【理解】Benchmark 压力测试

视频简介　为了准确地得到 RocketMQ 节点所能够承受的并发访问量，可以通过 Benchmark 提供的工具进行压力测试。本视频通过实例为读者讲解 Benchmark 提供的相关压力测试工具的使用。

在 RocketMQ 服务正式上线之前需要对服务进行压力测试，RocketMQ 提供了 Benchmark 压力测试工具包，其中有两个主要的操作命令：producer.sh（生产者压力测试）、consumer.sh（消费者压力测试），如图 8-12 所示。下面通过具体的步骤演示这两个操作命令的使用。

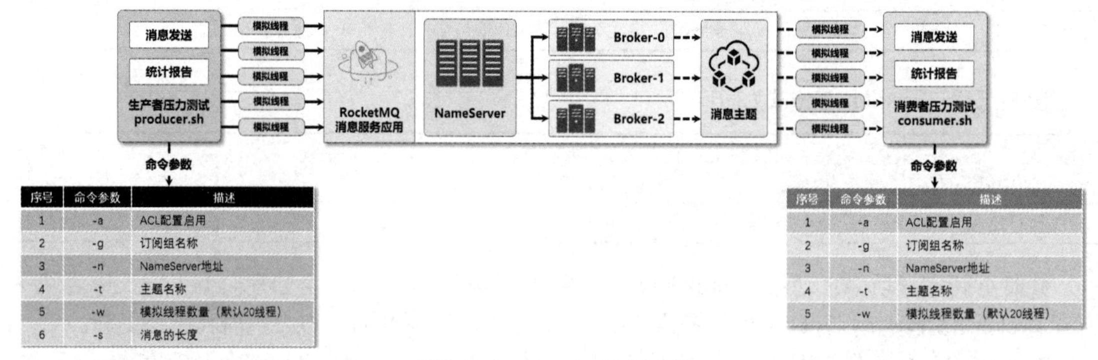

图 8-12　RocketMQ 压力测试

（1）【rocketmq-server 主机】为了使用该 Shell 脚本程序，首先需要对其进行授权。

```
chmod 777 -R /usr/local/rocketmq/benchmark/
```

（2）【rocketmq-server 主机】启动消费端，同时需要设置主题名称、NameServer 地址、订阅组名称。

```
/usr/local/rocketmq/benchmark/consumer.sh -t TopicBenchmark -n rocketmq-server:9876 -g muyan-group
```

程序执行结果

```
Current Time: 1625463607232          // 当前时间戳
TPS: 59883                           // 消费端TPS（系统吞吐量）
FAIL: 0                              // 消费失败的总和
AVG(B2C) RT(ms): 30879.416           // Broker到Consumer的平均响应时间（毫秒）
AVG(S2C) RT(ms): 30878.315           // NameServer到Consumer的平均响应时间（毫秒）
MAX(B2C) RT(ms): 36469               // Broker到Consumer的最大响应时间（毫秒）
MAX(S2C) RT(ms): 36460               // NameServer到Consumer的最大响应时间（毫秒）
```

（3）【rocketmq-server 主机】启动生产端，同时需要设置正确的主题名称与 NameServer 地址。

```
/usr/local/rocketmq/benchmark/producer.sh -t TopicBenchmark -n rocketmq-server:9876
```

程序执行结果

```
Current Time: 1625463611832          // 当前时间戳
Send TPS: 26213                      // 生产端TPS（系统吞吐量）
Max RT(ms): 685                      // 最大响应时间（毫秒）
Average RT(ms): 2.160                // 平均响应时间（毫秒）
Send Failed: 0                       // 发送失败的请求总和
Response Failed: 40                  // 返回失败的总响应数
```

（4）【rocketmq-server 主机】在压力测试期间可以通过 RocketMQ 控制台观察服务访问量，如图 8-13 所示。

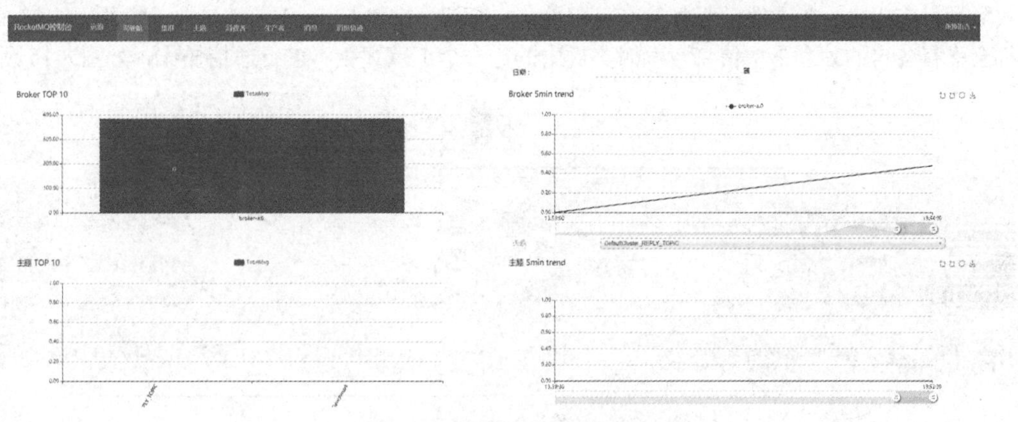

图 8-13　RocketMQ 控制台

（5）【rocketmq-server 主机】Benchmark 应用程序一旦启动，需要通过 shutdown.sh 脚本关闭相关进程。

关闭生产者	`/usr/local/rocketmq/benchmark/shutdown.sh producer`
关闭消费者	`/usr/local/rocketmq/benchmark/shutdown.sh consumer`

8.2　RocketMQ 实现架构

RocketMQ 核心概念

视频名称　0807_【掌握】RocketMQ 核心概念

视频简介　不同的消息组件有不同的处理架构。本视频为读者讲解 RocketMQ 所涉及的架构，并解释核心模块的主要作用。

RocketMQ 是基于队列模型来实现消息首发功能的应用组件，而为了保障服务的稳定运行，一般都会以集群的形式进行服务部署，如图 8-14 所示。在整个应用集群中 RocketMQ 有 4 个核心结构：NameServer、Broker、Producer、Consumer。每一个核心结构都可能是一套完整的服务集群。

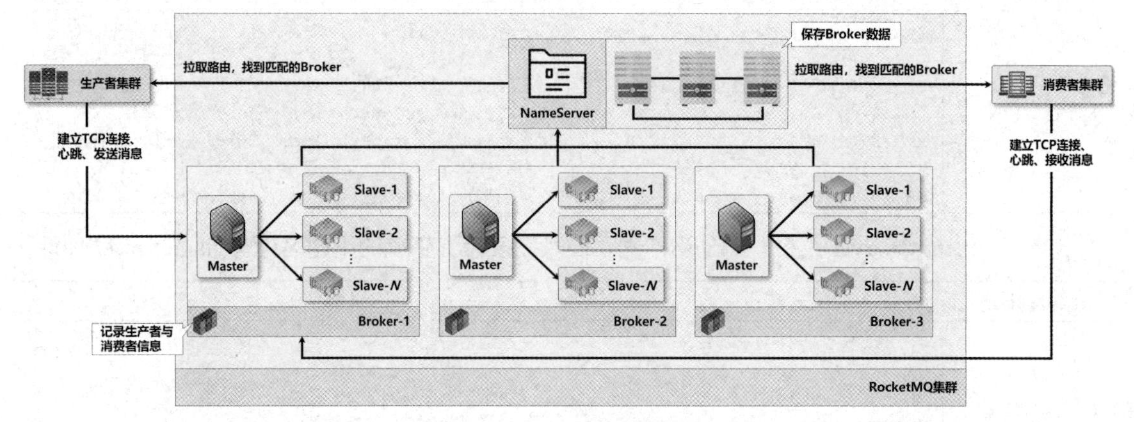

图 8-14　RocketMQ 集群架构

1. NameServer 集群

NameServer 集群保存整个 RocketMQ 集群中的核心数据（主题数据、Broker 数据），生产者与消费者都需要通过 NameServer 提供的路由信息来与相应的 Broker 连接，完成消息的生产与消费处理。在实际项目应用中，为了保证服务的高可用性，一般都会搭建 NameServer 集群，集群中的每一个节点都会保存各自的独立路由信息，同时集群中的每一个节点互相不进行通信操作，如图 8-15 所示。

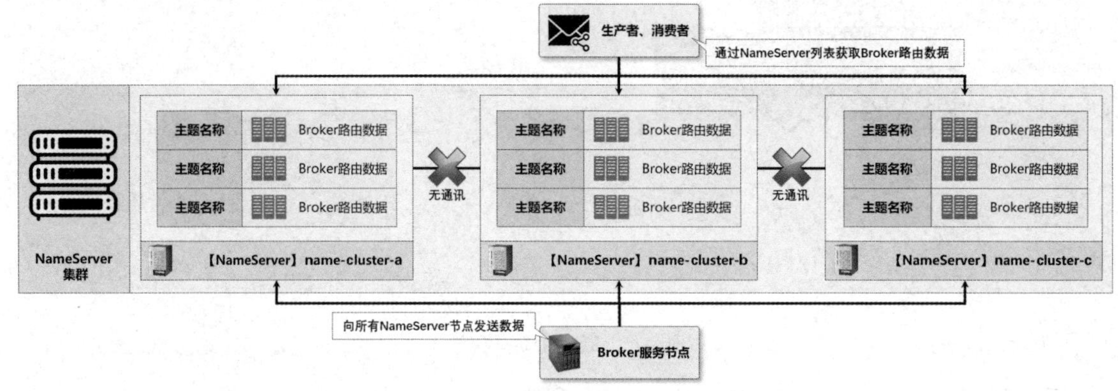

图 8-15　NameServer 集群

NameServer 集群中的每一个节点都保存相同的配置数据，这样就避免了数据一致性的处理操作，也减少了服务器数据同步所带来的性能损耗。由于所有的节点保存相同的配置数据，因此每一个 Broker 启动时需要向所有的 NameServer 节点发送相同的注册信息。这样当某一个节点发生故障时，生产者或消费者可以自动找到集群中的其他节点来获取 Broker 路由数据，从而实现 NameServer 的高可用性。

2. Broker 集群

Broker 集群是 RocketMQ 的核心概念，负责消息的发送、消费以及持久化存储。每一个 Broker 节点启动时都会自动向对应的 NameServer 发送其所保存的主题以及服务路由数据，这样生产者与消费者就可以根据其操作的主题与指定的 Broker 建立连接，从而实现消息的生产与发送处理，如图 8-16 所示。

提示：长连接模式的应用。

生产者与消费者会通过 NameServer 获取 Broker 路由信息，这样就建立了一个主题订阅的模式。而为了可以及时获取路由信息，就需要与 NameServer 创建一个长连接。

所有的 Broker 每 30 秒会向 NameServer 发送一次心跳数据，NameServer 利用这些心跳数据来及时实现主题更新，并判断 Broker 是否存活。为了提高这种心跳数据的传输效率，需要在 Broker 与 NameServer 之间建立一个长连接。

NameServer 每 10 秒扫描一次存活的 Broker 链接，若某一个 Broker 在 2 分钟内都没有发送心跳数据，则认为该 Broker 下线，同时 NameServer 会自动断开与该 Broker 的连接，并调整主题与 Broker 的对应关系，但是 Broker 死机的信息是不会主动发送给生产者与消费者链接的。

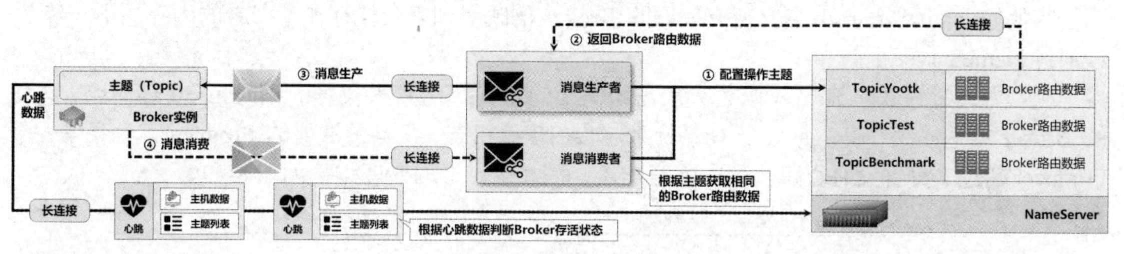

图 8-16　Broker 连接

注意：主题过多会导致“假死”问题。

Broker 每次向 NameServer 发送心跳数据时，都会传输所有的主题列表。一旦在 Broker 中创建了过多的主题（比如几万个主题），则可能会导致每次心跳传输的数据量过大，造成传输失败的情况，而这样就有可能导致 NameServer 误认为 Broker 死机。

随着项目中的业务处理功能的完善，以及访问量的增加，某些消息主题也必然会带来更多的消息内容。为了可以及时、有效地处理更多的消息数据，可以在 RocketMQ 中引入 Broker 集群，其中的 Broker 节点全部为一个主题服务，而不同的 Broker 中可以有不同的队列实现消息的传输，如图 8-17 所示。

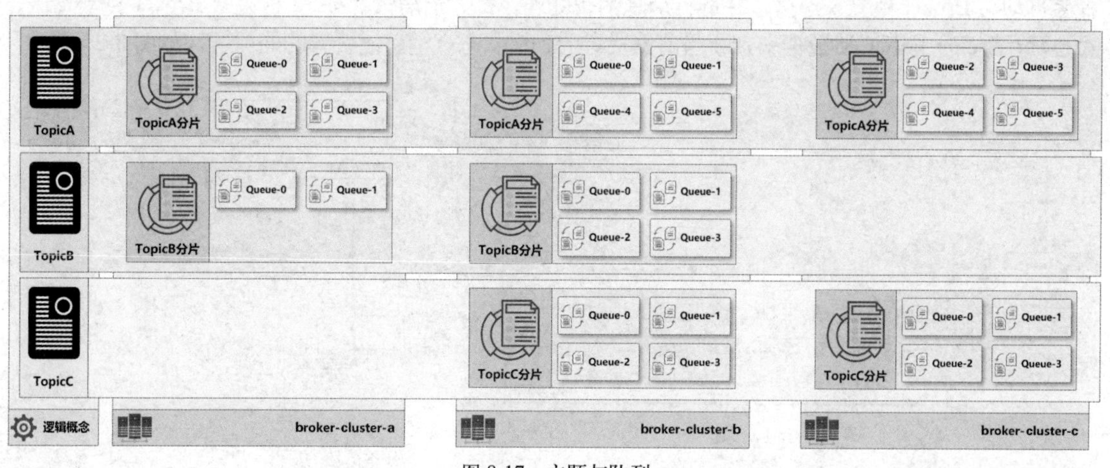

图 8-17　主题与队列

Broker 在整个 RocketMQ 运行架构中起着消息数据保存的重要作用，如果某一个 Broker 集群中的节点出现了故障，那么会造成消息数据的丢失。为了保证 Broker 节点的高可用性，需要引入主从架构（一个 Master 节点对应若干个 Slave 节点），如图 8-18 所示。为了保证数据的安全，主从架构中的每一个 Broker 节点都保存相同的数据。

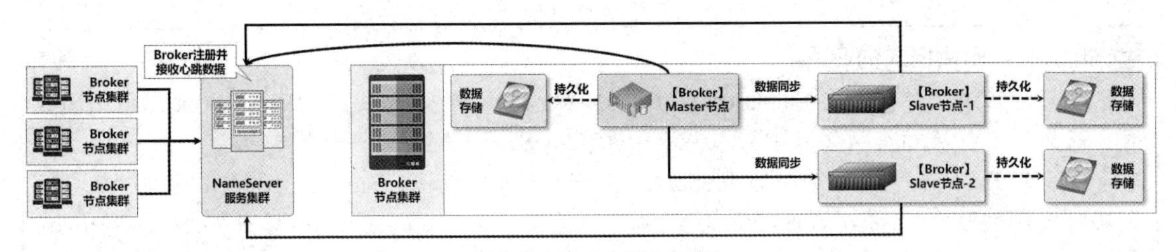

图 8-18　Broker 节点集群

3. Producer 集群

Producer 集群提供集群消息生产者，这些生产者拥有相同的分组，每一个生产者在启动时都需要与 NameServer 建立长连接。如果有一个 NameServer 节点出现了问题，则生产者会根据服务列表连接其他的 NameServer 节点。所有获取的 Broker 的映射列表会保存在生产者的内存之中，每 30 秒通过 NameServer 更新一次。

生产者通过获取的 Broker 映射列表与所需要的 Broker 建立长连接，同时每 30 秒向建立连接的 Broker 发送一次心跳。每一个 Broker 也会每 10 秒扫描一次当前注册的 Producer，如果某一个 Producer 超过 2 分钟没有心跳，则会断开连接。消息生产者的处理结构如图 8-19 所示。

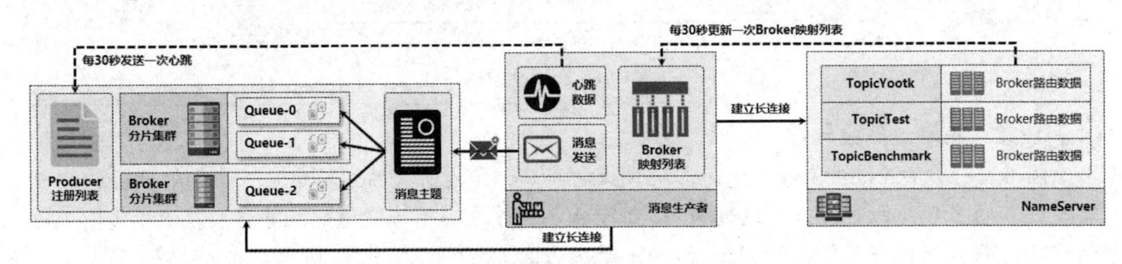

图 8-19　消息生产者

4. Consumer 集群

为了提高消息的消费处理性能，可以设置同一分组的多个消息消费者，每一个消费者都需要与 NameServer 保持长连接，这样才可以及时抓取最新的 Broker 映射列表。获得 Broker 映射列表后消费者会自动与 Broker 建立长连接，并进行消息消费处理。每一个消费者每 30 秒会向 Broker 自动发送一次心跳，Broker 依靠心跳数据来判断当前消费者的存活状态。消息消费者的处理结构如图 8-20 所示。

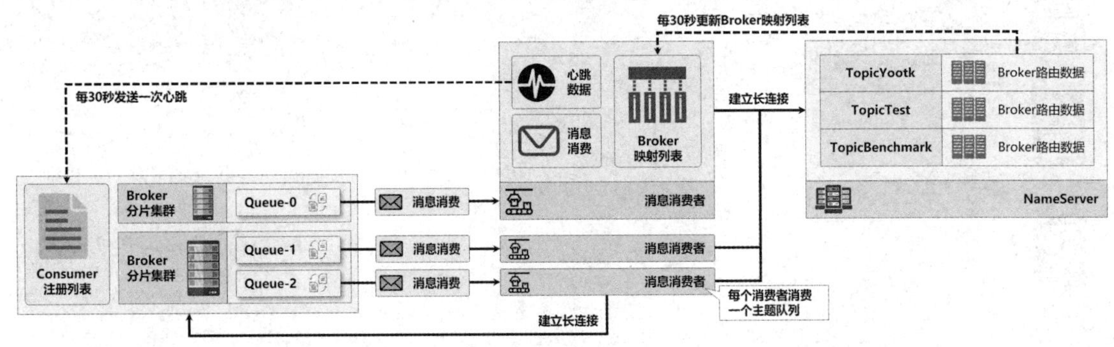

图 8-20　消息消费者

考虑不同的场景需要，在进行消息消费时一般会有广播消费与集群消费两种处理模式，这两种模式的特点如下。

- 广播消费：若干个消费者消费同一个主题下的所有队列消息，每个消费者获得同样的消息，如图 8-21 所示。
- 集群消费：在一个主题中由若干个不同的消费者分担消费。例如，现在一个消息主题中有

3 个队列，这样就可以启动 3 个消费者，每个消费者消费不同队列的消息数据。

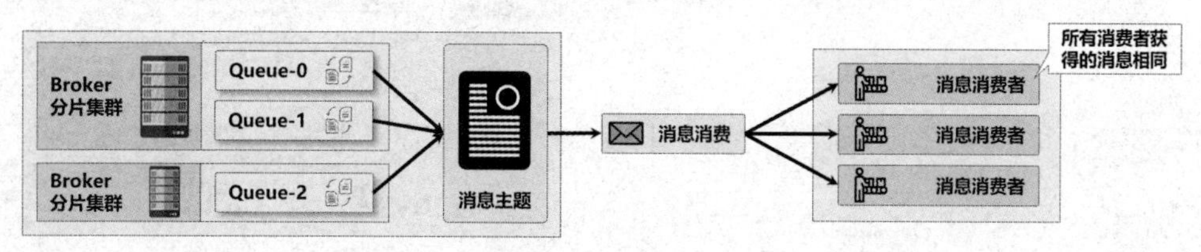

图 8-21　广播消费

8.2.1　Remoting 通信模块

Remoting 通信模块

视频名称　0808_【理解】Remoting 通信模块

视频简介　RocketMQ 消息组件需要基于长连接的模式与 NameServer 或 Broker 实现通信，而这些通信机制的实现都是基于 Netty 完成的。本视频通过 RocketMQ 项目中的 remoting 子模块的源代码分析 Netty 通信的核心处理机制。

RocketMQ 中的 RPC 通信是基于 Netty 框架实现的。为了便于所有的通信操作管理，RocketMQ 项目中定义了一个 rocketmq-remoting 子模块，而该项目中的其他子模块（如 common、broker、namesrv 等）都需要通过此模块完成通信。打开 remoting 模块的依赖库配置文件（pom.xml），可以发现里面引入了 Netty 依赖库，如图 8-22 所示。

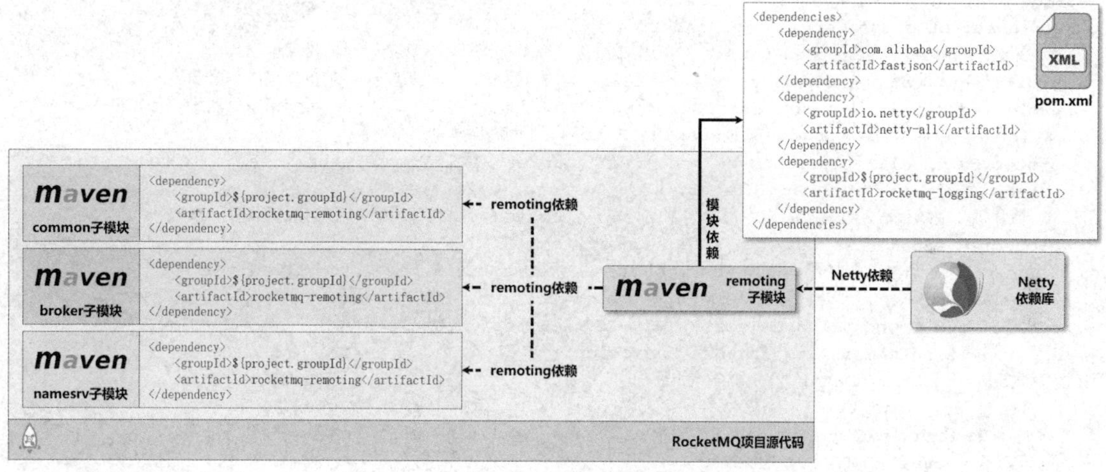

图 8-22　remoting 依赖结构

为了实现客户端与服务端之间高效的数据请求与接收，RocketMQ 消息队列自定义了通信协议，并在 Netty 的基础之上扩展了通信模块，包括消息协议格式、消息编解码器、消息通信模式（同步、异步、单向）。在 remoting 子模块中为了便于所有服务的统一管理，还定义了一个 RemotingService 的接口，该接口源代码如下。

```
package org.apache.rocketmq.remoting;
public interface RemotingService {
    void start();                                    // 服务开启
    void shutdown();                                 // 服务关闭
    void registerRPCHook(RPCHook rpcHook);           // 注册RPC钩子处理类
}
```

在网络通信过程中，RemotingService 的实现分为服务端（RemotingServer 接口）与客户端（RemoteClient 接口）两个处理部分，而具体的服务实现则基于 Netty 组件库开发。图 8-23 给出了 RemotingService 接口的继承关系。

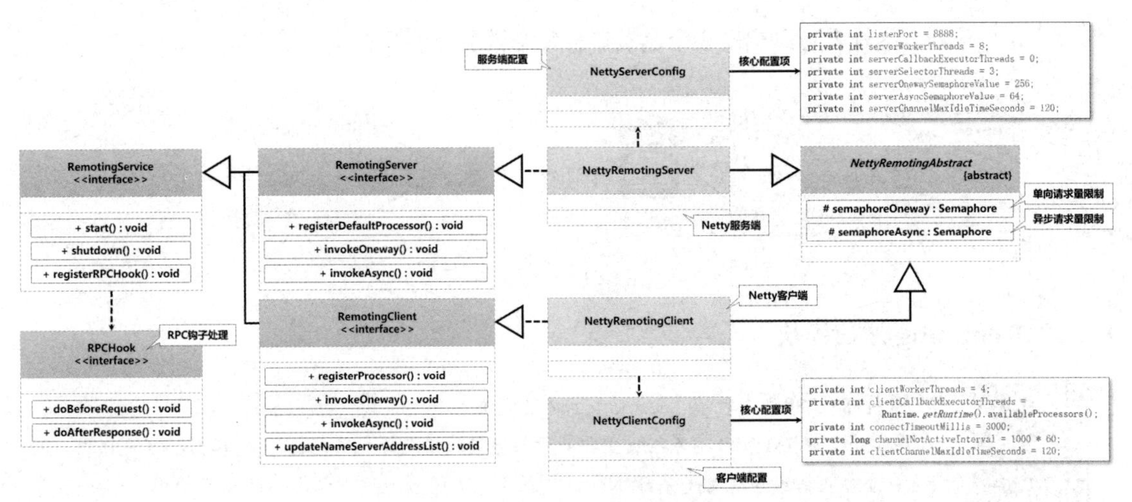

图 8-23　RemotingService 接口的继承关系

RemotingService 的两个具体实现子类（NettyRemotingServer、NettyRemotingClient）定义时都默认继承了一个公共抽象类（NettyRemotingAbstract），该类主要实现了远程通信模块。该类的核心结构如下所示。

范例：NettyRemotingAbstract 类核心结构

```
package org.apache.rocketmq.remoting.netty;
public abstract class NettyRemotingAbstract {
   // 此处只列出NettyRemotingAbstract类的核心结构，完整代码请参考RocketMQ项目
   protected final Semaphore semaphoreOneway;                     // 单向请求信号量
   protected final Semaphore semaphoreAsync;                      // 异步请求信号量
   // 保存正在进行中的请求缓存，并通过ResponseFuture接收响应
   protected final ConcurrentMap<Integer /* opaque */, ResponseFuture> responseTable =
      new ConcurrentHashMap<Integer, ResponseFuture>(256);
   // 保存每个请求代码的处理器，对于每个传入的请求可以通过此集合找到对应的处理器来进行请求处理
   protected final HashMap<Integer/* request code */,
               Pair<NettyRequestProcessor, ExecutorService>> processorTable =
      new HashMap<Integer, Pair<NettyRequestProcessor, ExecutorService>>(64);
   public void processMessageReceived(ChannelHandlerContext ctx,
        RemotingCommand msg) throws Exception {                   // 消息接收处理
      final RemotingCommand cmd = msg;                            // 远程操作类型
      if (cmd != null) {
         switch (cmd.getType()) {                                 // 类型匹配
            case REQUEST_COMMAND:                                 // 请求操作
               processRequestCommand(ctx, cmd);                   // 使用Netty实现请求处理
               break;
            case RESPONSE_COMMAND:                                // 响应操作
               processResponseCommand(ctx, cmd);                  // 使用Netty实现响应处理
               break;
            default:
               break;
         }
      }
   }
   public void scanResponseTable() {                              // 扫描超时请求并使其过期
      final List<ResponseFuture> rfList = new LinkedList<ResponseFuture>(); // 未过期的请求
      // 获取当前所有处理请求的迭代对象（这些请求对象都是在执行请求时保存的）
      Iterator<Entry<Integer, ResponseFuture>> it =
               this.responseTable.entrySet().iterator();          // 获取Iterator接口实例
      while (it.hasNext()) {                                      // 迭代处理
         Entry<Integer, ResponseFuture> next = it.next();         // 获取待响应请求
         ResponseFuture rep = next.getValue();                    // 获取异步请求响应
         if ((rep.getBeginTimestamp() + rep.getTimeoutMillis() + 1000)
```

```
                <= System.currentTimeMillis()) {          // 请求超时判断
            rep.release();                                 // 请求释放
            it.remove();                                   // 列表删除
            rfList.add(rep);                               // 请求保存
          }
        }
        for (ResponseFuture rf : rfList) {                 // 有效请求处理迭代
          try { executeInvokeCallback(rf);                 // 执行请求处理
          } catch (Throwable e) {}
        }
    }
    public RemotingCommand invokeSyncImpl(final Channel channel,
        final RemotingCommand request, final long timeoutMillis) throws … {
     // 发送同步访问请求，并等待请求响应
    }
    public void invokeAsyncImpl(final Channel channel, final RemotingCommand request,
         final long timeoutMillis, final InvokeCallback invokeCallback) throws … {
        // 发送异步访问请求，并通过响应监听的方式等待响应
    }
    public void invokeOnewayImpl(final Channel channel, final RemotingCommand request,
         final long timeoutMillis) throws … {
        // 单向消息处理，此时不需要进行响应监听
    }
}
```

通过源代码的定义可以发现，NettyRemotingAbstract 类主要实现了请求与响应的处理操作，它在每次发出消息时（同步、异步）都会将该请求操作的响应保存在 responseTable 的集合之中。为了防止消息堆积，它会每秒通过 scanResponseTable()方法进行一次 responseTable 回收，并通过 executeInvokeCallback()方法实现响应接收。

RocketMQ 的服务端是由 NettyRemotingServer 子类实现的，该类明确定义了服务端所使用的 Reactor 处理模型。下面首先来观察此实现类的核心源代码。

范例：NettyRemotingServer 核心源代码

```
package org.apache.rocketmq.remoting.netty;
public class NettyRemotingServer extends NettyRemotingAbstract
              implements RemotingServer {                          // 服务端通信实现类
   // 由于NettyRemotingServer类的源代码较多，此处只重点列出Reactor线程模型相关的源代码
   private final ServerBootstrap serverBootstrap;                  // Netty核心类
   private final EventLoopGroup eventLoopGroupSelector;            // 子线程池
   private final EventLoopGroup eventLoopGroupBoss;                // 主线程池
   private final NettyServerConfig nettyServerConfig;              // Netty默认配置
   private final ExecutorService publicExecutor;                   // 业务线程池
   private final ChannelEventListener channelEventListener;
   private DefaultEventExecutorGroup defaultEventExecutorGroup; // 耗时任务线程池
   public NettyRemotingServer(final NettyServerConfig nettyServerConfig,
      final ChannelEventListener channelEventListener) {
      // 调用父类构造，传递单向信号量与异步处理信号量内容
      super(nettyServerConfig.getServerOnewaySemaphoreValue(),
              nettyServerConfig.getServerAsyncSemaphoreValue());
      this.serverBootstrap = new ServerBootstrap();                // 对象实例化
      this.nettyServerConfig = nettyServerConfig;                  // 接收Netty默认配置
      this.channelEventListener = channelEventListener;            // Channel处理
      // 通过Runtime类获取当前主机配置的内核线程数量，该数量将作为任务处理的线程池大小
      int publicThreadNums = nettyServerConfig.getServerCallbackExecutorThreads();
      if (publicThreadNums <= 0) {                                 // 没有获取成功
         publicThreadNums = 4;                                     // 默认4个线程池
      }
      this.publicExecutor = Executors.newFixedThreadPool(publicThreadNums,
              new ThreadFactory() { … });                          // 创建业务线程池
      if (useEpoll()) {                                            // 事件驱动
         this.eventLoopGroupBoss = new EpollEventLoopGroup(1,
```

```
                new ThreadFactory() {});                        // 主线程池
        this.eventLoopGroupSelector = new EpollEventLoopGroup(nettyServerConfig
                .getServerSelectorThreads(), new ThreadFactory() { … });   // 子线程池
    } else {                                                    // 其他模式
        this.eventLoopGroupBoss = new NioEventLoopGroup(1,
                        new ThreadFactory() {…});               // 主线程池
        this.eventLoopGroupSelector = new NioEventLoopGroup(nettyServerConfig
                .getServerSelectorThreads(), new ThreadFactory() { … });   // 子线程池
    }
    loadSslContext();
  }
  private boolean useEpoll() {                                  // 判断是否为Epoll模式
    return RemotingUtil.isLinuxPlatform()
        && nettyServerConfig.isUseEpollNativeSelector()
        && Epoll.isAvailable();
  }
  @Override
  public void start() {                                         // 服务启动
    this.defaultEventExecutorGroup = new DefaultEventExecutorGroup(
        nettyServerConfig.getServerWorkerThreads(),
        new ThreadFactory() { … });                             // 耗时任务线程池
    prepareSharableHandlers();                                  // 实例化公共组件类
    ServerBootstrap childHandler = this.serverBootstrap.group(
            this.eventLoopGroupBoss, this.eventLoopGroupSelector)
          .channel(useEpoll() ? EpollServerSocketChannel.class :
            NioServerSocketChannel.class)                       // 配置通道类型
          .option(ChannelOption.SO_BACKLOG, 1024)               // 连接等待队列长度
          .option(ChannelOption.SO_REUSEADDR, true)             // 允许重复使用本地地址
          .option(ChannelOption.SO_KEEPALIVE, false)            // 心跳检测
          .childOption(ChannelOption.TCP_NODELAY, true)         // 禁用Nagle算法，小数据传输
          .childOption(ChannelOption.SO_SNDBUF,
            nettyServerConfig.getServerSocketSndBufSize())      // 发送缓冲大小
          .childOption(ChannelOption.SO_RCVBUF,
            nettyServerConfig.getServerSocketRcvBufSize())      // 接收缓冲大小
          .localAddress(new InetSocketAddress(
            this.nettyServerConfig.getListenPort()))            // 监听端口
          .childHandler(new ChannelInitializer<SocketChannel>() {
            @Override
            public void initChannel(SocketChannel ch) throws Exception {
              ch.pipeline()
                .addLast(defaultEventExecutorGroup, …)          // 执行链
                .addLast(defaultEventExecutorGroup, encoder,…);// 执行链
            }
          });
    // 后续程序代码略
  }
  // 本类中定义的invokeSync()、invokeAsync()、invokeOneway()等方法体内部调用了父类实现，略
  @Override
  public ExecutorService getCallbackExecutor() {    // NettyRemotingAbstract类使用
    return this.publicExecutor;                     // 返回任务执行线程池
  }
}
```

通过以上源代码，可以清楚地发现在构造方法中实例化了 3 个线程池：eventLoopGroupBoss（接收用户请求的主线程池）、eventLoopGroupSelector（工作处理的子线程池）、publicExecutor（业务线程池）。而后在服务启动时又实例化了耗时（复杂且时间不可控的）任务线程池（defaultEventExecutorGroup），如图 8-24 所示。

RocketMQ 服务端之所以引入 4 个线程池的处理模式，其核心目的在于将通信业务与核心业务有效分离，将简单的任务交由通信处理部分来完成，而所有耗时的任务统一由耗时任务线程池进行处理，如图 8-25 所示。这样的设计可以提高整个应用的通信效率，并使 RocketMQ 的整体处理性能更高。

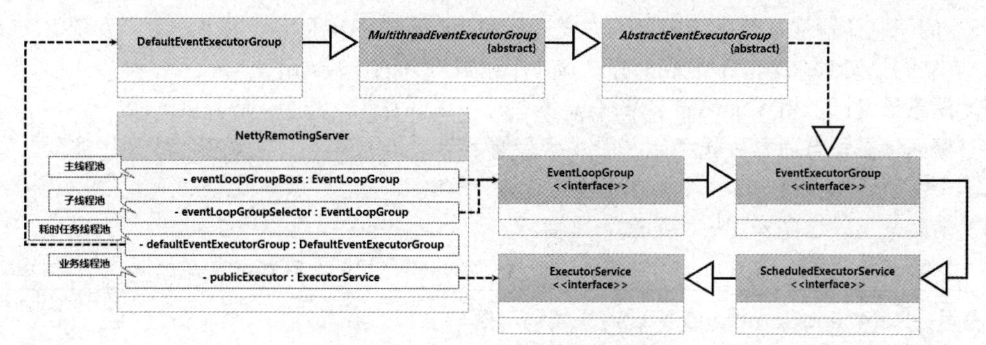

图 8-24 NettyRemotingServer 线程池

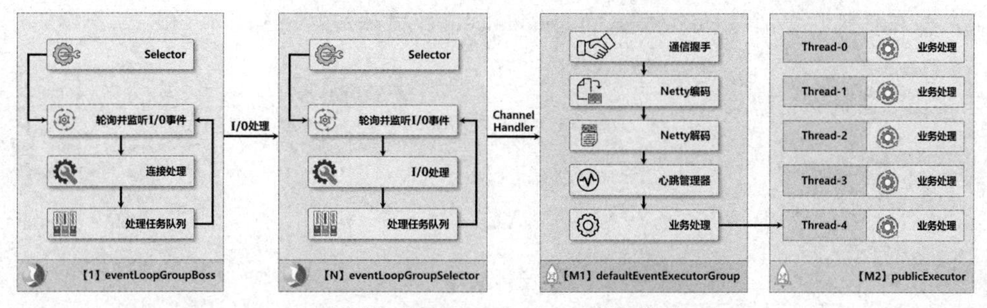

图 8-25 NettyRemotingServer 线程模型

提示："1+*N*+M1+M2" Reactor 多线程模型。

传统的 Reactor 编程模型中往往只采用"1 + *N*"的模式，即 1 个主线程负责连接处理、*N* 个子线程负责 I/O 处理。RocketMQ 对这一模式进行了扩展，在 *N* 个子线程之后有 M1 线程池负责处理所有的耗时任务，同时每一个具体的任务交由 M2 线程池完成。

8.2.2 消息结构

视频名称 0809_【理解】消息结构

视频简介 消息是 RocketMQ 之中主要的数据传输内容。本视频为读者讲解消息协议的组成结构，并通过消息的类源代码介绍消息的组成内容以及继承结构。

消息（Message）是整个 RocketMQ 之中最核心的处理逻辑单元之一。生产者生产完成的消息经过网络传输后才可以被消费端消费，为了保证消息传输的正确性，就需要按照图 8-26 所示的结构进行通信协议的定义。消息生产者根据指定的消息结构进行信息编码，而消息消费者根据同样的消息结构进行信息解码。

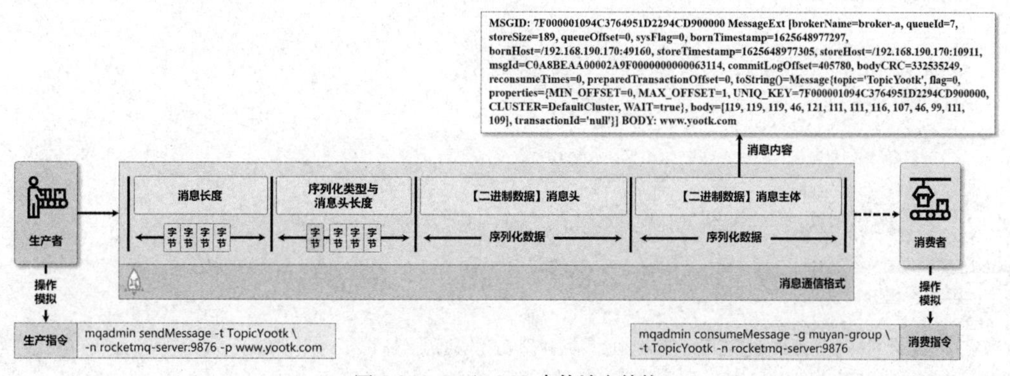

图 8-26 RocketMQ 中的消息结构

所有的消息传输都是基于二进制的方式进行的。通过图 8-26 所示的结构，可以发现 RocketMQ 将一个完整的消息划分为 4 个组成部分，这 4 个组成部分的作用如下。

- **消息长度**：占用 4 个存储字节（int 类型）。
- **序列化类型与消息头长度**：占用 4 个存储字节，其中第 1 个字节表示序列化类型，后面 3 个字节表示长度。
- **消息头**：经过序列化后的消息头数据。
- **消息主体**：经过序列化后的消息数据，在 RocketMQ 中使用 Message 类保存消息数据。

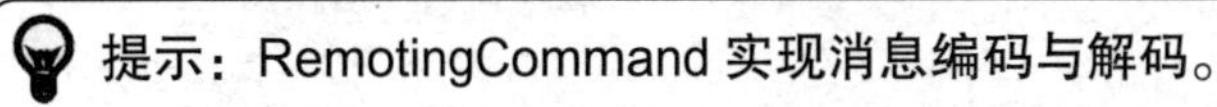

RocketMQ 之中的所有消息在进行传输之前都需要通过 RemotingCommand.encode()方法进行编码，而在接收时要使用 RemotingCommand.decode()方法进行解码，如图 8-27 所示。

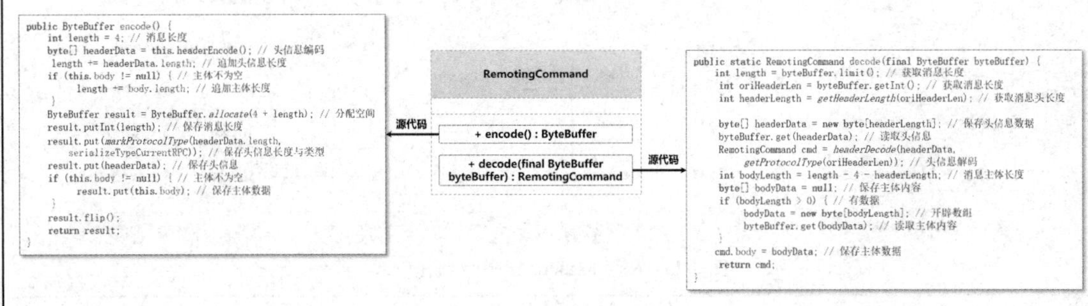

图 8-27 消息编码与解码

读者可以根据 encode()与 decode()源代码并结合图 8-26 所给出的消息结构，清楚地观察到消息编码与解码的实现过程。

在使用 mqadmin 进行消息生产与消费模拟时可以清楚地发现，每一个消息被消费时都会保存大量的数据信息，而这些信息都是由 Message 类定义的。为便于读者理解消息的完整结构，下面通过该类的源代码来进行详细说明。

范例：打开 common 子模块中的 Message 类源代码

```java
package org.apache.rocketmq.common.message;
public class Message implements Serializable {
   private static final long serialVersionUID = 8445773977080406428L;
   private String topic;                                   // 消息的主题
   private int flag;                                       // 网络通信层标记
   private Map<String, String> properties;                 // Message参数信息
   private byte[] body;                                    // 消息的主体数据
   private String transactionId;                           // 事务消息ID
   public void setTags(String tags) {                      // 设置消息标签
      this.putProperty(MessageConst.PROPERTY_TAGS, tags);
   }
   public void setKeys(Collection<String> keys) {          // 设置消息业务Key
      StringBuffer sb = new StringBuffer();
      for (String k : keys) {
         sb.append(k);
         sb.append(MessageConst.KEY_SEPARATOR);
      }
      this.setKeys(sb.toString().trim());
   }
   public void setDelayTimeLevel(int level) {              // 消息延迟处理级别
      this.putProperty(MessageConst.PROPERTY_DELAY_TIME_LEVEL, String.valueOf(level));
   }
   // 在同步刷盘的机制下是否需要等待数据落地才返回消息成功发送的标记
   public void setWaitStoreMsgOK(boolean waitStoreMsgOK) {
```

```
        this.putProperty(MessageConst.PROPERTY_WAIT_STORE_MSG_OK,
                Boolean.toString(waitStoreMsgOK));
    }
}
```

org.apache.rocketmq.common.message.Message 类只是定义了消息的核心组成结构（主题、内容），而消息的一些附加内容（业务 Key、标签）则采用了属性的形式来配置。但是通过消费端和 RocketMQ 控制台进行消息查看时，会发现消费端接收的消息所包含的内容远比 Message 类定义的要多，而这些信息实际上都保存在了 MessageExt 子类之中。下面继续观察 MessageExt 子类的源代码。

范例：打开 common 子模块中的 MessageExt 子类

```
package org.apache.rocketmq.common.message;
public class MessageExt extends Message {                  // 消息信息扩展类
    private static final long serialVersionUID = 5720810158625748049L;
    private String brokerName;                            // Broker名称
    private int queueId;                                  // 消息队列ID
    private int storeSize;                                // 消息在Broker中的存储大小
    private long queueOffset;                             // 消息在队列中的偏移量
    private int sysFlag;                                  // 系统标志的开关状态
    private long bornTimestamp;                           // 消息创建时间戳
    private SocketAddress bornHost;                       // 生产者地址
    private long storeTimestamp;                          // 消息储存时间戳
    private SocketAddress storeHost;                      // 消息存储地址
    private String msgId;                                 // 消息ID
    private long commitLogOffset;                         // 提交日志偏移量
    private int bodyCRC;                                  // 消息内容CRC校验值
    private int reconsumeTimes;                           // 重试消费次数
    private long preparedTransactionOffset;               // 事务消息相关字段
}
```

图 8-28 所示为消息类的继承结构。通过 MessageExt 子类的定义可以发现，在 RocketMQ 中所传递的每一个消息都会被清楚地记录消息存放的位置（brokerName）、消息队列的 ID（queueId）、生产者信息（主机地址、生产日期）、消息 ID（msgId）等核心内容，而这些内容在传输、存储或消费过程中会根据需要进行更新。

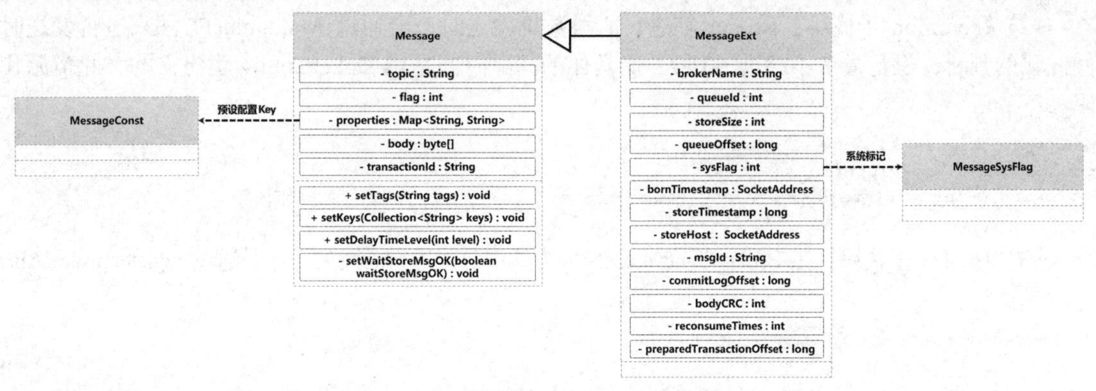

图 8-28　Message 与 MessageExt 的继承结构

8.2.3　心跳检测

视频名称　0810_【理解】心跳检测

视频简介　RocketMQ 除了核心的消息内容之外，还需要定时进行心跳数据的发送，以实现心跳检查。本视频通过 RocketMQ 的源代码分析心跳机制中的发送与检查实现。

在 RocketMQ 集群之中的许多节点都需要网络通信的支持，例如，Broker 需要向 NameServer 定期发送心跳以更新主题列表与路由信息，生产者与消费者又需要向 Broker 定期发送心跳以确定存活状态，如图 8-29 所示。

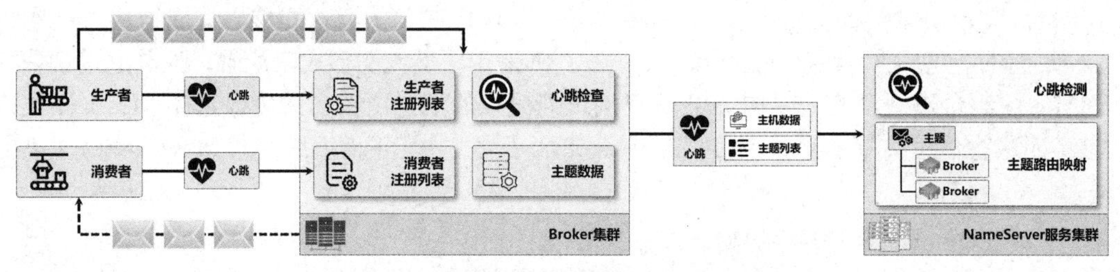

图 8-29 心跳检测

心跳机制是 TCP 中用于检查长连接的一种手段，客户端向服务端发送心跳的主要目的是报告自己的存活状态，而服务端需要维持长连接的客户端列表，则可以根据心跳的状态来进行此列表的动态更新。下面以 Broker 与 NameServer 之间的心跳处理为例进行说明。

（1）【broker 子模块】Broker 的主要功能是由 BrokerController 类定义的，而在 BrokerController 中的心跳发送是启用一个定时调度的线程池完成的。这一点可以通过 BrokerController.start()方法观察到。

```
package org.apache.rocketmq.broker;
public class BrokerController {
   public void start() throws Exception {                        // 其他代码略
      this.scheduledExecutorService.scheduleAtFixedRate(new Runnable() {
         @Override
         public void run() {
            try {
               BrokerController.this.registerBrokerAll(true, false,
                     brokerConfig.isForceRegister());            // Broker注册操作
            } catch (Throwable e) { }
         }
      }, 1000 * 10, Math.max(10000, Math.min(brokerConfig.getRegisterNameServerPeriod(), 60000)),
TimeUnit.MILLISECONDS);                                          // 定时调度线程池
   }
}
```

（2）【common 子模块】Broker 在进行心跳数据发送时，会通过 Math.min()方法来进行发送时间间隔的判断，但是最多不超过 60 秒。而具体的间隔时间是由 BokerConfig 类定义的，此类源代码如下。

```
package org.apache.rocketmq.common;
public class BrokerConfig {
   private int registerNameServerPeriod = 1000 * 30;             // 30秒间隔
}
```

（3）【broker 子模块】在 BrokerController.start()方法中每次进行心跳发送都是通过 registerBrokerAll()方法完成的。

```
package org.apache.rocketmq.broker;
public class BrokerController {
   public synchronized void registerBrokerAll(final boolean checkOrderConfig,
            boolean oneway, boolean forceRegister) {             // 同步处理
      TopicConfigSerializeWrapper topicConfigWrapper = this.getTopicConfigManager()
         .buildTopicConfigSerializeWrapper();                    // 获取主题配置序列化包装类实例
      // 判断当前要注册的Broker是否拥有读写权限
      if (!PermName.isWriteable(this.getBrokerConfig().getBrokerPermission())
         || !PermName.isReadable(this.getBrokerConfig().getBrokerPermission())) {
         ConcurrentHashMap<String, TopicConfig> topicConfigTable =
               new ConcurrentHashMap<String, TopicConfig>();     // 保存主题配置
         for (TopicConfig topicConfig : topicConfigWrapper.getTopicConfigTable()
                  .values()) {                                   // 迭代主题集合
```

```
            TopicConfig tmp = new TopicConfig(topicConfig.getTopicName(),
                topicConfig.getReadQueueNums(), topicConfig.getWriteQueueNums(),
                this.brokerConfig.getBrokerPermission());              // 获取主题配置项
            topicConfigTable.put(topicConfig.getTopicName(), tmp);     // 保存主题信息
        }
        topicConfigWrapper.setTopicConfigTable(topicConfigTable);      // 保存主题集合
    }
    if (forceRegister || needRegister(this.brokerConfig.getBrokerClusterName(),
        this.getBrokerAddr(),
        this.brokerConfig.getBrokerName(),
        this.brokerConfig.getBrokerId(),
        this.brokerConfig.getRegisterBrokerTimeoutMills())) {          // 判断Broker是否需要注册
        doRegisterBrokerAll(checkOrderConfig, oneway, topicConfigWrapper); // 注册
    }
  }
}
```

BrokerController.registerBrokerAll()方法首先会获取当前 Broker 中所有配置的主题信息（TopicConfig）。由于该信息最终都需要通过序列化处理，因此将获取的主题集合保存在 TopicConfigSerializeWrapper 对象实例之中，以实现心跳数据中主题数据的传输。

（4）【broker 子模块】最终 Broker 的心跳处理是由 BrokerController.doRegisterBrokerAll()方法完成的。

```
package org.apache.rocketmq.broker;
public class BrokerController {
   private void doRegisterBrokerAll(boolean checkOrderConfig, boolean oneway,
      TopicConfigSerializeWrapper topicConfigWrapper) {
      List<RegisterBrokerResult> registerBrokerResultList = this.brokerOuterAPI
         .registerBrokerAll(                                        // 获取Broker注册数据
            this.brokerConfig.getBrokerClusterName(),               // 集群名称
            this.getBrokerAddr(),                                   // Broker地址
            this.brokerConfig.getBrokerName(),                      // Broker名称
            this.brokerConfig.getBrokerId(),                        // Broker ID
            this.getHAServerAddr(),                                 // Broker分片集群地址
            topicConfigWrapper,                                     // 主题配置包装
            this.filterServerManager.buildNewFilterServerList(),    // 过滤配置
            oneway,                                                 // 单向信息标记
            this.brokerConfig.getRegisterBrokerTimeoutMills(),      // 超时时间
            this.brokerConfig.isCompressedRegister());              // 是否压缩注册
         // 其他代码略
   }
}
```

该方法会将所有接收的 Broker 数据通过 BrokerOuterAPI.registerBrokerAll()方法进行处理，而这也是实现 Broker 心跳数据向 NameServer 发送的最终方法。

（5）【broker 子模块】BrokerOuterAPI.registerBrokerAll()实现发送心跳数据源代码。

```
package org.apache.rocketmq.broker.out;
public class BrokerOuterAPI {
   public List<RegisterBrokerResult> registerBrokerAll(         // 发送心跳数据
      final String clusterName,                                 // 集群名称
      final String brokerAddr,                                  // Broker地址
      final String brokerName,                                  // Broker名称
      final long brokerId,                                      // Broker ID
      final String haServerAddr,                                // HA服务地址
      final TopicConfigSerializeWrapper topicConfigWrapper,     // 主题配置项
      final List<String> filterServerList,                      // 服务过滤
      final boolean oneway,                                     // 单向消息
      final int timeoutMills,                                   // 注册超时时间
      final boolean compressed) {                               // 是否压缩
      final List<RegisterBrokerResult> registerBrokerResultList =
            new CopyOnWriteArrayList<>();                       // 保存注册结果
      List<String> nameServerAddressList = this.remotingClient
            .getNameServerAddressList();                        // 获取NameServer地址
```

```
        if (nameServerAddressList != null &&
                nameServerAddressList.size() > 0) {                  // 判断是否存在NameServer地址
            final RegisterBrokerRequestHeader requestHeader =
                    new RegisterBrokerRequestHeader();                   // Broker注册头信息
            requestHeader.setBrokerAddr(brokerAddr);                     // Broker地址
            requestHeader.setBrokerId(brokerId);                         // Broker ID
            requestHeader.setBrokerName(brokerName);                     // Broker名称
            requestHeader.setClusterName(clusterName);                   // 集群名称
            requestHeader.setHaServerAddr(haServerAddr);                 // HA服务地址
            requestHeader.setCompressed(compressed);                     // 压缩配置
            RegisterBrokerBody requestBody = new RegisterBrokerBody();   // Broker注册主体
            requestBody.setTopicConfigSerializeWrapper(topicConfigWrapper); // 主题数据
            requestBody.setFilterServerList(filterServerList);           // 服务过滤
            final byte[] body = requestBody.encode(compressed);          // 数据编码
            final int bodyCrc32 = UtilAll.crc32(body);                   // CRC校验
            requestHeader.setBodyCrc32(bodyCrc32);                       // 设置CRC头信息
            final CountDownLatch countDownLatch =
                    new CountDownLatch(nameServerAddressList.size());    // 计数同步
            // Broker在发送心跳时，需要向所有的NameServer节点发送，以达到数据存储一致的目的
            for (final String namesrvAddr : nameServerAddressList) {     // NameServer迭代
                brokerOuterExecutor.execute(new Runnable() {             // 注册线程池处理
                    @Override
                    public void run() {                                  // Broker注册
                        try {
                            // 通过RemotingCommand类提供的invokeOneway()方法发送Broker数据
                            RegisterBrokerResult result = registerBroker(namesrvAddr, oneway,
                              timeoutMills, requestHeader, body);        // 服务注册
                            if (result != null) {                        // 有结果返回
                                registerBrokerResultList.add(result);    // 保存处理结果
                            }
                        } catch (Exception e) {
                        } finally { countDownLatch.countDown(); }        // 减少计数
                    }
                });
            }
            try {
                countDownLatch.await(timeoutMills, TimeUnit.MILLISECONDS); // 线程等待
            } catch (InterruptedException e) { }
        }
        return registerBrokerResultList;                                 // 返回注册结果集合
    }
}
```

通过此源代码可以清楚地发现，在 Broker 进行数据注册前需要配置 Broker 注册的头信息以及主体信息（编码为二进制数据）。RocketMQ 集群中会有多个 NameServer，为了保持 NameServer 数据的一致性，需要向每一个 NameServer 节点发送 Broker 数据。根据以上分析，可以得出图 8-30 所示的类结构模型。

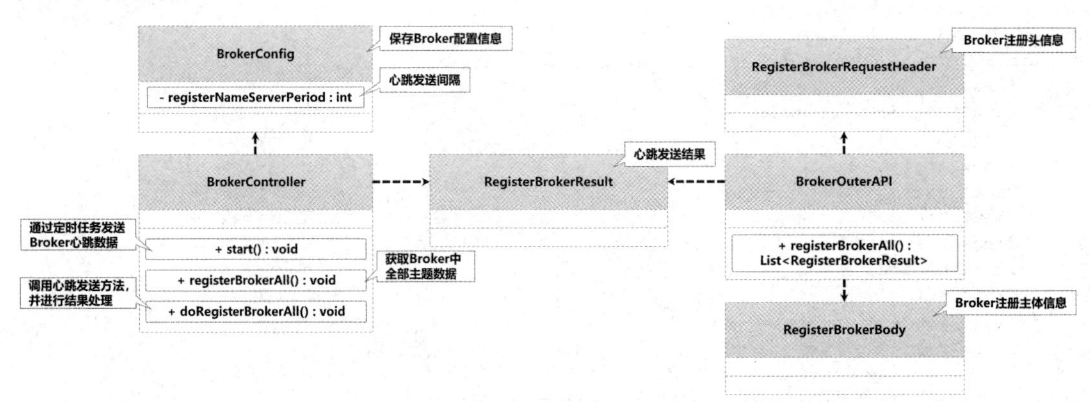

图 8-30　Broker 心跳发送类结构模型

（6）【namesrv 子模块】Broker 中的心跳数据被发送到 NameServer，而后由 RouteInfoManager.registerBroker()方法实现心跳数据的接收与解析。下面观察此方法的源代码。

```
package org.apache.rocketmq.namesrv.routeinfo;
public class RouteInfoManager {
   public RegisterBrokerResult registerBroker(                    // Broker注册
      final String clusterName,                                   // 集群名称
      final String brokerAddr,                                    // Broker地址
      final String brokerName,                                    // Broker名称
      final long brokerId,                                        // Broker ID
      final String haServerAddr,                                  // HA服务地址
      final TopicConfigSerializeWrapper topicConfigWrapper, // 主题数据
      final List<String> filterServerList,                        // 过滤配置
      final Channel channel) {                                    // 数据通道
      RegisterBrokerResult result = new RegisterBrokerResult(); // Broker注册结果
      try {
         try {
            this.lock.writeLock().lockInterruptibly();      // 写锁
            Set<String> brokerNames = this.clusterAddrTable
                  .get(clusterName);                              // 获取Broker名称
            if (null == brokerNames) {                            // 未获取Broker名称
               brokerNames = new HashSet<String>();         // 开辟新集合
               this.clusterAddrTable.put(clusterName, brokerNames); // 保存Broker名称
            }
            brokerNames.add(brokerName);                          // 保存Broker名称
            boolean registerFirst = false;                        // 第一个注册标记
            BrokerData brokerData = this.brokerAddrTable.get(brokerName); // 数据
            if (null == brokerData) {                             // 数据不为空
               registerFirst = true;                              // 修改注册标记
               brokerData = new BrokerData(clusterName, brokerName,
                     new HashMap<Long, String>());          // 保存Broker数据
               this.brokerAddrTable.put(brokerName, brokerData); // 保存Broker地址信息
            }
            Map<Long, String> brokerAddrsMap = brokerData.getBrokerAddrs();
            // 切换Slave节点到Master节点，首先通过NameServer删除第二个配置项（<1, IP:PORT>）
            // 随后在集合首位增加一个新的配置项（<0, IP:PORT>），相同“IP:PORT”只能保留一条记录
            Iterator<Entry<Long, String>> it = brokerAddrsMap.entrySet().iterator();
            while (it.hasNext()) {                                // 数据迭代
               Entry<Long, String> item = it.next();        // 获取地址项
               if (null != brokerAddr && brokerAddr.equals(item.getValue())
                     && brokerId != item.getKey()) {        // Broker地址重复
                  it.remove();                                    // 数据删除
               }
            }
            // 根据Broker ID保存新的Broker地址，并返回原始对应的Broker地址
            String oldAddr = brokerData.getBrokerAddrs().put(brokerId, brokerAddr);
            registerFirst = registerFirst || (null == oldAddr); // 修改标记
            if (null != topicConfigWrapper && MixAll.MASTER_ID == brokerId) {
               if (this.isBrokerTopicConfigChanged(brokerAddr,
                     topicConfigWrapper.getDataVersion()) || registerFirst) {
                  ConcurrentMap<String, TopicConfig> tcTable = topicConfigWrapper
                     .getTopicConfigTable();                      // 获取主题数据
                  if (tcTable != null) {
                     for (Map.Entry<String, TopicConfig> entry : tcTable.entrySet()) {
                        this.createAndUpdateQueueData(brokerName,
                              entry.getValue());                  // 创建或更新队列
                     }
                  }
               }
            }
            BrokerLiveInfo prevBrokerLiveInfo = this.brokerLiveTable.put(brokerAddr,
               new BrokerLiveInfo(                                // Broker存活信息
```

```
                System.currentTimeMillis(),             // 时间戳
                topicConfigWrapper.getDataVersion(),    // 数据版本
                channel,                                // NIO通道
                haServerAddr));                         // HA服务地址
            // 后续代码，略
        } finally { this.lock.writeLock().unlock(); }
    } catch (Exception e) { }
    return result;
  }
}
```

（7）【namesrv 子模块】除了 Broker 注册处理之外，在 NameServer 中还需要对存活的 Broker 进行定时检测，而这个检测任务的处理是由 NamesrvControler()构造方法定义的。

```
package org.apache.rocketmq.namesrv;
public class NamesrvController {
  public NamesrvController(NamesrvConfig namesrvConfig,
          NettyServerConfig nettyServerConfig) {
    // 此处只列出部分核心代码，其他代码略
    this.scheduledExecutorService.scheduleAtFixedRate(new Runnable() { // 定时任务
      @Override
      public void run() {                          // 扫描存活的Broker
        NamesrvController.this.routeInfoManager.scanNotActiveBroker();
      }
    }, 5, 10, TimeUnit.SECONDS);                   // 每10秒检查一次
    return true;
  }
}
```

（8）【namesrv 子模块】在存活检查时不活跃的 Broker 会被删除，而这是由 RouteInfoManager. scanNotActiveBroker()方法完成的。下面打开此方法的源代码并观察。

```
package org.apache.rocketmq.namesrv.routeinfo;
public class RouteInfoManager {
  public void scanNotActiveBroker() {                         // 扫描不存活的Broker
    Iterator<Entry<String, BrokerLiveInfo>> it = this.brokerLiveTable
          .entrySet().iterator();                             // 存活Broker集合迭代
    while (it.hasNext()) {                                    // Broker迭代
      Entry<String, BrokerLiveInfo> next = it.next();         // 获取Broker存活信息
      long last = next.getValue().getLastUpdateTimestamp();  // 获取最后一次更新时间戳
      if ((last + BROKER_CHANNEL_EXPIRED_TIME) <
            System.currentTimeMillis()) {                     // 当前Broker已经失效
        RemotingUtil.closeChannel(next.getValue().getChannel()); // 通道关闭
        it.remove();                                          // 删除Broker
        this.onChannelDestroy(next.getKey(), next.getValue().getChannel());
      }
    }
  }
}
```

Broker 的存活检查是基于对每一个 Broker 的最后一次心跳时间戳的判断实现的，Broker 失效后会从存活列表中被删除，Broker 对应的通道被关闭。

8.2.4　数据存储

数据存储

视频名称　0811_【掌握】数据存储

视频简介　RocketMQ 需要对消息数据进行存储。本视频为读者分析 Broker 中数据存储目录的作用，同时分析 CommitLog 与 Queue 之间的消费处理流程。

每当生产者进行消息生产时，所有的消息数据都会保存在 Broker 节点之中，而为了便于消息

数据的存储，往往会通过 broker.conf 文件定义消息的存储目录。这样一旦服务启动，开发者往往会见到图 8-31 所示的目录结构，该目录中一般会有 3 个重要的配置文件目录。

- commitlog：提交日志目录，所有生产者生产的消息数据全部保存在该目录之中，该目录中的文件会随着消息量的增加而增多。考虑到性能问题，每一个数据存储文件只有 1GB 大小。同时为便于数据操作，commitlog 目录中的文件名称采用“上一文件名称 + 物理偏移量（Offset）”的规则进行创建，这样就可以在消费时根据当前 commitLogOffset 来确定要使用的数据文件名称。
- config：配置保存目录，用户所创建的主题、消费组等信息都在此目录中存储。
- index：保存消息索引，此索引可以实现 MessageKey 的数据查询。

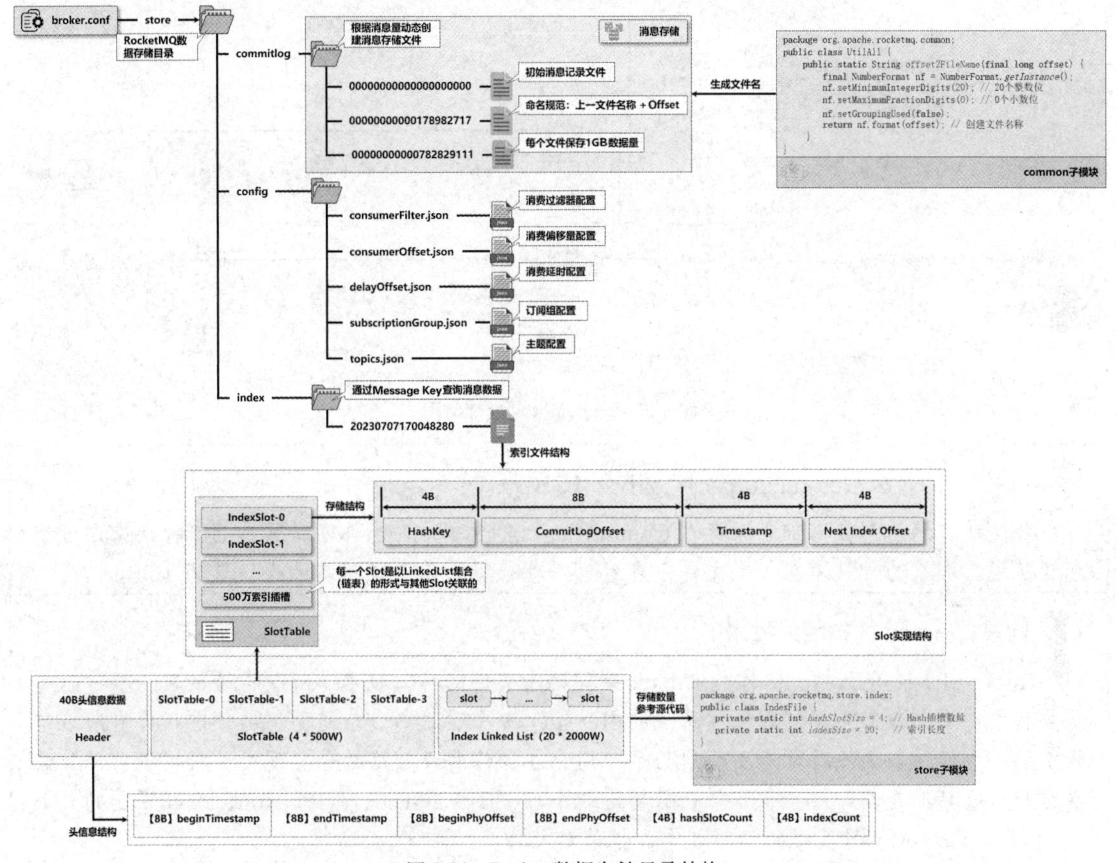

图 8-31 Broker 数据存储目录结构

提示：IndexFile 文件的补充说明。

IndexFile 的主要目的是根据 MessageKey 实现数据的查询，而为了便于获取消息内容，在每次存储时都会保存对应的 CommitLogOffset，这样就便于定位 CommitLog 文件中的数据内容。每一个 IndexFile 本质上都是一个链表的集合，一共可以保存 2000 万个数据索引（4 个 SlotTable）。

为了便于数据的快速查找，每一个索引文件都会包含头信息，该信息的组成结构如下。

① beginTimestamp：long 数据类型（长度为 8 位），第一个索引的消息保存在 Broker 的时间戳。

② endTimestamp：long 数据类型（长度为 8 位），最后一个索引的消息保存在 Broker 时间戳。

③ beginPhyOffset：long 数据类型（长度为 8 位），第一个索引的消息 commitLog 偏移量。

④ endPhyOffset：long 数据类型（长度为 8 位），最后一个索引的消息 commitLog 偏移量。

⑤ hashSlotCount：int 数据类型（长度为 4 位），占用的插槽位数。

⑥ indexCount：int 数据类型（长度为 4 位），构建的索引个数。

本次演示的重点在于 CommitLog 文件的作用解析，至于 IndexFile 具体操作流程以及其他细节，有兴趣的读者可以自行参考 store 子模块的源代码。

RocketMQ 的生产者进行数据生产时，为了实现与消费者之间的隔离，会将所有生产的消息写入 CommitLog（消息存储日志数据）文件，而且若干个主题的数据会写入同一个 CommitLog 文件。在进行最终消费处理时，需要通过特定的处理操作将每一个已经写入的消息转为消息队列，这样消费者就可以实现消费。为了防止重复消费，在每次消费时还会通过一个 consumerOffset 记录每一个消息队列中的当前消费位置。其处理结构如图 8-32 所示。

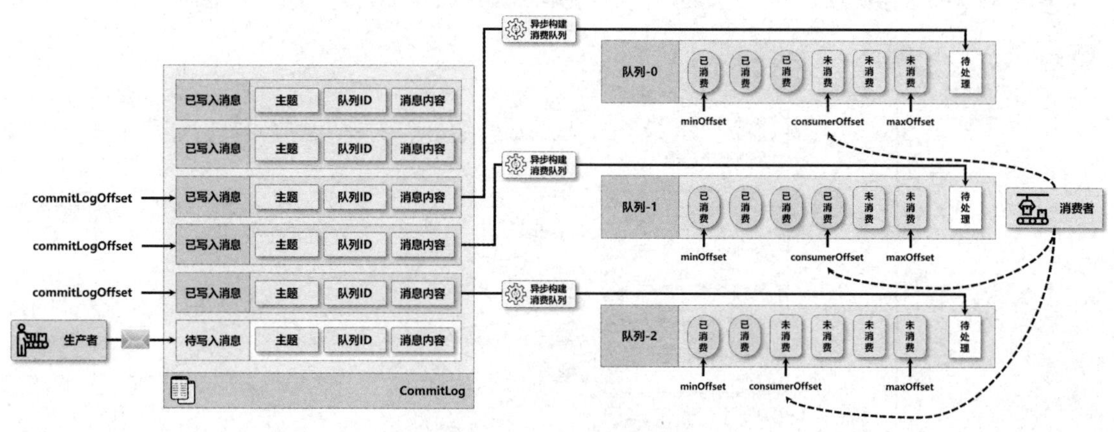

图 8-32 CommitLog 处理结构

提问：为什么 CommitLog 文件大小为 1GB？

既然所有的消息数据都会保存在 CommitLog 文件中，并且也都可以通过 Offset 快速定位消息的位置，那么为什么不在一个文件中多存储一些数据，而仅仅将其大小设置为 1GB 呢？

回答：Mmap 内存映射技术。

RocketMQ 使用了 Mmap 作为 Broker 读写磁盘文件的核心技术，所以在进行数据操作时不需要将文件中的数据先复制到操作系统的内核 I/O 缓冲区之中，只需要将用户进程私有地址空间中的一块区域与文件对象建立映射关系。用户在操作内存数据时会实现文件数据操作，只需要进行一次数据复制即可。对于容量较大的文件（一般在 2GB 以内，而 RocketMQ 限制的大小为 1GB），采用 Mmap 实现的读写效率和性能都非常高。

Java 提供的 NIO 开发包提供了 MappedByteBuffer 类，结合 FileChannel 提供的 map()方法即可实现映射关系的建立，如图 8-33 所示。该内存不属于 JVM 内存，所以不受 GC（Garbage Cllection，垃圾回收）机制的控制，但会受操作系统虚拟内存大小的限制。

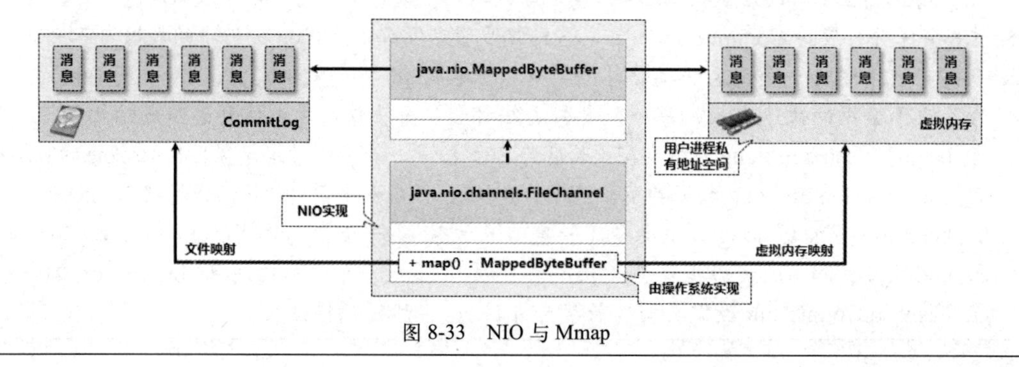

图 8-33 NIO 与 Mmap

在 RocketMQ 设计中，每一个 Broker 都会有一个对应的 MessageStore 对象实例，通过此对象实例可实现消息的存储。而每一个 MessageStore 中都会保存一个 CommitLog 对象实例，并由 CommitLog 维护一个 MappedFileQueue（映射文件队列）。每一个队列中会维护多个 MappedFile（映射到文件系统中的数据存储文件），其结构如图 8-34 所示。

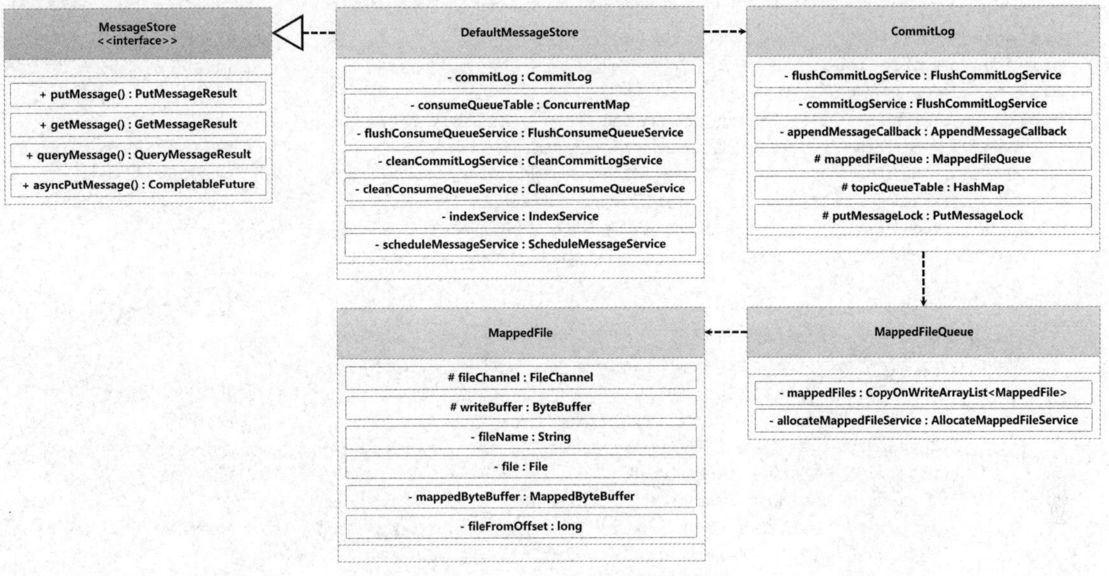

图 8-34　CommitLog 源代码实现类结构

MessageStore 接口定义了消息存储与读取的处理方法的标准，而后具体的实现由 DefaultMessageStore 子类来完成。为了便于读者理解消息的存储操作，下面对 DefaultMessageStore 类中 putMessage() 方法的源代码进行分析。

范例：【store 子模块】DefaultMessageStore 源代码

```
package org.apache.rocketmq.store;
public class DefaultMessageStore implements MessageStore {
   @Override
   public PutMessageResult putMessage(MessageExtBrokerInner msg) {      // 消息存放
      PutMessageStatus checkStoreStatus = this.checkStoreStatus();      // 消息存放状态
      if (checkStoreStatus != PutMessageStatus.PUT_OK) {                // 错误状态
         return new PutMessageResult(checkStoreStatus, null);           // 返回错误消息
      }
      PutMessageStatus msgCheckStatus = this.checkMessage(msg);         // 消息检查
      if (msgCheckStatus == PutMessageStatus.MESSAGE_ILLEGAL) {         // 消息出错
         return new PutMessageResult(msgCheckStatus, null);             // 返回错误消息
      }
      long beginTime = this.getSystemClock().now();                     // 开始时间戳
      PutMessageResult result = this.commitLog.putMessage(msg);         // 由CommitLog类处理
      long elapsedTime = this.getSystemClock().now() - beginTime;       // 耗时统计
      if (elapsedTime > 500) {                                          // 时间超过500毫秒
         log.warn("not in lock …", elapsedTime, msg.getBody().length); // 警告日志
      }
      this.storeStatsService.setPutMessageEntireTimeMax(elapsedTime);   // 耗时记录
      if (null == result || !result.isOk()) {                           // 操作失败
         this.storeStatsService.getPutMessageFailedTimes()
               .incrementAndGet();                                      // 失败次数增长
      }
      return result;
   }
}
```

在 DefaultMessageStore 子类中实现的 putMessage()方法会首先对消息状态进行检查，而后通过

CommitLog 对象实例实现数据的保存，如果发现保存的时间过长（超过 500 毫秒）则会进行警告日志的记录。下面再重点来观察 CommitLog 类中 putMessage()方法的源代码。

范例：【store 子模块】CommitLog 类源代码

```
package org.apache.rocketmq.store;
public class CommitLog {
    protected final PutMessageLock putMessageLock;                          // 存储锁
    private volatile long beginTimeInLock = 0;                              // 锁标记
    private final AppendMessageCallback appendMessageCallback;              // 保存回调
    public PutMessageResult putMessage(final MessageExtBrokerInner msg) { // 消息存储
        msg.setStoreTimestamp(System.currentTimeMillis());                  // 消息存储时间戳
        msg.setBodyCRC(UtilAll.crc32(msg.getBody()));                       // CRC校验
        AppendMessageResult result = null;                                  // 消息追加结果
        // 获取StoreStatsService对象实例，以实现消息统计信息的处理
        StoreStatsService storeStatsService = this.defaultMessageStore
                .getStoreStatsService();
        String topic = msg.getTopic();                                      // 获取消息主题
        int queueId = msg.getQueueId();                                     // 获取队列ID
        final int tranType = MessageSysFlag.getTransactionValue(
                msg.getSysFlag());                                          // 获取事务类型
        if (tranType == MessageSysFlag.TRANSACTION_NOT_TYPE                 // 没有类型
                || tranType == MessageSysFlag.TRANSACTION_COMMIT_TYPE) {    // 事务提交类型
            if (msg.getDelayTimeLevel() > 0) {                              // 延迟交货
                if (msg.getDelayTimeLevel() > this.defaultMessageStore
                    .getScheduleMessageService().getMaxDelayLevel()) {      // 延迟判断
                    msg.setDelayTimeLevel(this.defaultMessageStore
                    .getScheduleMessageService().getMaxDelayLevel());
                }
                topic = TopicValidator.RMQ_SYS_SCHEDULE_TOPIC;              // 获得系统任务主题
                queueId = ScheduleMessageService.delayLevel2QueueId(
                    msg.getDelayTimeLevel());                               // 获取指定队列
                MessageAccessor.putProperty(msg, MessageConst.PROPERTY_REAL_TOPIC,
                    msg.getTopic());                                        // 备用主题
                MessageAccessor.putProperty(msg, MessageConst.PROPERTY_REAL_QUEUE_ID,
                    String.valueOf(msg.getQueueId()));                      // 备用队列
                msg.setPropertiesString(MessageDecoder.messageProperties2String(
                    msg.getProperties()));                                  // 属性配置
                msg.setTopic(topic);                                        // 保存主题
                msg.setQueueId(queueId);                                    // 保存队列
            }
        }
        InetSocketAddress bornSocketAddress = (InetSocketAddress) msg.getBornHost();
        if (bornSocketAddress.getAddress() instanceof Inet6Address) {
            msg.setBornHostV6Flag();                                        // 发送地址
        }
        InetSocketAddress storeSocketAddress = (InetSocketAddress) msg.getStoreHost();
        if (storeSocketAddress.getAddress() instanceof Inet6Address) {
            msg.setStoreHostAddressV6Flag();                                // 存储地址
        }
        long elapsedTimeInLock = 0;                                         // 保存锁定耗时
        MappedFile unlockMappedFile = null;                                 // 文件锁定
        MappedFile mappedFile = this.mappedFileQueue.getLastMappedFile();// 获取MappedFile
        putMessageLock.lock();                                              // 类型由依赖配置决定
        try {      // 写入操作需要进行锁定，为了便于统计锁定耗时，首先获取锁定前的时间戳
            long beginLockTimestamp = this.defaultMessageStore.getSystemClock().now();
            this.beginTimeInLock = beginLockTimestamp;                      // 时间戳保存
            msg.setStoreTimestamp(beginLockTimestamp);                      // 全局有序存储
            if (null == mappedFile || mappedFile.isFull()) {                // CommitLog文件已满
                mappedFile = this.mappedFileQueue.getLastMappedFile(0);     // 创建新映射文件
            }
            if (null == mappedFile) {                                       // 对象为空
                log.error("create mapped file1 error ...");                 // 错误日志
```

```
            beginTimeInLock = 0;                                    // 修改锁标记
            return new PutMessageResult(
                PutMessageStatus.CREATE_MAPEDFILE_FAILED, null);
        }
        result = mappedFile.appendMessage(msg, this.appendMessageCallback); // 消息保存
        switch (result.getStatus()) {                               // 结果判断
            case PUT_OK:                                            // 保存成功
                break;                                              // 结束判断
            case END_OF_FILE:                                       // 文件已满
                unlockMappedFile = mappedFile;                      // 锁定当前文件
                // 创建新文件并进行消息数据重新写入
                mappedFile = this.mappedFileQueue.getLastMappedFile(0);
                if (null == mappedFile) {                           // 文件创建失败
                    log.error("create mapped ... ");                // 错误日志
                    beginTimeInLock = 0;                            // 修改锁定标记
                    return new PutMessageResult(PutMessageStatus
                            .CREATE_MAPEDFILE_FAILED, result);      // 消息保存失败
                }
                result = mappedFile.appendMessage(msg, this.appendMessageCallback);
                break;
            case MESSAGE_SIZE_EXCEEDED:                             // 文件过大
            case PROPERTIES_SIZE_EXCEEDED:                          // 属性过多
                beginTimeInLock = 0;                                // 修改锁定标记
                return new PutMessageResult(PutMessageStatus.MESSAGE_ILLEGAL, result);
            case UNKNOWN_ERROR:                                     // 未知错误
                beginTimeInLock = 0;                                // 修改锁定标记
                return new PutMessageResult(PutMessageStatus.UNKNOWN_ERROR, result);
            default:
                beginTimeInLock = 0;                                // 修改锁定标记
                return new PutMessageResult(PutMessageStatus.UNKNOWN_ERROR, result);
        }
        elapsedTimeInLock = this.defaultMessageStore.getSystemClock().now() -
                beginLockTimestamp;                                 // 耗时统计
        beginTimeInLock = 0;                                        // 修改锁定标记
    } finally { putMessageLock.unlock(); }                          // 解除锁定
    if (elapsedTimeInLock > 500) {                                  // 超时警告
        log.warn("[NOTIFYME]putMessage in lock ...");               // 警告日志
    }
    if (null != unlockMappedFile && this.defaultMessageStore.getMessageStoreConfig()
            .isWarmMapedFileEnable()) {
        this.defaultMessageStore.unlockMappedFile(unlockMappedFile);// 文件解锁
    }
    PutMessageResult putMessageResult = new PutMessageResult(
            PutMessageStatus.PUT_OK, result);                       // 保存成功
    storeStatsService.getSinglePutMessageTopicTimesTotal(
            msg.getTopic()).incrementAndGet();                      // 数据统计
    storeStatsService.getSinglePutMessageTopicSizeTotal(topic)
            .addAndGet(result.getWroteBytes());                     // 数据统计
    handleDiskFlush(result, putMessageResult, msg);                 // 数据刷盘
    handleHA(result, putMessageResult, msg);                        // HA处理
    return putMessageResult;                                        // 返回保存结果
  }
}
```

通过源代码可以发现，所有的消息在进行存储时，会自动保存当前的时间戳。如果是普通消息（非二阶段事务消息），则会根据配置的主题和队列进行消息的保存。如果是二阶段消息，则先使用一个系统主题进行保存，而普通消息会通过 MappedFile 类提供的 appendMessage()方法进行保存。下面观察此类的源代码。

范例：【store 子模块】MappedFile 类源代码

```
package org.apache.rocketmq.store;
public class MappedFile extends ReferenceResource {
```

```
    protected final AtomicInteger wrotePosition = new AtomicInteger(0); // 写入位置
    protected int fileSize;                                          // 文件大小
    public AppendMessageResult appendMessagesInner(final MessageExt messageExt,
                final AppendMessageCallback cb) {
        assert messageExt != null;                                   // 消息不为空
        assert cb != null;                                           // 有Callback实例
        int currentPos = this.wrotePosition.get();                   // 获取写入位置
        if (currentPos < this.fileSize) {                            // 未超过文件大小
            ByteBuffer byteBuffer = writeBuffer != null ? writeBuffer.slice() :
                    this.mappedByteBuffer.slice();
            byteBuffer.position(currentPos);                         // 写入位置
            AppendMessageResult result;                              // 写入结果
            // 最终的消息写入需要采用二进制的方式处理，而具体由AppendMessageCallback子类实现
            // 该接口子类为CommitLogFile.DefaultAppendMessageCallback
            if (messageExt instanceof MessageExtBrokerInner) {       // 消息类型判断
                result = cb.doAppend(this.getFileFromOffset(), byteBuffer,
                    this.fileSize - currentPos, (MessageExtBrokerInner) messageExt);
            } else if (messageExt instanceof MessageExtBatch) {      // 消息类型判断
                result = cb.doAppend(this.getFileFromOffset(), byteBuffer,
                    this.fileSize - currentPos, (MessageExtBatch) messageExt);
            } else {
                return new AppendMessageResult(AppendMessageStatus.UNKNOWN_ERROR);
            }
            this.wrotePosition.addAndGet(result.getWroteBytes()); // 数据写入
            this.storeTimestamp = result.getStoreTimestamp();     // 存储时间戳
            return result;                                           // 返回追加结果
        }
        log.error("MappedFile.appendMessage ...");                   // 错误日志
        return new AppendMessageResult(AppendMessageStatus.UNKNOWN_ERROR);
    }
}
```

通过此源代码可以发现，所有的 MappedFile 类内部都会维持一个写入消息位置的属性。这样就可以依据此位置进行数据的顺序式写入，而写入的具体实现是由 AppendMessageCallback 接口所提供的 doAppend()方法完成的。该方法会依据 NIO 提供的 ByteBuffer 类将消息对象中的数据写入 CommitLog 文件。类关联结构如图 8-35 所示。

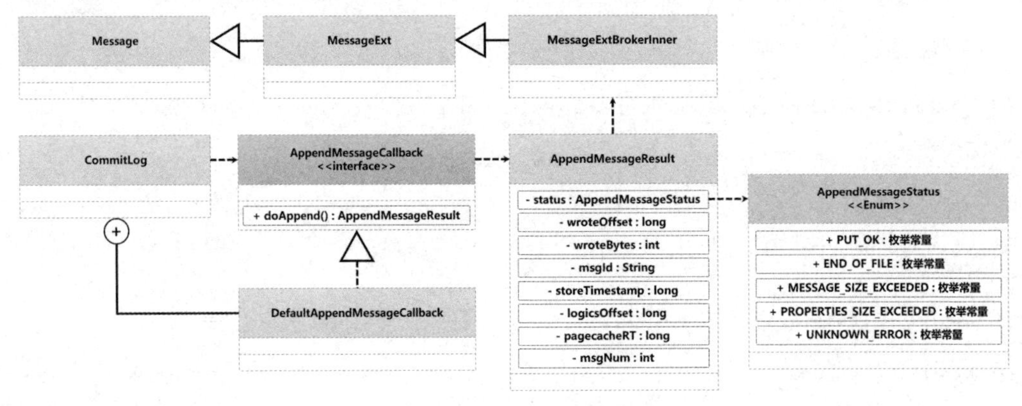

图 8-35　数据写入实现类关联结构

8.2.5　数据刷盘

视频名称　0812_【掌握】数据刷盘

视频简介　数据刷盘是 RocketMQ 提供的重要存储性能优化的实现。本视频为读者讲解数据刷盘的两种处理形式，同时分析 PageCache 以及核心源代码。

生产者发送到 Broker 中的数据会首先保存在 Broker 主机的 CommitLog 文件之中，随后才会交给消息队列进行消费处理。而 Broker 在进行消息接收时，会首先将接收的消息内容保存在内存之中，而后通过内存向磁盘写入数据，如图 8-36 所示，这样才可以实现最终的消息刷盘处理。

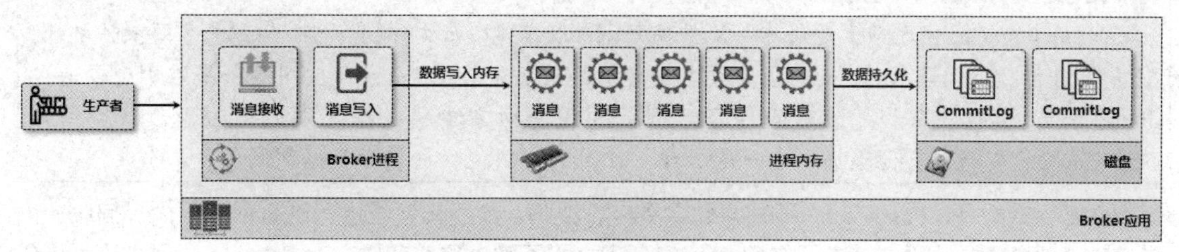

图 8-36　Broker 消息数据持久化

为了保证消息组件中的数据在 Broker 死机时不会出现全部丢失的问题，最终的消息内容肯定要保存在磁盘之中。而为了实现这样的数据存储，并在 RocketMQ 中实现数据刷盘操作，可采用“同步刷盘”与“异步刷盘”两种机制。

（1）同步刷盘

在接收消息并将消息持久化到磁盘后，Broker 才会发送一个应答响应（ACK）给生产者，如图 8-37 所示。同步刷盘是一种可靠性较高的形式，但是会有性能问题，一般可以应用在一些严谨的业务应用领域（如金融、保险等可靠性要求较高的领域）。

（2）异步刷盘

异步刷盘主要利用了操作系统提供的 PageCache（页面缓存）的优势，只要将接收的消息写入 PageCache 即可向生产端返回应答响应，最终由操作系统负责将 PageCache 中的数据写入磁盘文件，如图 8-38 所示。此种方式减少了读写延迟，提高了消息组件的性能和吞吐量，并且不会造成主线程的阻塞。

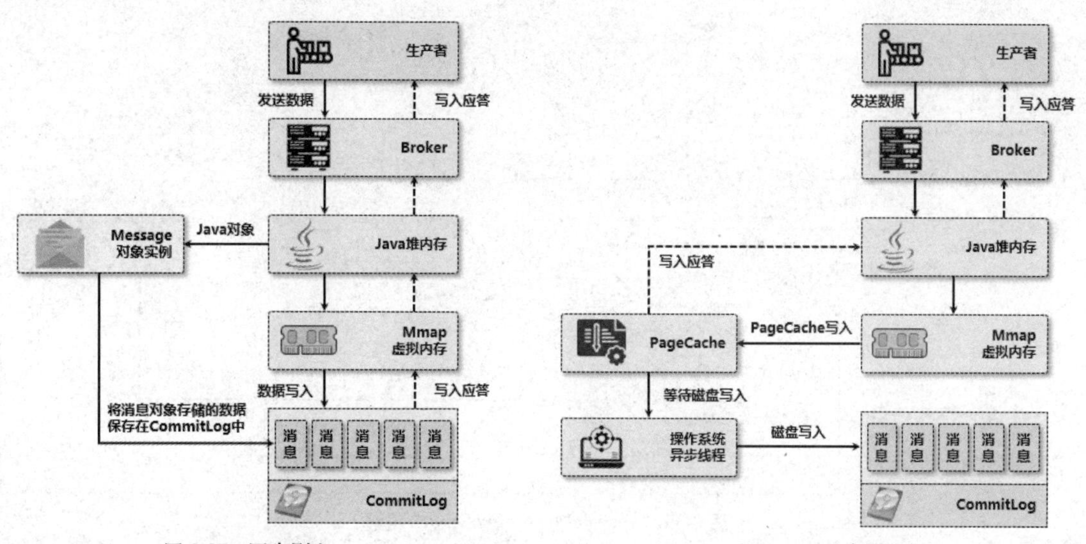

图 8-37　同步刷盘

图 8-38　异步刷盘

> **提示：PageCache 机制。**
>
> PageCache 机制是指操作系统对文件缓存的支持机制，操作系统会将一部分内存作为 PageCache，主要用于提高文件的读写性能，使得程序对文件的顺序读写速度接近于内存的读写速度。

RocketMQ 在进行异步刷盘操作时，将 Mmap 映射到了操作系统的 PageCache，这样在进行刷盘操作时就只需要将数据复制到磁盘文件之中。这样的机制既提高了主线程的处理性能，又提升了I/O 的处理性能（避免多次数据复制所带来的性能损耗）。

提示：PageCache 数据读取操作。

利用 PageCache 之前需要进行一次数据复制，以实现磁盘文件的存储。但是除了写操作之外，还需要进行数据的读操作。此时首先要判断读取的数据是否在 PageCache 中，如果在则可以通过 PageCache 读取；而如果此时的 PageCache 中没有所需要的数据，则通过磁盘加载数据到 PageCache，邻近的其他数据块也一起加载，此时的加载过程也属于数据复制。

需要注意的是，PageCache 机制并非完全没有缺陷，当操作系统进行脏页回写、内存回收、内存 SWAP 操作时，它会带来较大的消息读写延迟。为了解决这些问题，RocketMQ 采用了多种优化技术，如内存预分配、mlock 系统调用、文件预热等。下面分别来看这些机制的实现特点。

1. 内存预分配

消息的写入处理需要通过 CommitLog.putMessage()方法来完成，而在使用此方法时，CommitLog 类会先通过消息文件队列（MappedFileQueue）获取一个映射文件（MappedFile），获取的 MappedFile 对象实例就是由一个后台的分配映射文件服务 AllocateMappedFileService 线程提前准备好的，如图 8-39 所示。Broker 服务启动后，它就会通过 AllocateMappedFileService.run()方法不断运行，只要队列存在请求，就会分配请求（AllocateRequest）对象，进行 MappedFile 映射文件的创建和预分配工作。这一点可以通过 AllocateMappedFileService.mmapOperation()方法的源代码观察到。

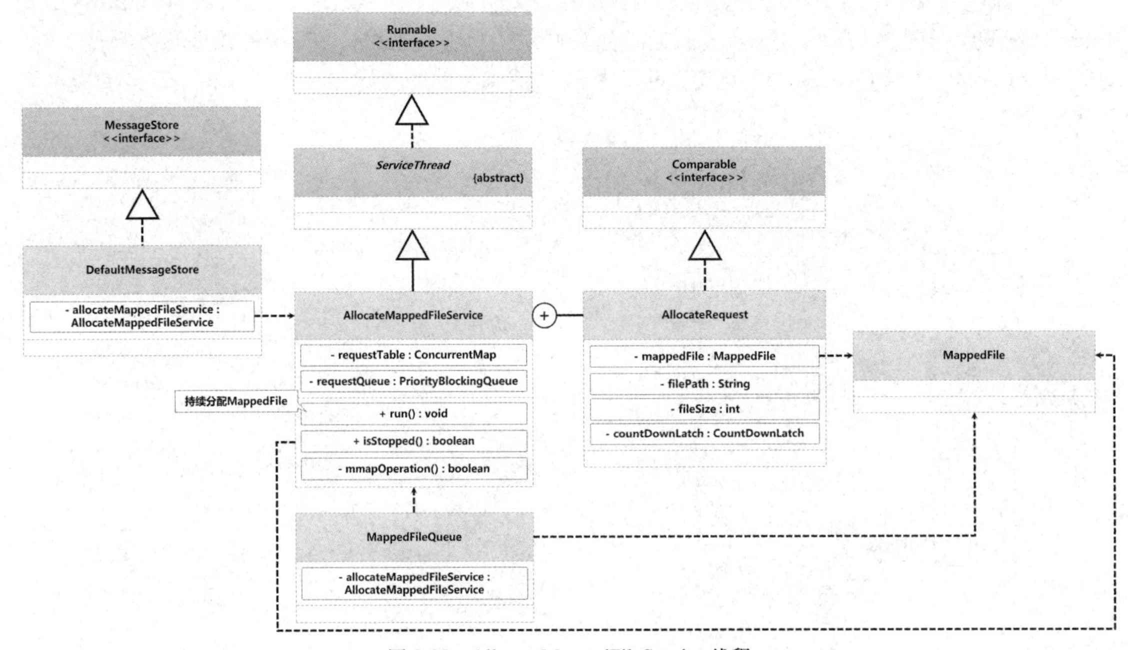

图 8-39　AllocateMappedFileService 线程

范例：【store 子模块】AllocateMappedFileService 类源代码

```
package org.apache.rocketmq.store;
public class AllocateMappedFileService extends ServiceThread {
   private static int waitTimeOut = 1000 * 5;                       // 分配超时
   private ConcurrentMap<String, AllocateRequest> requestTable =
      new ConcurrentHashMap<String, AllocateRequest>();             // 分配表
   private PriorityBlockingQueue<AllocateRequest> requestQueue =
      new PriorityBlockingQueue<AllocateRequest>();                 // 分配队列
   private volatile boolean hasException = false;                   // 分配异常记录
```

```
// 处理请求并返回MappedFile对象实例，MappedFileQueue.getLastMappedFile()会调用此方法
public MappedFile putRequestAndReturnMappedFile(String nextFilePath,
      String nextNextFilePath, int fileSize) {
    int canSubmitRequests = 2;                                    // 分配请求数量
    if (this.messageStore.getMessageStoreConfig()
            .isTransientStorePoolEnable()) {                      // 配置判断
        if (this.messageStore.getMessageStoreConfig()
                .isFastFailIfNoBufferInStorePool() && BrokerRole.SLAVE !=
                this.messageStore.getMessageStoreConfig().getBrokerRole()) {
            canSubmitRequests = this.messageStore.getTransientStorePool()
                .availableBufferNums() - this.requestQueue.size();
        }
    }
    // 分配请求，保存下一个文件路径与文件大小
    AllocateRequest nextReq = new AllocateRequest(nextFilePath, fileSize); // 分配请求
    boolean nextPutOK = this.requestTable.putIfAbsent(
            nextFilePath, nextReq) == null;                       // 保存分配记录
    if (nextPutOK) {                                              // 集合保存成功
        if (canSubmitRequests <= 0) {
            log.warn("[NOTIFYME]TransientStorePool is not enough..."); // 警告信息
            this.requestTable.remove(nextFilePath);               // 删除分配请求
            return null;                                          // 返回null
        }
        boolean offerOK = this.requestQueue.offer(nextReq);       // 添加请求队列
        if (!offerOK) {  log.warn("never expected here..."); }    // 警告日志
        canSubmitRequests--;                                      // 减少请求个数
    }
    // 分配请求，保存下一个文件路径与文件大小
    AllocateRequest nextNextReq = new AllocateRequest(nextNextFilePath, fileSize);
    boolean nextNextPutOK = this.requestTable.putIfAbsent(
            nextNextFilePath, nextNextReq) == null;               // 保存分配记录
    if (nextNextPutOK) {                                          // 保存成功
        if (canSubmitRequests <= 0) {
            log.warn("[NOTIFYME]TransientStorePool is not enough..."); // 警告日志
            this.requestTable.remove(nextNextFilePath);           // 删除分配请求
        } else {
            boolean offerOK = this.requestQueue.offer(nextNextReq); // 添加请求队列
            if (!offerOK) { log.warn("never expected here..."); } // 警告日志
        }
    }
    if (hasException) {                                           // 产生异常
        log.warn("service has exception…");                       // 警告日志
        return null;                                              // 返回null
    }
    AllocateRequest result = this.requestTable.get(nextFilePath); // 从集合获取分配请求
    try {
        if (result != null) {                                     // 有请求对象
            boolean waitOK = result.getCountDownLatch().await(waitTimeOut,
                TimeUnit.MILLISECONDS);
            if (!waitOK) {                                        // 分配失败
                log.warn("create mmap timeout ...");              // 警告日志
                return null;                                      // 返回null
            } else {                                              // 分配成功
                this.requestTable.remove(nextFilePath);           // 删除请求
                return result.getMappedFile();                    // 返回MappedFile
            }
        } else {
            log.error("find preallocate mmap failed...");         // 警告日志
        }
    } catch (InterruptedException e) {}
    return null;                                                  // 返回null
}
public void run() {
```

```
        while (!this.isStopped() && this.mmapOperation()) {}          // 不停分配
    }
    private boolean mmapOperation() {                                  // 不停分配
        boolean isSuccess = false;                                     // 成功标记
        AllocateRequest req = null;                                    // 分配请求
        try {
            req = this.requestQueue.take();                            // 获取分配请求
            AllocateRequest expectedRequest = this.requestTable.get(
                    req.getFilePath());                                // 获取分配数据
            if (null == expectedRequest) {                             // 请求不为空
                log.warn("this mmap request expired...");              // 警告日志
                return true;                                           // 下一轮分配
            }
            if (expectedRequest != req) {                              // 请求不匹配
                log.warn("never expected here...");                    // 警告日志
                return true;                                           // 下一轮分配
            }
            if (req.getMappedFile() == null) {                         // 没有MappedFile实例
                long beginTime = System.currentTimeMillis();           // 开始时间戳
                MappedFile mappedFile;                                 // MappedFile对象
                if (messageStore.getMessageStoreConfig()
                        .isTransientStorePoolEnable()) {               // 启用临时存储池
                    try {
                        mappedFile = ServiceLoader.load(MappedFile.class)
                            .iterator().next();                        // 获取MappedFile
                        // 映射文件初始化，传入分配的文件路径、文件大小
                        mappedFile.init(req.getFilePath(), req.getFileSize(),
                            messageStore.getTransientStorePool());
                    } catch (RuntimeException e) {                     // 出现异常
                        log.warn("Use default implementation.");
                        mappedFile = new MappedFile(req.getFilePath(), req.getFileSize(),
                            messageStore.getTransientStorePool());     // 实例化新对象
                    }
                } else {                                               // 未启用临时存储池
                    mappedFile = new MappedFile(req.getFilePath(),
                            req.getFileSize());                        // 实例化新对象
                }
                long elapsedTime = UtilAll.computeElapsedTimeMilliseconds(beginTime);
                if (elapsedTime > 10) {                                // 耗时判断
                    int queueSize = this.requestQueue.size();          // 获取队列长度
                    log.warn("create mappedFile spent ...");           // 警告日志
                }
                if (mappedFile.getFileSize() >= this.messageStore.getMessageStoreConfig()
                    .getMappedFileSizeCommitLog() &&
                    this.messageStore.getMessageStoreConfig().isWarmMapedFileEnable()) {
                    mappedFile.warmMappedFile(this.messageStore.
                            getMessageStoreConfig().getFlushDiskType(),
                        this.messageStore.getMessageStoreConfig()
                            .getFlushLeastPagesWhenWarmMapedFile());   // 映射文件预热
                }
                req.setMappedFile(mappedFile);                         // 保存MappedFile实例
                this.hasException = false;                             // 未发生异常
                isSuccess = true;                                      // 分配成功
            }
        } catch (InterruptedException e) {                             // 中断异常
            log.warn(this.getServiceName() + " interrupted, possibly by shutdown.");
            this.hasException = true;                                  // 发生异常
            return false;
        } catch (IOException e) {                                      // I/O异常
            log.warn(this.getServiceName() + " service has exception. ", e);
            this.hasException = true;                                  // 发生异常
            if (null != req) {
                requestQueue.offer(req);                               // 队列保存
                try { Thread.sleep(1); } catch (InterruptedException ignored) {}
```

```
            }
        } finally {
            if (req != null && isSuccess)
                req.getCountDownLatch().countDown();                    // 减少等待计数
        }
        return true;                                                    // 下一轮分配
    }
}
```

通过 AllocateMappedFileService 类的源代码可以发现，所有的分配请求都是通过 putRequestAndReturnMappedFile()方法接收的，而所接收的全部分配请求会保存在请求分配等待队列（requestQueue）与请求分配保存集合（requestTable）之中。这样 run()方法在处理时，就会针对队列中的分配请求进行 MappedFile 分配。如果此时 MappedFile 文件大小超过了预定的存储大小，则会进行 MappedFile 的预分配，下次再获取时就可以避免 MappedFile 创建分配所产生的时间延迟。

2. mlock（内存锁定）系统调用

将 Broker 进程所使用的空间锁定在物理内存之中，可以防止其被交换到 SWAP 空间，这样就可以尽可能多地使用物理内存，从而提升读写性能。

3. 文件预热

虽然可以通过 mlock 锁定物理内存，但是有可能其中的一些分页（虚拟内存系统用来存储逻辑地址和物理地址之间映射的数据结构）是写时复制的，这样就会出现内存碎片，所以就需要对每个内存页面写入一个假的数据（例如，预先写入一些随机数字到 Mmap，映射内存空间）。

除此之外，在通过 Mmap 进行内存映射之后，操作系统只是建立了虚拟内存地址与物理地址的映射表，实际上并不会加载任何数据信息到内存之中。当程序要进行数据访问时，操作系统会检查该数据是否已经在内存之中，如果不在，则发出一次缺页中断，而如果文件较大，有可能在出现较多的缺页中断后才会将数据加载到内存，所以 RocketMQ 在实现 Mmap 内存映射的同时会进行 madvise 系统调用（告诉系统内核将所映射的内存页面提升为大页面）。这样可以在 Mmap 内存映射后将对应文件尽可能多地预加载到内存之中，从而达到内存预热的效果，如图 8-40 所示。

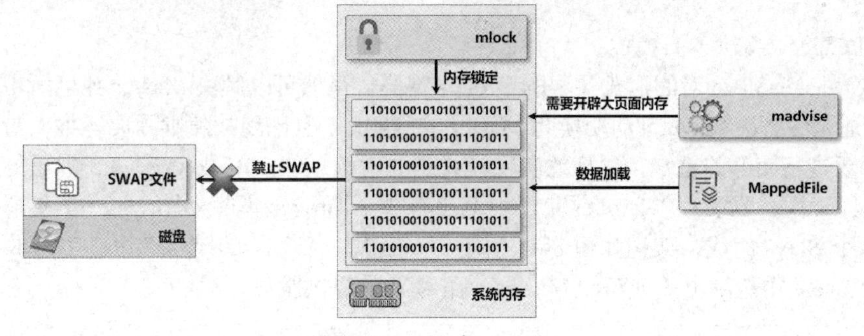

图 8-40　mlock 与 madvise

8.3　RocketMQ 集群服务

RocketMQ 集群服务概述

视频名称　0813_【掌握】RocketMQ 集群服务概述

视频简介　稳定与高效是消息组件最重要的保障，RocketMQ 可以基于集群服务的方式实现多 Broker 的节点配置。本视频为读者分析 Broker 集群的 4 种实现方式，同时给出本次 RocketMQ 集群服务搭建所需要的服务主机信息。

在实际的项目运行环境中，需要随时考虑到 RocketMQ 服务性能以及服务的稳定性，所以在生产环境中一般都会采用集群环境实现 RocketMQ 服务部署，即需要提供多个 NameServer 实例以及多个

Broker 的分片集群。由于 NameServer 采用独立的设计，因此在集群之中只要有一台 NameServer 提供服务即可实现路由信息的存储，而 Broker 集群模式较为烦琐，在 RocketMQ 中有 4 种 Broker 集群部署方案。

（1）多 Master 模式

每一个 Broker 分片为单一 Master 节点，在一个 Broker 集群中由若干个 Master 组成存储分片，如图 8-41 所示。采用此种集群设计方案时，服务运维较为方便，也可以得到较好的性能。但是在某一个 Broker 节点死机后，该节点中未被消费的消息在节点恢复前将无法被消费。

（2）多 Master-Slave 异步复制模式

在 Broker 分片节点中，每一个 Master 配置一个 Slave 集群对。在进行数据同步时，由 Master 向 Slave 发出一个异步的数据复制信号，如图 8-42 所示。此方案有短暂的数据延迟（毫秒级延迟）。如果其中一个 Master 出现问题，可以自动切换到 Slave 进行消费处理（Slave 无法自动切换为 Master），消息的可靠性较高，性能与方案（1）相同。

（3）多 Master-Slave 同步写入模式

每一个 Broker 分片节点中都存在一个 Master-Slave 对，在数据写入时会向 Master 和 Slave 同步写入，只有全部写成功后才返回写入成功的信息，如图 8-43 所示。这样可以保证数据的可用性，但是在 Master 节点死机后，Slave 无法自动切换为 Master。

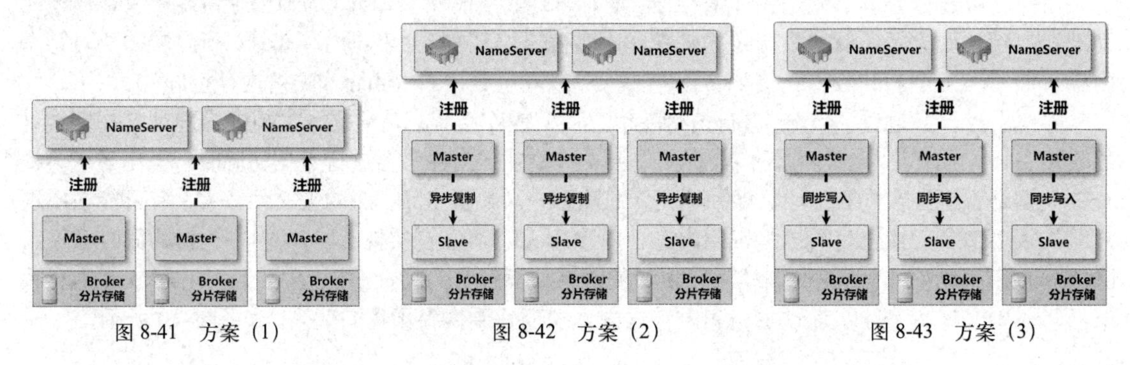

图 8-41 方案（1） 图 8-42 方案（2） 图 8-43 方案（3）

（4）DLedger（多副本）模式

利用多 Master-Slave 对的形式实现的 Broker 集群，虽然可以解决 Master 死机所带来的消息消费问题，但是无法解决 Slave 自动切换为 Master 的问题。一旦出现故障则需要运维人员进行手动切换，并且需要进行集群的重启，这样的实现模式是非常烦琐的。RocketMQ 4.5 以后的版本提供了 DLedger 集群架构，要求一个 Broker 集群中至少有 3 个 Broker 集群分片，每一个集群分片由一个 Master 和两个 Slave 组成，架构如图 8-44 所示。一旦某一个 Master 节点死机，DLedger 会自动从剩下的两个 Slave 中选举出一个新的 Master 来继续对外提供服务。

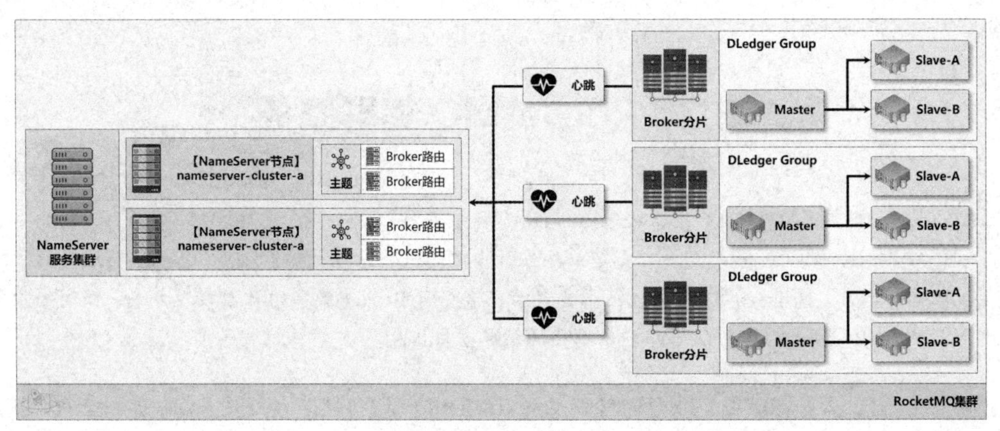

图 8-44 方案（4）

考虑到实际生产环境下的 RocketMQ 应用，本次将采用方案（4）实现服务搭建；再结合 NameServer 节点的高可用，本次将通过 11 台服务主机来实现 RocketMQ 集群搭建，这些主机的作用如表 8-11 所示。

表 8-11 RocketMQ 集群主机

序号	主机名称	IP 地址	描述
01	nameserver-cluster-a	192.168.190.176	独立的 NameServer 进程，保存 Broker 路由信息
02	nameserver-cluster-b	192.168.190.177	独立的 NameServer 进程，保存 Broker 路由信息
03	broker-cluster-a-node-1	192.168.190.180	Broker 进程，某一分片的 Master 节点
04	broker-cluster-a-node-2	192.168.190.181	Broker 进程，某一分片的 Slave 节点
05	broker-cluster-a-node-3	192.168.190.182	Broker 进程，某一分片的 Slave 节点
06	broker-cluster-b-node-1	192.168.190.183	Broker 进程，某一分片的 Master 节点
07	broker-cluster-b-node-2	192.168.190.184	Broker 进程，某一分片的 Slave 节点
08	broker-cluster-b-node-3	192.168.190.185	Broker 进程，某一分片的 Slave 节点
09	broker-cluster-c-node-1	192.168.190.186	Broker 进程，某一分片的 Master 节点
10	broker-cluster-c-node-2	192.168.190.187	Broker 进程，某一分片的 Slave 节点
11	broker-cluster-c-node-3	192.168.190.188	Broker 进程，某一分片的 Slave 节点

为便于服务的配置连接，建议修改 RocketMQ 集群中各个主机节点（虚拟机主机以及开发者主机）的 hosts 文件，添加本次所要使用的全部主机项，这样就可以通过主机名称实现服务调用。

范例：修改 hosts 配置文件

打开主机配置文件

```
vi /etc/hosts
```

添加主机配置列表

```
192.168.190.176 nameserver-cluster-a
192.168.190.177 nameserver-cluster-b
192.168.190.180 broker-cluster-a-node-1
192.168.190.181 broker-cluster-a-node-2
192.168.190.182 broker-cluster-a-node-3
192.168.190.183 broker-cluster-b-node-1
192.168.190.184 broker-cluster-b-node-2
192.168.190.185 broker-cluster-b-node-3
192.168.190.186 broker-cluster-c-node-1
192.168.190.187 broker-cluster-c-node-2
192.168.190.188 broker-cluster-c-node-3
```

8.3.1 NameServer 集群

NameServer 集群

视频名称 0814_【掌握】NameServer 集群

视频简介 NameServer 提供了所有服务的映射数据项存储，是 RocketMQ 中最为重要的服务组件。本视频将介绍通过虚拟机实现 NameServer 集群搭建。

NameServer 是 RocketMQ 之中的核心组件，其完整地保留了每一个主题对应的 Broker 路由信息。在 RocketMQ 中为了防止单一 NameServer 死机所带来的服务集群瘫痪，一般会配置若干个 NameServer 节点。考虑到路由数据存储的性能问题，每一个 NameServer 彼此独立，并且会保存相同的路由数据。客户端在连接集群时，一般都会将 NameServer 所有的节点地址全部进行定义（不同地址之间使用“;”进行分隔），这样当某一个 NameServer 发生故障时，可以自动切换到其他 NameServer，如图 8-45 所示。

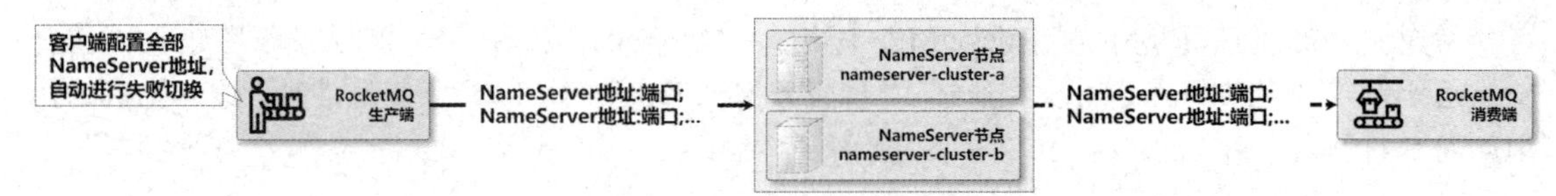

图 8-45　连接 NameServer 集群

考虑到服务配置的简化应用，本次将讲解搭建拥有两个节点的 NameServer 集群（实际的生产环境下可以根据需要添加若干个 NameServer）。下面通过具体的步骤进行配置讲解。

（1）【nameserver-cluster-a 主机】将 RocketMQ 开发包上传到该主机，保存父路径为/var/ftp。

（2）【nameserver-cluster-a 主机】将上传完成的 rocketmq-all-4.9.0-bin-release.zip 包解压缩到"/usr/local"目录。

```
unzip /var/ftp/rocketmq-all-4.9.0-bin-release.zip -d /usr/local/
```

（3）【nameserver-cluster-a 主机】为便于配置，将更改后的目录更名为 rocketmq。

```
mv /usr/local/rocketmq-all-4.9.0-bin-release/ /usr/local/rocketmq
```

（4）【nameserver-cluster-a 主机】在 rocketmq 目录中创建日志存储目录。

```
mkdir /usr/local/rocketmq/logs
```

（5）【nameserver-cluster-a 主机】打开 runserver.sh 文件，修改 NameServer 相关的 JVM 启动参数。

打开配置文件	`vi /usr/local/rocketmq/bin/runserver.sh`
修改 JVM 参数	`JAVA_OPT="${JAVA_OPT} -server -Xms1g -Xmx1g -Xmn1g -XX:MetaspaceSize=128m -XX:MaxMetaspaceSize=320m"`

（6）【nameserver-cluster-a 主机】将配置完成的 rocketmq 目录复制到 nameserver-cluster-b 主机。

```
scp -r /usr/local/rocketmq nameserver-cluster-b:/usr/local/
```

（7）【nameserver-cluster-*主机】分别启动两台 NameServer 服务器。

```
nohup /usr/local/rocketmq/bin/mqnamesrv > /usr/local/rocketmq/logs/rocketmq-namesrv.log 2>&1 &
```

（8）【nameserver-cluster-*主机】NameServer 启动后占用 9876 端口，修改防火墙配置。

添加访问规则	`firewall-cmd --zone=public --add-port=9876/tcp --permanent`
配置重新加载	`firewall-cmd --reload`

8.3.2　Broker 集群

Broker 集群

视频名称　0815_【掌握】Broker 集群

视频简介　Broker 是最终实现消息处理的核心单元，本视频将通过九台服务器实现三组 Broker 分片集群，并通过具体的配置步骤实现服务的搭建以及控制台改造。

考虑到可维护性的特点，本次实现 Broker 集群采用的是 DLeger 多副本模式，这样就需要通过 Broker 配置文件的方式定义每一个 DLegerGroup；在每一个 DLegerGroup 中必须存在三个 Broker 节点，这样就需要为每一个节点设置一个 ID 标记，并通过此标记来绑定与之相关的主机地址。配置的核心结构如图 8-46 所示。

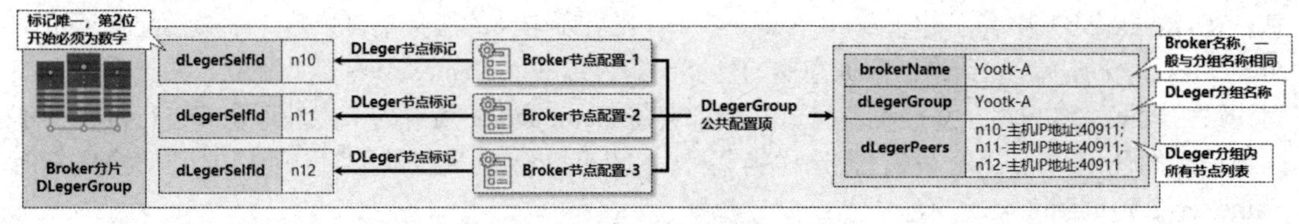

图 8-46　DLeger 多副本模式的核心结构

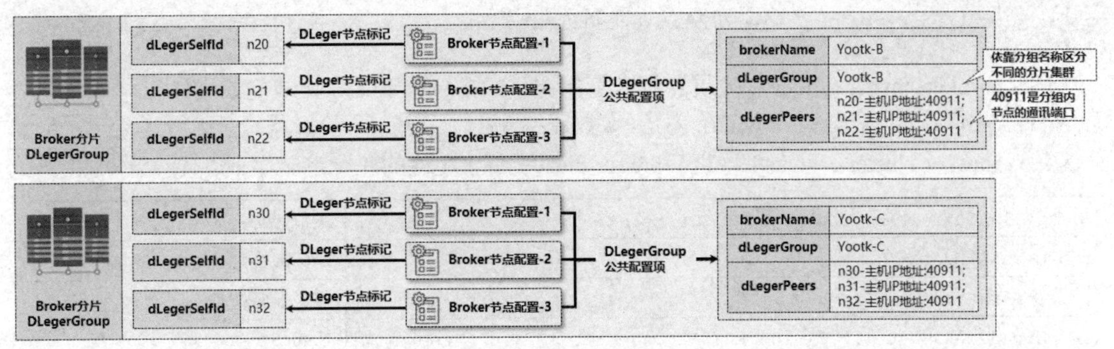

图 8-46 DLeger 多副本模式的核心结构（续）

注意：DLegerID 有格式要求。

在配置 DLeger 模式时，所定义的“dLegerSelfId”属性项的第一位可以是任意的字符，但是第二位开始必须使用数字，这一点可以参考 RocketMQ 源代码中 store 子模块提供的 DLedgerCommitLog 类。

范例：【store 子模块】DLedgerCommitLog 类核心定义

```
package org.apache.rocketmq.store.dledger;
public class DLedgerCommitLog extends CommitLog {
   public DLedgerCommitLog(final DefaultMessageStore store) { // 构造方法
      // DLedger的相关配置项注入，代码略
      id = Integer.valueOf(dLedgerConfig.getSelfId()
               .substring(1)) + 1;                                // 第2位开始转为数字
   }
}
```

如果没有按照要求进行配置，则在启动时会出现“NumberFormatException”。在“dLegerPeers”选项配置时，采用的格式为“dLegerSelfId-主机 IP 地址:通信端口”，使用的也是 dLegerSelfId 配置项。

为了便于读者理解 Broker 集群搭建，本次将依据图 8-46 所示的结构实现 9 台 Broker 服务器的三组 DLegerGroup 的分配，具体实现的操作步骤如下。

（1）【broker-cluster-*-node-*主机】由于 Broker 需要存储消息数据，所以创建一个数据存储目录。

```
mkdir -p /usr/data/rocketmq/store/commitlog
```

（2）【broker-cluster-a-node-1 主机】将 RocketMQ 开发包上传到服务器之中，保存父路径为“/var/ftp”。

（3）【broker-cluster-a-node-1 主机】将上传完成的 rocketmq-all-4.9.0-bin-release.zip 组件包解压缩到“/usr/local”目录中。

```
unzip /var/ftp/rocketmq-all-4.9.0-bin-release.zip -d /usr/local/
```

（4）【broker-cluster-a-node-1 主机】为便于配置将更改后的目录更名为 rocketmq。

```
mv /usr/local/rocketmq-all-4.9.0-bin-release/ /usr/local/rocketmq
```

（5）【broker-cluster-a-node-1 主机】创建一个 Broker 日志存储目录。

```
mkdir -p /usr/local/rocketmq/logs
```

（6）【broker-cluster-a-node-1 主机】调整 BrokerServer 的内存配置。

打开配置文件	`vi /usr/local/rocketmq/bin/runbroker.sh`
修改 JVM 参数	`JAVA_OPT="${JAVA_OPT} -server -Xms1g -Xmx1g -Xmn1g"`

（7）【broker-cluster-a-node-1 主机】此时需要进行 dledger 集群的配置。在 RocketMQ 中给出了“/conf/dledger/*.conf”配置参考文件，本处可以通过复制已有文件的方式进行 Broker 的配置定义。首先实现文件复制操作。

```
cp /usr/local/rocketmq/conf/dledger/broker-n0.conf /usr/local/rocketmq/conf/dledger/broker-cluster.conf
```

（8）【broker-cluster-a-node-1 主机】打开复制完成的 broker-cluster.conf 配置文件。

```
vi /usr/local/rocketmq/conf/dledger/broker-cluster.conf
```

（9）【broker-cluster-a-node-1 主机】broker-cluster.conf 文件的配置定义如下。

`brokerClusterName = YootkRocketMQCluster`	集群名称
`brokerName=Yootk-A`	Broker 名称
`listenPort=30911`	服务监听端口
`namesrvAddr=nameserver-cluster-a:9876;nameserver-cluster-b:9876`	NameServer 列表
`storePathRootDir=/usr/data/rocketmq/store/`	Broker 数据存储目录
`storePathCommitLog=/usr/data/rocketmq/store/commitlog`	CommitLog 存储目录
`enableDLegerCommitLog=true`	启用 DLedger 模式
`dLegerGroup=Yootk-A`	DLedger 组名称
`dLegerPeers=n10-192.168.190.180:40911;` `n11-192.168.190.181:40911;n12-192.168.190.182:40911`	集群节点信息，主要实现 DLedger 内部通信
`dLegerSelfId=n10`	节点 ID，与 dLegerPeers 名称一致
`sendMessageThreadPoolNums=16`	消息发送的线程数量

（10）【broker-cluster-a-node-1 主机】将当前配置好的 RocketMQ 目录复制到其余的 Broker 主机。

复制到“broker-cluster-a-node-2”主机	`scp -r /usr/local/rocketmq broker-cluster-a-node-2:/usr/local/`
复制到“broker-cluster-a-node-3”主机	`scp -r /usr/local/rocketmq broker-cluster-a-node-3:/usr/local/`
复制到“broker-cluster-b-node-1”主机	`scp -r /usr/local/rocketmq broker-cluster-b-node-1:/usr/local/`
复制到“broker-cluster-b-node-2”主机	`scp -r /usr/local/rocketmq broker-cluster-b-node-2:/usr/local/`
复制到“broker-cluster-b-node-3”主机	`scp -r /usr/local/rocketmq broker-cluster-b-node-3:/usr/local/`
复制到“broker-cluster-c-node-1”主机	`scp -r /usr/local/rocketmq broker-cluster-c-node-1:/usr/local/`
复制到“broker-cluster-c-node-2”主机	`scp -r /usr/local/rocketmq broker-cluster-c-node-2:/usr/local/`
复制到“broker-cluster-c-node-3”主机	`scp -r /usr/local/rocketmq broker-cluster-c-node-3:/usr/local/`

（11）【broker-cluster-a-node-*主机】修改第一个 Broker 分片集群中各个节点的 broker-cluster.conf 配置文件。

修改“broker-cluster-a-node-2”主机配置	`brokerName=Yootk-A` `dLegerSelfId=n11`
修改“broker-cluster-a-node-3”主机配置	`brokerName=Yootk-A` `dLegerSelfId=n12`

（12）【broker-cluster-b-*主机】修改第二个 Broker 分片集群中各个节点的 broker-cluster.conf 配置文件。

修改“broker-cluster-b-node-1”主机配置	`brokerName=Yootk-B`
	`dLegerGroup=Yootk-B`
	`dLegerPeers=n20-192.168.190.183:40911;`
	`n21-192.168.190.184:40911;n22-192.168.190.185:40911`
	`dLegerSelfId=n20`

修改“broker-cluster-b-node-2”主机配置	`brokerName=Yootk-B`
	`dLegerGroup=Yootk-B`
	`dLegerPeers=n20-192.168.190.183:40911;`
	`n21-192.168.190.184:40911;n22-192.168.190.185:40911`
	`dLegerSelfId=n21`

修改“broker-cluster-b-node-3”主机配置	brokerName=Yootk-B dLegerGroup=Yootk-B dLegerPeers=n20-192.168.190.183:40911; n21-192.168.190.184:40911;n22-192.168.190.185:40911 dLegerSelfId=n22

（13）【broker-cluster-c-*主机】修改第三个 Broker 分片集群中各个节点的 broker-cluster.conf 配置文件。

修改“broker-cluster-c-node-1”主机配置	brokerName=Yootk-C dLegerGroup=Yootk-C dLegerPeers=n30-192.168.190.186:40911; n31-192.168.190.187:40911;n32-192.168.190.188:40911 dLegerSelfId=n30
修改“broker-cluster-c-node-2”主机配置	brokerName= Yootk-C dLegerGroup= Yootk-C dLegerPeers=n30-192.168.190.186:40911; n31-192.168.190.187:40911;n32-192.168.190.188:40911 dLegerSelfId=n31
修改“broker-cluster-c-node-3”主机配置	brokerName= Yootk-C dLegerGroup= Yootk-C dLegerPeers=n30-192.168.190.186:40911; n31-192.168.190.187:40911;n32-192.168.190.188:40911 dLegerSelfId=n32

（14）【broker-cluster-*-*主机】修改防火墙配置，添加服务端口访问权限。

添加访问端口	firewall-cmd --zone=public --add-port=30909/tcp --permanent firewall-cmd --zone=public --add-port=30911/tcp --permanent firewall-cmd --zone=public --add-port=40911/tcp --permanent
重新加载配置	firewall-cmd --reload

（15）【broker-cluster-*-*主机】启动所有 Broker 服务进程。

```
nohup /usr/local/rocketmq/bin/mqbroker -c /usr/local/rocketmq/conf/dledger/broker-cluster.conf >
/usr/local/rocketmq/logs/rocketmq-broker.log 2>&1 &
```

（16）【rocketmq-console 源代码】此时已经成功启动了 RocketMQ 服务集群，如果想通过 RocketMQ 控制台观察集群状态，则需要修改“rocketmq-console”子模块源代码中的 application.yml 配置文件。

```
rocketmq.config.namesrvAddr=nameserver-cluster-a:9876;nameserver-cluster-b:9876
```

（17）【RocketMQ 控制台】通过控制台查看集群信息，最终结果如图 8-47 所示。

分片	编号	地址	版本	生产消息TPS	消费消息TPS	昨日生产总数	昨日消费总数	今天生产总数	今天消费总数	操作
Yootk-C	0(master)	192.168.190.187:30911	V4_9_0	0.00	0.00	0	0	0	0	状态 配置
Yootk-C	31(slave)	192.168.190.186:30911	V4_9_0	0.00	0.00	0	0	0	0	状态 配置
Yootk-C	33(slave)	192.168.190.188:30911	V4_9_0	0.00	0.00	0	0	0	0	状态 配置
Yootk-B	0(master)	192.168.190.184:30911	V4_9_0	0.00	0.00	0	0	0	0	状态 配置
Yootk-B	21(slave)	192.168.190.183:30911	V4_9_0	0.00	0.00	0	0	0	0	状态 配置
Yootk-B	23(slave)	192.168.190.185:30911	V4_9_0	0.00	0.00	0	0	0	0	状态 配置
Yootk-A	0(master)	192.168.190.181:30911	V4_9_0	0.00	0.00	0	0	0	0	状态 配置
Yootk-A	11(slave)	192.168.190.180:30911	V4_9_0	0.00	0.00	0	0	0	0	状态 配置
Yootk-A	13(slave)	192.168.190.182:30911	V4_9_0	0.00	0.00	0	0	0	0	状态 配置

图 8-47 RocketMQ 集群列表

（18）【RocketMQ 控制台】通过控制台创建一个新的主题，同时配置该主题存储的集群以及 Broker 名称，如图 8-48 所示。

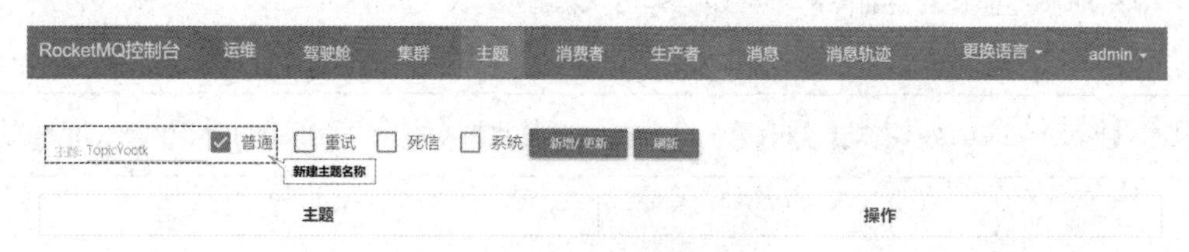

图 8-48 新建主题

（19）【RocketMQ 控制台】主题建立完成后还需要进行存储配置，需要选择消息集群以及主题分片，如图 8-49 所示。

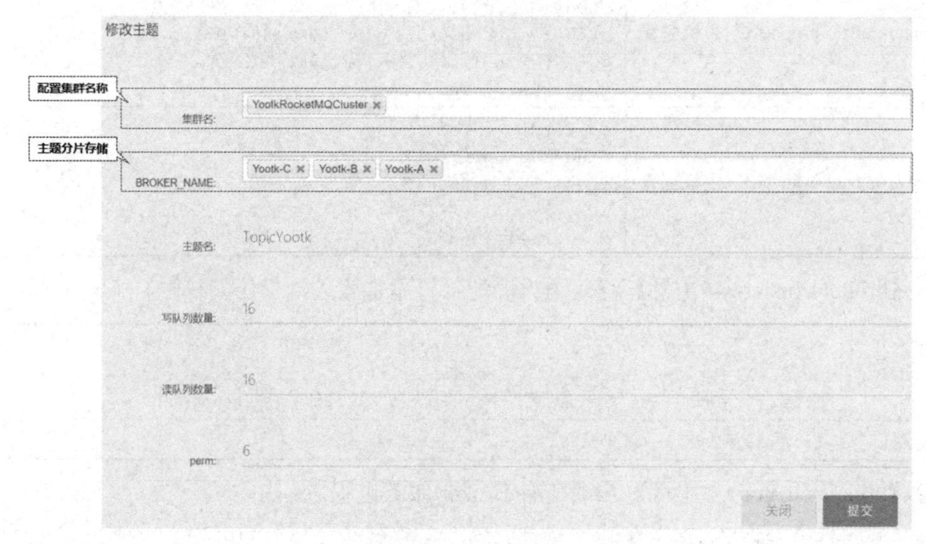

图 8-49 存储配置

（20）【RocketMQ 控制台】通过控制台模拟消息的发送，可以发现每一条消息会随机分配到不同的 Broker 分片之中。

8.4 RocketMQClient 程序开发

RocketMQClient 基本使用

视频名称 0816_【掌握】RocketMQClient 基本使用

视频简介 RocketMQ 应用服务搭建完成之后，在实际生产环境下，需要通过程序来实现生产端与消费端。本视频介绍通过 Apache 官方给出的 RocketMQ 相关依赖实现消息服务的调用。

在现实的开发中，RocketMQ 中的生产者与消费者都需要通过具体的业务来实现消息的生产与消费处理逻辑。为了便于相关程序的编写，Apache 提供了 RocketMQClient 依赖支持库，开发者根据指定类库的使用原则即可方便地实现消息的生产与发送。下面通过具体的代码来介绍如何实现这一基础操作。

（1）【microcloud 项目】创建一个 rocketmq 子模块，随后修改 build.gradle 追加所需依赖。

```
project('rocketmq') {                                              // 子模块
    dependencies {
        implementation('org.apache.rocketmq:rocketmq-client:4.9.0')   // RocketMQ依赖
```

```
        implementation('org.apache.rocketmq:rocketmq-acl:4.9.0')  // RocketMQACL依赖
    }
}
```

（2）【rocketmq 子模块】定义消息消费者。

```
package com.yootk.rocketmq;
public class MessageConsumer {
    public static final String NAME_SERVER_LIST = "rocketmq-server:9876";
    private static final String CONSUMER_GROUP = "muyan-group";  // 消费组
    private static final String TOPIC = "TopicYootk";            // 主题名称
    public static final String ACCESSKEY = "RocketMQMuyan";      // 用户名
    public static final String SECRETKEY = "helloyootk";         // 密码
    public static void main(String[] args) throws Exception {
        RPCHook clientHook = new AclClientRPCHook(
                new SessionCredentials(ACCESSKEY, SECRETKEY));  // ACL认证
        DefaultMQPushConsumer consumer = new DefaultMQPushConsumer(CONSUMER_GROUP,
          clientHook, new AllocateMessageQueueAveragely());      // 推模式消费端
        consumer.setNamesrvAddr(NAME_SERVER_LIST);               // NameServer列表
        consumer.subscribe(TOPIC, "*");                          // 匹配所有标签
        consumer.registerMessageListener(new MessageListenerConcurrently() { // 消息监听
            @Override
            public ConsumeConcurrentlyStatus consumeMessage(List<MessageExt> msgs,
                                               ConsumeConcurrentlyContext context) {
                System.out.printf("【%s】接收新的消息：%s %n",
                        Thread.currentThread().getName(), msgs);    // 消息输出
                return ConsumeConcurrentlyStatus.CONSUME_SUCCESS;  // 消费成功标记
            }
        });
        consumer.start();                        // 启动消费端
    }
}
```

由于 RocketMQ 已经启用了 ACL 安全控制，因此在消费端连接时需要通过 SessionCredentials 对象保存认证数据，随后再基于 RPCHook 接口实例包装后传递给 DefaultMQPushConsumer 消费端处理类。这样只需要设置好 NameServer 主机地址、主题名称、匹配表达式以及消息监听接口，就可以在每次收到消息时自动调用 MessageListenerConcurrently 接口所提供的 consumeMessage()方法获取消息内容。RocketMQ 消费处理的类关联结构如图 8-50 所示。

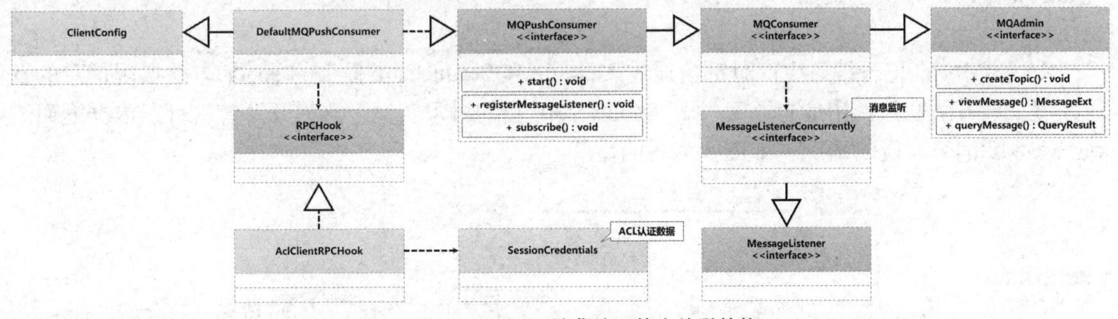

图 8-50　RocketMQ 消费处理的类关联结构

（3）【rocketmq 子模块】定义消息生产端。

```
package com.yootk.rocketmq;
public class MessageProducer {
    public static final String NAME_SERVER_LIST = "rocketmq-server:9876";
    public static final String PRODUCER_GROUP = "muyan-group";   // 消费组
    public static final String TOPIC = "TopicYootk";             // 主题
    public static final String ACCESSKEY = "RocketMQMuyan";      // 用户名
    public static final String SECRETKEY = "helloyootk";         // 密码
    public static void main(String[] args) throws Exception {
        RPCHook clientHook = new AclClientRPCHook(
```

```
                new SessionCredentials(ACCESSKEY, SECRETKEY));  // ACL认证
        DefaultMQProducer producer = new DefaultMQProducer(
                PRODUCER_GROUP, clientHook);                    // 创建消息生产者
        producer.setNamesrvAddr(NAME_SERVER_LIST);              // NameServer地址
        producer.start();                                       // 启动消息生产者
        for (int x = 0; x < 100; x++) {                         // 循环发送消息
            Message msg = new Message(TOPIC, "沐言科技: www.yootk.com"
                    .getBytes(RemotingHelper.DEFAULT_CHARSET)); // 创建消息处理机制
            SendResult result = producer.send(msg);             // 发送信息
            System.out.printf("【消息发送】%s%n", result);       // 接收消息返回结果
        }
        producer.shutdown();                                    // 关闭消息发送者
    }
}
```

消息发送时首先需要通过 RPCHook 实现认证数据的封装，同时设置好正确的 NameServer 地址，这样就可以将所发送的内容与主题封装在 Message 对象实例之中，而后会有系统生成一些扩展的消息内容。随后利用 DefaultMQProducer 类提供的 send()方法实现消息发送，并且在每次发送完成后都会返回一个 SendResult 结果集对象以保存消息回执。本程序的类关联结构如图 8-51 所示。

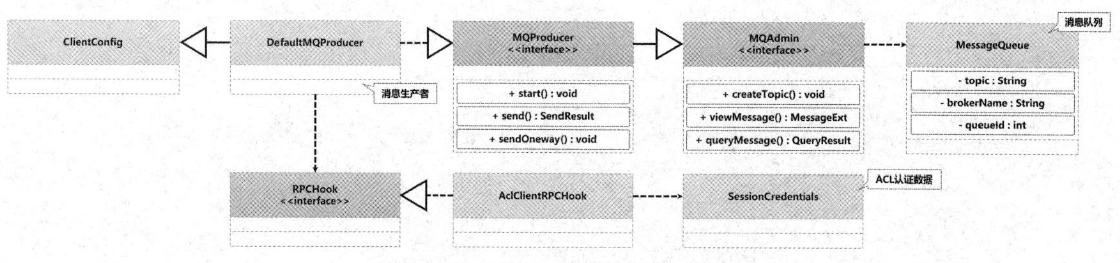

图 8-51 RocketMQ 消息发送的关联结构

8.4.1 消息生产模式

消息生产模式

视频名称 0817_【掌握】消息生产模式

视频简介 RocketMQ 提供了 3 种不同的消息生产模式。本视频为读者详细地解释这 3 种消息生产模式的特点以及具体实现。

RocketMQ 在进行消息生产时都会通过 DefaultMQProducer 类提供的 send()方法实现消息的发送处理。该方法在被调用后会返回一个 SendResult 类的对象实例，这样开发者就可以依据返回的状态来判断消息发送的情况，如图 8-52 所示。

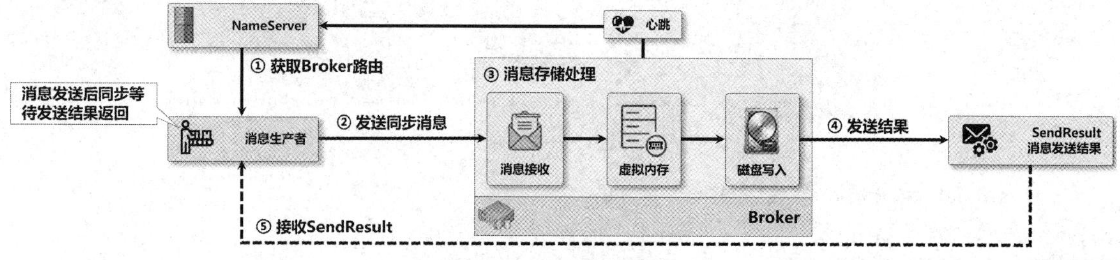

图 8-52 send()发送消息

但是在默认情况下，send()方法采用的是同步发送，即在消息发送后，整个线程将进入阻塞状态，一直持续到 Broker 返回最终的响应状态才继续进行下一消息的发送。当 Broker 消息过多，或者网络延迟较高时，程序就会出现较大的性能问题。所以 RocketMQ 除了同步发送之外，还提供异步消息发送和单向消息发送机制。

异步消息发送是在每次进行消息发送时将所发送的消息交由异步线程池处理，而此时的消息生产线程就可以按照同样的逻辑进行其他消息的发送。最终所有发送到 Broker 的消息全部由异步线程池进行发送，并且会有一个单独的线程进行消息发送结果的接收与处理（由 SendCallback 回调接口实现 Broker 响应处理），处理流程如图 8-53 所示。

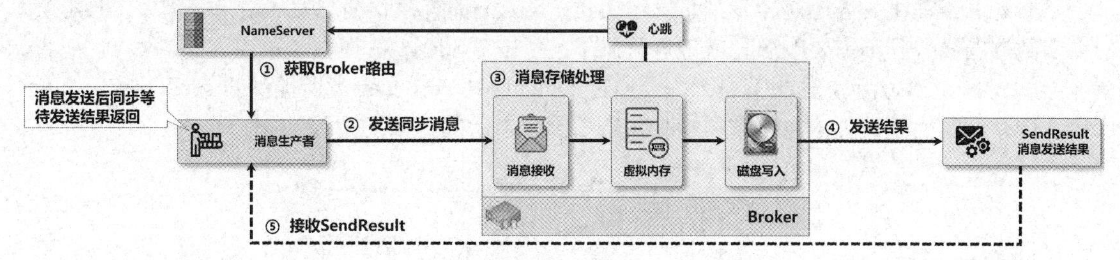

图 8-53 异步消息发送

范例：【rocketmq 子模块】异步消息发送

```
package com.yootk.rocketmq;
public class MessageProducer {
   public static final String NAME_SERVER_LIST = "rocketmq-server:9876";// NameServer
   public static final String PRODUCER_GROUP = "muyan-group";          // 消费组
   public static final String TOPIC = "TopicYootk";                    // 主题
   public static final String ACCESSKEY = "RocketMQMuyan";             // 用户名
   public static final String SECRETKEY = "helloyootk";                // 密码
   public static void main(String[] args) throws Exception {
      RPCHook clientHook = new AclClientRPCHook(
            new SessionCredentials(ACCESSKEY, SECRETKEY));             // ACL认证
      DefaultMQProducer producer = new DefaultMQProducer(
            PRODUCER_GROUP, clientHook);                               // 创建消息生产者
      producer.setNamesrvAddr(NAME_SERVER_LIST);                       // NameServer地址
      producer.start();                                                // 启动消息生产者
      Message msg = new Message(TOPIC, "沐言科技：www.yootk.com"
            .getBytes(RemotingHelper.DEFAULT_CHARSET));                // 创建消息
      final CountDownLatch countDownLatch = new CountDownLatch(1);  // 发送等待
      System.out.println("【消息发送】当前线程：" +
               Thread.currentThread().getName());                      // 信息输出
      producer.send(msg, new SendCallback() {                          // 异步接收
         @Override
         public void onSuccess(SendResult sendResult) {                // 消息发送成功
            countDownLatch.countDown();                                // 减少同步计数
            System.out.printf("【消息发送成功 - %s】%s%n", Thread.currentThread()
                .getName(), sendResult);                               // 输出成功信息
         }
         @Override
         public void onException(Throwable e) {                        // 消息发送失败
            countDownLatch.countDown();                                // 减少同步计数
            System.out.printf("【消息发送失败 - %s】%s%n", Thread.currentThread()
                .getName(), e.getMessage());                           // 输出失败信息
         }
      });
      countDownLatch.await();                                          // 线程等待
      producer.shutdown();                                             // 关闭消息发送者
   }
}
```

程序执行结果	【消息发送】当前线程：main 【消息发送成功 - NettyClientPublicExecutor_1】SendResult [sendStatus=SEND_OK, msgId=7F0000012EDC2437C6DC5F3D08940000, offsetMsgId=C0A8BEAA00002A9F000000000083A6A, messageQueue=MessageQueue [topic=TopicYootk, brokerName=broker-a, queueId=6], queueOffset=81]

通过当前程序的执行结果可以清楚地发现，消息的发送是由主线程负责处理的，而消息结果的处理是由一个专门的线程池来进行的。在本程序中通过 SendCallback 接口实现了 Broker 消息数据的响应处理，当消息发送成功时会自动回调 onSuccess()方法，而消息发送失败也会自动回调 onException()方法。在实际的开发中采用异步消息发送的模式可以在大规模消息生产中获得相对较好的处理性能。

不管是同步消息发送还是异步消息发送模式，实际上都有接收消息发送结果的需要，但如果面对海量数据发送并且不需要响应时就不再适用了。所以 RocketMQ 提供了单向（One-way）消息，操作结构如图 8-54 所示。这类消息只需要通过 DefaultMQProducer 类所提供的 sendOneway()方法即可实现。同时由于该类消息不需要接收响应，因此方法的返回类型为 void。

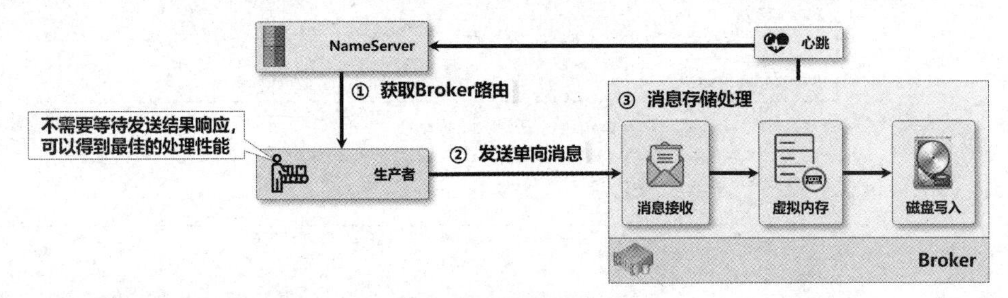

图 8-54　单向消息

范例：【rocketmq 子模块】发送单向消息

```
package com.yootk.rocketmq;
public class MessageProducer {
    public static final String NAME_SERVER_LIST = "rocketmq-server:9876"; // NameServer
    public static final String PRODUCER_GROUP = "muyan-group";    // 消费组
    public static final String TOPIC = "TopicYootk";              // 主题
    public static final String ACCESSKEY = "RocketMQMuyan";       // 用户名
    public static final String SECRETKEY = "helloyootk";          // 密码
    public static void main(String[] args) throws Exception {
        RPCHook clientHook = new AclClientRPCHook(
                new SessionCredentials(ACCESSKEY, SECRETKEY));      // ACL认证
        DefaultMQProducer producer = new DefaultMQProducer(
                PRODUCER_GROUP, clientHook);                        // 创建消息生产者
        producer.setNamesrvAddr(NAME_SERVER_LIST);                  // NameServer地址
        producer.start();                                           // 启动消息生产者
        Message msg = new Message(TOPIC, "沐言科技：www.yootk.com"
                .getBytes(RemotingHelper.DEFAULT_CHARSET));         // 创建消息处理机制
        producer.sendOneway(msg);                                   // 发送单向消息
        producer.shutdown();                                        // 关闭消息发送者
    }
}
```

此程序使用了 sendOneway()方法来实现单向消息的发送，当消息发送后消费端可以获得发送的消息数据，而生产端在发送完成后即可结束线程，不需要等待 Broker 响应。

8.4.2　消费模式

消费模式

视频名称　0818_【掌握】消费模式

视频简介　RocketMQ 的消费端有两种消费模式。本视频将为读者讲解这两种消费模式的工作原理以及具体的代码实现。

在 RocketMQ 中消费端需要通过指定的主题连接到消息队列，而后才可以实现消息的消费处理。而在消费时需要通过 MessageListener 启动消息监听，这样每当有新的消息发送过来时，就可以进行消息处理。RocketMQ 有推与拉两种消费模式，这两种消费模式的特点如下。

- 推模式：消费端与 Broker 建立连接后，每当服务端有消息时，会将消息推送给消费端，如图 8-55 所示。

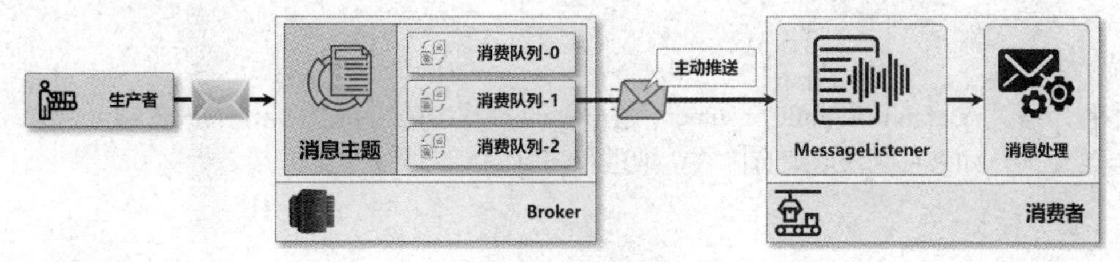

图 8-55 推模式

提示：推模式有连接上限。

推模式中，服务端会随时向消费端推送新的消息内容，这样在设计时就必须建立一个长连接，以避免重复“连接—关闭”所带来的性能损耗。当消费端较多时，这必然会造成服务端的性能下降。

- 拉模式：消费端以轮询的方式不断对当前连接 Broker 的消费队列进行检查，如果有新的消息则拉取到消费端，如图 8-56 所示。

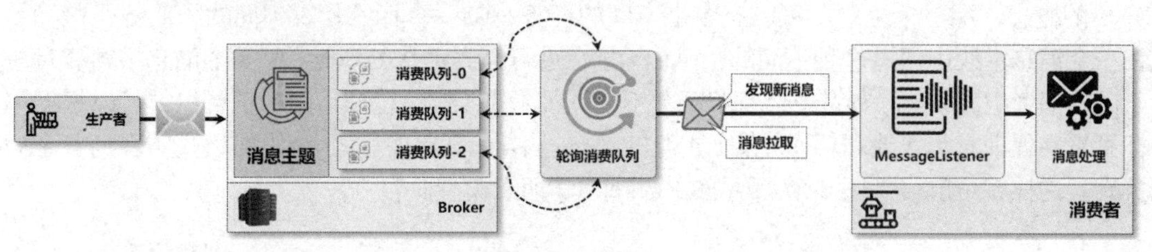

图 8-56 拉模式

提示：拉模式与长轮询。

拉模式采用的是全部消息队列的轮询检查模式，由于不需要长连接，因此使用拉模式可以实现更多的消费端连接。但考虑到服务通信的性能问题，在没有新消息时，服务端会将当前的消费端连接挂起，当有新消息时才会对此连接进行响应，而后重复这一操作以不断实现新的消息消费处理。

范例：【rocketmq 子模块】使用拉模式

```java
package com.yootk.rocketmq;
public class MessageConsumer {
    public static final String NAME_SERVER_LIST = "rocketmq-server:9876";
    private static final String CONSUMER_GROUP = "muyan-group";  // 消费组
    private static final String TOPIC = "TopicYootk";             // 主题名称
    public static final String ACCESSKEY = "RocketMQMuyan";       // 用户名
    public static final String SECRETKEY = "helloyootk";          // 密码
    public static void main(String[] args) throws Exception {
        RPCHook clientHook = new AclClientRPCHook(
                new SessionCredentials(ACCESSKEY, SECRETKEY));     // ACL认证
        DefaultLitePullConsumer consumer = new DefaultLitePullConsumer(
                    CONSUMER_GROUP, clientHook);                  // 拉模式消费端
        consumer.setNamesrvAddr(NAME_SERVER_LIST);                // NameServer列表
        consumer.subscribe(TOPIC, "*");                           // 匹配所有标签
        consumer.start();                                         // 启动消费端
        boolean running = true;                                   // 定义一个循环控制变量
        while (running) {                                         // 循环拉取消息
            List<MessageExt> messages = consumer.poll();          // 拉取数据
            if (messages != null && messages.size() > 0) {        // 获取消息
                for (Message msg : messages) {                    // 迭代消息集合
                    System.out.printf("%s%n", msg);               // 消息输出
```

```
                }
            }
        }
        consumer.shutdown();                                    // 关闭消费端
    }
}
```

本程序通过 DefaultLitePullConsumer 类实现了拉模式开发。程序启动后消费端会持续地进行 Broker 的轮询，如果可以获取消息内容，则进行消息输出。

8.4.3　业务标签

业务标签

视频名称　0819_【掌握】业务标签

视频简介　为了实现更加详细的子业务消息的处理逻辑，可以基于消息主题实现进一步的划分，所以 RocketMQ 提供了标签的概念。本视频为读者详细解释 RocketMQ 中 Topic、Group、Tag 之间的关联，并通过实例演示标签的使用。

大部分的消息组件都以消息的主题（Topic）实现不同的业务功能划分，这样一来就有可能出现许多有关联的业务运行在不同的主题之中。RocketMQ 为了解决这一设计问题，提出了标签（Tag）的概念，即一个主题中可以有多个不同的标签，每一个标签代表不同的子业务，这样就可以在一个消息主题内传递多种不同的消息内容。在进行消息消费处理时，同一个消息主题会对应多个不同的消费组，每个消费组中会有若干个不同的消费者实例，这样不同的消费组就可以根据不同的标签来实现消息消费处理；而当某个主题的消息过多时，也可以基于分组的模式实现消费者的动态扩充。这样就可以得到图 8-57 所示的 RocketMQ 消息模型。

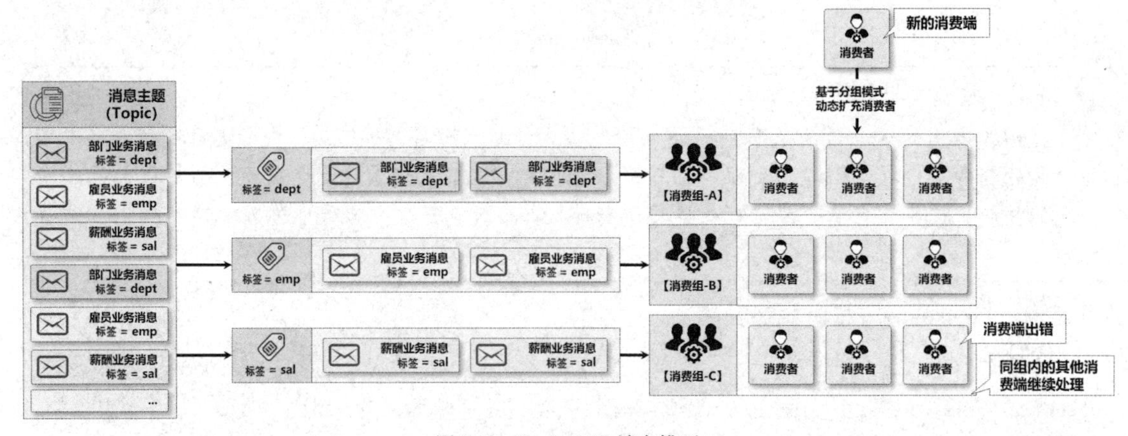

图 8-57　RocketMQ 消息模型

如果想在消息处理中使用标签进行子业务消息的处理，那么需要在消息发送时明确设置标签的内容（通过 Message 实例存储标签信息），而在消费时可以通过 subscribe()方法设置匹配的标签名称。下面来观察具体的代码实现。

（1）【rocketmq 子模块】消息标签的设置可以通过 Message 的构造方法进行定义，本次将依据循环发送 10 条消息数据，并且依据循环的次数为消息设置不同的标签（两个标签的名称分别为“dept”与“emp”）。

```
package com.yootk.rocketmq;
public class MessageProducer {
    public static final String NAME_SERVER_LIST = "rocketmq-server:9876"; // NameServer
    public static final String PRODUCER_GROUP = "muyan-group"; // 消费组
    public static final String TOPIC = "TopicYootk";          // 主题
    public static final String ACCESSKEY = "RocketMQMuyan";   // 用户名
    public static final String SECRETKEY = "helloyootk";      // 密码
```

```
    public static void main(String[] args) throws Exception {
        RPCHook clientHook = new AclClientRPCHook(
                new SessionCredentials(ACCESSKEY, SECRETKEY)); // ACL认证
        DefaultMQProducer producer = new DefaultMQProducer(
                PRODUCER_GROUP, clientHook);                          // 创建消息生产者
        producer.setNamesrvAddr(NAME_SERVER_LIST);                    // NameServer地址
        producer.start();                                             // 启动消息生产者
        Message msg = null;                                           // 保存消息内容
        for (int x = 0; x < 10; x++) {                                // 循环发送消息
            if (x % 2 == 0) {                                         // 创建消息时设置标签为dept
                msg = new Message(TOPIC, "dept", ("【部门消息 - " + x +
                    "】沐言科技：www.yootk.com").getBytes(RemotingHelper.DEFAULT_CHARSET));
            } else {                                                  // 创建消息时设置标签为emp
                msg = new Message(TOPIC, "emp", ("【部门消息 - " + x +
                    "】沐言科技：www.yootk.com").getBytes(RemotingHelper.DEFAULT_CHARSET));
            }
            SendResult result = producer.send(msg);                  // 发送信息
            System.out.printf("【消息发送】%s%n", result);             // 接收消息返回结果
        }
        producer.shutdown();                                          // 关闭消息发送者
    }
}
```

（2）【rocketmq 子模块】在进行消息接收时，如果要接收全部标签的消息，则可以使用“*”进行匹配；如果要接收指定标签的消息，则可以写上具体的标签名称。首先创建部门消费处理类。

```
package com.yootk.rocketmq.tag;
public class DeptMessageConsumer {
    public static final String NAME_SERVER_LIST = "rocketmq-server:9876";// NameServer
    private static final String CONSUMER_GROUP = "muyan-group-dept"; // 消费组
    private static final String TOPIC = "TopicYootk";             // 主题名称
    public static final String ACCESSKEY = "RocketMQMuyan";       // 用户名
    public static final String SECRETKEY = "helloyootk";          // 密码
    public static void main(String[] args) throws Exception {
        RPCHook clientHook = new AclClientRPCHook(
                new SessionCredentials(ACCESSKEY, SECRETKEY));     // ACL认证
        DefaultMQPushConsumer consumer = new DefaultMQPushConsumer(CONSUMER_GROUP,
                clientHook, new AllocateMessageQueueAveragely()); // 推模式消费端
        consumer.setNamesrvAddr(NAME_SERVER_LIST);                 // NameServer列表
        consumer.subscribe(TOPIC, "dept");                         // 指定标签
        // 消息监听与接收代码略
    }
}
```

（3）【rocketmq 子模块】创建雇员消费处理类，其整体结构与部门消费处理类相同，但是需要设置不同的分组与标签。

```
package com.yootk.rocketmq.tag;
public class DeptMessageConsumer {
    // 考虑到篇幅问题，此处只列出修改部分，其余代码与部门消费端的相同
    private static final String CONSUMER_GROUP = "muyan-group-dept"; // 消费组
    public static void main(String[] args) throws Exception {
        consumer.subscribe(TOPIC, "emp");                              // 指定标签
        // 消息监听与接收代码略
    }
}
```

消费端开发完成后再分别启动，这样两个消费者就会监听各自标签范围内的消息数据。生产者生产消息后，会自动依据标签的定义找到不同的消费组，从而实现消息消费处理。

提示：消费多个标签。

在某一个标签的消息量没有这么大，或者业务独立性不是这么强的情况下，可以考虑让一个消费者消费多个标签的消息，中间使用“||”分隔。例如，现在某一个消费端要求同时可以消费

dept 与 emp 消息，则在编写订阅表达式时，可以采用如下方式定义。

范例：同时订阅多个标签信息

```
consumer.subscribe(TOPIC, "dept||emp");     // 多个订阅标签
```

根据此时配置的订阅表达式，该消费端启动后会同时接收同一主题下的“dept”与“emp”两个标签的消息内容。

8.4.4　消息识别码

消息识别码

视频名称　0820_【掌握】消息识别码

视频简介　为进一步实现业务消息的标记，RocketMQ 提供了消息识别码的功能。本视频为读者分析消息识别码的作用以及具体应用。

生产者生产完成消息后，会为每一个消息自动分配一个消息 ID，这样管理员就可以依据指定的消息 ID 实现消息数据的查询。但是仅仅凭借消息 ID 是很难与具体的业务进行联系的，所以 RocketMQ 为了便于业务消息的区分，特意设置了一个专属的消息识别码（keys）。利用此消息识别码可以非常方便地实现消息的定位，以解决消息丢失的问题，开发者可以在消息生产时通过 Message 类的构造方法设置该字段的内容。

范例：生产消息并设置 keys

```
package com.yootk.rocketmq;
public class MessageProducer {
   // RocketMQ连接所需要的相关配置项，略
   public static void main(String[] args) throws Exception {
      RPCHook clientHook = new AclClientRPCHook(
           new SessionCredentials(ACCESSKEY, SECRETKEY));          // ACL认证
      DefaultMQProducer producer = new DefaultMQProducer(
           PRODUCER_GROUP, clientHook);                            // 创建消息生产者
      producer.setNamesrvAddr(NAME_SERVER_LIST);                   // NameServer地址
      producer.start();                                            // 启动消息生产者
      Message msg = null;                                          // 保存消息内容
      for (int x = 0; x < 10; x++) {                               // 循环发送消息
         if (x % 2 == 0) {                                         // 设置标签与keys
            String keys = "yootk-dept-keys-" + Math.random();      // 自定义keys
            msg = new Message(TOPIC, "dept", keys,
                 ("【部门消息 - " + x + "】沐言科技：www.yootk.com")
                    .getBytes(RemotingHelper.DEFAULT_CHARSET)); // 创建消息
         } else {                                                  // 设置标签与keys
            String keys = "yootk-emp-keys-" + Math.random();       // 自定义keys
            msg = new Message(TOPIC, "emp", keys,
                  ("【部门消息 - " + x + "】沐言科技：www.yootk.com")
                    .getBytes(RemotingHelper.DEFAULT_CHARSET)); // 创建消息
         }
         SendResult result = producer.send(msg);                   // 发送信息
         System.out.printf("【消息发送】%s%n", result);            // 接收消息返回结果
      }
      producer.shutdown();                                         // 关闭消息发送者
   }
}
```

本程序通过自定义的简单算法规则实现了 keys 的生成（格式为“yootk-业务 TAG-keys-随机数”），同时将每一个生成的 keys 保存在了 Message 消息对象之中。这样在消息发送完成后，开发者就可以通过 RocketMQ 控制台根据 keys 实现消息查询，如图 8-58 所示。

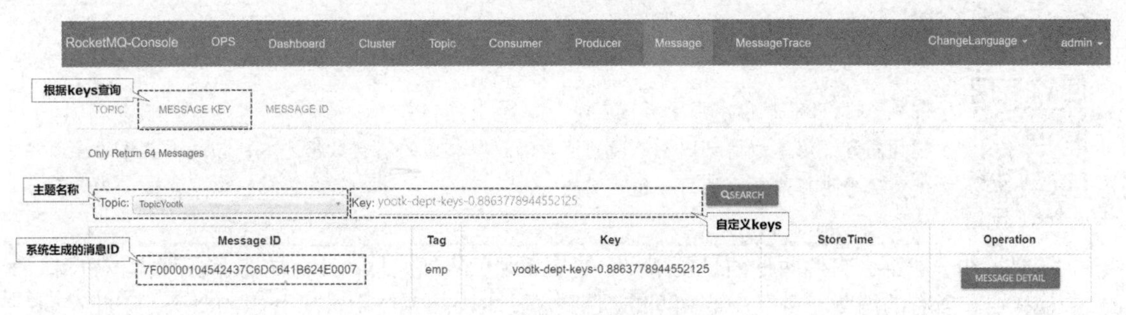

图 8-58　消息查询

8.4.5　NameSpace

视频名称　0821_【理解】NameSpace

视频简介　NameSpace 是一种基于命名空间实现消息分类管理的操作机制。本视频为读者讲解 NameSpace 的作用，以及如何通过代码实现生产者与消费者的 NameSpace 定义。

RocketMQ 提供了非常丰富的消息管理机制，除了之前使用过的标签与识别码之外，还提供了命名空间（NameSpace）的分类处理，即利用命名空间隔离不同消息的生产者与消费者，只有在同一命名空间之中的生产者与消费者之间才可以实现消息的传递。命名空间如图 8-59 所示。

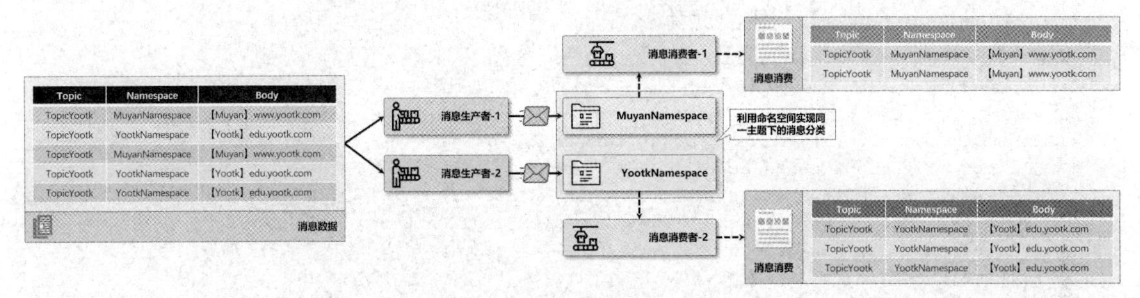

图 8-59　命名空间

RocketMQ 命名空间的设置操作，需要通过生产端（DefaultMQProducer）和消费端（DefaultMQPushConsumer）处理类中的 setNamespace()方法完成。

范例：设置命名空间

消息生产端	`producer.setNamespace("MuyanNamespace");　　// 定义命名空间`
消息消费端	`consumer.setNamespace("MuyanNamespace");　　// 定义命名空间`

命名空间设置完成后，只有在命名空间相同的情况下，生产端发送的消息才可以被消费端获取，这样也可以实现所需子业务的消息隔离。

8.5　消息处理模式

RocketMQ 除了可以实现基础的消息生产与消费处理之外，还有多种消息的处理模式，如消息广播、消息排序、延迟消息、消息过滤、消息批处理以及日志消息处理等机制。本节为读者依次讲解这几种消费模式的特点与使用。

8.5.1 消息广播

消息广播

视频名称 0822_【掌握】消息广播

视频简介 消息广播可以使不同的消费者获取相同的消息数据。本视频通过具体的应用实例为读者讲解消息广播与分组之间的关系。

消息组件一般都会支持两种消息模型：一种是点对点队列消息；另一种是广播消息。在广播消息的模式中，所有的消费者可以获取相同的消息数据。RocketMQ 引入了分组的概念，因此在同一分组下的消费者将共同消费相同的内容（每个消费者消费不同的数据），如图 8-60 所示。为便于理解，下面通过具体的实例操作进行说明。

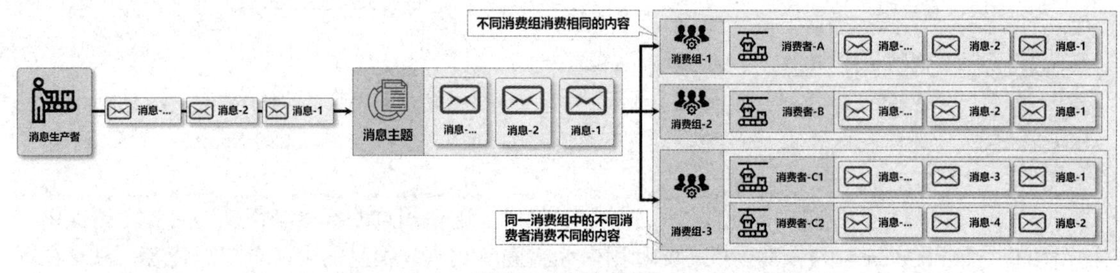

图 8-60 消息广播

（1）【rocketmq 子模块】创建消费者 A。

```
package com.yootk.rocketmq.broadcast;
public class BroadcastMessageConsumerA {
    // RocketMQ连接所需要的相关配置项，略
    public static void main(String[] args) throws Exception {
        RPCHook clientHook = new AclClientRPCHook(
                new SessionCredentials(ACCESSKEY, SECRETKEY));         // ACL认证
        DefaultMQPushConsumer consumer = new DefaultMQPushConsumer(CONSUMER_GROUP,
                clientHook, new AllocateMessageQueueAveragely());      // 推模式消费端
        consumer.setNamesrvAddr(NAME_SERVER_LIST);                     // NameServer列表
        consumer.subscribe(TOPIC, "*");                                // 匹配全部标签
        consumer.registerMessageListener(new MessageListenerConcurrently() { // 消息监听
            @Override
            public ConsumeConcurrentlyStatus consumeMessage(List<MessageExt> msgs,
                                                ConsumeConcurrentlyContext context) {
                for (MessageExt msg : msgs) {                          // 消息迭代
                    System.out.printf("【消费者-A】QueueID = %s, 消息ID = %s, " +
                            "消息TAG = %s、CommitLogOffset = %s, 消息内容 = %s %n",
                        msg.getQueueId(), msg.getMsgId(), msg.getTags(),
                         msg.getCommitLogOffset(), new String(msg.getBody()));
                }
                return ConsumeConcurrentlyStatus.CONSUME_SUCCESS;      // 消费成功的标记
            }
        });
        consumer.start();                                              // 启动消费端
    }
}
```

（2）【rocketmq 子模块】创建消费者 B（只列出核心实现）。

定义消息分组	`private static final String CONSUMER_GROUP = "broadcast-group-b";`

（3）【rocketmq 子模块】创建消费者 C1（只列出核心实现）。

定义消息分组	`private static final String CONSUMER_GROUP = "broadcast-group-c";`

（4）【rocketmq 子模块】创建消费者 C2（只列出核心实现）。

定义消息分组	`private static final String CONSUMER_GROUP = "broadcast-group-c";`

（5）【rocketmq 子模块】创建消息生产者。

```
package com.yootk.rocketmq.broadcast;
public class BroadcastMessageProducer {
    // RocketMQ连接所需要的相关配置项，略
    public static void main(String[] args) throws Exception {
        RPCHook clientHook = new AclClientRPCHook(
                new SessionCredentials(ACCESSKEY, SECRETKEY));      // ACL认证
        DefaultMQProducer producer = new DefaultMQProducer(
                PRODUCER_GROUP, clientHook);                         // 创建消息生产者
        producer.setNamesrvAddr(NAME_SERVER_LIST);                   // NameServer地址
        producer.start();                                            // 启动消息生产者
        for (int x = 0; x < 10; x++) {                               // 循环发送消息
            Message msg = new Message(TOPIC, ("【" + String.format("%03d", x) +
                    "】沐言科技：www.yootk.com").getBytes(RemotingHelper.DEFAULT_CHARSET));
            producer.sendOneway(msg);                                // 发送信息
        }
        producer.shutdown();                                         // 关闭消息发送者
    }
}
```

此时将启动 4 个消费者，同时这 4 个消费者会处于不同的消费组之中。这样在消息生产者启动后，不同的消费组将获得相同的消息数据，从而实现消息广播。

8.5.2　消息排序

消息排序

视频名称　0823_【掌握】消息排序

视频简介　RocketMQ 中的消息生产与消费都依赖于消息队列（Message Queue），为了实现消息的顺序处理，RocketMQ 提供了顺序消费支持。本视频将通过具体的实例分析为读者讲解 MessageQueueSelector 与 MessageListenerOrderly 接口。

在一般情况下如果用户要进行消息的发送，一般都会基于 RocketMQ 默认的处理形式来实现消息队列的分配。这样当有多个消费者进行同一主题消息消费时，就会出现随机选择消息队列，从而导致消息消费顺序混乱，如图 8-61 所示。

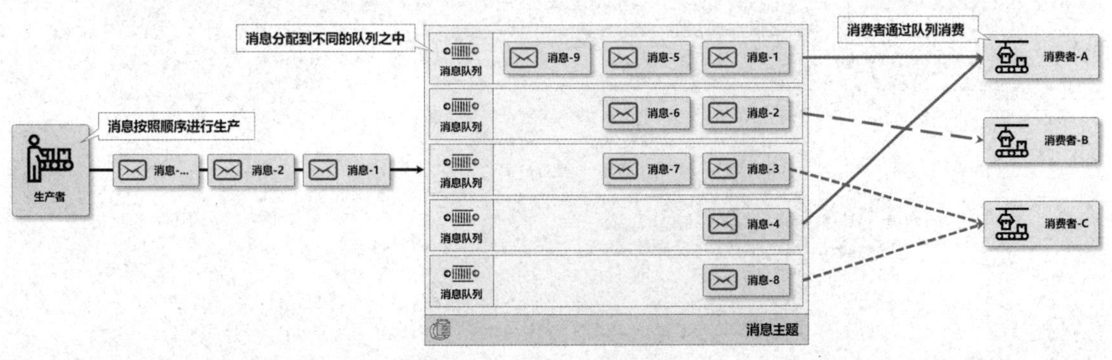

图 8-61　消息消费顺序混乱

要实现消息的顺序消费，一般需要在生产端和消费端上分别进行处理。首先生产端需要自己控制消息队列的选择（MessageQueueSelector 接口实现），而后消费端则可以根据顺序消费（MessageListenerOrderly 接口实现）。这样每一个消费者就可以根据顺序来实现不同消息队列的消费处理。下面通过具体的步骤来实现这一功能。

（1）【rocketmq 子模块】生产消息并选择消息队列。

```
package com.yootk.rocketmq.orders;
public class OrderMessageProducer {
    // RocketMQ连接所需要的相关配置项，略
    public static final String TAGS[] = new String[]{
            "tagA", "tagB", "tagC", "tagD", "tagE"};                // 业务标签
    public static void main(String[] args) throws Exception {
        RPCHook clientHook = new AclClientRPCHook(
                new SessionCredentials(ACCESSKEY, SECRETKEY));     // ACL认证
        DefaultMQProducer producer = new DefaultMQProducer(
                PRODUCER_GROUP, clientHook);                        // 创建消息生产者
        producer.setNamesrvAddr(NAME_SERVER_LIST);                  // NameServer
        producer.start();                                           // 启动消息生产者
        for (int x = 0; x < 100; x++) {                             // 循环发送消息
            Message msg = new Message(TOPIC, TAGS[x % TAGS.length],
                    ("【" + String.format("%03d", x) + "】沐言科技：www.yootk.com")
                            .getBytes(RemotingHelper.DEFAULT_CHARSET)); // 创建消息
            producer.sendOneway(msg, new MessageQueueSelector() {   // 队列选择器
                @Override
                public MessageQueue select(List<MessageQueue> mqs,
                    Message msg, Object arg) {
                    int index = ((Integer) arg) % mqs.size();       // 获取设置的ID
                    return mqs.get(index);                          // 选择指定队列
                }
            }, x % 5);                                              // 单向消息
        }
        producer.shutdown();                                        // 关闭消息发送者
    }
}
```

本程序实现了一个单向消息的发送，在进行消息发送时，通过 MessageQueueSelector 接口对象实例，实现了消息队列的选择。这样就具备了不同消息队列顺序消费消息的前提。

（2）【rocketmq 子模块】定义消息消费者 A。

```
package com.yootk.rocketmq.orders;
public class OrderMessageConsumerA {
    // RocketMQ连接所需要的相关配置项，略
    public static void main(String[] args) throws Exception {
        RPCHook clientHook = new AclClientRPCHook(
                new SessionCredentials(ACCESSKEY, SECRETKEY));     // ACL认证
        DefaultMQPushConsumer consumer = new DefaultMQPushConsumer(CONSUMER_GROUP,
                clientHook, new AllocateMessageQueueAveragely()); // Push消费端
        consumer.setNamesrvAddr(NAME_SERVER_LIST);              // NameServer列表
        consumer.subscribe(TOPIC, "tagA||tagC");                // 指定tag
        consumer.registerMessageListener(new MessageListenerOrderly() { // 顺序监听
            @Override
            public ConsumeOrderlyStatus consumeMessage(
                    List<MessageExt> msgs, ConsumeOrderlyContext context) {
                for (MessageExt msg : msgs) {                   // 不同消费者使用不同的标记
                    System.out.printf("【消费者-A】QueueID = %s，消息ID = %s，" +
                            "消息TAG = %s、CommitLogOffset = %s，消息内容 = %s %n",
                            msg.getQueueId(), msg.getMsgId(), msg.getTags(),
                            msg.getCommitLogOffset(), new String(msg.getBody()));
                }
                return ConsumeOrderlyStatus.SUCCESS;            // 消费成功标记
            }
        });
        consumer.start();                                       // 启动消费端
    }
}
```

（3）【rocketmq 子模块】定义消费者 B（只列出核心代码）。

定义消息分组	`private static final String CONSUMER_GROUP = "muyan-group-b"; // 消费组`
设置标签表达式	`consumer.subscribe(TOPIC, "tagB||tagE");`

（4）【rocketmq 子模块】定义消费者 C（只列出核心代码）。

定义消息分组	`private static final String CONSUMER_GROUP = "muyan-group-c"; // 消费组`
设置标签表达式	`consumer.subscribe(TOPIC, "tagD");`

此时一共创建了 3 个消费者，同时不同的消费者匹配了不同的消息标签，最重要的是每一个消费端在进行消息监听时是通过 MessageListenerOrderly 接口实现的，这样就可以根据当前的消息队列来实现消息的顺序消费。

8.5.3　延迟消息

延迟消息

视频名称　0824_【掌握】延迟消息

视频简介　为了实现合理的业务与消息之间的关联，RocketMQ 支持延迟消息的功能。本视频为读者讲解延迟消息设计存在的意义，同时通过具体的实例讲解延迟消息的开发以及 Broker 相关配置项的作用。

生产者创建完成消息后（构建 Message 对象实例），在默认情况下只要调用了 send()方法就会进行发送处理。但是在一些情况下，有些消息有可能需要延迟一段时间发出。例如，用户创建了一个购物订单后，可能又会在不久后取消订单，为了应对这种情况，可以延迟 10 分钟向仓库系统发出与该订单有关的仓储变更以及订单配送的消息，如图 8-62 所示。

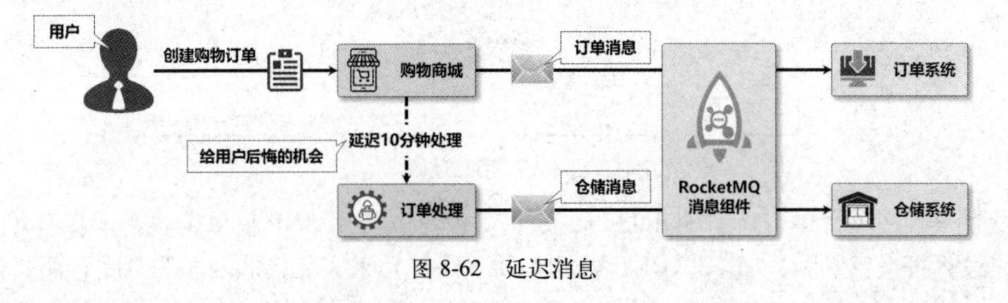

图 8-62　延迟消息

范例：【rocketmq 子模块】定义延迟消息

```
package com.yootk.rocketmq;
public class MessageProducerDelay {
   // RocketMQ连接所需要的相关配置项，略
   public static void main(String[] args) throws Exception {
      RPCHook clientHook = new AclClientRPCHook(
           new SessionCredentials(ACCESSKEY, SECRETKEY));    // ACL认证
      DefaultMQProducer producer = new DefaultMQProducer(
           PRODUCER_GROUP, clientHook);                        // 创建消息生产者
      producer.setNamesrvAddr(NAME_SERVER_LIST);               // NameServer地址
      producer.start();                                        // 启动消息生产者
      Message msg = new Message(TOPIC, "沐言科技：www.yootk.com"
            .getBytes(RemotingHelper.DEFAULT_CHARSET));        // 创建消息处理机制
      msg.setDelayTimeLevel(2);                                // 消息延迟5s发出
      producer.sendOneway(msg);                                // 消息发送
      producer.shutdown();                                     // 关闭消息发送者
   }
}
```

本程序在创建 Message 对象实例后通过 setDelayTimeLevel()方法设置了消息发送的延迟时间级别为 2（5 秒），这样消息发送到 Broker 5 秒后消费端才能够收到该消息。

提示：RocketMQ 不支持自定义延迟时间。

虽然 RocketMQ 提供了延迟消息的机制，但是该机制是不能够由用户自定义延迟时间的，只能使用固定的时间间隔。在默认情况下 RocketMQ 固定的时间间隔配置为（s=秒，m=分，h=时）：

```
1s 5s 10s 30s 1m 2m 3m 4m 5m 6m 7m 8m 9m 10m 20m 30m 1h 2h
```

如果希望延迟 5s 发送消息，则延迟级别应该设置为 2。同样的道理，要想延迟 30s 发送，则延迟级别设置为 4 即可。要想修改延迟级别，可以通过 messageDelayLevel 配置项进行 Broker 定义。

8.5.4 消息过滤

消息过滤

视频名称 0825_【掌握】消息过滤

视频简介 在 RocketMQ 中可以通过标签实现不同子业务的消息逻辑处理，而除了此机制外还有更加灵活的消息过滤机制，即 SQL92 过滤。本视频为读者分析 SQL92 过滤的作用，同时讲解 MessageSelector 实现 SQL92 过滤的程序。

RocketMQ 中可以对生产或消费的数据执行一些 SQL92 过滤逻辑的处理，核心的流程在于，消息发送时根据需要设置相关的业务属性内容，而在消费端进行消费时，就可以根据消息生产者所发送的属性内容按照 SQL92 标准进行判断，如果满足条件则可以进行消费，否则将无法消费，如图 8-63 所示。

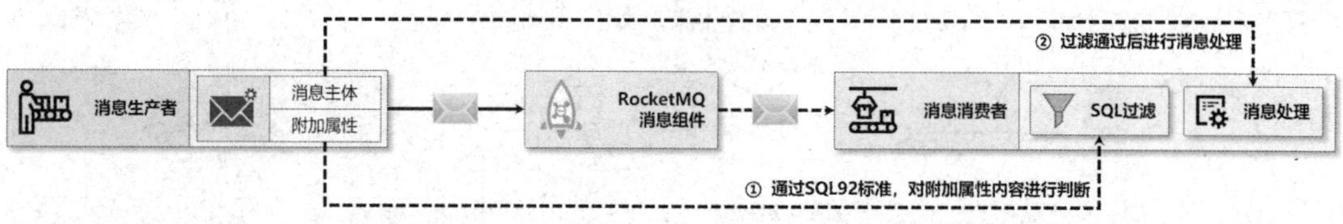

图 8-63 消息过滤

开发者一旦在项目中引入了 SQL92 过滤支持，就可以使用 SQL 提供的各类运算符（比较运算符、逻辑运算符、空判断、LIKE、BETWEEN…AND 等）来实现属性内容的判断，而具体的判断操作则可以通过 MessageSelector 类来实现。这一机制不能与标签机制混用，并且要求在 Broker 上配置启用才可以生效，下面讲解具体的实现步骤。

（1）【rocketmq-server 主机】修改 broker.conf 配置文件，增加启用 SQL92 过滤支持的配置项。

打开配置文件	`vi /usr/local/rocketmq/conf/broker.conf`
启用过滤支持	`enablePropertyFilter=true`

（2）【rocketmq-server 主机】配置完成后重新启动 Broker 服务进程。

```
nohup /usr/local/rocketmq/bin/mqbroker -c /usr/local/rocketmq/conf/broker.conf > /usr/local/
rocketmq/logs/rocketmq-broker.log 2>&1 &
```

（3）【rocketmq 子模块】消费端需要通过 MessageSelector.bySql()方法编写判断表达式，满足判断条件则可以接收消息。

```
package com.yootk.rocketmq;
public class MessageConsumer {
    // RocketMQ连接所需要的相关配置项，略
    public static void main(String[] args) throws Exception {
        RPCHook clientHook = new AclClientRPCHook(
                new SessionCredentials(ACCESSKEY, SECRETKEY));              // ACL认证
        DefaultMQPushConsumer consumer = new DefaultMQPushConsumer(CONSUMER_GROUP,
                clientHook, new AllocateMessageQueueAveragely());          // 推模式消费端
        consumer.setNamesrvAddr(NAME_SERVER_LIST);                         // NameServer列表
```

```
        // 如果要基于标签判断，则使用MessageSelector.byTag("tag表达式")方法
        // 本次是通过SQL92表达式判断所传递的level属性范围是否在8 ~ 12
        consumer.subscribe(TOPIC, MessageSelector.bySql("level BETWEEN 8 AND 12"));
        consumer.registerMessageListener(new MessageListenerConcurrently() { // 消息监听
            @Override
            public ConsumeConcurrentlyStatus consumeMessage(List<MessageExt> msgs,
                                                ConsumeConcurrentlyContext context) {
                System.out.printf("【%s】接收到新的消息：%s %n",
                        Thread.currentThread().getName(), msgs);      // 消息输出
                return ConsumeConcurrentlyStatus.CONSUME_SUCCESS;  // 消费成功标记
            }
        });
        consumer.start();                                          // 启动消费端
    }
}
```

（4）【rocketmq 子模块】生产端在发送消息时可以通过 Message 类提供的 putUserProperty()方法进行附加属性设置。

```
package com.yootk.rocketmq;
public class MessageProducer {
    // RocketMQ连接所需要的相关配置项，略
    public static void main(String[] args) throws Exception {
        RPCHook clientHook = new AclClientRPCHook(
                new SessionCredentials(ACCESSKEY, SECRETKEY));    // ACL认证
        DefaultMQProducer producer = new DefaultMQProducer(
                PRODUCER_GROUP, clientHook);                     // 创建消息生产者
        producer.setNamesrvAddr(NAME_SERVER_LIST);               // NameServer地址
        producer.start();                                        // 启动消息生产者
        Message msg = new Message(TOPIC, "沐言科技：www.yootk.com"
                .getBytes(RemotingHelper.DEFAULT_CHARSET));     // 创建消息
        msg.putUserProperty("level", "10");                      // 设置属性内容
        producer.sendOneway(msg);                                // 消息发送
        producer.shutdown();                                     // 关闭消息发送者
    }
}
```

此时生产端启动后会将消息发送到 Broker 之中，同时由于附加属性的存在，只有匹配消费端定义表达式的消息才可以被成功消费。

8.5.5 消息批处理

消息批处理

视频名称 0826_【掌握】消息批处理

视频简介 面对大规模消息生产的应用环境，为了提高传输效率可以通过批量方式进行处理。本视频为读者分析批量消息的设计意义，并通过具体代码介绍如何实现。

在消息组件生产端进行消息发送时，一般都是通过 DefaultMQProducer 类所提供的 send()方法实现，而该方法在每次发送时都只会发送一条消息。这样在进行大批量消息发送时，就会形成图 8-64 所示的状况。

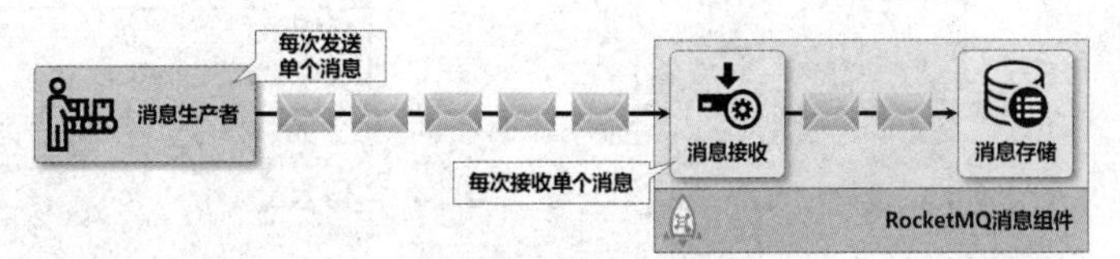

图 8-64　接收单个消息

在图 8-64 所示的处理中，每次只能够接收单个消息，同时还要持续进行批量消息的接收，这

样服务的处理性能肯定是不足的。所以最佳的做法是，将要发送的大批量消息拆分为不同的部分(每部分的大小限定在 4MB 以内)，这样就可以将一大批完整的消息分割为若干个不同批次，实现消息的批量发送，如图 8-65 所示。

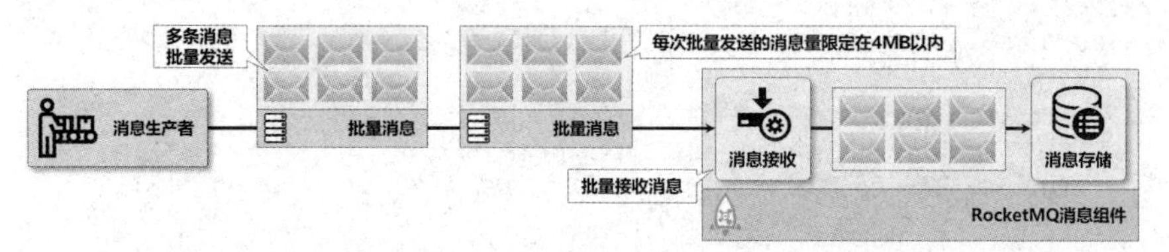

图 8-65　批量发送消息

范例：批量发送消息

```
package com.yootk.rocketmq;
public class MessageProducer {
    // RocketMQ连接所需要的相关配置项，略
    public static void main(String[] args) throws Exception {
        RPCHook clientHook = new AclClientRPCHook(
             new SessionCredentials(ACCESSKEY, SECRETKEY));   // ACL认证
        DefaultMQProducer producer = new DefaultMQProducer(
             CONSUMER_GROUP, clientHook);                      // 创建消息生产者
        producer.setNamesrvAddr(NAME_SERVER_LIST);             // NameServer地址
        producer.start();                                      // 启动消息生产者
        List<Message> messages = new ArrayList<>();            // 消息集合
        for (int x = 0; x < 100; x++) {                        // 循环发送消息
            Message msg = new Message(TOPIC, "沐言科技：www.yootk.com"
                .getBytes(RemotingHelper.DEFAULT_CHARSET));    // 创建消息处理机制
            messages.add(msg);                                 // 保存消息
        }
        SendResult result = producer.send(messages);           // 批量发送
        System.out.printf("【批量发送结果】发送状态：%s %n", result.getSendStatus());
        producer.shutdown();                                   // 关闭消息发送者
    }
}
```

程序执行结果	【批量发送结果】发送状态：SEND_OK

本程序首先将要发送的消息数据保存在 List 集合之中，随后就可以按照原始的形式，利用 DefaultMQProducer 类所提供的 send()方法将集合中所保存的全部消息数据发送到 Broker。虽然以上方式可以实现消息的批量发送，但是如果开发者在集合中保存的数据过多（超过 4MB），那么也是无法实现消息正常发送的。

要想解决集合过大而导致消息发送失败的问题，最佳的做法是创建一个集合拆分工具类。该工具类可以将一个完整的集合按照其保存消息的长度拆分为若干个子集合(每个子集合的大小不超过 4MB，如果遇见超大的消息则可以进行消息丢弃)，而后通过迭代的形式每次将子集合的消息批量发送到 Broker，如图 8-66 所示。

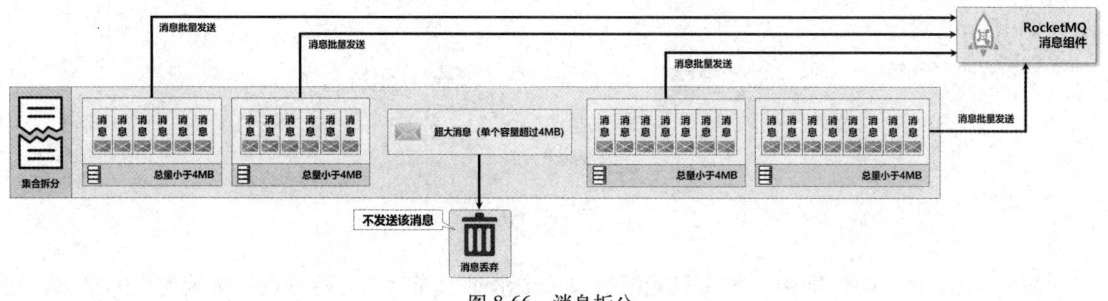

图 8-66　消息拆分

范例：定义集合拆分工具

```
package com.yootk.rocketmq;
public class MessageListSplitter implements Iterator<List<Message>> {
   private static final int SIZE_LIMIT = 1024 * 1024 * 4;                    // 每个集合的大小
   private List<Message> messages;                                           // 保存消息数据
   private int currentIndex;                                                 // 当前数据量
   public MessageListSplitter(List<Message> messages) {                      // 发送消息集合
      this.messages = messages;                                              // 集合保存
   }
   @Override
   public boolean hasNext() {                                                // 继续迭代
      return this.currentIndex < messages.size();                            // 未超过保存个数
   }
   @Override
   public List<Message> next() {                                             // 获取子集合
      int startIndex = this.getBeginIndex();                                 // 获取开始索引
      int nextIndex = startIndex;                                            // 集合结束索引
      int totalSize = 0;                                                     // 发送总长度
      for (; nextIndex < this.messages.size(); nextIndex++) {                // 循环获取集合数据
         Message message = this.messages.get(nextIndex);                     // 获取消息
         int messageSize = this.calcMessageSize(message);                    // 计算单个消息长度
         if (messageSize + totalSize > SIZE_LIMIT) {                         // 超过长度上限
            break;                                                           // 中断循环
         } else {                                                            // 未超过上限
            totalSize += messageSize;                                        // 保存长度
         }
      }
      List<Message> subMessageList = this.messages.subList(startIndex, nextIndex);
      this.currentIndex = nextIndex;                                         // 修改当前索引
      return subMessageList;                                                 // 返回单次发送子集合
   }
   private int getBeginIndex() {                                             // 获取截取索引
      Message currentMessage = messages.get(this.currentIndex);              // 获取当前消息
      int messageSize = calcMessageSize(currentMessage);                     // 计算当前消息长度
      while(messageSize > SIZE_LIMIT) {                                      // 单个消息超过总长度
         this.currentIndex += 1;                                             // 下一条消息
         Message message = this.messages.get(this.currentIndex);             // 获取消息
         messageSize = calcMessageSize(message);                             // 计算消息长度
      }
      return this.currentIndex;                                              // 返回数据索引
   }
   private int calcMessageSize(Message message) {                            // 计算单个消息长度
      int tempSize = message.getTopic().length() +
                     message.getBody().length;                               // 当前消息长度
      Map<String, String> properties = message.getProperties();              // 获取消息属性
      for (Map.Entry<String, String> entry : properties.entrySet()) {        // 属性迭代
         tempSize += entry.getKey().length() + entry.getValue().length();    // 计算属性长度
      }
      tempSize = tempSize + 20;                                              // 日志长度（20字节）
      return tempSize;                                                       // 返回单个消息长度
   }
}
```

本程序的构造方法首先需要接收要发送的全部消息集合，而后为了方便生产者获取子集合数据（批量发送的消息内容），本类实现了 Iterator 父接口。这样每次调用 next()方法时可以获取一个子集合，同时该子集合的消息量也是通过消息的长度（消息内容、附加属性、日志信息）动态计算得来的。

范例：生产端通过拆分集合实现批量消息发送

```
package com.yootk.rocketmq;
public class MessageProducer {
```

```
    // RocketMQ连接所需要的相关配置项，略
    public static void main(String[] args) throws Exception {
        // 与消息生产者有关的重复代码，略
        List<Message> messages = new ArrayList<>();                        // 消息集合
        for (int x = 0; x < 100; x++) {                                    // 循环发送消息
            Message msg = new Message(TOPIC, "沐言科技：www.yootk.com"
                .getBytes(RemotingHelper.DEFAULT_CHARSET));                // 创建消息处理机制
            messages.add(msg);                                             // 保存消息
        }
        MessageListSplitter messageSplitter = new MessageListSplitter(messages);
        while (messageSplitter.hasNext()) {                                // 集合拆分
            SendResult result = producer.send(messageSplitter.next());     // 批量发送
            System.out.printf("【批量发送结果】发送状态：%s %n", result.getSendStatus());
        }
        producer.shutdown();                                               // 关闭消息发送者
    }
}
```

此时的生产者会将要发送的消息集合保存在 MessageListSplitter 工具类对象实例之中，这样每次调用 hasNext()就可以获取部分集合，并利用 send()方法实现批量发送（发送数据的大小不超过 4MB）。

8.5.6　日志消息处理

日志消息处理

视频名称　0827_【掌握】日志消息处理

视频简介　RocketMQ 虽然没有提供像 Kafka 那样优秀的日志采集性能，但是在一些日志量较小的环境下，也可以整合已有的日志组件，实现日志采集的功能。本视频通过具体的实例讲解 Logback 日志的采集处理。

在项目运行过程中，日志是非常重要的跟踪手段，所以 RocketMQ 提供了单向消息类型，同时为了便于用户实现日志记录，也提供了专属的“rocketmq-logappender”依赖库。这样就可以将之与已有的日志组件结合在一起，将所有通过日志组件记录的日志消息发送到 RocketMQ，实现日志采集的功能，采集架构如图 8-67 所示。下面通过具体步骤演示日志采集处理功能的实现。

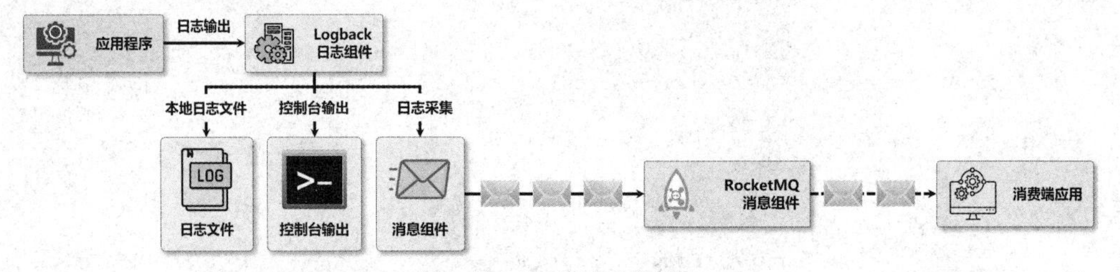

图 8-67　日志消息采集架构

（1）【microcloud 项目】为便于读者观察日志消息操作机制，本次演示将创建一个 rocketmq-logback 子模块，该子模块主要通过 Logback 向 RocketMQ 发出日志消息。修改 build.gradle 配置文件，引入 RocketMQ 日志支持库。

```
project('rocketmq-logback') {  // 子模块
    dependencies {
        implementation('org.apache.rocketmq:rocketmq-client:4.9.0')        // RocketMQ依赖
        implementation('org.apache.rocketmq:rocketmq-acl:4.9.0')           // ACL依赖
        implementation('org.apache.rocketmq:rocketmq-logappender:4.9.0')  // 日志依赖
        implementation('ch.qos.logback:logback-core:1.2.4')               // Logback日志组件
        implementation('org.slf4j:slf4j-api:1.7.32')                      // slf4j标准
        implementation('ch.qos.logback:logback-classic:1.2.4')            // Logback实现类
    }
}
```

（2）【rocketmq-logback 子模块】在 src/main/resources 源代码目录中创建 logback.xml 配置文件。

```
<?xml version="1.0" encoding="UTF-8"?>
<configuration>
   <property name="LOG_HOME" value="c:/log" />                      <!-- 日志保存目录 -->
   <appender name="STDOUT"
       class="ch.qos.logback.core.ConsoleAppender">                 <!-- 控制台输出 -->
      <Encoding>UTF-8</Encoding>                                     <!-- 日志编码 -->
      <layout class="ch.qos.logback.classic.PatternLayout">          <!-- 日志格式 -->
         <pattern>%d{HH:mm:ss.SSS} [%thread] %-5level %logger{50} - %msg%n</pattern>
      </layout>
   </appender>
   <appender name="FILE"
         class="ch.qos.logback.core.rolling.RollingFileAppender"><!-- 本地日志文件 -->
      <Encoding>UTF-8</Encoding>                                     <!-- 日志编码 -->
      <!-- 定义日志文件的生成结构，将每天的日志保存在一个文件之中，最多保留30天日志 -->
      <rollingPolicy class="ch.qos.logback.core.rolling.TimeBasedRollingPolicy">
         <FileNamePattern>${LOG_HOME}/rocketmq.%d{yyyy-MM-dd}.log</FileNamePattern>
         <MaxHistory>30</MaxHistory>
      </rollingPolicy>
      <layout class="ch.qos.logback.classic.PatternLayout">          <!-- 日志格式 -->
         <pattern>%d{HH:mm:ss.SSS} [%thread] %-5level %logger{50} - %msg%n</pattern>
      </layout>
   </appender>
   <!-- 定义RocketMQ连接的相关属性内容，注意：在RocketMQLogback中没有ACL配置支持 -->
   <appender name="RocketMQAppender"
         class="org.apache.rocketmq.logappender.logback.RocketmqLogbackAppender">
      <tag>logback</tag>                                             <!-- 消息标签 -->
      <topic>TopicLogback</topic>                                    <!-- 消息主题 -->
      <producerGroup>logback-group</producerGroup>                   <!-- 消息分组 -->
      <nameServerAddress>rocketmq-server:9876</nameServerAddress>    <!-- NameServer -->
      <layout><pattern>%date %p %t - %m%n</pattern></layout>         <!-- 消息格式 -->
   </appender>
   <appender name="RocketMQAsyncAppender"
         class="ch.qos.logback.classic.AsyncAppender">               <!-- 异步消息 -->
      <queueSize>1024</queueSize>                                    <!-- 阻塞队列长度 -->
      <discardingThreshold>80</discardingThreshold>                  <!-- 丢弃阈值 -->
      <maxFlushTime>2000</maxFlushTime>                              <!-- 刷新时间 -->
      <neverBlock>true</neverBlock>                                  <!-- 异步处理 -->
      <appender-ref ref="RocketMQAppender"/>                         <!-- 配置引用 -->
   </appender>
   <logger name="com.yootk" level="debug"/>                          <!-- 日志级别 -->
   <root level="DEBUG">                                              <!-- 日志级别 -->
      <appender-ref ref="RocketMQAppender" />                        <!-- 消息日志 -->
      <appender-ref ref="STDOUT" />                                  <!-- 控制台日志 -->
      <appender-ref ref="FILE" />                                    <!-- 文件日志 -->
   </root>
</configuration>
```

此时的日志配置中定义了与 RocketMQ 相关的配置项，而消息保存的主题名称为 TopicLogback，这样消费端就需要配置与之相同的主题名称才可以收到消息数据。

> **提示：RocketmqLogbackAppender 不支持 ACL。**
>
> 虽然 RocketMQ 提供了日志采集支持依赖库，但是在 RocketmqLogbackAppender 配置类中并没有提供 ACL 相关配置项。由于日志采集一般都是系统内部的工作，因此可以用 ACL 白名单的方式实现日志消息的正确发送。

（3）【rocketmq-logback 子模块】创建一个日志输出类。

```
package com.yootk.rocketmq.logback;
public class LogbackProducer {
   private final static Logger LOGGER =
         LoggerFactory.getLogger(LogbackProducer.class);          // 获取日志对象实例
   public static void main(String[] args) {
```

```
        LOGGER.info("【INFO】沐言科技：www.yootk.com");          // 普通日志
        LOGGER.error("【ERROR】沐言科技：www.yootk.com");        // 错误日志
        LOGGER.debug("【DEBUG】沐言科技：www.yootk.com");        // 调试日志
    }
}
```

程序执行结果	INFO　com.yootk.rocketmq.logback.LogbackProducer - 【INFO】沐言科技：www.yootk.com ERROR　com.yootk.rocketmq.logback.LogbackProducer - 【ERROR】沐言科技：www.yootk.com DEBUG　com.yootk.rocketmq.logback.LogbackProducer - 【DEBUG】沐言科技：www.yootk.com

程序运行之后，日志内容会在控制台显示，也会在本地日志文件中进行存储，同时被保存在 RocketMQ 中以实现日志采集。

8.5.7　事务消息

事务消息简介

视频名称　0828_【掌握】事务消息简介

视频简介　RocketMQ 提供了事务消息处理的机制，这个机制可以有效地解决分布式状态下的事务处理问题。本视频为读者分析分布式事务的实现意义，并通过具体的操作演示事务消息的执行与回调检查。

RocketMQ 消息组件的最大的特点在于对分布式事务的有效支持，分布式状态下微服务的数据更新操作可以实现 ACID 设计原则。例如，现在有部门微服务和雇员微服务两个不同的服务接口，需要实现一个新部门及雇员创建的业务处理，那么为了保证数据更新的正确性，就需要引入分布式事务处理，如图 8-68 所示。

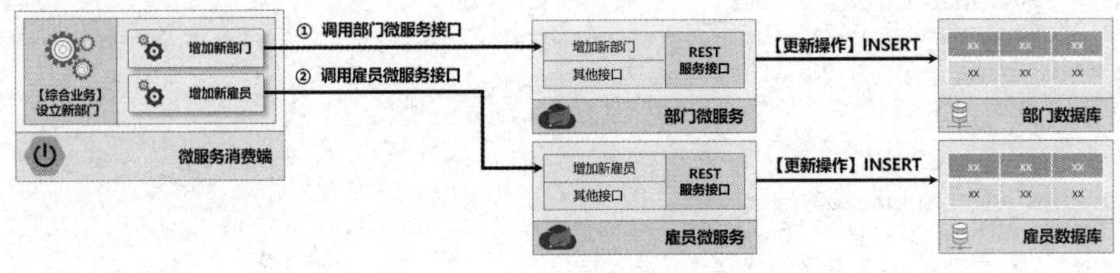

图 8-68　分布式事务处理

提示：本部分以事务消息概念为主。

事务消息最大的特点在于二阶段提交，而这一操作可以很好地应用于分布式事务的开发。考虑到阿里巴巴公司已经提供了 Seata 应用组件，所以此处仅以 RocketMQ 事务消息的概念为主。有兴趣的读者可以在学习完 Seata 之后尝试将其与 RocketMQ 编码融合，以实现高性能的分布式事务应用开发。

为了解决此类事务处理需要，RocketMQ 提供了 HalfMessage（半消息）的处理机制。该机制属于二阶段提交方式，即发送到 Broker 中的消息数据暂时不能被消费端消费，而是等待进一步提交的指令。如果给出了 COMMIT 指令，则该消息允许被消费端消费；如果给出了 ROLLBACK 指令，则该消息将被删除。事务消息处理流程如图 8-69 所示。每一个操作阶段的具体功能如下。

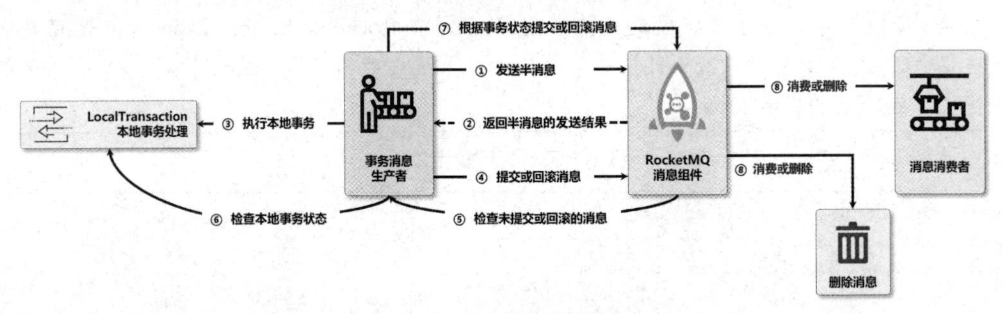

图 8-69　事务消息处理流程

（1）消息的生产者以同步的形式发送一个事务消息 Half Message 到 Broker，表示需要进行分布式事务处理。

（2）Broker 收到事务消息后，会将其保存在一个名称为 RMQ_SYS_TRANS_HALF_TOPIC 的内置默认半消息主题之中，写入成功后给生产者返回成功状态。

（3）在本地消息生产端可以获取该事务消息的事务 ID（TransactionID），同时进行本地事务处理。

（4）如果本地事务成功处理，则提交“COMMIT”状态；如果失败，则提交“ROLLBACK”状态；如果处理超时或某些特殊状况导致无法正常提交，则可以先返回“UNKNOW”状态（会触发 Broker 事务回查机制）。

（5）为防止事务处理超时所带来的问题，所有返回“UNKNOW”状态的消息会定期被 Broker 回查。若生产者执行本地事务超过 6 秒，则进行第一次事务回查。总共回查 15 次（回查间隔 60 秒），在每次回查时将此事务消息再一次写入“RMQ_SYS_TRANS_HALF_TOPIC”主题。

（6）在本地事务状态回查时，可以根据当前事务消息的 ID 来获取本地事务的处理状态。

（7）回查完成后会对所有状态为“COMMIT”及“ROLLBACK”的消息再次进行提交或回滚处理。

（8）最终状态为“COMMIT”的消息可以被消费者消费，最终状态为“ROLLBACK”的消息将被删除。

为了便于事务消息的实现处理，RocketMQ 提供了 TransactionMQProducer 事务消息的生产者处理类。当通过该类提供的 sendMessageInTransaction()方法进行消息发送时，会将一个半消息发送到 Broker，随后需要等待本地的事务处理完成，这样就需要通过 TransactionListener 接口进行本地事务执行与回查检测的具体实现。该接口提供了两个重要的事务处理方法。

- executeLocalTransaction()：执行本地事务，即每一个独立的业务单元都进行本地事务的调用。
- checkLocalTransaction()：事务消息回查，即处理那些未及时完成或处理超时的本地事务。

因为不同的项目有不同的业务功能，所以事务处理的方式也会有所不同。这样就需要开发者根据自身业务的需要去实现 TransactionListener 接口，而后将此接口实例绑定在 TransactionMQProducer 实例化对象之中，以实现对事务消息的控制，其实现结构如图 8-70 所示。下面将通过一系列的具体操作模拟事务消息的处理。

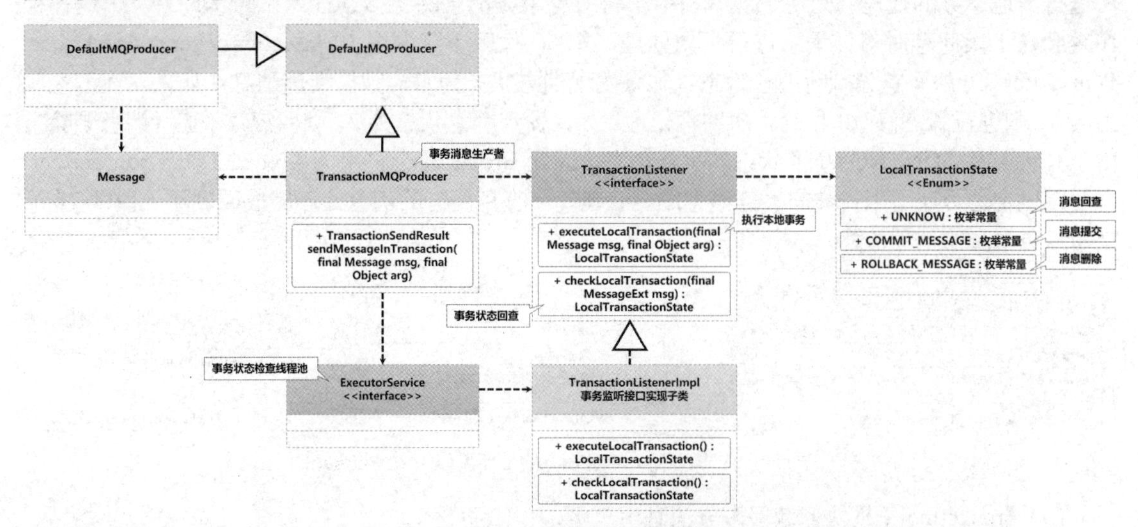

图 8-70　事务消息类实现结构

（1）【rocketmq-server 主机】为便于观察可以修改回查时间。本次将间隔时间修改为 10s。

打开配置文件	`vi /usr/local/rocketmq/conf/broker.conf`
增加新配置项	`transactionCheckInterval=10000`

（2）【rocketmq 子模块】创建 TransactionListener 接口子类并实现业务消息的状态处理。

```
package com.yootk.rocketmq.transaction;
public class TransactionListenerImpl implements TransactionListener {
   private AtomicInteger transactionIndex = new AtomicInteger(0);       // 操作计数
   // 【模拟】保存每一个事务消息对应的业务执行状态（key = TransactionId、value = 业务状态标记）
   private ConcurrentHashMap<String, Integer> localTransMap = new ConcurrentHashMap<>();
   @Override                                                            // 执行本地业务处理
   public LocalTransactionState executeLocalTransaction(Message msg, Object arg) {
      int value = transactionIndex.getAndIncrement();                   // 模拟业务处理
      int status = value % 3;                                           // 模拟业务计算
      this.localTransMap.put(msg.getTransactionId(), status);           // 保存业务状态
      // 业务处理存在操作耗时，所以将业务处理结果以回查的方式进行处理（回查通过后提交）
      return LocalTransactionState.UNKNOW;                              // 回查处理
   }
   @Override                                                            // 事务消息回查
   public LocalTransactionState checkLocalTransaction(MessageExt msg) {
      Integer status = this.localTransMap.get(msg.getTransactionId()); // 获取业务状态
      if (null != status) {                                             // 存在业务状态
         switch (status) {                                              // 状态判断
            case 0:                                                     // 状态判断
               int value = transactionIndex.getAndIncrement();          // 模拟业务处理
               status = value % 3;                                      // 模拟业务计算
               this.localTransMap.put(msg.getTransactionId(), status);// 保存业务状态
               return LocalTransactionState.UNKNOW;                     // 回查处理
            case 1:                                                     // 状态判断
               return LocalTransactionState.COMMIT_MESSAGE;             // 消息提交
            case 2:                                                     // 状态判断
               return LocalTransactionState.ROLLBACK_MESSAGE;           // 消息回滚
         }
      }
      return LocalTransactionState.COMMIT_MESSAGE;                      // 消息提交
   }
}
```

生产者将半消息发送到 Broker 之后，就会调用 TransactionListener.executeLocalTransaction()方法进行本地事务的处理。为便于观察，本次将所有事务消息的状态设置为“UNKNOW”，这样程序就会在下次回查时对该消息进行再次处理。第 1 次回查（调用 TransactionListener.checkLocalTransaction()方法）首先判断消息状态：如果发现消息状态为 1，则提交该消息；如果发现消息状态为 0，则进行状态修改，留给下次回调处理；如果发现消息状态为 2，则回滚该消息（消息删除）。由于此时还有“UNKNOWN”状态的事务消息，因此 Broker 还会再次调用 checkLocalTransaction()方法以确定是否还有未提交的事务消息需要处理，一直到所有的消息处理完成或删除。图 8-71 给出了事务消息的处理流程。

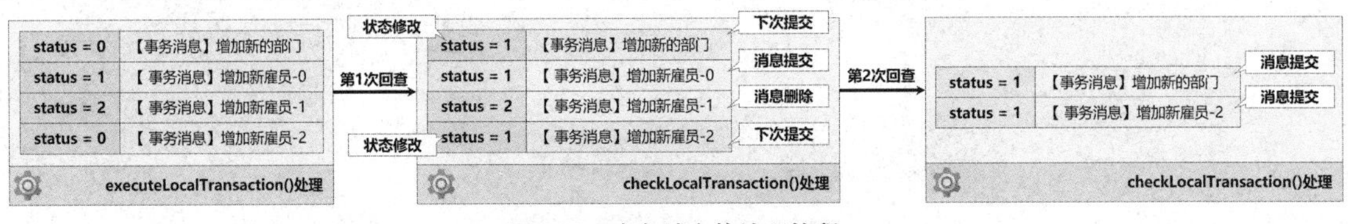

图 8-71　事务消息的处理流程

（3）【rocketmq 子模块】编写事务消息生产者。

```
package com.yootk.rocketmq.transaction;
public class TransactionMessageProducer {
   public static final String NAME_SERVER_LIST = "rocketmq-server:9876";    // NameServer
   public static final String PRODUCER_GROUP = "muyan-transaction-group";   // 生产组
   public static final String TOPIC = "TopicTransaction";                   // 主题
   public static final String ACCESSKEY = "RocketMQMuyan";                  // 用户名
```

```
    public static final String SECRETKEY = "helloyootk";              // 密码
    public static void main(String[] args) throws Exception {
        RPCHook clientHook = new AclClientRPCHook(
            new SessionCredentials(ACCESSKEY, SECRETKEY));             // ACL认证
        TransactionListener transactionListener = new TransactionListenerImpl();
        TransactionMQProducer producer = new TransactionMQProducer(
                PRODUCER_GROUP, clientHook);                           // 消息生产者
        producer.setNamesrvAddr(NAME_SERVER_LIST);                     // NameServer地址
        ExecutorService executorService = Executors.newFixedThreadPool(8) ; // 线程池
        producer.setExecutorService(executorService);                  // 事务状态检查线程池
        producer.setTransactionListener(transactionListener);          // 事务监听
        producer.start();                                              // 启动生产者
        sendDeptMessage(producer);                                     // 增加部门消息
        sendEmpMessage(producer);                                      // 增加雇员消息
        TimeUnit.MINUTES.sleep(Long.MAX_VALUE);                        // 等待事务消息处理
        producer.shutdown();                                           // 关闭生产者
    }
    public static void sendDeptMessage(TransactionMQProducer producer) throws Exception {
        Message msg = new Message(TOPIC, "dept",
                "【DEPT】增加新部门事务".getBytes(RemotingHelper.DEFAULT_CHARSET));
        TransactionSendResult result = producer.sendMessageInTransaction(
                msg, "DeptProvider");                                  // 发送事务消息
        TimeUnit.MICROSECONDS.sleep(10);                               // 模拟调用间隔
    }
    public static void sendEmpMessage(TransactionMQProducer producer) throws Exception {
        for (int x = 0; x < 3; x++) {
            Message msg = new Message(TOPIC, "emp",
                    ("【EMP】增加新雇员事务 - " + x).getBytes(RemotingHelper.DEFAULT_CHARSET));
            TransactionSendResult result = producer.sendMessageInTransaction(
                    msg, "EmpProvider");                               // 发送事务消息
            TimeUnit.MICROSECONDS.sleep(10);                           // 模拟调用间隔
        }
    }
}
```

此时程序一共发出了 4 条事务消息（对应部门业务以及雇员业务），而后发出的消息会交由 TransactionListener 接口子类进行处理。同时为了便于事务状态检查，还需要为生产者提供一个检查线程池实例。

8.6 本章概览

1．RocketMQ 是基于 Java 开发的消息组件，最早由阿里巴巴公司开发，现在已交由 Apache 基金会更新维护。

2．RocketMQ 是一种介于 RabbitMQ 与 Kafka 之间的消息组件，其性能比 RabbitMQ 更强，在日志采集功能实现中的处理性能低于 Kafka，但是比 Kafka 可靠。

3．RocketMQ 4.4.0 后的版本提供了 ACL 安全机制，可以基于此机制实现用户认证与授权操作管理，使得第三方平台的接入更加便捷。

4．由于配置环境的原因，RocketMQ 并没有直接提供可用管理控制台。开发者可以根据自身需要，对控制台源代码进行编译与部署，以实现服务管理的需求。

5．RocketMQ 中有 4 个核心部分：NameServer（保存主题路由数据）、Broker（消息处理）、Producer（生产者）、Consumer（消费者）。每一个核心部分都可能以集群的形式在应用中呈现。

6．RocketMQ 中所传递的消息由 4 个部分组成：消息长度、序列化类型与消息头长度、消息头、消息主体。

7．RocketMQ 中的所有 Broker 需要向 NameServer 注册并发送心跳数据，每次的心跳数据包含

最新的路由信息。NameServer 根据心跳检测判断 Broker 存活状态，而生产端与消费端也需要与 Broker 建立心跳检测机制，以便 Broker 确定其存活状态。

8．RocketMQ 中的所有数据都保存在 CommitLog 文件之中，每一个 CommitLog 文件的大小限制为 1GB，这样便于实现内存映射，提高文件的保存性能。

9．RocketMQ 提供了同步刷盘与异步刷盘两种机制，其中异步刷盘可以利用内存映射以及操作系统提供的 PageCache 缓存技术实现异步数据写入。

10．为了提高 NameServer 处理性能，不同的 NameServer 彼此独立，并且会保存相同的数据，这样就避免了数据同步所带来的性能损耗问题。

11．RocketMQ 集群中 Broker 有 4 种集群形式：多 Master 模式、多 Master-Slave 异步复制模式、多 Master-Slave 同步写入模式、DLedger（多副本）模式。其中 DLedger 模式主要采用的是集群方案，并且提供了失败切换支持。

12．RocketMQ 客户端进行消息开发时，可以利用主题、业务标签、识别码以及命名空间实现不同业务的划分与管理，实际开发中要根据自身的需求进行技术选用。

13．RocketMQ 中的消费端是以分组的形式进行管理的，在同一组中的多个消费端可以实现队列消息的处理，而不同分组的消费端可以实现广播消息处理。

14．RocketMQ 支持 SQL92 过滤标准，而在使用之前需要手动修改 Broker 配置项（enablePropertyFilter=true）开启。

15．为了提高消息的传输与处理性能，可以基于批处理的方式进行消息的批量发送，每次发送的长度不超过 4MB。

16．RocketMQ 可以与 Logback 日志组件整合实现日志消息的采集处理。

17．事务消息基于二阶段消息处理模式，会先进行半消息的发送。如果本地事务处理完成后的结果为“COMMIT”状态，则会将该消息提交到消费端；如果结果为“ROLLBACK”，则会进行消息丢弃。

18．事务消息处理中会对消息状态为“UNKNOW”的消息进行回查处理，默认的回查间隔为 60 秒，回查 15 次，这些配置都可以在 Broker 中进行修改。

第 9 章

微服务辅助技术

本章学习目标

1. 掌握 Seata 分布式事务组件的实现原理，可以熟练地搭建 Seata 服务应用；

2. 掌握 Seata 提供的 4 种事务处理模式（AT、XT、TCC 以及 Saga），并可以使用这些模式实现分布式事务开发；

3. 理解 Spring Cloud Stream 主要作用与实现架构，并可以基于 RocketMQ 实现 Spring Cloud Stream 应用；

4. 理解 Spring Cloud Config 的作用，并可以结合 Nacos 实现微服务配置管理与服务更新；

5. 理解 Spring Cloud Bus 的主要作用，并可以结合 Spring Cloud Config 实现服务的动态更新管理。

微服务实现了完整业务的拆分，随之而来的就是分布式事务的实现需要。本章将为读者全面讲解 Seata 分布式事务组件的实现原理与具体应用，还将介绍基于 RocketMQ 实现 Spring Cloud Stream 标准化开发，以及在 Netflix 套件中的 Spring Cloud Config 动态配置技术。

9.1 Spring Cloud Stream

Spring Cloud Stream 简介

视频名称 0901_【理解】Spring Cloud Stream 简介

视频简介 Spring Cloud Stream 是一种事件驱动型的设计框架，可以轻松地以标准的结构实现微服务业务的整合处理。本视频从传统的消息组件操作问题出发，讲解消息驱动型微服务框架的主要作用，并分析其核心的组成架构。

在项目开发中，为了方便解决不同服务之间的业务通信，往往会基于消息组件进行平台业务整合。这样不仅可以降低不同模块之间的业务耦合，也可以避免高峰流量所带来的安全隐患，如图 9-1 所示。

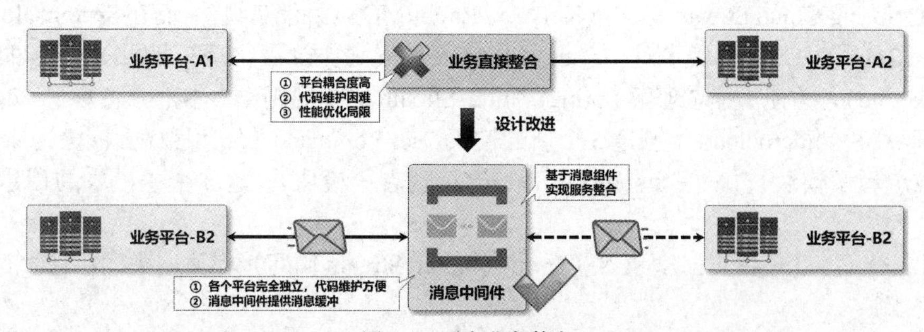

图 9-1　平台业务整合

既然确定了基于消息组件的业务整合方案，那么随后就需要在项目中搭建消息服务。由于现代项目开发中可以选择的消息组件很多，如 RabbitMQ、Kafka、RocketMQ 等，因此在不同的项目中有可能使用不同的消息组件。然而随着项目的不断完善，就有可能形成多种消息组件混用的情况，如图 9-2 所示。

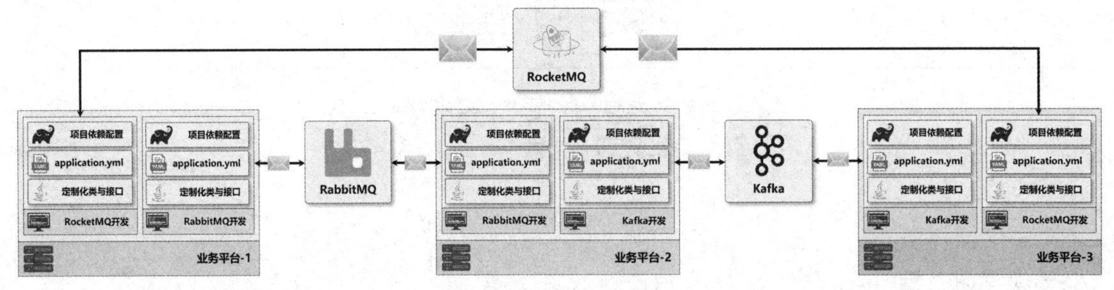

图 9-2　消息组件混用

为了解决消息组件使用的统一性问题，Spring Cloud 提供了 Spring Cloud Stream 开发框架，该框架可用于构建高度可扩展的基于事件驱动的微服务，其实现架构如图 9-3 所示。

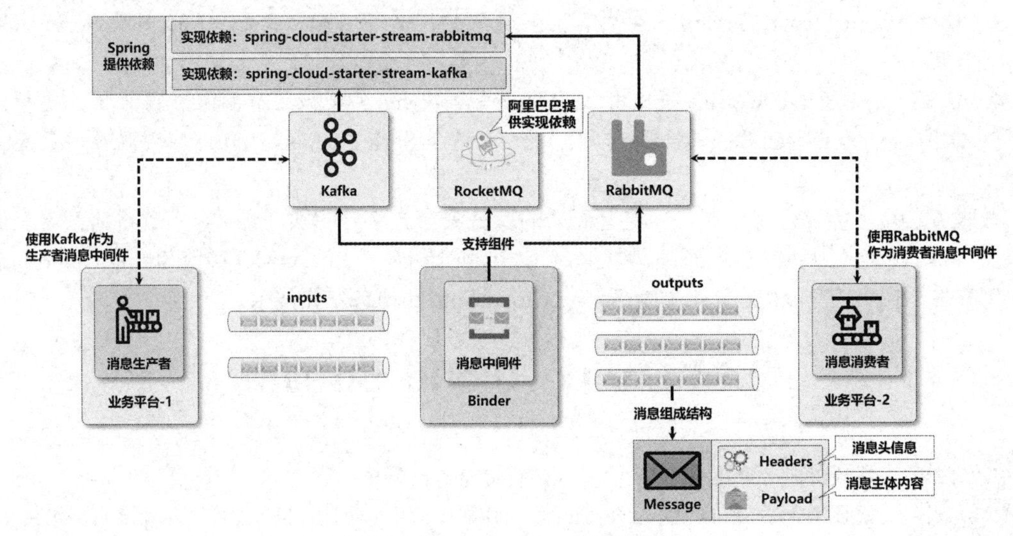

图 9-3　Spring Cloud Stream 实现架构

利用 Spring Cloud Stream 可以对消息组件实现进一步的封装，使其开发代码更具通用性。这样就彻底解决了开发人员对某一具体消息中间件的依赖，从而使开发人员可以“无感知地”使用各种消息中间件进行开发。最重要的是，可以方便地实现消息中间件的混用，例如，生产者使用 Kafka，而消费者使用 RabbitMQ。

最初的 Spring Cloud Stream 只支持 Kafka 与 RabbitMQ 两种消息组件，而在 Spring Cloud Alibaba 引入之后，RocketMQ 也提供了对 Spring Cloud Stream 的支持（实现依赖：spring-cloud-starter-stream-rocketmq）。为便于读者理解 Spring Cloud Stream 的实现操作，本部分将基于 RocketMQ 进行讲解，同时将在 microcloud 中创建 3 个新的模块：scs-command（公共程序子模块）、scs-producer（Stream 生产者子模块）、scs-consumer（Stream 消费者子模块）。这 3 个子模块的创建与配置步骤如下。

（1）【microcloud 项目】创建 SCS（Spring Cloud Stream 简写）相关模块，并修改 build.gradle 配置模块依赖。

```
project('scs-common') {}                                    // 公共子模块
project('scs-producer') {                                   // 生产端子模块
    dependencies {
```

```
        implementation('com.alibaba.cloud:spring-cloud-starter-stream-rocketmq:2021.1')
        implementation(project(':scs-common'))                   // 引入公共子模块
    }
}
project('scs-consumer') {                                         // 消费端子模块
    dependencies {
        implementation('com.alibaba.cloud:spring-cloud-starter-stream-rocketmq:2021.1')
        implementation(project(':scs-common'))                   // 引入公共子模块
    }
}
```

（2）【scs-common 子模块】Spring Cloud Stream 最主要的特点是可以直接实现对象的传输，生产者会自动将其转为 JSON 数据（Jackson 实现转换），而后消费者在获取消息数据时也可以自动将其反序列化为具体的对象实例，如图 9-4 所示。为便于该类的管理与维护，下面将在公共子模块创建一个部门传输类。

```
package com.yootk.common.dto;
import lombok.Data;
import java.io.Serializable;
@Data                                                     // Lombok结构生成注解
public class DeptDTO implements Serializable {            // 部门数据传输类
    private Long deptno;                                  // 部门编号
    private String dname;                                 // 部门名称
    private String loc;                                   // 部门位置
}
```

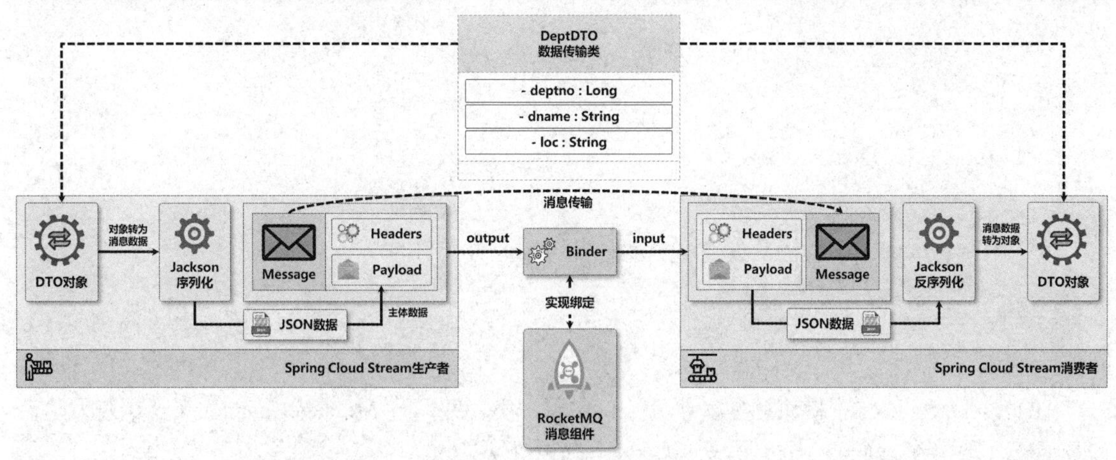

图 9-4　数据传输类

（3）【scs-common 子模块】由于此模块需要被生产端与消费端引用，因此需要对当前模块进行编译，命令为 gradle build。

9.1.1　SCS 消息生产者

SCS 消息生产者

视频名称　0902_【理解】SCS 消息生产者

视频简介　Spring Cloud Stream 提供了统一的生产者消息处理规范。本视频将为读者讲解基于 RocketMQ 的 SCS 生产端服务配置，并通过具体的代码介绍如何实现消息发送处理。

Spring Cloud Stream 是一套标准的消息驱动程序模型，这就意味着只要开发者根据模型编写出了程序代码，那么这套代码就可以在已支持的消息组件之中任意移植。本次演示将介绍基于图 9-5 所示的结构实现一个消息生产者。

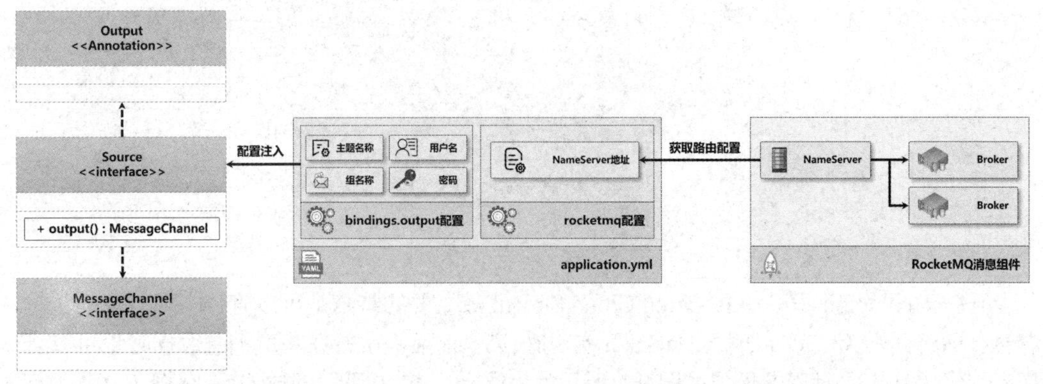

图 9-5　Spring Cloud Stream 输出配置的结构

在进行 Spring Cloud Stream 输出配置时，需要在“spring.cloud.stream.bindings.[名称]”配置项下进行，而这些配置都是为了构建 MessageChannel 的接口实例，只要存在该接口实例就可以实现消息的发送。下面通过具体的开发实例，介绍如何实现消息生产者的开发。

（1）【scs-producer 子模块】修改 application.yml 文件，并配置 RocketMQ 服务连接信息。

```
spring:                                                 # Spring配置
  cloud:                                                # Cloud配置项
    stream:                                             # SCS配置项
      rocketmq:                                         # 整合RoketMQ
        binder:                                         # 服务绑定
          name-server: rocketmq-server:9876             # NameServer地址
      bindings:                                         # 操作绑定
        output:                                         # 输出配置
          destination: TopicSSC                         # 消息主题
          group: scs-producer-group                     # 生产组
          access-key: RocketMQMuyan                     # 用户名
          secret-key: helloyootk                        # 密码
```

（2）【scs-producer 子模块】创建一个消息发送的业务接口。

```
package com.yootk.service;
public interface IDeptMessageService {                          // 业务接口
    public void sendMessage(DeptDTO dept);                      // 发送部门消息
}
```

（3）【scs-producer 子模块】创建业务接口实现子类，并通过 MessageChannel 实现消息发送。

```
package com.yootk.service.impl;
@Service
public class DeptMessageServiceImpl implements IDeptMessageService {     // 业务接口实现子类
    @Autowired
    private MessageChannel output;                                       // 消息通道
    @Override
    public void sendMessage(DeptDTO dept) {
        this.output.send(MessageBuilder.withPayload(dept).build());      // 发送消息
    }
}
```

（4）【scs-producer 子模块】消息传输需要将对象转为 JSON 数据，所以需要创建一个 Jackson 提供的 ObjectMapper 实例。

```
package com.yootk.config;
@Configuration
public class ProducerConfig {                                            // 生产者配置类
    @Bean
    public ObjectMapper getJacksonObjectMapper() {                       // 对象序列化
        return new ObjectMapper();
    }
}
```

（5）【scs-producer 子模块】创建程序启动类，并设置启用 Stream 输出配置。

```
package com.yootk;
@SpringBootApplication
@EnableBinding(Source.class)                                    // 配置绑定
public class StartRocketMQStreamProducerApplication {
    public static void main(String[] args) {
        SpringApplication.run(StartRocketMQStreamProducerApplication.class, args);
    }
}
```

（6）【scs-producer 子模块】创建一个测试类，实现消息发送。

```
package com.yootk.test;
@ExtendWith(SpringExtension.class)                              // JUnit 5测试工具
@WebAppConfiguration                                            // 启动Web配置
@SpringBootTest(classes = StartRocketMQStreamProducerApplication.class) // 启动类
public class TestDeptMessageService {
    @Autowired
    private IDeptMessageService messageService;                 // 注入业务实例
    @Test
    public void testSend() throws Exception{
        DeptDTO dept = new DeptDTO();                           // 实例化DTO对象
        dept.setDeptno(99L);                                    // DTO对象属性设置
        dept.setDname("沐言科技教学研发部");                    // DTO对象属性设置
        dept.setLoc("北京");                                    // DTO对象属性设置
        this.messageService.sendMessage(dept);                  // 消息发送
        TimeUnit.SECONDS.sleep(20);                             // 等待发送完成
    }
}
```

（7）【RocketMQ 控制台】消息发送完成后，可以通过 RocketMQ 控制台进行消息查看，如图 9-6 所示。

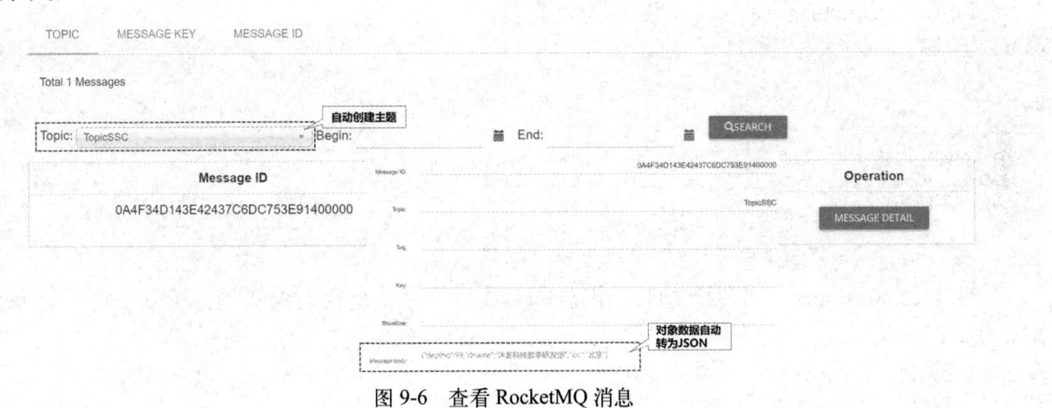

图 9-6　查看 RocketMQ 消息

9.1.2　SCS 消息消费者

SCS 消息消费者

视频名称　0903_【理解】SCS 消息消费者

视频简介　Spring Cloud Stream 消息消费者依然采用监听的形式实现消息数据的接收。本视频将介绍基于 RocketMQ 实现消费端的配置与消息接收。

Spring Cloud Stream 消息接收依然遵循着监听的操作特点，只要设置好监听的主题，每当有消息时，就可以通过 SubscribableChannel 订阅通道进行接收。如果传输的是一个对象，也会基于 Jackson 组件自动将 JSON 数据转为指定的对象实例。在使用 Spring Cloud Stream 实现消费端的标准开发时，需要通过 Input 配置项来实现 Sink 通道的定义，程序的实现结构如图 9-7 所示。下面通过具体的操作步骤介绍消费端代码。

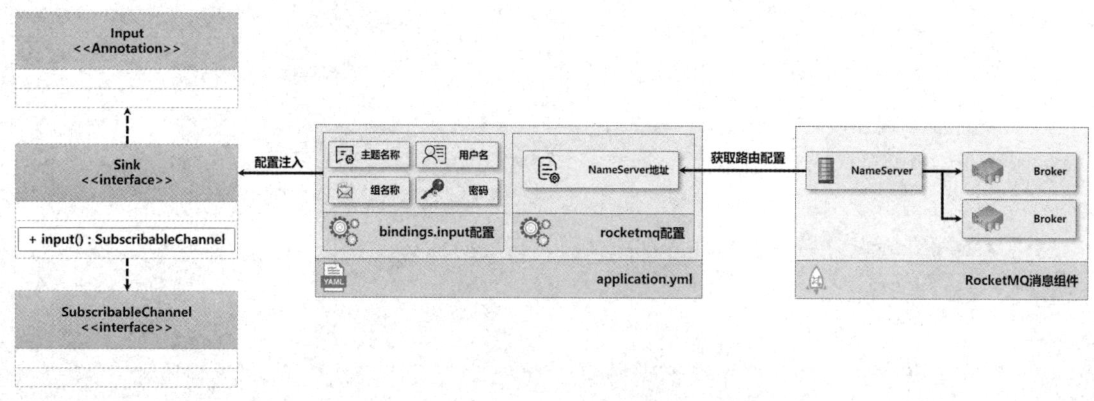

图 9-7　Spring Cloud Stream 消费端的实现结构

（1）【scs-consumer 子模块】创建 application.yml 配置文件，并进行消息输入源配置。

```
spring:                                                    # Spring配置
  cloud:                                                   # Cloud配置项
    stream:                                                # SCS配置项
      rocketmq:                                            # 整合RoketMQ
        binder:                                            # 服务绑定
          name-server: rocketmq-server:9876                # NameServer地址
      bindings:                                            # 操作绑定
        input:                                             # 配置默认消费通道
          destination: TopicSSC                            # 消息主题
          group: scs-consumer-group                        # 消费组
          access-key: RocketMQMuyan                        # 用户名
          secret-key: helloyootk                           # 密码
```

（2）【scs-consumer 子模块】创建消息监听类。

```
package com.yootk.service;
@Component                                                 // Bean注册
@Slf4j                                                     // Lombok日志注解
public class DeptMessageListener {                         // 消息监听类
    @StreamListener(Sink.INPUT)                            // 监听通道
    public void receive(Message<DeptDTO> message) {        // 接收消息
        log.info(message.toString());                      // 日志输出
    }
}
```

（3）【scs-consumer 子模块】接收的消息需要反序列化为指定对象实例，所以创建一个 Jackson 配置类。

```
package com.yootk.config;
@Configuration
public class ConsumerConfig {                              // 消费端配置类
    @Bean
    public ObjectMapper getJacksonObjectMapper() {         // 消息反序列化处理
        return new ObjectMapper();
    }
}
```

（4）【scs-consumer 子模块】创建消费端启动类，并绑定订阅配置。

```
package com.yootk;
@SpringBootApplication
@EnableBinding(Sink.class)                                 // 配置绑定
public class StartRocketMQStreamConsumerApplication {
    public static void main(String[] args) {
        SpringApplication.run(StartRocketMQStreamConsumerApplication.class, args);
    }
}
```

消费端应用启动后会自动进行消息监听，一旦有消息发送过来就通过日志的形式进行消息的记录，同时该程序代码模型可以任意切换到所有支持 Spring Cloud Stream 标准的消息组件中。

9.1.3　消费过滤

视频名称　0904_【理解】消费过滤

视频简介　虽然 Spring Cloud Stream 采用更加标准的方式实现了消息驱动的通信模型，但是不同的消息组件有其自身的消息分类。本视频介绍对已有的代码进行改造，使其可以同时绑定多个不同的消费通道，并实现不同消息的过滤。

RocketMQ 消息组件除了提供基本的消息生产与消费支持之外，还可以在一个消息主题中根据消息标签（Tag 过滤）以及消息识别码（keys 过滤）、消息附加属性（SQL92 过滤）实现不同子业务的消息消费处理。Spring Cloud Stream 提供了这种 RocketMQ 特殊实现（其他消息组件整合时未必支持），可以基于消息头进行该操作的配置实现，如图 9-8 所示。下面通过具体步骤对已有的程序代码进行修改。

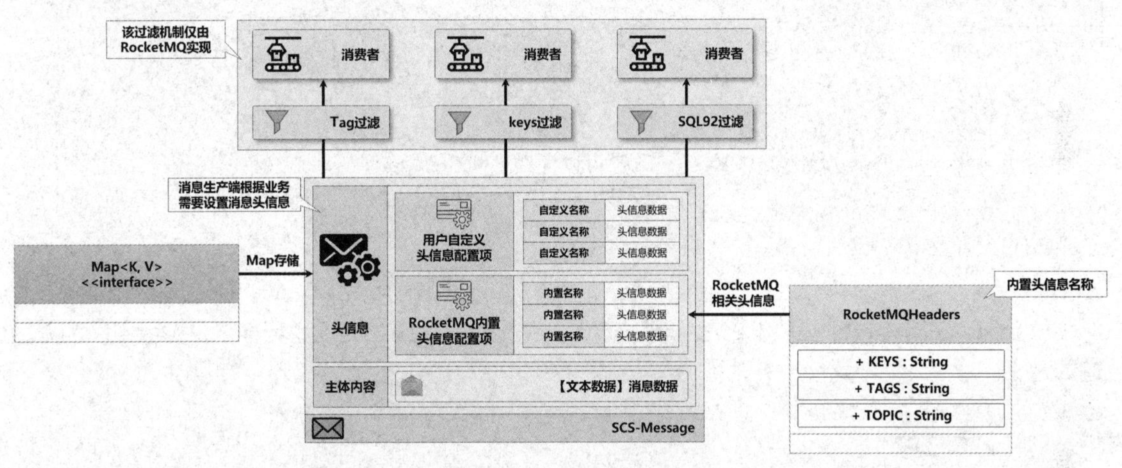

图 9-8　SCS 消费过滤支持

（1）【scs-producer 子模块】修改 DeptMessageServiceImpl 业务实现类，实现子业务消息发送。

```
package com.yootk.service.impl;
@Service
public class DeptMessageServiceImpl implements IDeptMessageService { // 业务实现类
    @Autowired
    private MessageChannel output;                              // 消息通道
    @Override
    public void sendMessage(DeptDTO dept) {
        {   // 发送第一条消息，该消息将传递标签
            Map<String, Object> headers = new HashMap<>();      // 头信息
            headers.put("author", "李兴华");                    // 自定义头信息
            headers.put("李兴华编程训练营", "edu.yootk.com");   // 自定义头信息
            headers.put(RocketMQHeaders.TAGS, "dept");          // 添加标签
            dept.setLoc("Tag消息");                             // 追加操作标记
            Message message = MessageBuilder.createMessage(dept,
                    new MessageHeaders(headers));               // 创建消息
            this.output.send(message);                          // 创建并发送消息
        }
        {   // 发送第二条消息，该消息将传递keys
            Map<String, Object> headers = new HashMap<>();      // 头信息
            headers.put("author", "李兴华");                    // 自定义头信息
            headers.put("李兴华编程训练营", "edu.yootk.com");   // 自定义头信息
            headers.put(RocketMQHeaders.KEYS, "dept-keys");     // 添加识别码
```

```
        dept.setLoc("Keys消息");                                    // 追加操作标记
        Message message = MessageBuilder.createMessage(dept,
              new MessageHeaders(headers));                         // 创建消息
        this.output.send(message);                                  // 创建并发送消息
    }
    {   // 发送第三条消息，该消息将传递附加属性，实现SQL92过滤操作
        Map<String, Object> headers = new HashMap<>();              // 头信息
        headers.put("author", "李兴华");                             // 自定义头信息
        headers.put("李兴华编程训练营", "edu.yootk.com");              // 自定义头信息
        headers.put("level", "10");                                 // 自定义头信息
        dept.setLoc("SQL92消息");                                    // 追加操作标记
        Message message = MessageBuilder.createMessage(dept,
              new MessageHeaders(headers));                         // 创建消息
        this.output.send(message);                                  // 创建并发送消息
    }
  }
}
```

（2）【scs-consumer 子模块】创建一个 YootkSink 消费接口，用于绑定 3 个不同的消息通道。

```
package com.yootk.channel;
public interface YootkSink {                                        // 自定义消费通道
    String INPUT_TAG = "inputTag";                                  // 通道标记常量
    String INPUT_KEYS = "inputKeys";                                // 通道标记常量
    String INPUT_SQL92 = "inputSQL92";                              // 通道标记常量
    @Input(INPUT_TAG)                                               // 消费通道标记
    SubscribableChannel inputTag();
    @Input(INPUT_KEYS)                                              // 消费通道标记
    SubscribableChannel inputKeys();
    @Input(INPUT_SQL92)                                             // 消费通道标记
    SubscribableChannel inputSQL92();
}
```

（3）【scs-consumer 子模块】修改 application.yml 文件，根据各类子业务标记，实现不同消费通道的定义。

```
spring:                                                             # Spring配置
  cloud:                                                            # Cloud配置项
    stream:                                                         # SCS配置项
      rocketmq:                                                     # 整合RoketMQ
        binder:                                                     # 服务绑定
          name-server: rocketmq-server:9876                         # NameServer地址
        bindings:                                                   # 通道绑定配置
          inputTag:                                                 # 消费通道名称
            consumer:                                               # 根据标签头信息匹配
              tags: dept||emp                                       # 匹配表达式
          inputKeys:                                                # 消费通道名称
            consumer:                                               # 根据keys头信息匹配
              kes: dept-keys                                        # 匹配内容
          inputSQL92:                                               # 消费通道名称
            consumer:                                               # 根据SQL92头信息匹配
              sql: 'level BETWEEN 8 AND 12'                         # SQL表达式
      bindings:                                                     # 操作绑定
        inputTag:                                                   # 配置消费通道
          destination: TopicSSC                                     # 消息主题
          group: scs-consumer-group-tag                             # 消费组
          access-key: RocketMQMuyan                                 # 用户名
          secret-key: helloyootk                                    # 密码
        inputKeys:                                                  # 配置消费通道
          destination: TopicSSC                                     # 消息主题
          group: scs-consumer-group-keys                            # 消费组
          access-key: RocketMQMuyan                                 # 用户名
          secret-key: helloyootk                                    # 密码
        inputSQL92:                                                 # 配置消费通道
          destination: TopicSSC                                     # 消息主题
```

```
      group: scs-consumer-group-sql92                        # 消费组
      access-key: RocketMQMuyan                              # 用户名
      secret-key: helloyootk                                 # 密码
```

（4）【scs-consumer 子模块】修改消息监听类，定义 3 个消息监听方法，用于实现 3 种子业务消息的匹配。

```
package com.yootk.service;
@Component                                                   // Bean注册
@Slf4j                                                       // Lombok日志注解
public class DeptMessageListener {                           // 消息监听
    @StreamListener(YootkSink.INPUT_TAG)                     // 监听通道
    public void receiveTag(Message<DeptDTO> message) {       // 接收消息
        log.info("【receiveTag()】头信息：{}、主体数据：{}",
            message.getHeaders(), message.getPayload());     // 日志输出
    }
    @StreamListener(YootkSink.INPUT_KEYS)                    // 监听通道
    public void receiveKeys(Message<DeptDTO> message) {      // 接收消息
        log.info("【receiveKeys()】头信息：{}、主体数据：{}",
            message.getHeaders(), message.getPayload());     // 日志输出
    }
    @StreamListener(YootkSink.INPUT_SQL92)                   // 监听通道
    public void receiveSQL92(Message<DeptDTO> message) {     // 接收消息
        log.info("【receiveSQL92()】头信息：{}、主体数据：{}",
            message.getHeaders(), message.getPayload());     // 日志输出
    }
}
```

（5）【scs-consumer 子模块】由于此时自定义了 YootkSink 通道类，因此要修改启动类中的注解定义。

```
package com.yootk;
@SpringBootApplication
@EnableBinding(YootkSink.class)                              // 绑定自定义通道配置类
public class StartRocketMQStreamConsumerApplication {
    public static void main(String[] args) {
        SpringApplication.run(StartRocketMQStreamConsumerApplication.class, args);
    }
}
```

消费端程序启动后，会启动 3 个不同的消费通道，而后结合 application.yml 中的绑定配置，就可以根据标签、keys、SQL92 标准的条件，实现各自业务消息的接收。

9.2 Spring Cloud Config

Spring Cloud Config 简介

视频名称 0905_【了解】Spring Cloud Config 简介

视频简介 Spring Cloud Config 提供了一种动态配置的解决方案。本视频为读者分析这种动态配置处理的主要作用，并介绍基于 GitLab 实现服务仓库的构建。

所有的项目都不是一成不变的。一些不断进行业务完善的项目需要进行代码的大量变更，因此在每次修改后都需要重新部署服务。而有些项目可能只是修改了一个配置文件项，在这样的情况下还需要重复进行服务的部署就有些不方便了，所以 Spring Cloud 套件提供了 Spring Cloud Config 动态配置技术，如图 9-9 所示。

在 Spring Cloud Config 开发中，可以将一些允许产生变动的配置项保存在 SVN 或 Git 这样的版本控制工具之中，而要使用 Spring Cloud Config 微服务，必须通过一个配置服务器进行配置加载。每当需要进行配置变更时，可以直接修改版本仓库中的配置内容，从而实现最终配置的动态更新。所以，为了便于配置管理，首先要在 GitLab 中创建一个新的仓库，仓库名称为“microcloud”，如图 9-10 所示。

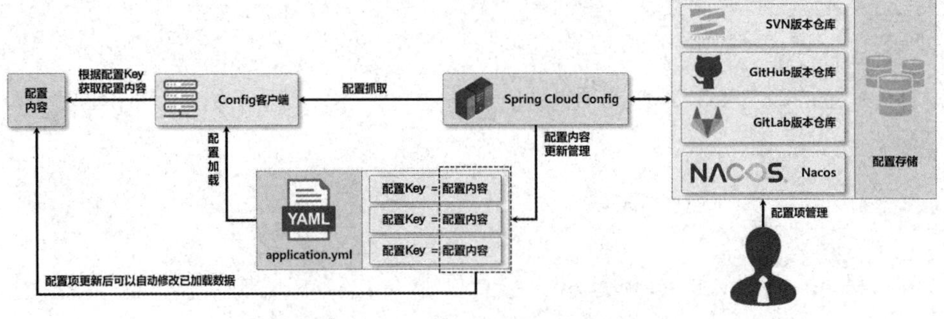

图 9-9 Spring Cloud Config 动态配置

图 9-10 创建 GitLab 仓库

提示：本次将基于 GitLab 讲解。

虽然 Spring Cloud Config 支持多种不同的数据存储形式，但是考虑到现在的版本都是以 Git 为主，所以推荐使用 GitHub 和 GitLab。又考虑到网络访问的稳定性以及配置安全性的问题，本书推荐使用 GitLab 来进行配置管理。不会配置 GitLab 的读者可以参考本套丛书中关于 Java 项目构建与管理的书自行学习。

仓库创建完成之后需要向仓库中保存配置文件，此时就需要开发者通过本地仓库实现配置文件的创建与管理，同时这些文件都要统一提交到 GitLab 的仓库之中，其开发结构如图 9-11 所示。下面通过具体的操作来演示本地仓库的创建、配置文件定义以及代码提交操作。

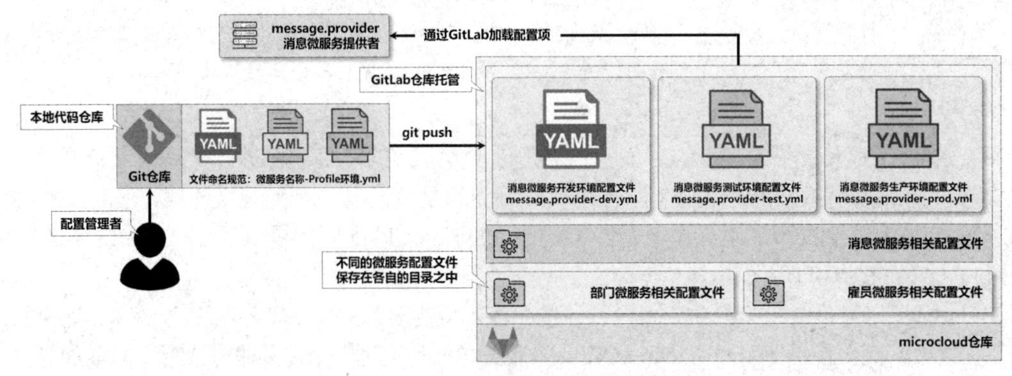

图 9-11 GitLab 仓库托管开发结构

（1）【GitBash 命令】为了便于通过 SSH 实现与 GitLab 的连接，需要在创建 SSH 密钥对时使用“-m”定义密钥格式，随后将生成的公钥信息配置到 GitLab 之中。

```
ssh-keygen -m PEM
```

（2）【本地 Git 仓库】在本地任意磁盘中创建一个新的目录，名称为“cloud-config”。

（3）【本地 Git 仓库】初始化本地 GIT 仓库。

```
git init
```

（4）【本地 Git 仓库】在 cloud-config 目录中创建 provider-message 目录，在该目录中创建 3 个配置文件，文件内容如下。

message.provider-dev.yml	yootk:　　# 自定义属性 message:　　# 自定义属性 flag: dev　　# 环境标记 content: 沐言科技：www.yootk.com　　# 消息内容
message.provider-test.yml	yootk:　　# 自定义属性 message:　　# 自定义属性 flag: test　　# 环境标记 content: 李兴华高薪就业编程训练营：edu.yootk.com　　# 消息内容
message.provider-prod.yml	yootk:　　# 自定义属性 message:　　# 自定义属性 flag: prod　　# 环境标记 content: 课程资源下载：www.yootk.com/resources　　# 消息内容

（5）【本地 Git 仓库】将当前目录中的 3 个配置文件添加到 Git 暂存库之中。

```
git add
```

（6）【本地 Git 仓库】将暂存库中保存的代码提交到版本库之中。

```
git commit -m "Create Message Config Files"
```

（7）【本地 Git 仓库】建立本地仓库与远程 GitLab 仓库之间的连接。

```
git remote add origin git@gitlab-server:muyan/microcloud.git
```

（8）【本地 Git 仓库】将本地代码提交到 GitLab 仓库之中。

```
git push -u origin master
```

（9）【远程 GitLab 仓库】代码提交完成后，访问 GitLab 仓库，可以发现图 9-12 所示的内容。

图 9-12　GitLab 仓库列表

9.2.1　Spring Cloud Config 服务端

Spring Cloud Config 服务端

视频名称　0906_【了解】Spring Cloud Config 服务端

视频简介　Spring Cloud Config 服务端需要与 GitLab 或 GitHub 连接，这样才可以实现配置抓取。本视频通过具体的实现操作，讲解配置服务器的搭建以及 SSH 连接配置。

所有的核心配置项都保存在了 GitLab 仓库之中，而要想进行该仓库的配置访问，就需要搭建一个 Config 服务端应用，在该应用中需要明确配置 GitLab 的仓库地址、SSH 私钥、仓库目录等。由于该应用也是一个 Spring Cloud 微服务，因此应该统一向 Nacos 注册，而后就可以在调用时基于 Nacos 注册名称来实现服务调用。项目的设计架构如图 9-13 所示。下面通过具体的步骤来介绍 Spring Cloud Config 服务端的开发。

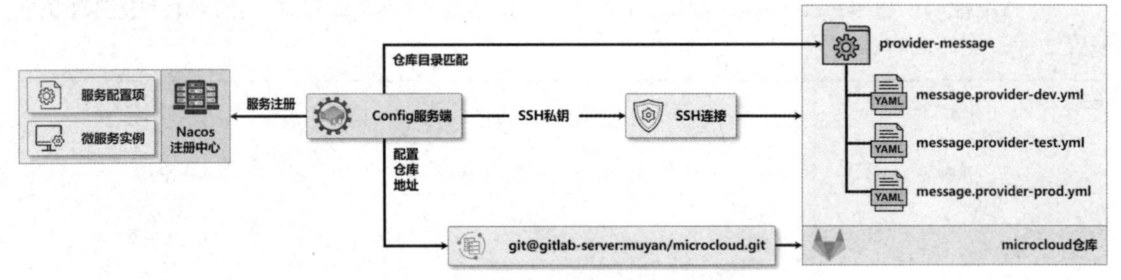

图 9-13 Spring Cloud Config 服务端设计结构

（1）【microcloud 项目】创建 config-server-7501 子模块，同时修改 build.gradle 配置文件添加模块依赖。

```
project('config-server-7501') {                              // 子模块
    dependencies {
        implementation('org.springframework.boot:spring-boot-starter-web')
        implementation('org.springframework.cloud:spring-cloud-config-server')
        implementation('de.codecentric:spring-boot-admin-starter-client:2.3.0')
        implementation('org.springframework.boot:spring-boot-starter-actuator')
        implementation('com.alibaba.cloud:' +
                'spring-cloud-starter-alibaba-nacos-discovery') {
            exclude group: 'com.alibaba.nacos', module: 'nacos-client'
        }
        implementation('com.alibaba.cloud:' +
                'spring-cloud-starter-alibaba-nacos-config') {
            exclude group: 'com.alibaba.nacos', module: 'nacos-client'
        }
        implementation('com.alibaba.nacos:nacos-client:2.0.0')
    }
}
```

（2）【config-server-7501 子模块】创建 application.yml 配置文件，定义 GitLab 连接配置项。

```
server:                                                 # 服务端配置
  port: 7501                                            # 7501端口监听
spring:                                                 # Spring配置
  application:                                          # 应用配置
    name: config.server                                 # 应用名称
  cloud:                                                # Spring Cloud配置
    nacos:                                              # Nacos注册中心
      discovery:                                        # 发现服务
        weight: 10                                      # 服务权重
        username: nacos                                 # 用户名
        password: nacos                                 # 密码
        service: ${spring.application.name}             # Nacos服务名称
        server-addr: nacos-server:8848                  # 服务地址
        max-retry: 1                                    # 重试次数
        namespace: 650fab32-c7dc-4ae1-8ac4-2dbdefd7e617 # 命名空间ID
        group: MICROCLOUD_GROUP                         # 注册分组
        cluster-name: YootkCluster                      # 集群名称
        metadata:                                       # 元数据
          version: 1.0                                  # 自定义数据项
          company: 沐言科技                              # 自定义数据项
          url: www.yootk.com                            # 自定义数据项
          author: 李兴华                                 # 自定义数据项
    config:                                             # Spring Cloud Config配置
```

```
    server:                                              # Config服务端配置
      git:                                               # Git连接配置
        uri: git@gitlab-server:muyan/microcloud.git      # Git地址
        default-label: master                            # 分支名称
        search-paths:                                    # 仓库目录匹配
          - provider-*                                   # 匹配以“provider-”开头的目录
        private-key:
          -----BEGIN RSA PRIVATE KEY-----
          用户通过sshkey-gen所生成的本地密钥内容，略
          -----END RSA PRIVATE KEY-----
```

（3）【config-server-7501 子模块】创建 bootstrap.yml 配置文件定义与 Nacos 连接的配置项（此部分代码与前面代码相同，略）。

（4）【config-server-7501 子模块】创建服务启动类。

```
package com.yootk;
@SpringBootApplication
@EnableDiscoveryClient
@EnableConfigServer                                         // 配置服务注解
public class StartConfigServerApplication7501 {
   public static void main(String[] args) {
      SpringApplication.run(StartConfigServerApplication7501.class, args); // 应用启动
   }
}
```

（5）【Nacos 控制台】向 Nacos 中注册一个名称为“config.server”的服务实例，这样就需要在 Nacos 中添加一个新的配置项，名称为“config.server.properties”，如图 9-14 所示。

Data Id	Group	归属应用:	操作
dept.provider.properties	MICROCLOUD_GROUP		详情 | 示例代码 | 编辑 | 删除 | 更多
gateway.config	MICROCLOUD_GROUP		详情 | 示例代码 | 编辑 | 删除 | 更多
emp.provider.properties	MICROCLOUD_GROUP		详情 | 示例代码 | 编辑 | 删除 | 更多
config.server.properties	MICROCLOUD_GROUP		详情 | 示例代码 | 编辑 | 删除 | 更多

新增Nacos配置项

图 9-14　新增 Nacos 配置项

（6）【本地系统】启动 ConfigServer 服务应用，同时为了便于访问修改本地的 hosts 文件，添加新的主机映射名称。

```
127.0.0.1   config-server-7501
```

（7）【Postman】配置完成后可以通过 ConfigServer 应用读取 GitLab 中的配置文件，在进行配置读取时，应该采用“{label}/{application.name}-{profile}.yml”的格式进行访问，如下所示。

读取开发环境配置	config-server-7501:7501/master/message.provider-dev.yml
读取测试环境配置	config-server-7501:7501/master/message.provider-test.yml
读取生产环境配置	config-server-7501:7501/master/message.provider-prod.yml

虽然此时的配置文件保存在 provider-message 目录之中，但是由于配置了目录的匹配模式项 search-paths，开发者只需要输入“{application.name}-{profile}.yml”（不需要输入目录名称）即可实现配置加载。

9.2.2　Spring Cloud Config 客户端

Spring Cloud Config 客户端

视频名称　0907_【了解】Spring Cloud Config 客户端

视频简介　配置服务端是为了便于实现 GitLab 仓库中配置项的抓取。本视频介绍创建一个客户端微服务，并通过微服务名称实现配置项的抓取。

Spring Cloud Config 只提供了配置存储仓库的配置与加载功能，而要真正将其应用在整个微服务的实现架构之中，还需要在项目中引入 Spring Cloud Config 客户端，如图 9-15 所示。

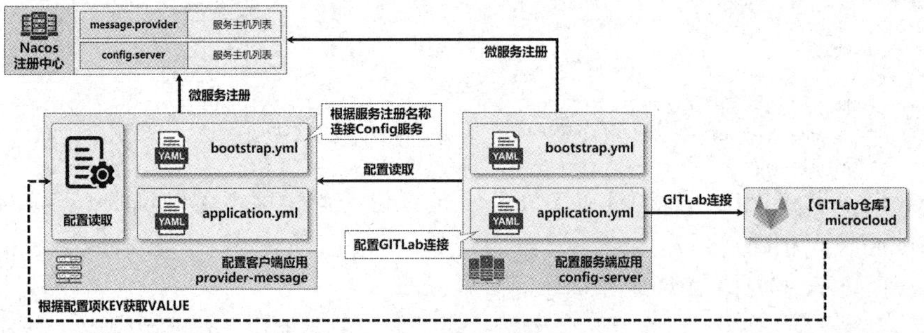

图 9-15　Spring Cloud Config 客户端

此时的 Config 客户端就是一个最终要进行业务功能提供的微服务，但是某些配置项需要通过 GitLab 中的配置文件来实现加载，这样就需要在客户端的 bootstrap.yml 配置文件中定义 Config 服务端的连接信息（服务地址、profile 名称、应用名称），而后在 Config 客户端应用启动时就可以自动实现所需配置的加载。下面通过具体的操作步骤介绍如何实现这一功能。

（1）【microcloud 项目】创建 provider-message-8201 子模块，随后修改 build.gradle 配置文件添加所需依赖项。

```
project('provider-message-8201') {
    dependencies {
        // 此处依赖与config-server-7501的类似，需要配置Nacos相关依赖、Spring Boot启动依赖
        // 并将spring-cloud-config-server依赖更换为spring-cloud-starter-config依赖
        implementation('org.springframework.cloud:spring-cloud-starter-config')
    }
}
```

（2）【provider-message-8201 子模块】创建 application.yml 配置文件，该文件的核心配置项如下。

```
server:                                   # 服务端配置
  port: 8201                              # 8201端口监听
spring:                                   # Spring配置
  application:                            # 应用配置
    name: message.provider                # 应用名称
  cloud:                                  # Spring Cloud配置
    nacos:                                # Nacos相关配置，略
```

此时配置的“message.provider”与 GitLab 仓库中保存的配置文件的名称相同，这样在进行配置文件加载时，就可以根据当前应用名称来实现自动匹配。

（3）【provider-message-8201 子模块】创建 bootstrap.yml 配置文件，该文件核心配置项如下。

```
spring:                                   # Spring配置
  application:                            # 应用配置
    name: message.provider                # 【重复配置】应用名称
  cloud:                                  # Spring Cloud配置
    config:                               # Spring Cloud Config配置
      name: ${spring.application.name}    # 资源名称
      profile: dev                        # 环境名称
      label: master                       # 分支名称
      discovery:                          # 发现服务
        enabled: true                     # 服务启用
        service-id: config.server         # Nacos注册名称
    nacos:                                # Nacos注册中心
      config:                             # Nacos相关配置，略
    discovery:                            # Nacos发现服务相关配置，略
```

在 Config 客户端进行配置加载时，可以通过 spring.cloud.config.uri 的配置项（http://config-server-7501:7501）定义配置服务加载地址，但是这样一旦 Config 服务端出现问题，将导致整个微服务集

群不可用。所以此时的程序将基于 Nacos 注册中心的注册服务名称来实现配置加载，这样就需要在 bootstrap.yml 中定义 Nacos 发现服务的相关配置。

（4）【provider-message-8201 子模块】创建一个 MessageAction 控制器类，实现配置项的读取。

```
package com.yootk.provider.action;
@RestController
@RequestMapping("/provider/message/*")                    // 映射父路径
public class MessageAction {
    @Value("${yootk.message.flag}")                        // 获取配置项内容
    private String flag;
    @Value("${yootk.message.content}")                     // 获取配置项内容
    private String content;
    @RequestMapping("config")                              // 映射子路径
    public Object config() {
        Map<String, Object> result = new HashMap<>();      // 保存返回结果
        result.put("flag", this.flag);                     // 保存结果项
        result.put("content", this.content);               // 保存结果项
        return result;
    }
}
```

（5）【provider-message-8201 子模块】创建消息微服务应用启动类。

```
package com.yootk.provider;
@SpringBootApplication
@EnableDiscoveryClient
public class StartMessageApplication8201 {
    public static void main(String[] args) {
        SpringApplication.run(StartMessageApplication8201.class, args); // 应用启动
    }
}
```

（6）【Nacos 控制台】由于此时新增加了一个微服务，因此需要在 Nacos 中添加 message.provider.properties 配置项，参考本书第 3 章。

（7）【本地系统】为便于服务访问，修改本地系统的 hosts 文件。

```
127.0.0.1   provider-message-8201
```

（8）【Postman 测试】在 message.provider 应用启动时后台会自动出现服务抓取的日志信息，随后可以通过 Postman 进行接口测试，如图 9-16 所示。

图 9-16　Postman 接口测试

9.2.3　Spring Cloud Bus

视频名称　0908_【了解】Spring Cloud Bus

视频简介　使用 Spring Cloud Config 的核心目的在于可以动态实现配置项的加载，而这一功能需要基于 Spring Cloud Bus 并结合消息组件完成。本视频在已有的 RocketMQ 服务的基础之上介绍如何实现配置的动态刷新操作。

此时已经成功地实现了 Spring Cloud Config 中的服务端与客户端的开发架构，但是在实际的开发中保存在 GitLab 仓库中的配置文件是极有可能被管理员更新的，而一旦更新，肯定需要 Config 客户端的动态内容加载。为了实现这一配置更新机制，Spring Cloud Config 提供了 Spring Cloud Bus 技术，开发者只需要在应用中结合消息组件即可实现配置的动态抓取。其实现架构如图 9-17 所示。

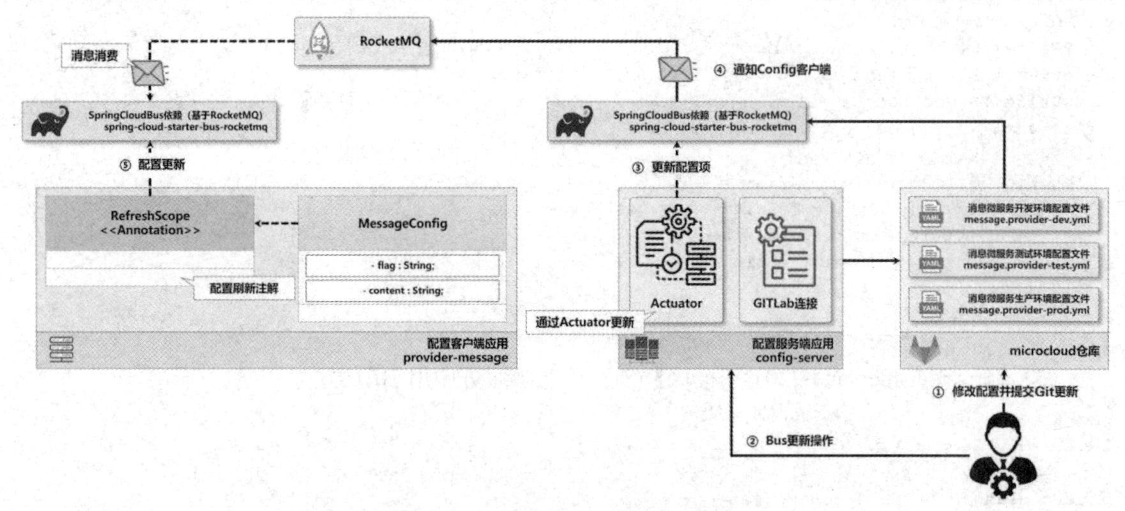

图 9-17　Spring Cloud Bus 的实现架构

Spring Cloud Bus 技术需要依赖于消息组件，本次可以基于 RocketMQ 实现，在整个实现过程中最重要的一项就是需要在动态更新的属性类中使用@RefreshScope 注解进行声明。下面通过具体的操作步骤介绍如何实现配置更新。

（1）【microcloud 项目】修改 build.gradle 配置文件，为 config-server-7501 与 provider-message-8201 两个子模块添加 spring-cloud-starter-bus-rocketmq 依赖库。

```
implementation('com.alibaba.cloud:spring-cloud-starter-bus-rocketmq')
```

（2）【config-server-7501、provider-message8201 子模块】修改 application.yml 配置文件，追加 RocketMQ 的整合配置项。

```
spring:                                                    # Spring配置
  application:                                             # 应用配置
    name: config.server                                    # 应用名称
  cloud:                                                   # Spring Cloud配置
    stream:                                                # 配置消息组件
      rocketmq:                                            # RocketMQ配置
        binder:                                            # 通道绑定
          name-server: rocketmq-server:9876                # NameServer地址
          access-key: RocketMQMuyan                        # 用户名
          secret-key: helloyootk                           # 密码
```

（3）【provider-message-8201 子模块】建立一个 MessageConfig 配置类，用于实现动态配置加载操作。

```
package com.yootk.provider.vo;
@Data                                                      // 自动生成类结构
@Component                                                 // Bean注册
@RefreshScope                                              // 动态加载
public class MessageConfig {
   @Value("${yootk.message.flag}")                         // 获取配置项内容
   private String flag;
   @Value("${yootk.message.content}")                      // 获取配置项内容
   private String content;
}
```

（4）【provider-message-8201 子模块】修改 MessageAction 类，通过 MessageConfig 实例返回配置数据。

```
package com.yootk.provider.action;
@RestController
@RequestMapping("/provider/message/*")                         // 映射父路径
public class MessageAction {
    @Autowired
    private MessageConfig messageConfig;                       // 消息配置类
    @RequestMapping("config")                                  // 映射子路径
    public Object config() {                                   // 控制层处理方法
        Map<String, Object> result = new HashMap<>();          // 保存返回结果
        result.put("flag", this.messageConfig.getFlag());      // 保存结果项
        result.put("content", this.messageConfig.getContent()); // 保存结果项
        return result;                                         // 返回配置结果
    }
}
```

（5）【config-server-7501 子模块】Spring Cloud Bus 需要 Actuator 的支持，所以此时需要修改 application.yml 配置文件，开放 Actuator 访问端口。

```
management:                                                    # Actuator配置
  endpoints:                                                   # 终端管理
    web:                                                       # Web终端
      exposure:                                                # 暴露刷新服务端点
        include: 'bus-refresh'                                 # Actuator刷新端口
```

（6）【Git 本地仓库】为了可以观察到配置的动态更新，下面可以通过 Git 本地仓库实现 message.provider-dev.yml 配置文件的更新与代码提交的操作。

修改文件内容	yootk:　　# 自定义属性 　message:　　# 自定义属性 　　flag: dev　　# 环境标记 　　content: 沐言科技（yootk.com） — 李兴华老师　　消息内容
代码提交到暂存库	git add .
代码提交到版本库	git commit -m "Modify message.provider-dev.yml File"
发布到 GitLab 仓库	git push -u origin master

（7）【Postman 测试】GitLab 仓库中的配置项修改完成之后，消息微服务中的配置项默认是不会发生任何改变的，需要开发者手动进行刷新，刷新地址为“config-server-7501:7501/ actuator/bus-refresh”，请求类型为“POST”。

（8）【Postman 测试】Config 服务端刷新完成后，对应的 Config 客户端中的选项也会发生改变，重新进行消息微服务的调用，执行结果如图 9-18 所示。

图 9-18　Config 客户端自动更新

9.2.4　Spring Cloud Config 整合 Nacos

Spring Cloud
Config整合Nacos

视频名称　0909_【掌握】Spring Cloud Config 整合 Nacos

视频简介　Spring Cloud Alibaba 针对 Spring Cloud Config 提供了更加简化的处理机制。本视频为读者分析传统的 Spring Cloud Config 的缺陷，并通过具体的代码介绍实现基于 Nacos 的配置动态管理操作。

Spring Cloud Config 属于早期的 Spring Cloud Netflix 配置的一种动态管理机制，但是其在实现的过程中需要开发者进行大量的配置（Config 相关依赖配置、Bus 依赖配置）以及服务接入（消息组件、Git 或 SVN 仓库）处理，管理人员也必须有版本控制工具的使用经验。这样一来在实际的项目中，它就很难得到切实有效的应用，所以一般不建议采用此技术实现管理。

但是这种动态配置管理的机制在实际的项目中依然会用到，所以 Spring Cloud Alibaba 套件提供了 Nacos 解决方案，即开发者可以基于 Nacos 来实现配置项的管理，并且可以通过 Nacos 控制台基于可视化的方案来实现配置的更新（Nacos 配置项支持监听，可以直接修改，避免了消息组件的操作），如图 9-19 所示。下面将介绍对已有的消息微服务进行修改，实现 Nacos 下的配置动态更新操作。

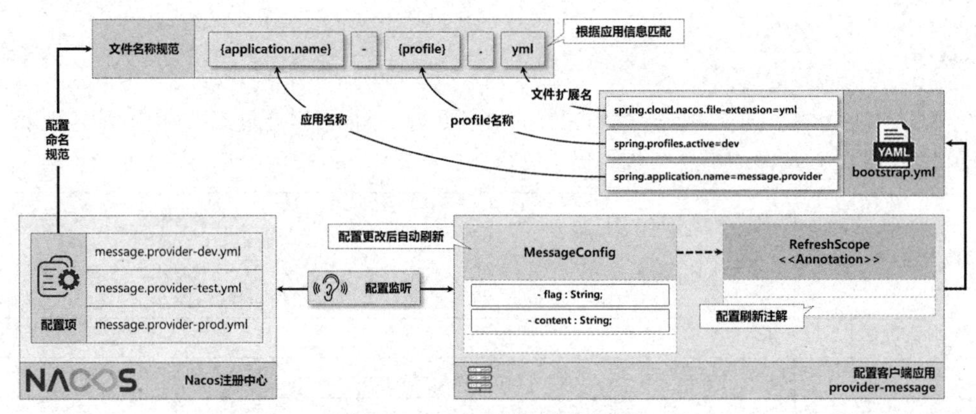

图 9-19　Nacos 实现配置动态管理

（1）【Nacos 控制台】将当前项目中所使用的 3 个配置文件以配置项的形式定义在 Nacos 注册中心之中，如图 9-20 所示。

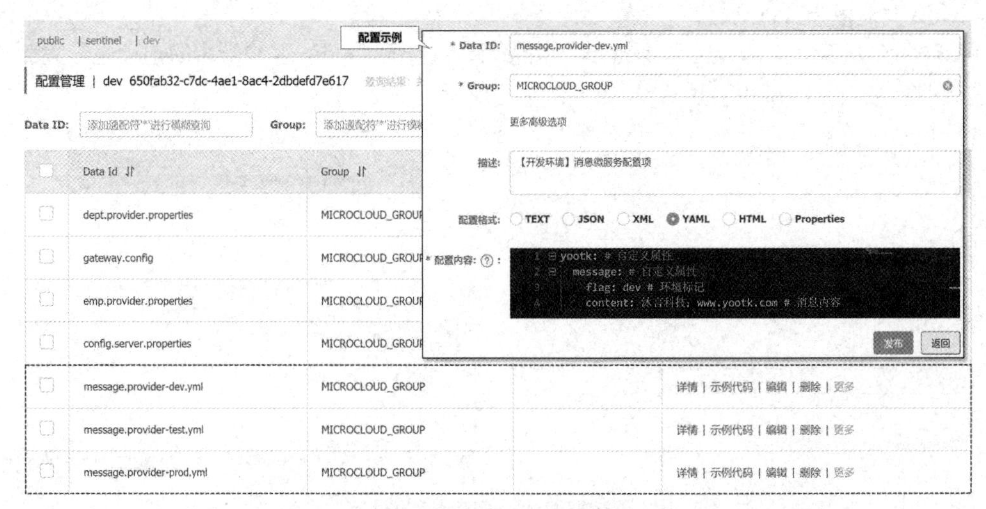

图 9-20　定义 Nacos 配置项

（2）【microcloud 项目】修改 build.gradle 配置文件，删除与 Spring Cloud Config 相关的依赖库。

（3）【provider-message-8201 子模块】修改 bootstrap.yml 配置文件。

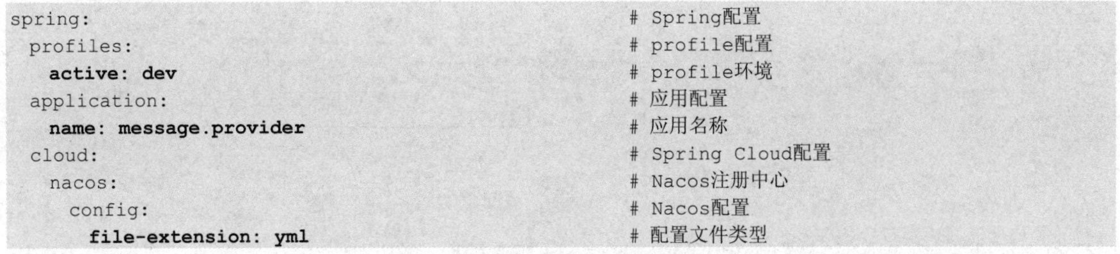

```
spring:                                   # Spring配置
  profiles:                               # profile配置
    active: dev                           # profile环境
  application:                            # 应用配置
    name: message.provider                # 应用名称
  cloud:                                  # Spring Cloud配置
    nacos:                                # Nacos注册中心
      config:                             # Nacos配置
        file-extension: yml               # 配置文件类型
```

程序启动后将根据应用信息找到与之关联的配置项，管理员可以通过 Nacos 注册中心实现可视化的配置修改，从实现难度以及易用性来讲都要比“原始”的 Spring Cloud Config 更加方便。

9.3 Seata 分布式事务组件

Seata 分布式事务简介

视频名称 0910_【掌握】Seata 分布式事务简介

视频简介 分布式的业务中心要实现统一的事务调度，必须基于二阶段提交的模式来完成。本视频为读者分析分布式事务与微服务之间的关联，同时通过具体的案例分析 Seata 业务层分布式事务处理的特点、实现流程以及实现分类。

在微服务的实现架构系统之中，不同的业务会被拆分到不同的微服务实例之中，这样会彻底丧失传统数据库事务的支持。在进行数据更新时，每一个微服务都仅仅与当前关联的数据库实现事务处理，不同的微服务之间没有事务上的联系。一旦出现错误的数据就可能会出现违反 ACID 原则的操作，从而使整个业务数据混乱，导致业务处理出现问题，如图 9-21 所示。

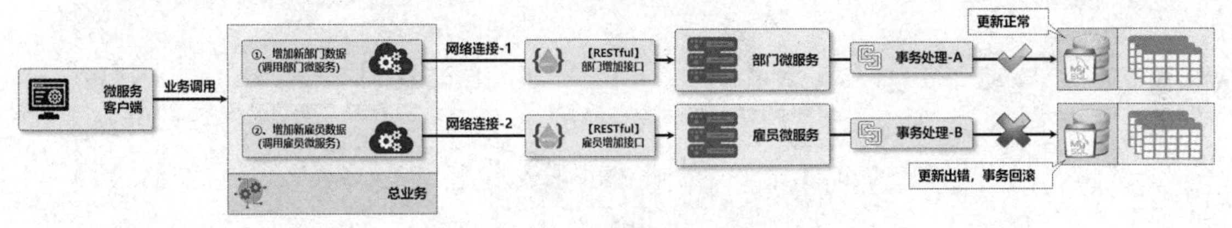

图 9-21 微服务事务处理

提示：关于 JTA 处理。

如果若干个数据库处于统一应用之中，那么依然可以通过容器内提供的 JTA 进行事务管理。而关于这一点的实现，本套丛书中关于 Spring Boot 的书已经为读者进行了完整的分析，对于还未掌握这种数据库二阶段提交概念的读者，笔者强烈建议先做知识补充。

由于不同的微服务对应不同的数据库事务，因此不同的事务之间不会有任何关联，而要想解决此类问题，就需要引入一个第三方的协调机制。每一个微服务的事务由该协调组件基于二阶段提交处理的模式进行统一调度，在第一阶段（Prepare 阶段）执行本地的数据库事务处理，在第二阶段（Commit 阶段）针对第一阶段的执行结果来决定最终是否要提交或回滚事务，如图 9-22 所示，而这就是 Seata 的主要作用。

Seata 是由阿里巴巴公司推出的针对业务层分布式事务的解决方案，在实现过程中只依赖单台数据库的事务处理能力。在 Seata 框架中一个分布式事务包含 3 种角色。

- Transaction Coordinator（TC）：事务协调器，维护全局事务运行状态，负责协调并驱动全局事务的提交或回滚。

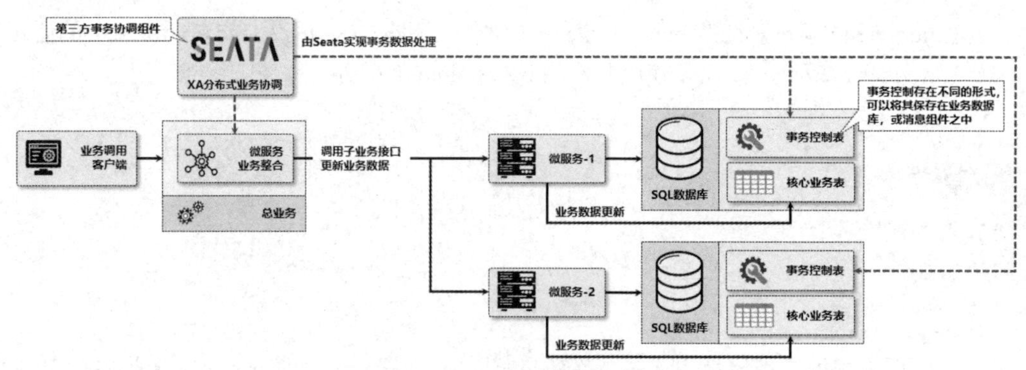

图 9-22　Seata 操作模式（AT 模式）

- Transaction Manager（TM）：控制全局事务的边界，负责开启全局事务，并最终做出全局提交或回滚的决议。
- Resource Manager（RM）：控制分支（本地数据库）事务，负责分支事务的注册与状态汇报，并接收事务协调器的指令，驱动分支事务的提交和回滚。

其中 TM 是一个分布式事务中核心的发起者与终结者，而 TC 负责维护分布式事务的运行状态。由于微服务集群业务中会有大量的子业务微服务，因此需要通过 RM 实现本地事务的运行，如图 9-23 所示。所有的本地事务都需要向 TC 进行注册，而后执行第一阶段的事务处理操作，并将最终的事务处理结果发送给 TC。TC 将汇总全部的事务处理结果，对全局事务做出提交或回滚的决议。最后 RM 根据 TC 的决议结果来执行本地事务第二阶段事务处理。

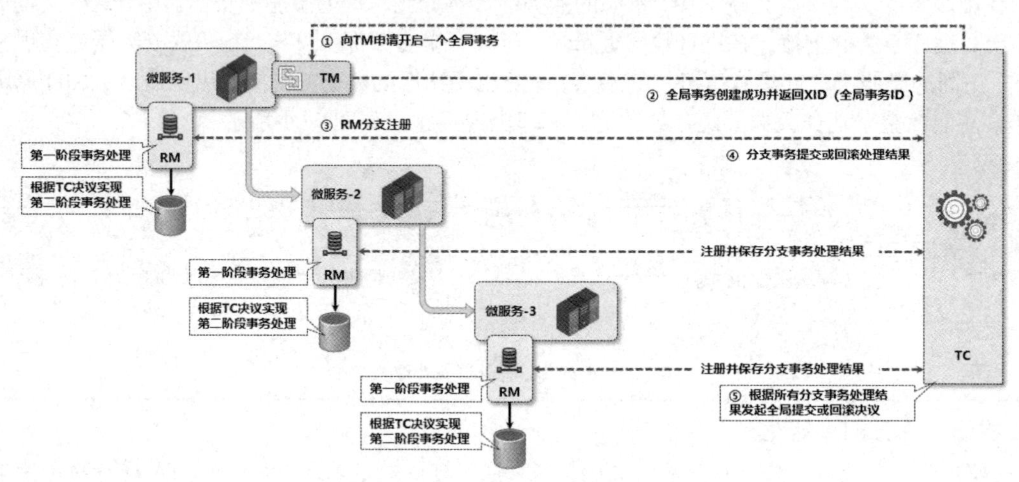

图 9-23　Seata 处理流程

Seata 为了解决分布式事务处理操作，一共有 4 种模式：AT 模式（2019 年 1 月开源）、TCC 模式（2019 年 3 月开源）、Saga 模式（2019 年 8 月开源）和 XA 模式（2019 年 12 月开源）。下面将对这些模式展开讲解。

9.3.1　雇员微服务

雇员微服务

视频名称　0911_【掌握】雇员微服务

视频简介　要想实现完整的分布式事务处理，必须启用新的微服务提供者。本视频介绍依据已有的部门业务模型搭建雇员微服务，以构成分布式事务实现场景。

为了便于读者观察分布式事务的操作过程，下面将模仿已有的部门微服务创建一个新的雇员微服务，并提供部门增加的业务接口。其实现结构如图 9-24 所示。

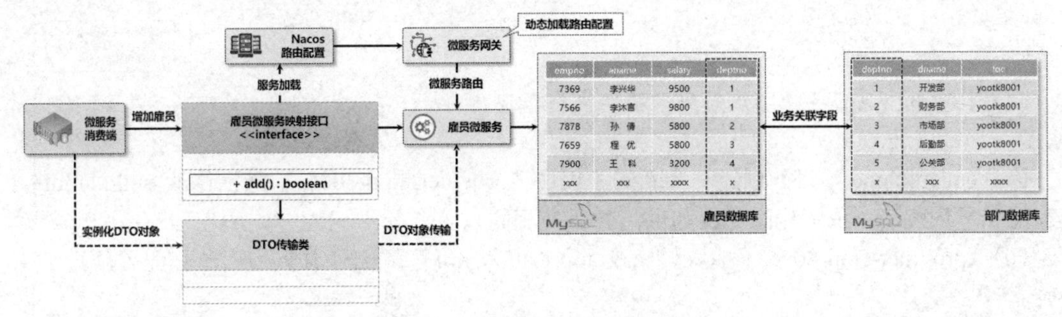

图 9-24　创建雇员微服务的实现结构

雇员微服务将主要实现一个雇员增加的 REST 接口，但是要想完成这一功能，还需要根据需要进行创建雇员数据库、Feign 接口定义与实现、网关路由配置等操作。下面通过具体的操作步骤进行讲解。

提示：雇员微服务与部门微服务环境相同。

要想更好地展示分布式事务的具体实现，需要基于前面的方式开发雇员微服务。由于该微服务依然是通过 MyBatis/MyBatisPlus 开发，也需要向 Nacos 注册，并支持 Sentinel 管理，因此在本模块的实现过程中将为读者列出核心代码。如果读者对于这些核心代码的使用不是很理解，建议回顾本书前面的内容，或者参考随书附赠的完整项目代码。

（1）【SQL 脚本】雇员数据库创建脚本。

```
DROP DATABASE IF EXISTS yootk8006;
CREATE DATABASE yootk8006 CHARACTER SET UTF8;
USE yootk8006;
CREATE TABLE emp (
   empno     BIGINT,
   ename     VARCHAR(50),
   salary    DOUBLE,
   deptno     BIGINT,
   CONSTRAINT pk_empno PRIMARY KEY(empno)
) ENGINE=InnoDB DEFAULT CHARSET=utf8;
INSERT INTO emp(empno, ename, salary, deptno) VALUES (7369, '李兴华', 9500, 1);
INSERT INTO emp(empno, ename, salary, deptno) VALUES (7566, '李沐言', 9800, 1);
INSERT INTO emp(empno, ename, salary, deptno) VALUES (7878, '孙倩', 5800, 2);
INSERT INTO emp(empno, ename, salary, deptno) VALUES (7888, '王寨塞', 6800, 2);
INSERT INTO emp(empno, ename, salary, deptno) VALUES (7659, '程优', 5800, 3);
INSERT INTO emp(empno, ename, salary, deptno) VALUES (7900, '王科', 3200, 4);
INSERT INTO emp(empno, ename, salary, deptno) VALUES (7839, '金拓', 4900, 5);
INSERT INTO emp(empno, ename, salary, deptno) VALUES (7869, '郭竹音', 8900, 5);
COMMIT;
```

（2）【common-api 子模块】创建一个 EmpDTO 数据传输类。

```
package com.yootk.common.dto;
@Data                                                        // Lombok结构生成注解
public class EmpDTO implements Serializable {                // 雇员数据传输类
   private Long empno;                                       // 雇员编号
   private String ename;                                     // 雇员姓名
   private Double salary;                                    // 雇员工资
   private Long deptno;                                      // 部门编号
}
```

（3）【common-api 子模块】创建雇员业务接口。

```
package com.yootk.service;
@FeignClient(value="microcloud.gateway",configuration = FeignConfig.class)
public interface IEmpService {                               // 雇员业务接口
   /**
    * 增加雇员数据
    * @param dept 雇员传输类实例
```

```
    * @return 雇员增加成功返回true，否则返回false
    */
   @PostMapping("/provider/emp/add")                    // Feign接口映射路径
   public boolean add(EmpDTO dept);
}
```

（4）【common-api 子模块】对当前模块进行编译。

```
gradle clean build
```

（5）【microcloud 项目】创建一个新的子模块“provider-emp-8006”，随后修改 build.gradle，为其配置所需依赖库（依赖库的配置与部门微服务相同）。

（6）【provider-emp-8006 子模块】修改 application.yml 配置文件（只列出核心配置项）。

```
server:                                                  # 服务端配置
 port: 8006                                              # 8006端口监听
spring:                                                  # Spring配置
 application:                                            # 应用配置
   name: emp.provider                                    # 应用名称
 datasource:                                             # 数据源配置
   url: jdbc:mysql://localhost:3306/yootk8006            # 连接地址
```

（7）【provider-emp-8006 子模块】部门与 emp 数据表映射的 VO 类。

```
package com.yootk.provider.vo;
@TableName("emp")                                        // 映射数据表
@Data                                                    // Lombok结构生成注解
public class Emp {
   @TableId                                              // 设置ID映射
   private Long empno;                                   // 映射empno字段
   private String ename;                                 // 映射ename字段
   private double salary;                                // 映射salary字段
   private Long deptno;                                  // 映射deptno字段
}
```

（8）【provider-emp-8006 子模块】创建 IEmpDAO 数据层操作接口。

```
package com.yootk.provider.dao;
@Mapper                                                  // MyBatis映射
public interface IEmpDAO extends BaseMapper<Emp> {}      // 数据层接口
```

（9）【provider-emp-8006 子模块】创建 EmpServiceImpl 业务实现子类。

```
package com.yootk.provider.service.impl;
@Service
public class EmpServiceImpl implements IEmpService {     // 业务实现类
   @Autowired
   private IEmpDAO empDAO;                               // 注入IEmpDAO接口实例
   @Override
   public boolean add(EmpDTO emp) {
      Emp empVO = new Emp();                             // DAO对象
      DeepBeanUtils.copyProperties(emp, empVO);          // 对象复制
      return this.empDAO.insert(empVO) > 0;              // 数据保存
   }
}
```

（10）【provider-emp-8006 子模块】创建 EmpAction 控制层类，发布 REST 接口。

```
package com.yootk.provider.action;
@RestController                                          // REST控制器
@RequestMapping("/provider/emp/*")                       // 父路径
public class EmpAction {
   @Autowired
   private IEmpService empService;                       // 注入业务接口实例
   @PostMapping("add")
   public Object add(@RequestBody EmpDTO empDTO) {
      return this.empService.add(empDTO);                // 部门信息添加
   }
}
```

（11）【本地系统】修改 hosts 配置文件，追加雇员微服务的地址映射。

```
127.0.0.1   provider-emp-8006
```

（12）【Nacos 控制台】在 Nacos 中为雇员微服务添加一个配置项（配置项名称为“emp.provider.properties”，保存分组为“MICROCLOUD_GROUP”），如图 9-25 所示。

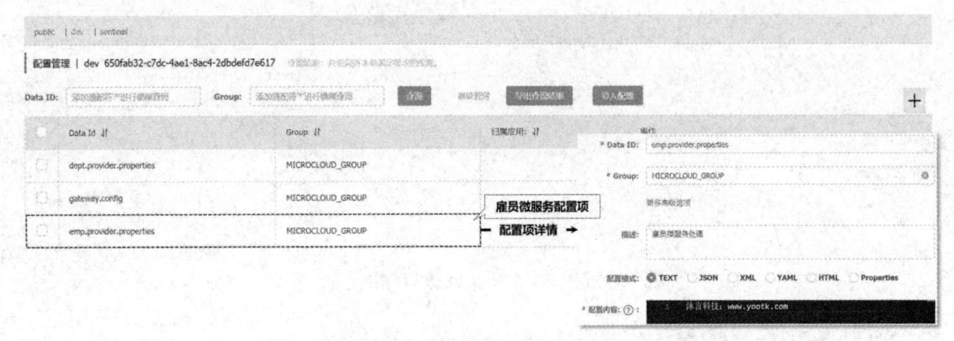

图 9-25　添加雇员微服务配置项

（13）【Nacos 控制台】修改 gateway.config 配置项，追加雇员微服务的路由配置。

```
[ // 此处只列出新增的路由配置项，已有的路由配置项代码略
  {
    "id": "emp",                                              // 路由配置项ID
    "uri": "lb://emp.provider",                               // 路由映射地址
    "order": 1,                                               // 定义路由执行顺序
    "predicates": [                                           // 配置路由断言
      { "name": "Path",                                       // 定义断言路径
        "args": { "pattern": "/provider/emp/**" }             // 断言匹配模式
      }  ]  } ]
```

（14）【Nacos 控制台】全部微服务实例启动后，可以通过 Nacos 控制台查询服务注册信息，如图 9-26 所示。

图 9-26　Nacos 已注册微服务

（15）【Spring Boot Admin 控制台】由于所有的微服务都会向 Spring Boot Admin 微服务注册，因此也可以打开该控制台，观察微服务的运行状态，如图 9-27 所示。

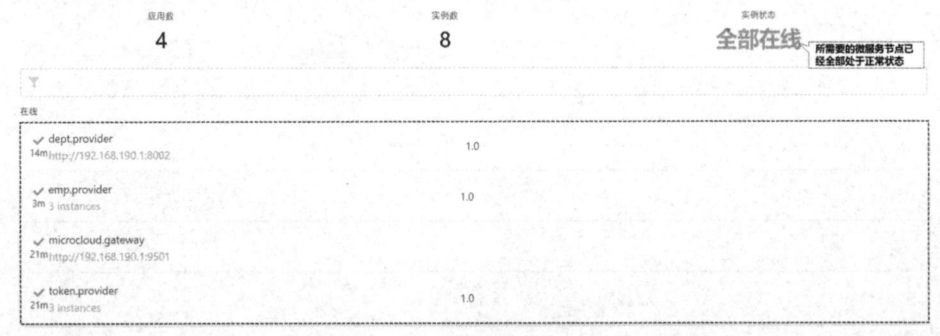

图 9-27　查询微服务运行状态

（16）【Postman 接口测试】随后通过 Postman 测试当前 REST 接口是否可以实现雇员数据添加，

测试结果如图 9-28 所示。

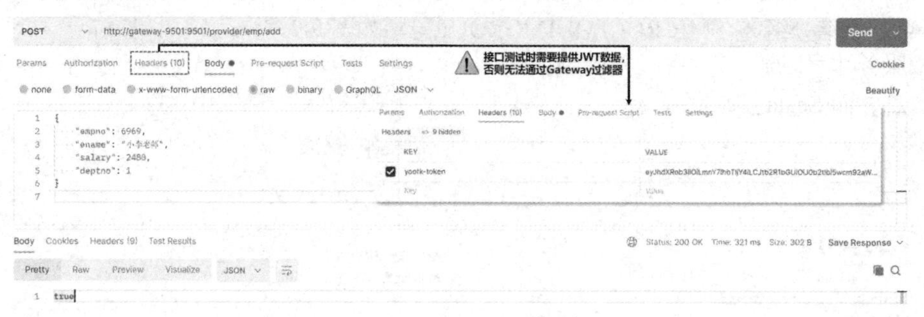

图 9-28　Postman 接口测试

9.3.2　Seata 服务安装与配置

Seata 服务安装与配置

视频名称　0912_【掌握】Seata 服务安装与配置

视频简介　Seata 是公布在 GitHub 上的一套完整开源项目。本视频通过 GitHub 为读者演示组件的获取操作，并依据实例演示 Seata 的配置，以及与 Nacos 的集成。

开发者如果想在项目中引入 Seata 组件进行分布式事务处理，则可以登录 Seata 官方站点，如图 9-29 所示，根据官方站点提供的文档资料学习并使用 Seata。

图 9-29　Seata 官方站点

官方站点给出了 Seata 项目托管的 GitHub 地址，单击链接即可得到 Seata 的源代码以及打包后的部署应用程序，如图 9-30 所示。

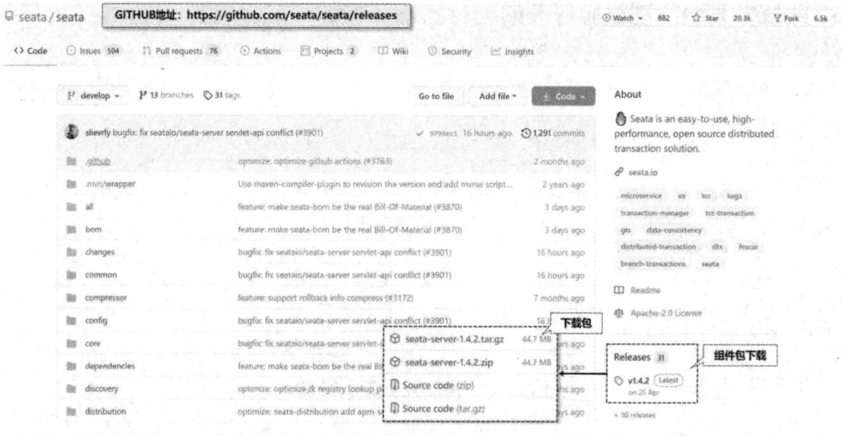

图 9-30　下载 Seata 组件

Seata 本质上属于一个服务应用，所以要想让其他微服务可以正确地使用 Seata 服务应用，需要将 Seata 应用在 Nacos 中进行注册（通过 registry.conf 配置文件实现），这样其他参与分布式事务处理的微服务才可以通过 Nacos 获取 Seata 应用的信息。所有通过 Seata 实现的事务处理需要存储相关的信息，而 Seata 在实现存储时也较为灵活，可以根据需要选择文件、数据库或 Redis 实现存储（通过 file.conf 配置文件实现），在配置完成后还需要将 Seata 相关配置项保存在 Nacos 之中，这样才可以构建起一个完整的 Seata 应用，如图 9-31 所示。下面通过具体的操作步骤介绍 Seata 服务的安装与配置。

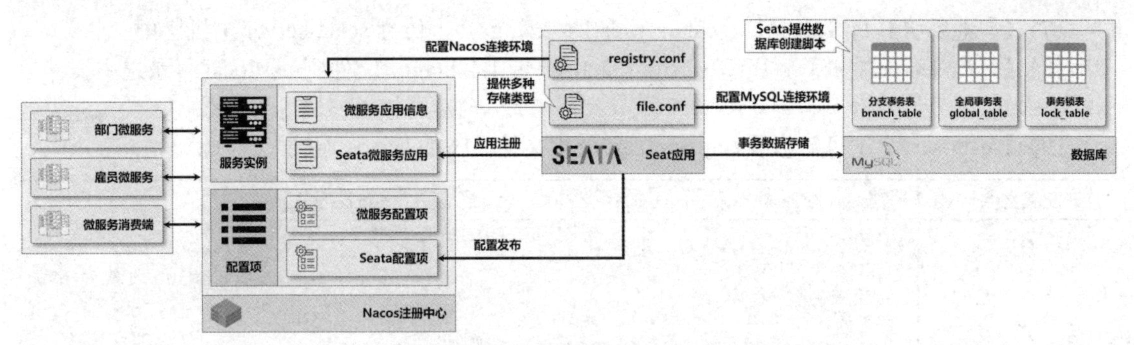

图 9-31　Seata 服务安装与配置

（1）【本地系统】将 Seata 数据库的创建脚本上传到 Linux 主机的“/var/ftp”目录之中，Seata 所提供的数据库创建脚本保存在 Seata 源代码目录之中，如图 9-32 所示。

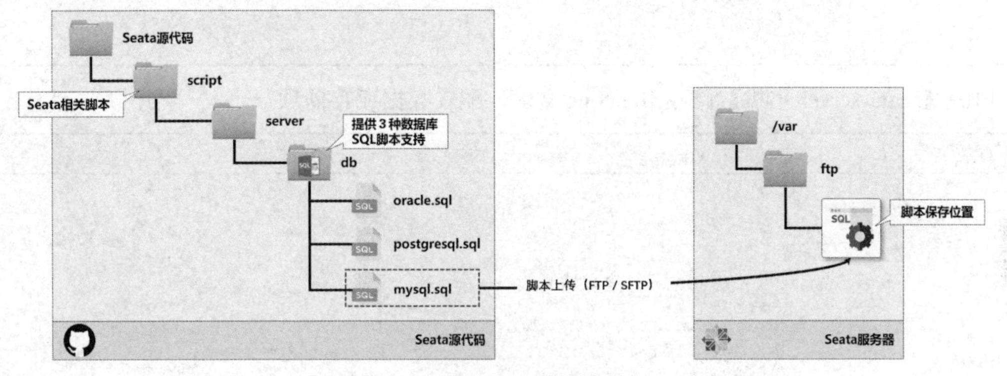

图 9-32　上传 Seata 数据库脚本

（2）【seata-server 主机】启动本机的 MySQL 服务进程。

```
service mysqld start
```

（3）【seata-server 主机】启动 MySQL 客户端。

```
/usr/local/mysql/bin/mysql -uroot -pmysqladmin
```

（4）【seata-server 主机】在 MySQL 中创建 Seata 数据库。

```
CREATE DATABASE seata CHARACTER SET UTF8 ;
USE seata;
```

（5）【seata-mysql 主机】执行 Seata 源代码中所给出的 MySQL 数据库创建脚本，执行完成后，会自动为用户创建“branch_table”“global_table”及“lock_table”3 张数据表。

```
source /var/ftp/mysql.sql
```

（6）【nacos 控制台】Seata 需要向 Nacos 进行注册，而为了服务的注册管理，需要通过 Nacos 控制台创建一个新的命名空间 seata（ID：af4c2724-87ca-4f23-aa50-b60c4b70134c），实现 Seata 相关信息的存储，如图 9-33 所示。

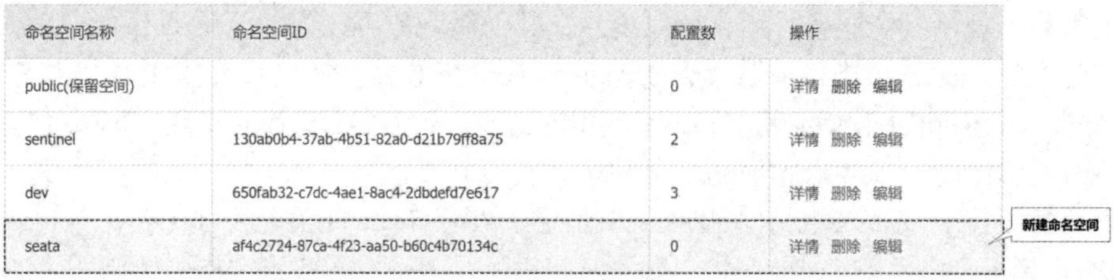

命名空间名称	命名空间ID	配置数	操作
public(保留空间)		0	详情 删除 编辑
sentinel	130ab0b4-37ab-4b51-82a0-d21b79ff8a75	2	详情 删除 编辑
dev	650fab32-c7dc-4ae1-8ac4-2dbdefd7e617	3	详情 删除 编辑
seata	af4c2724-87ca-4f23-aa50-b60c4b70134c	0	详情 删除 编辑

新建命名空间

图 9-33　新建命名空间

（7）【本地系统】将 Seata 包（seata-server-1.4.2.tar.gz）上传到 seata-server 主机之中。

（8）【seata-server 主机】解压缩/var/ftp/seata-server-1.4.2.tar.gz 文件到/usr/local 目录之中。

```
tar xzvf /var/ftp/seata-server-1.4.2.tar.gz -C /usr/local/
```

（9）【seata-server 主机】修改 registry.conf 配置文件，添加 Nacos 连接配置项。

打开配置文件

```
vi /usr/local/seata/seata-server-1.4.2/conf/registry.conf
```

编辑配置项

```
registry {
  type = "nacos"
  nacos {
    application = "seata-server"
    serverAddr = "nacos-server:8848"
    group = "SEATA_GROUP"
    namespace = "af4c2724-87ca-4f23-aa50-b60c4b70134c"
    cluster = "SeataCluster"
    username = "nacos"
    password = "nacos"
  } # 其他配置项，略
}
```

（10）【seata-server 主机】修改 file.conf 文件，配置数据库存储项。

打开配置文件

```
vi /usr/local/seata/seata-server-1.4.2/conf/file.conf
```

编辑配置项

```
store {
  mode = "db"
  db {
    datasource = "druid"
    dbType = "mysql"
    driverClassName = "com.mysql.jdbc.Driver"
    url = "jdbc:mysql://192.168.190.191:3306/seata?rewriteBatchedStatements=true"
    user = "root"
    password = "mysqladmin"
    minConn = 5
    maxConn = 10
    globalTable = "global_table"
    branchTable = "branch_table"
    lockTable = "lock_table"
    queryLimit = 10
    maxWait = 5000
  }
}
```

（11）【seata-server 主机】本次使用的 MySQL 版本为 8.x，所以需要移动 MySQL 驱动程序包到 lib 目录之中。

```
mv /usr/local/seata/seata-server-1.4.2/lib/jdbc/mysql-connector-java-8.0.19.jar /usr/local/
seata/seata-server-1.4.2/lib/
```

（12）【seata-server 主机】启动 Seata 服务。

```
nohup /usr/local/seata/seata-server-1.4.2/bin/seata-server.sh > /usr/local/seata/seata-server-
1.4.2/logs/seata.log 2>&1 &
```

（13）【seata-server 主机】Seata 服务启动后默认占用 8091 服务端口，修改防火墙规则。

添加新的防火墙规则	`firewall-cmd --zone=public --add-port=8091/tcp --permanent`
重新加载防火墙配置	`firewall-cmd --reload`

（14）【Nacos 控制台】Seata 服务启动后会自动向 Nacos 注册中心保存数据，如图 9-34 所示。

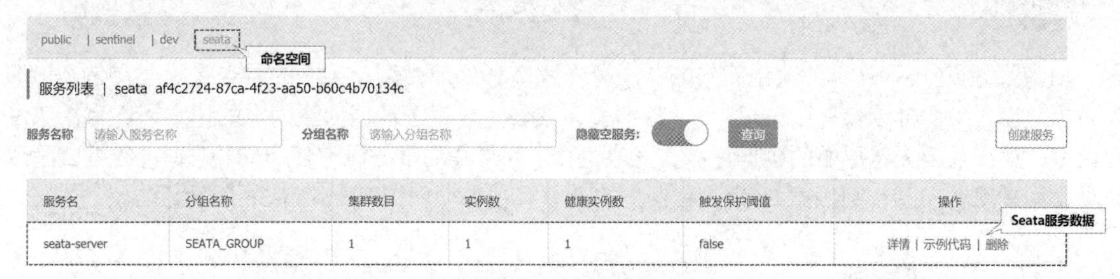

图 9-34　Seata 服务数据注册

（15）【seata-server 主机】除了服务数据之外，还需要向 Nacos 中添加配置数据文件（config.txt）。通过 Seata 源代码获取配置数据文件上传到/usr/local/seata/seata-server-1.4.2/目录之中，而后根据当前的配置进行修改。

打开配置文件	`vi /usr/local/seata/seata-server-1.4.2/config.txt`

修改部分配置	`service.vgroupMapping.my_test_tx_group=default`
	`store.mode=db`
	`store.db.url=jdbc:mysql://192.168.190.190:3306/seata?useUnicode=`
	`true&rewriteBatchedStatements=true`
	`store.db.user=root`
	`store.db.password=mysqladmin`

（16）【seata-server 主机】所有的配置项需要保存在 Nacos 之中，可以通过 Seata 源代码给出的处理程序（nacos-config.sh 文件）将 config.txt 的内容上传，该程序需要保存在/usr/local/seata/seata-server-1.4.2/bin 目录之中。

切换当前目录	`cd /usr/local/seata/seata-server-1.4.2/bin`
提交配置项	`sh /usr/local/seata/seata-server-1.4.2/bin/nacos-config.sh -h nacos-server -p 8848 \` `-t af4c2724-87ca-4f23-aa50-b60c4b70134c -g SEATA_GROUP -u nacos -w nacos`

（17）【Nacos 控制台】通过 Nacos 控制台查看 Seata 配置项，如图 9-35 所示。

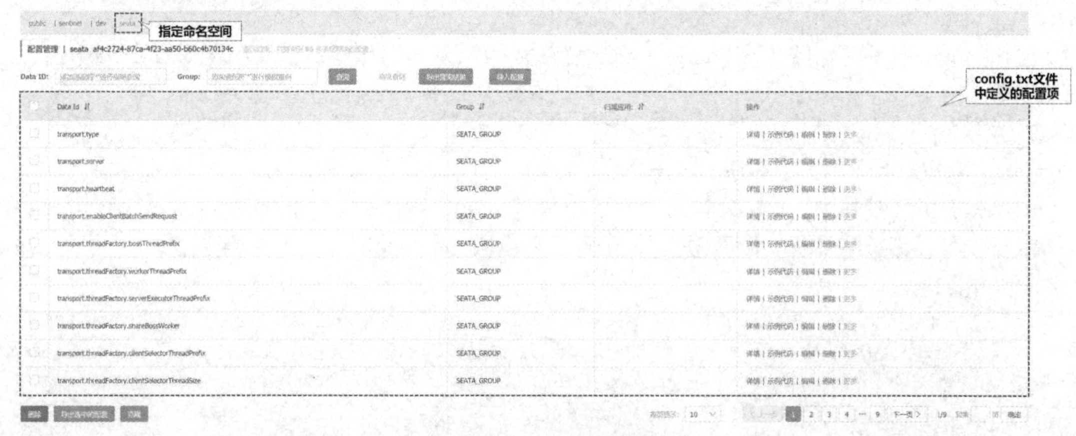

图 9-35　Seata 配置项

9.3.3　AT 模式

AT 模式

视频名称　0913_【掌握】AT 模式

视频简介　Seata 提供了 AT 模式来实现与传统 XA 模式的对接。本视频将为读者分析 AT 模式的操作特点，并通过代码介绍如何实现 AT 模式下的分布式事务管理。

AT 模式是基于 XA 事务演化而来的一个分布式事务处理模型，最大的特点是需要在每个微服务的数据库之中额外创建一张 undo_log 数据表。该表主要用于数据回滚内容的记录，在每次第一阶段事务提交前要先将原始的数据记录在此表之中。如果全局事务没有任何问题，则进行第二阶段提交；如果出现问题，则根据 undo_log 表中的内容实现数据回滚，如图 9-36 所示。

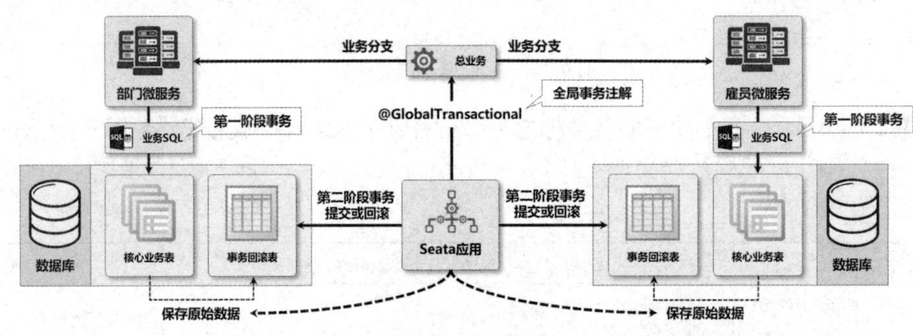

图 9-36　AT 模式

在 AT 模式下，开发者只需要根据自身的业务编写“业务 SQL”即可（第一阶段更新），而 Seata 框架会自动生成事务第二阶段的提交或回滚操作。为便于理解，下面通过具体的操作步骤进行开发演示。

（1）【数据库】在部门数据库和雇员数据库之中创建 undo_log 数据表，创建脚本如下。

```
CREATE TABLE IF NOT EXISTS 'undo_log'(
   'branch_id'      BIGINT(20)   NOT NULL COMMENT 'branch transaction id',
   'xid'            VARCHAR(100) NOT NULL COMMENT 'global transaction id',
   'context'        VARCHAR(128) NOT NULL COMMENT 'undo_log context,such as serialization',
   'rollback_info'  LONGBLOB     NOT NULL COMMENT 'rollback info',
   'log_status'     INT(11)      NOT NULL COMMENT '0:normal status,1:defense status',
   'log_created'    DATETIME(6)  NOT NULL COMMENT 'create datetime',
   'log_modified'   DATETIME(6)  NOT NULL COMMENT 'modify datetime',
   UNIQUE KEY 'ux_undo_log' ('xid', 'branch_id')
) ENGINE = InnoDB
  AUTO_INCREMENT = 1
  DEFAULT CHARSET = utf8 COMMENT ='AT transaction mode undo table';
```

（2）【microcloud 项目】修改 build.gradle 配置文件，为部门与雇员模块添加 Seata 依赖库。

```
implementation('com.alibaba.cloud:spring-cloud-starter-alibaba-seata:2021.1') {
   exclude group: 'io.seata', module: 'seata-spring-boot-starter'   // 排除旧依赖
}
implementation('io.seata:seata-spring-boot-starter:1.4.2')
```

（3）【微服务子模块】由于本次需要 3 个子模块（provider-dept-xx、provider-emp-xx、consumer-xx）共同参与分布式事务的处理，因此需要修改这 3 个模块中的 application.yml 配置文件，引入 Seata 相关配置项。

```
seata:                                                   # Seata配置
  application-id: seata-server                           # Seata应用名称
  tx-service-group: my_test_tx_group                     # 事务组
  service:                                               # 服务配置
    vgroup-mapping:                                      # 事务群组
      my_test_tx_group: SeataCluster                     # 集群名称
  config:                                                # 配置抓取
    nacos:                                               # Nacos存储
```

```
      server-addr: nacos-server:8848                    # Nacos地址
      namespace: af4c2724-87ca-4f23-aa50-b60c4b70134c   # 命名空间
      group: SEATA_GROUP                                # 配置组
      username: nacos                                   # 用户名
      password: nacos                                   # 密码
  registry:                                             # Seata注册配置
    type: nacos                                         # 注册类型
    nacos:                                              # Nacos配置
      application: seata-server                         # Seata应用名称
      server-addr: nacos-server:8848                    # Nacos地址
      group: SEATA_GROUP                                # 分组名称
      namespace: af4c2724-87ca-4f23-aa50-b60c4b70134c   # 命名空间
      username: nacos                                   # 用户名
      password: nacos                                   # 密码
      cluster: SeataCluster                             # 集群名称
```

（4）【consumer-springboot-80 子模块】在消费端进行部门与雇员微服务的业务整合。

```java
package com.yootk.consumer.action;
@RestController                                          // REST响应
@RequestMapping("/consumer/company/*")                   // 父路径
public class CompanyConsumerAction {
    @Autowired
    private IDeptService deptService;                    // 部门业务接口实例
    @Autowired
    private IEmpService empService;                      // 雇员业务接口实例
    @GetMapping("add")
    @GlobalTransactional                                 // 全局事务处理
    public Object add(DeptDTO dept, EmpDTO emp) {        // 数据增加
        Map<String, Object> result = new HashMap<>();    // 保存增加结果
        this.deptService.add(dept);                      // 调用部门微服务
        result.put("dept", dept);                        // 保存结果
        emp.setDeptno(dept.getDeptno());                 // 设置部门编号（随意）
        String ename = emp.getEname();                   // 获取雇员名称
        for (int x = 0; x < 3; x++) {                    // 添加3位雇员
            emp.setEmpno(emp.getEmpno() + x);            // 雇员编号
            emp.setEname(ename + " - " + x);             // 雇员姓名
            this.empService.add(emp);                    // 添加雇员
            result.put("emp-" + x, emp);                 // 保存结果
        }
        return result;
    }
}
```

此时的微服务实现了部门和雇员数据的添加业务（两个操作都属于更新操作）的功能调用。为了便于操作，此处通过循环的形式新增 3 位雇员（雇员编号累加），并进行 3 位雇员信息的添加（相当于发出 3 次更新业务），而为了保证整体业务的成功，在该方法上使用@GlobalTransactional 全局事务注解进行了标注，这样就会自动通过 Nacos 找到 Seata 应用，实现分布式事务控制。

（5）【Postman 测试】接口开发完成后，通过 Postman 工具来进行测试，图 9-37 所示为测试成功后的结果。而如果雇员编号出现了重复，则会通过 Seata 实现数据的回滚操作。

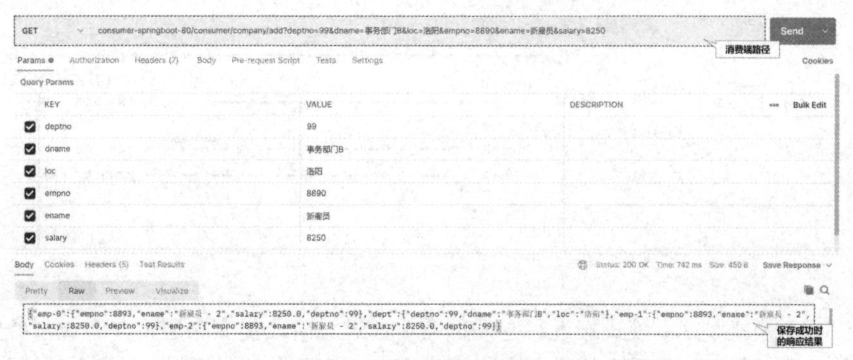

图 9-37　Postman 测试结果

从整个实现结构来讲，采用 AT 模式并不会对代码造成侵入，因此代码的结构性较好，开发者只需要进行一些简单的配置即可轻松地实现分布式事务。但是其处理性能较差，因此不会被应用在性能要求严苛的项目之中。

提示：XA 模式与 AT 模式类似。

在 Seata 之中还有一种 XA 控制模式，该模式是通过数据库的 XA 机制实现的。在此类模式中同样不需要开发者编写大量的事务处理逻辑，一切全部交由 Seata 处理。而实现其操作的过程中只需要将项目中使用的 DruidDataSource 更换为 DataSourceProxyXA 接口对象实例即可。

范例：更换 DataSource 实例

```
package com.yootk.provider.config;
import io.seata.rm.datasource.xa.DataSourceProxyXA;
@Configuration
public class XADataSourceConfiguration {
    @Bean("dataSourceProxy")
    public DataSource dataSource(DruidDataSource druidDataSource) {
        return new DataSourceProxyXA(druidDataSource); // 接收Druid实现类
    }
}
```

该模式的实现机制主要就是将 DataSource 实现类型更换为 Seata 提供的 DataSourceProxyXA（此类模式不需要 undo_log 数据表，完全由数据库实现）。考虑到代码实现的重复性以及知识的连续性，本书不再对此机制进行讲解，有兴趣的读者可以依据所学知识自行实现。

9.3.4　TCC 模式

TCC 模式

视频名称　0914_【掌握】TCC 模式

视频简介　TCC 模式是 AT 模式的手动化实现，可以获得较高的处理性能。本视频为读者分析 TCC 模式的实现流程，并通过具体的代码实现 TCC 分布式事务。

TCC（Try-Confirm/Cancel）模式是一种服务化的两阶段事务处理协议，在第一阶段（try 阶段）对操作资源进行先期判断与更新处理，在第二阶段根据 try 阶段的结果来决定当前的事务是否提交或回滚，如图 9-38 所示。

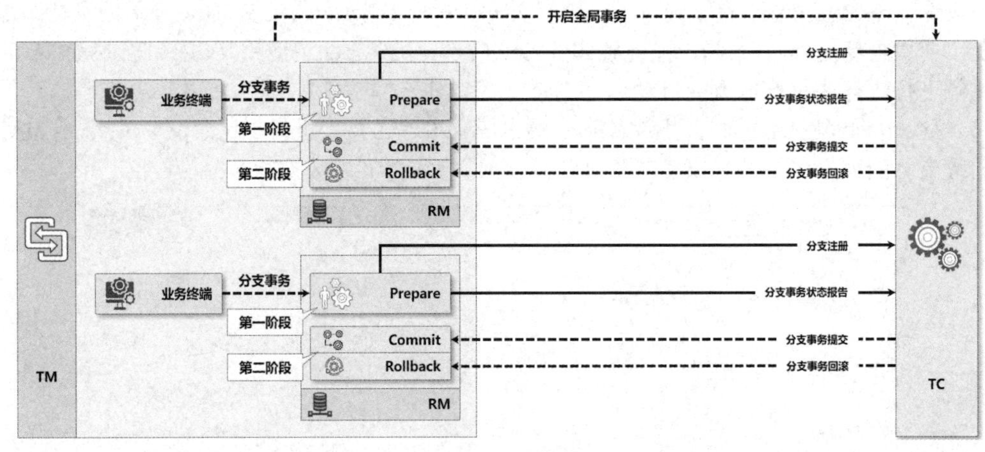

图 9-38　TCC 模式

本质上 TCC 模式与 AT 模式接近，它们都需要通过二阶段提交来进行分布式事务处理。可以将 TCC 简单地理解为 AT 的手动模式，允许开发者自定义两阶段处理逻辑，这样就可以避免依赖

undo_log 这类额外的事务数据表。但是 TCC 对业务的侵入性较大，需要开发者自行创建二阶段处理操作接口并手动实现，如图 9-39 所示。

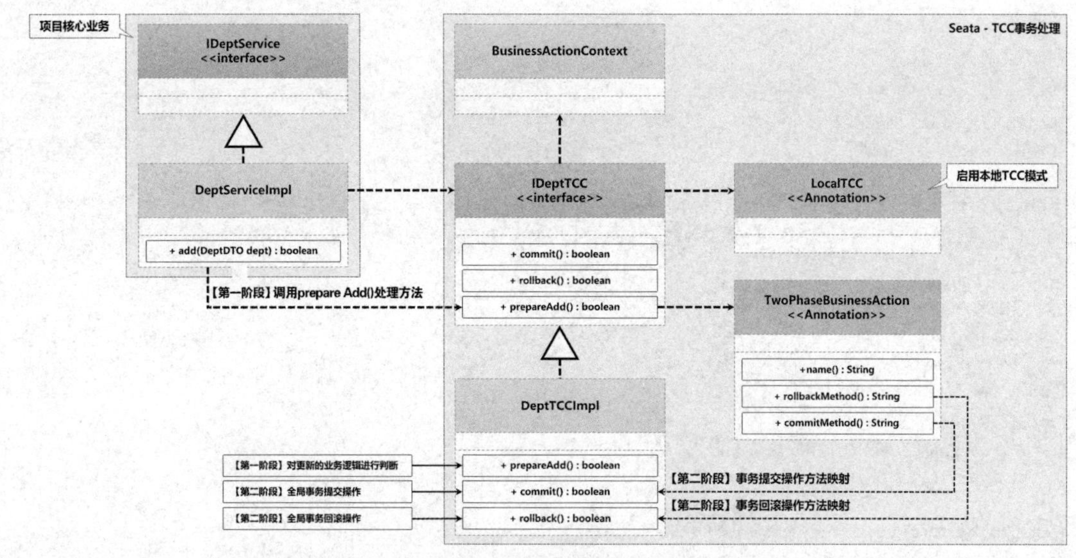

图 9-39　TCC 事务处理模型

图 9-39 给出了一个微服务的 TCC 事务处理模型，在该模型中，业务层不再直接进行数据层的操作，而是在业务层的基础之上嵌套了一个 TCC 层，并且在 TCC 层中明确定义二阶段的事务处理操作逻辑。为了帮助读者更好地理解，下面将对先前的微服务应用进行修改，具体实现步骤如下。

提示：接口的幂等性问题。

幂等性是接口稳定运行的一个重要标准，即同一个系统在相同的使用环境下，一次请求和多次重复请求对系统资源的影响是一致的。在现实开发中，网络的稳定性、服务的并发量等都有可能造成事务管理器对资源的重试处理，这样一个业务会被重复调用，为了解决这个问题，才需要在分布式事务中增加幂等性的控制（主要基于 xid 判断）。

（1）【common-api 子模块】在进行二阶段事务处理时，因为需要考虑接口调用的幂等性问题，所以需要对每次分布式事务的 xid 数据进行记录，这样就需要创建一个事务状态的存储类。

```
package com.yootk.common.util.tcc;
import java.util.Map;
import java.util.concurrent.ConcurrentHashMap;
public class TCCResultStore {                                      // TCC结果保存
   // 考虑到实际调用时多线程事务状态问题，需要创建一个xid的存储集合
   private static final Map<Class<?>, Map<String, String>> RESULT_MAP =
        new ConcurrentHashMap<Class<?>, Map<String, String>>(); // 状态存储
   /**
    * 分布式事务数据存储
    * @param tccClass 操作当前分布式事务的TCC处理类
    * @param xid 分布式ID存储，需要通过此ID实现二阶段提交或回滚
    * @param v 任意的标记内容
    */
   public static void setResult(Class<?> tccClass, String xid, String v) {
      Map<String, String> results = RESULT_MAP.get(tccClass);   // 获取xid存储集合
      if (results == null) {                                     // 当前集合为空
         synchronized (RESULT_MAP) {                             // 线程同步处理
            if (results == null) {                               // 判断集合状态
               results = new ConcurrentHashMap<>();              // 实例化Map集合
               RESULT_MAP.put(tccClass, results);                // 保存集合数据
            }
```

```
            }
        }
        results.put(xid, v);                                                        // 保存当前事务ID
    }
    /**
     * 获取xid数据
     * @param tccClass 操作的TCC类型
     * @param xid 当前的事务ID
     * @return 事务存储结果
     */
    public static String getResult(Class<?> tccClass, String xid) {
        Map<String, String> results = RESULT_MAP.get(tccClass);                     // 获取指定类型的集合
        if (results != null) {                                                      // 集合不为空
            return results.get(xid);                                                // 根据xid获取数据
        }
        return null;                                                                // 数据不存在返回null
    }
    /**
     * 删除当前的事务状态
     * @param tccClass 操作的TCC类型
     * @param xid 当前的事务ID
     */
    public static void removeResult(Class<?> tccClass, String xid) {
        Map<String, String> results = RESULT_MAP.get(tccClass);                     // 获取指定类型的集合
        if (results != null) {                                                      // 集合不为空
            results.remove(xid);                                                    // 删除xid数据
        }
    }
}
```

本程序类的核心流程在于，第一阶段进行事务预处理的时候，会在此类保存一个分布式事务的ID。而在第二阶段处理时，如果可以通过该xid获取数据，则表示当前的事务未进行第二阶段处理；如果其为空，则表示需要进行第二阶段处理。

（2）【provider-dept-800x 子模块】创建部门TCC操作接口，用于实现数据增加处理。

```
package com.yootk.provider.tcc;
@LocalTCC                                                                           // 必须开启本地事务
public interface IDeptTCC {
    @TwoPhaseBusinessAction(name = "deptTCCService",
        commitMethod = "commit", rollbackMethod = "rollback")                       // 二阶段事务注解
    public boolean prepareAdd(BusinessActionContext businessActionContext,
        @BusinessActionContextParameter(paramName = "dept") DeptDTO dept);   // 增加处理
    public boolean commit(BusinessActionContext businessActionContext);            // 二阶段提交
    public boolean rollback(BusinessActionContext businessActionContext);          // 二阶段回滚
}
```

（3）【provider-dept-800x 子模块】创建 IDeptTCC 接口子类，定义二阶段提交处理逻辑。

```
package com.yootk.provider.tcc.impl;
@Component                                                                          // 组件注册
@Slf4j                                                                              // 日志注解
public class DeptTCCImpl implements IDeptTCC {
    @Autowired
    private IDeptDAO deptDAO;                                                       // DAO接口
    @Override
    public boolean prepareAdd(BusinessActionContext businessActionContext,
                DeptDTO dept) {                                                     // try阶段处理
        log.info("【第一阶段】xid = {}、dept = {}",
                businessActionContext.getXid(), dept);                              // 日志输出
        if (dept.getDname() == null || "".equals(dept.getDname())) {               // 部门数据错误
            throw new RuntimeException("部门信息错误");                              // 异常抛出
        }
        TCCResultStore.setResult(getClass(),
```

```
                businessActionContext.getXid(), "d");              // 幂等性标识
        return true;
    }
    @Override
    public boolean commit(BusinessActionContext businessActionContext) {
        DeptDTO dept = ((JSONObject) businessActionContext.getActionContext("dept"))
                .toJavaObject(DeptDTO.class);                      // 获取操作数据
        log.info("【第二阶段】事务提交，xid = {}、dept = {}",
                businessActionContext.getXid(), dept);             // 日志输出
        if (TCCResultStore.getResult(getClass(),
                businessActionContext.getXid()) == null) {         // 防止重复提交
            return true;
        }
        Dept deptVO = new Dept();                                  // DAO对象
        DeepBeanUtils.copyProperties(dept, deptVO);                // 对象复制
        try {
            return this.deptDAO.insert(deptVO) > 0;                // 数据保存
        } finally {
            TCCResultStore.removeResult(getClass(),
                businessActionContext.getXid());                   // 删除标记
        }
    }
    @Override
    public boolean rollback(BusinessActionContext businessActionContext) {
        DeptDTO dept = ((JSONObject) businessActionContext.getActionContext("dept"))
                .toJavaObject(DeptDTO.class);                      // 获取操作数据
        log.info("【第二阶段】事务回滚，xid = {}、dept = {}",
                businessActionContext.getXid(), dept);             // 日志输出
        if (TCCResultStore.getResult(getClass(),
                businessActionContext.getXid()) == null) {         // 防止重复提交
            return true;
        }
        TCCResultStore.removeResult(getClass(),
                businessActionContext.getXid());                   // 删除标记
        return true;
    }
}
```

由于此时的业务逻辑较为简单，因此此处仅仅进行了部门名称是否为空的判断。在实际的开发中可能会有更加烦琐、复杂的业务逻辑，还有可能需要在 try 阶段实现一些数据的更新处理，那么此时就需要对旧数据进行记录，在发生事务回滚时，将这些旧数据重新保存在业务数据库之中。

（4）【provider-dept-800x 子模块】修改 DeptServiceImpl 业务实现子类，通过 IDeptTCC 接口实例实现数据增加。

```
package com.yootk.provider.service.impl;
@Service
public class DeptServiceImpl implements IDeptService {          // 业务实现类
    @Autowired
    private IDeptTCC deptTCC;                                  // 注入IDeptTCC实例
    // 此处只列出核心实现代码，其他重复代码略
    @Override
    public boolean add(DeptDTO dept) {
        return this.deptTCC.prepareAdd(new BusinessActionContext(), dept);
    }
}
```

（5）【provider-emp-8006 子模块】创建 IEmpTCC 事务接口。

```
package com.yootk.provider.tcc;
@LocalTCC
public interface IEmpTCC {
    @TwoPhaseBusinessAction(name = "empTCCService",
              commitMethod = "commit", rollbackMethod = "rollback")
```

```
    public boolean prepareAdd(BusinessActionContext businessActionContext,
        @BusinessActionContextParameter(paramName = "emp") EmpDTO emp);     // 增加处理
    public boolean commit(BusinessActionContext businessActionContext);      // 二阶段提交
    public boolean rollback(BusinessActionContext businessActionContext);    // 二阶段回滚
}
```

（6）【provider-emp-8006 子模块】创建 EmpTCCImpl 实现子类。

```
package com.yootk.provider.tcc.impl;
@Component                                                          // 组件注册
@Slf4j                                                              // 日志注解
public class EmpTCCImpl implements IEmpTCC {
    @Autowired
    private IEmpDAO empDAO;                                         // 注入IEmptDAO接口
    @Override
    public boolean prepareAdd(BusinessActionContext businessActionContext, EmpDTO emp) {
        log.info("【第一阶段】xid = {}、emp = {}",
                businessActionContext.getXid(), emp);               // 日志输出
        if (this.empDAO.selectById(emp.getEmpno()) != null) {       // 雇员编号已存在
            throw new RuntimeException("雇员信息已经存在。");        // 手动异常抛出
        }
        TCCResultStore.setResult(getClass(),
                businessActionContext.getXid(), "e");               // 幂等性标识
        return true;
    }
    @Override
    public boolean commit(BusinessActionContext businessActionContext) {
        EmpDTO emp = ((JSONObject) businessActionContext.getActionContext("emp"))
                .toJavaObject(EmpDTO.class);                        // 获取操作数据
        log.info("【第二阶段】事务提交，xid = {}、emp = {}",
                businessActionContext.getXid(), emp);               // 日志输出
        if (TCCResultStore.getResult(getClass(),
                businessActionContext.getXid()) == null) {          // 防止重复提交
            return true;
        }
        Emp empVO = new Emp();                                      // DAO对象
        DeepBeanUtils.copyProperties(emp, empVO);                   // 对象复制
        try {
            return this.empDAO.insert(empVO) > 0;                   // 数据保存
        } finally {
            TCCResultStore.removeResult(getClass(),
                businessActionContext.getXid());                    // 删除标记
        }
    }
    @Override
    public boolean rollback(BusinessActionContext businessActionContext) {
        EmpDTO emp = ((JSONObject) businessActionContext.getActionContext("emp"))
                .toJavaObject(EmpDTO.class);                        // 获取操作数据
        log.info("【第二阶段】事务回滚，xid = {}、emp = {}",
                businessActionContext.getXid(), emp);               // 日志输出
        if (TCCResultStore.getResult(getClass(),
                businessActionContext.getXid()) == null) {          // 防止重复提交
            return true;
        }
        TCCResultStore.removeResult(getClass(),
                businessActionContext.getXid());                    // 删除标记
        return true;
    }
}
```

（7）【provider-emp-8006 子模块】修改 EmpServiceImpl 业务实现类，通过 IEmpTCC 接口实例实现数据增加。

```
package com.yootk.provider.service.impl;
@Service
```

```
public class EmpServiceImpl implements IEmpService {                    // 业务实现类
    @Autowired
    private IEmpTCC empTCC;                                              // 注入IEmpTCC实例
    @Override
    public boolean add(EmpDTO emp) {
        return this.empTCC.prepareAdd(new BusinessActionContext(), emp);
    }
}
```

此时在部门和雇员微服务之中已经基于 TCC 的手动模式实现了分布式的事务处理，而在总业务逻辑中不需要做出任何修改，直接使用@GlobalTransactional 注解即可基于 Seata 实现分布式事务处理。最重要的是，此时的代码不需要对数据采用全局锁，允许多个事务同时实现数据操作，因此 TCC 的模式效率更高。

9.3.5 Saga 模式

Saga 模式

视频名称 0915_【理解】Saga 模式

视频简介 Saga 是一种跨平台的解决方案，可以很好地解决异构系统的分布式事务处理问题。本视频将为读者讲解 Saga 的主要作用、Saga 状态机设计器的服务构建，并介绍如何结合 Seata 实现分布式事务的处理。

在前面实现的 Seata 分布式事务 3 个操作模型中所使用的微服务全部可以根据开发者的需要进行修改。但是在一些特殊的环境下，如果要与一些封闭的系统，或者与老旧系统进行分布式业务对接，那么 AT、XA、TCC 模型将全部失效，如图 9-40 所示。为了解决这样的技术问题，Seata 引入了 Saga 模式。

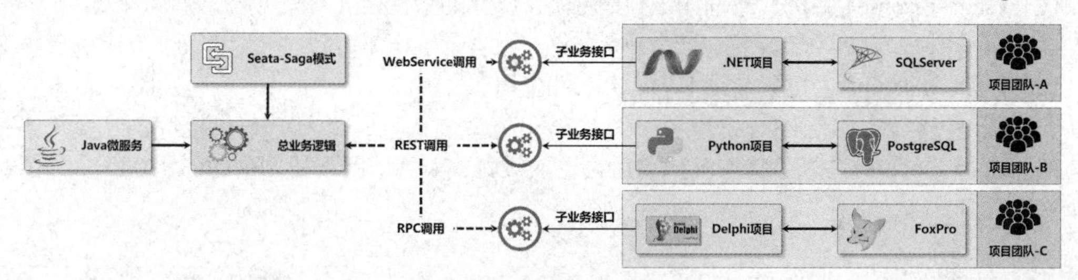

图 9-40 Saga 模式的产生背景

Saga 模式是 Seata 提供的长事务解决方案，提供了异构系统的事务统一处理模型。在 Saga 模式中，所有的子业务都不再直接参与整体事务的处理（只负责本地事务的处理），而是将其全部交由最终调用端来负责。而在进行总业务逻辑处理时，若某一个子业务出现问题，则自动补偿前面已经成功的其他参与者，这样一阶段的正向服务调用和二阶段的服务补偿处理全部由总业务开发实现，如图 9-41 所示。

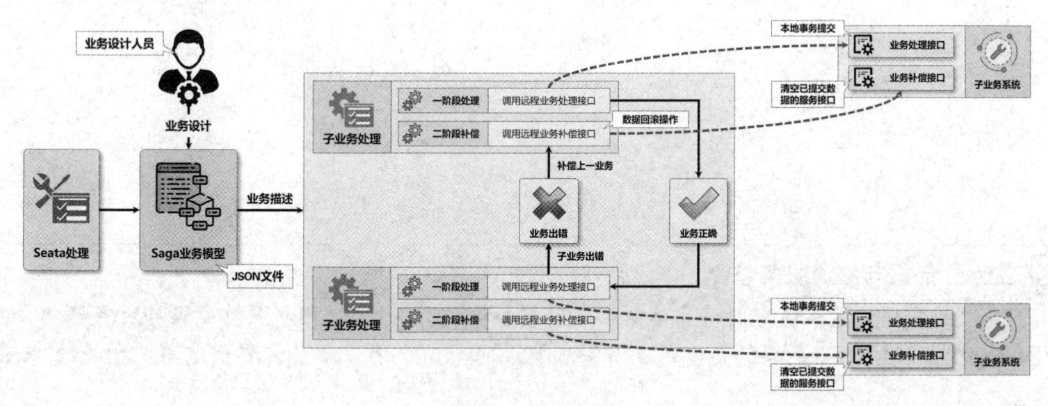

图 9-41 Saga 模型

提示：Saga 模型起源。

Saga 模型起源于 1987 年 Hector Garcia-Molina、Kenneth Salem 发表的论文 *Sagas*，异构分布式长事务相关概念在其中被提出。Seata 根据该论文将之转为现实可用的组件供开发者使用。

目前 Seata 提供的 Saga 模式是基于状态机引擎实现的，需要开发者手动进行 Saga 业务流程的绘制，并将其转换为 JSON 配置文件。而后程序运行时，将依据此配置文件实现业务处理以及服务补偿处理，Saga 状态图的绘制一般需要通过 Saga 状态机来实现。为了便于读者理解，下面将通过一系列的操作步骤，介绍 Saga 模式的应用。

提示：本节内容建议观看视频学习。

本次实现的 Saga 会涉及大量的开发步骤，而且 JSON 文件的生成操作步骤也相对烦琐。考虑到学习效果，笔者强烈建议读者跟随本书的配套视频进行学习，在视频讲解中也会有大量的 Seata 官方文档的参考。

另外本次所使用的 Saga 状态机需要基于 Node.JS 应用进行部署，对 Node.JS 或 NPM 不太理解的读者可以参考本套丛书中的相关图书进行学习。

（1）【state-machine 主机】通过 GitHub 下载 Seata 状态机设计工具代码。

```
git clone https://github.com/seata/seata.git
```

（2）【state-machine 主机】进入 Seata 状态机代码目录。

```
cd seata\saga\seata-saga-statemachine-designer
```

（3）【state-machine 主机】通过 NPM 安装项目所需的依赖库。

```
npm install
```

（4）【state-machine 主机】启动 Seata 状态机应用。

```
npm start
```

（5）【state-machine 主机】状态机启动后会进入设计窗口（访问路径：localhost:8080），如图 9-42 所示。

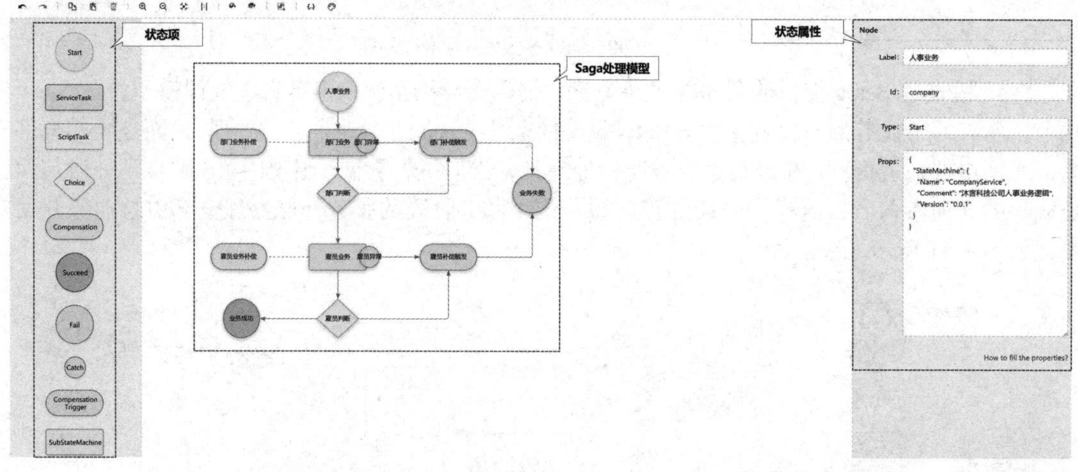

图 9-42　Saga 状态机

提示：部门与雇员服务整合。

本次的处理依然实现部门与雇员微服务的整合操作（此时的部门和雇员微服务没有任何分布式事务处理环境），而最终的整合是在消费端完成的，这样消费端就需要创建与之相关的 Bean 模型。

（6）【Saga 状态机设计工具】在 Saga 状态机中进行模型的构建，如图 9-43 所示。

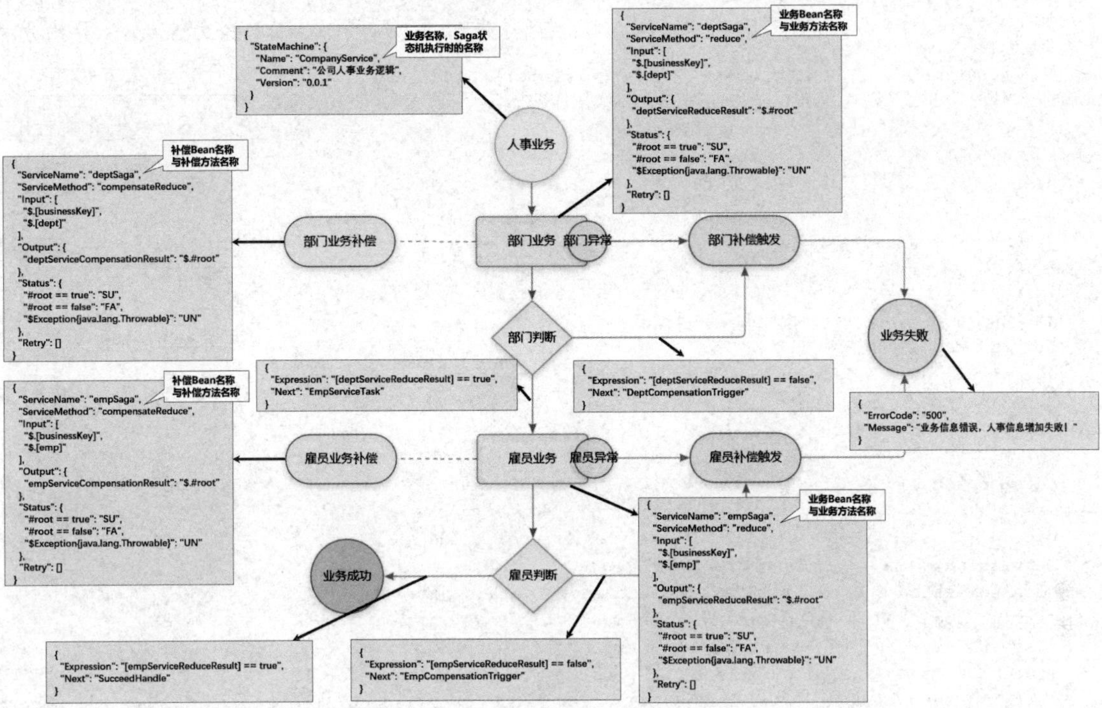

图 9-43　Saga 模型绘制

提示：Saga 模型与 Bean 模型。

本次进行的 Saga 模型绘制，除了状态的关系配置之外，实际上与程序之中注册的程序 Bean 和处理方法有关联，还涉及大量的 SpEL 应用。如果仅仅是观察此模型，对于初学者来说较难理解，建议参照本节视频进行学习。

（7）【consumer-springboot-80 子模块】将 Saga 状态机生成的 JSON 数据保存在 resource/saga/company-hr-add.json 之中。

（8）【数据库】此时所有的分布式事务处理全部交由消费端处理，这样就需要在消费端按照 Seata 要求进行数据库与数据表的创建，执行脚本如下。

```
DROP DATABASE IF EXISTS yootk80;
CREATE DATABASE yootk80 CHARACTER SET UTF8;
USE yootk80;
create table seata_state_machine_def(
   id              varchar(32)  not null comment 'id',
   name            varchar(128) not null comment 'name',
   tenant_id        varchar(32)  not null comment 'tenant id',
   app_name         varchar(32)  not null comment 'application name',
   type            varchar(20) comment 'state language type',
   comment_         varchar(255) comment 'comment',
   ver             varchar(16)  not null comment 'version',
   gmt_create       timestamp    not null comment 'create time',
   status          varchar(2)   not null comment 'status(AC:active|IN:inactive)',
   content          text comment 'content',
   recover_strategy varchar(16) comment 'transaction recover strategy(compensate|retry)',
   primary key (id)
);
create table seata_state_machine_inst(
   id                varchar(128) not null comment 'id',
   machine_id          varchar(32) not null comment 'state machine definition id',
```

```
    tenant_id           varchar(32) not null comment 'tenant id',
    parent_id           varchar(128) comment 'parent id',
    gmt_started         timestamp  not null comment 'start time',
    business_key        varchar(48) comment 'business key',
    start_params        text comment 'start parameters',
    gmt_end             timestamp comment 'end time',
    excep               blob comment 'exception',
    end_params          text comment 'end parameters',
    status              varchar(2) comment 'status(SU succeed|FA failed|UN unknown|SK skipped|RU
running)',
    compensation_status varchar(2) comment 'compensation status(SU succeed|FA failed|UN unknown|SK
skipped|RU running)',
    is_running          tinyint(1) comment 'is running(0 no|1 yes)',
    gmt_updated         timestamp  not null,
    primary key (id),
    unique key unikey_buz_tenant (business_key, tenant_id)
);
create table seata_state_inst(
    id                      varchar(48)  not null comment 'id',
    machine_inst_id           varchar(128)  not null comment 'state machine instance id',
    name                    varchar(128) not null comment 'state name',
    type                    varchar(20) comment 'state type',
    service_name             varchar(128) comment 'service name',
    service_method           varchar(128) comment 'method name',
    service_type             varchar(16) comment 'service type',
    business_key             varchar(48) comment 'business key',
    state_id_compensated_for varchar(50) comment 'state compensated for',
    state_id_retried_for      varchar(50) comment 'state retried for',
    gmt_started              timestamp   not null comment 'start time',
    is_for_update            tinyint(1) comment 'is service for update',
    input_params             text comment 'input parameters',
    output_params            text comment 'output parameters',
    status                  varchar(2)   not null comment 'status(SU succeed|FA failed|UN unknown|SK
skipped|RU running)',
    excep                   text comment 'exception',
    gmt_end                  timestamp comment 'end time',
    gmt_updated         timestamp  not null,
    primary key (id, machine_inst_id)
);
```

（9）【microcloud 项目】修改 build.gradle 配置文件，为 consumer-springboot-80 子模块引入 Druid 相关依赖。

```
implementation('mysql:mysql-connector-java:8.0.23')          // MySQL依赖
implementation('com.alibaba:druid:1.2.5')                    // Druid依赖
```

（10）【consumer-springboot-80 子模块】修改 application.yml 配置文件，引入 datasource 配置（核心项）如下。

```
spring:                                                        # Spring配置
  datasource:                                                  # 数据源配置
    type: com.alibaba.druid.pool.DruidDataSource               # 数据源类型
    driver-class-name: com.mysql.cj.jdbc.Driver                # 驱动程序类
    url: jdbc:mysql://localhost:3306/yootk80                   # 连接地址
```

（11）【consumer-springboot-80 子模块】编写 DruidConfig 配置类，用于创建 DruidDataSource 实例。

```
package com.yootk.consumer.config;
@Configuration
public class DruidConfig {
    @ConfigurationProperties(prefix = "spring.datasource")    // 配置项
    @Bean
    public DataSource druid(){
```

```
        return  new DruidDataSource();                              // DruidDataSource实例
    }
}
```

（12）【consumer-springboot-80 子模块】创建 StateMachineEngineConfig 配置类。

```
package com.yootk.consumer.config;
@Configuration
public class StateMachineEngineConfig {
    @Autowired
    private DataSource dataSource;                                  // 数据源
    @Bean
    public ThreadPoolExecutor getThreadExecutorConfig(){
        ThreadPoolTaskExecutor executor = new ThreadPoolTaskExecutor(); // 线程池定义
        executor.setCorePoolSize(1);                                // 核心线程数量
        executor.setMaxPoolSize(20);                                // 最大线程数量
        executor.setQueueCapacity(99999);                           // 延迟队列
        executor.setThreadNamePrefix("SAGA_ASYNC_EXEC_");           // 线程名称前缀
        executor.setRejectedExecutionHandler(new ThreadPoolExecutor.CallerRunsPolicy());
        executor.initialize();                                      // 初始化操作
        return executor.getThreadPoolExecutor();                    // 返回线程池
    }
    @Bean
    public DbStateMachineConfig dbStateMachineConfig(){             // 数据库配置
        DbStateMachineConfig stateMachineConfig = new DbStateMachineConfig(); // 存储配置
        stateMachineConfig.setDataSource(dataSource);               // 数据源配置
        Resource resource = new ClassPathResource("saga/company-hr-add.json"); // saga
        stateMachineConfig.setResources(new Resource[]{resource});  // 资源绑定
        stateMachineConfig.setEnableAsync(true);                    // 异步支持
        stateMachineConfig.setThreadPoolExecutor(getThreadExecutorConfig()); // 线程池定义
        return stateMachineConfig;
    }
    @Bean
    public ProcessCtrlStateMachineEngine stateMachineEngine(){      // 状态机引擎
        ProcessCtrlStateMachineEngine processCtrlStateMachineEngine =
                new ProcessCtrlStateMachineEngine();
        processCtrlStateMachineEngine.setStateMachineConfig(dbStateMachineConfig());
        return processCtrlStateMachineEngine;
    }
    @Bean
    public StateMachineEngineHolder stateMachineEngineHolder(){     // 状态机持有者
        StateMachineEngineHolder engineHolder = new StateMachineEngineHolder();
        engineHolder.setStateMachineEngine(stateMachineEngine());
        return engineHolder;
    }
}
```

（13）【consumer-springboot-80 子模块】创建 IDeptSaga 接口。

```
package com.yootk.consumer.saga;
public interface IDeptSaga {
    boolean reduce(String businessKey, DeptDTO dept);               // 业务处理
    boolean compensateReduce(String businessKey, DeptDTO dept);     // 业务补偿
}
```

（14）【consumer-springboot-80 子模块】创建 DeptSagaImpl 实现子类。

```
package com.yootk.consumer.saga.impl;
@Component("deptSaga")                                              // 与Saga模型中的Bean名称一致
@Slf4j                                                              // 日志定义
public class DeptSagaImpl implements IDeptSaga {
    @Autowired
    private IDeptService deptService;                               // Feign远程接口
    @Override
    public boolean reduce(String businessKey, DeptDTO dept) {       // 服务调用
        log.info("【部门业务处理】业务KEY：{}、部门信息：{}", businessKey, dept); // 日志输出
```

```
        if (dept.getDname() == null || "".equals(dept.getDname())) {
            throw new RuntimeException("部门名称不允许为空。");                    // 模拟异常
        }
        try {
            return this.deptService.add(dept);                                   // 业务调用
        } catch (Exception e) {
            return false;                                                        // 出错返回false
        }
    }
    @Override
    public boolean compensateReduce(String businessKey, DeptDTO dept) {       // 服务补偿
        log.info("【部门业务补偿】业务KEY：{}、部门信息：{}", businessKey, dept);  // 日志输出
        // 此处不再实现具体的业务补偿处理逻辑，实际中需要服务端提供补偿操作接口
        return true;
    }
}
```

（15）【consumer-springboot-80 子模块】创建 IEmpSaga 接口。

```
package com.yootk.consumer.saga;
public interface IEmpSaga {
    boolean reduce(String businessKey, EmpDTO emp);                       // 业务处理
    boolean compensateReduce(String businessKey, EmpDTO emp);             // 业务补偿
}
```

（16）【consumer-springboot-80 子模块】创建 EmpSagaImpl 实现子类。

```
package com.yootk.consumer.saga.impl;
@Component("empSaga")                                                    // 与Saga模型中的Bean名称一致
@Slf4j                                                                   // 日志定义
public class EmpSagaImpl implements IEmpSaga {
    @Autowired
    private IEmpService empService;                                      // Feign远程接口
    @Override
    public boolean reduce(String businessKey, EmpDTO emp) {              // 业务处理
        log.info("【雇员业务处理】业务Key：{}、雇员信息：{}", businessKey, emp);   // 日志输出
        try {
            return this.empService.add(emp);                             // 接口调用
        } catch (Exception e) {
            return false;                                                // 异常处理
        }
    }
    @Override
    public boolean compensateReduce(String businessKey, EmpDTO emp) {
        log.info("【雇员业务补偿】业务Key：{}。雇员信息：{}"。 businessKey, emp);   // 日志输出
        // 此处不再实现具体的业务补偿处理逻辑，实际操作中需要服务端提供补偿操作接口
        return true;
    }
}
```

（17）【consumer-springboot-80 子模块】修改 CompanyConsumerAction 类，利用状态机引擎 StateMachineEngine 类根据 company-hr-add.json 中配置的业务名称进行服务调用。

```
package com.yootk.consumer.action;
@RestController                                                  // REST响应
@RequestMapping("/consumer/company/*")                           // 父路径
@Slf4j
public class CompanyConsumerAction {
    @Autowired
    private StateMachineEngine stateMachineEngine;               // Saga状态机
    @GetMapping("add")
    public Object add(DeptDTO dept, EmpDTO emp) {                // 数据增加
        Map<String, Object> startParams = new HashMap<>(3);      // 参数存储
        String businessKey = String.valueOf(System.currentTimeMillis());
        startParams.put("businessKey", businessKey);             // 业务Key
```

```
        startParams.put("dept", dept);                          // 部门业务参数
        startParams.put("emp", emp);                            // 雇员业务参数
        StateMachineInstance inst = stateMachineEngine.startWithBusinessKey(
                "CompanyService", null, businessKey, startParams);
        if(ExecutionStatus.SU.equals(inst.getStatus())){        // Saga模型处理成功
            log.info("人事信息创建成功, XID = {}", inst.getId());
            return true;
        } else {                                                // Saga模型处理失败
            log.info("人事信息创建失败, XID = {}", inst.getId());
            return false;
        }
    }
}
```

以上操作完成后，如果所有的业务未出现错误，则/consumer/company/add 接口的最终响应结果是 true；如果有一个子业务出现错误，则响应结果为 false，同时会自动调用上一子业务中的业务补偿方法进行处理（雇员业务出错后会调用部门提供的业务补偿方法进行处理）。

9.4 本章概览

1．Spring Cloud Stream 技术的目的是通过一套统一的处理架构实现任意消息应用的整合。目前为止支持该技术的消息组件有 3 种：RabbitMQ、RocketMQ、Kafka。

2．每一种消息组件进行 Spring Cloud Stream 整合时，除了按照标准规范进行代码开发之外，还可以使用消息组件自己的一些操作模式，例如，RabbitMQ 可以实现路由消息，而 RocketMQ 可以实现消费过滤。

3．Spring Cloud Config 是一种基于动态配置管理的解决方案，早期的 Spring Cloud Netflix 套件需要开发者基于代码仓库（SVN、GitHub、GitLab）实现配置托管，而后通过 Spring Cloud Config 服务端实现与代码仓库的整合，并在客户端启动时进行配置项的读取。在此种操作机制下，如果需要进行配置更新，则必须基于消息组件与 Actuator 监控的 Spring Cloud Bus 技术来进行更新管理，实现过程过于烦琐。

4．Spring Cloud Alibaba 套件提供了 Nacos 配置项的监听支持，基于此种方式实现的配置管理不仅开发简单，管理员也可以基于 Web 配置页面实现配置项的管理。

5．Seata 是由阿里巴巴公司提供的分布式事务组件，基于二阶段处理操作模型，包括 AT 模式、XA 模式、TCC 模式与 Saga 模式。其中 AT 模式不需要进行业务侵入处理，XA 模式实现了基于数据库二阶段处理的分布式事务，而 TCC 模式则需要对代码的业务模型进行较大的侵入。

6．异构系统的分布式事务可以基于 Saga 模式实现。Saga 模式可以通过用户绘制的状态机来实现事务流程的处理。

第 10 章

服务跟踪

本章学习目标

1. 理解 Spring Cloud Sleuth 的主要作用与实现架构，并可以基于 Zipkin 实现微服务调用跟踪；

2. 理解 SkyWalking 全链路跟踪的配置与服务无侵入处理操作，并可以基于 ElasticSearch 实现数据存储；

3. 了解 Sentry 服务的使用，并可以通过其实现微服务错误日志记录的处理。

微服务的出现使得业务中心的开发与调用模式发生了极大的改变，伴随而来的就是微服务调用链的烦琐以及问题排查的困难，所以一个设计良好的微服务必然要有完善的服务跟踪机制。本章将在前面实现的微服务基础之上为读者讲解 Zipkin、SkyWalking、Sentry 等常用微服务监控工具的使用。

10.1 Spring Cloud Sleuth

Spring Cloud Sleuth 简介

视频名称 1001_【了解】Spring Cloud Sleuth 简介

视频简介 Spring Cloud Sleuth 是 Spring Cloud 内置的链路跟踪解决方案。本视频为读者分析链路跟踪的设计意义，并通过实例讲解 Spring Cloud Sleuth 的基本使用。

在微服务的设计架构之中，由于需要对微服务进行拆分，因此一个完整的业务处理最终就有可能调用若干个微服务来实现。如果想弄清楚微服务的调用轨迹，可以使用 Spring Cloud Sleuth 技术来实现链路跟踪处理。

Spring Cloud Sleuth 是为 Spring Cloud 提供的分布式微服务跟踪解决方案，建议将其与 Zipkin、HTrace 或 ELK 等日志追踪采集系统一起使用，这样可以很清楚地观察到某一个请求所经过的微服务节点，帮助使用者方便地厘清服务间的调用关系，如图 10-1 所示。每一次链路追踪都会提供一个 TraceID，而后每一个调用的节点都会生成一个 SpanID，利用 TraceID 的内容即可观察服务的调用轨迹。

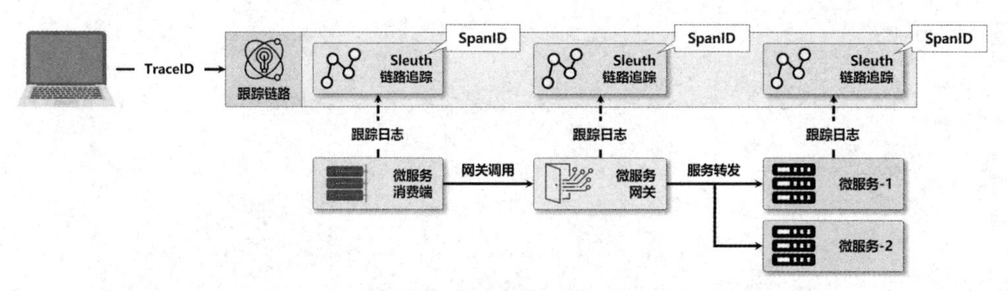

图 10-1　Spring Cloud Sleuth

范例：在需要进行监控的微服务（部门、网关、Token 以及消费端）中添加 Sleuth 依赖

```
implementation('org.springframework.cloud:spring-cloud-starter-sleuth')
```

日志信息输出	INFO [dept.provider,aed1c13cbf33c829,b20974ec440070a5,false] [] 19736 --- [nio-8002-exec-4]…

依赖添加完成之后，在每次进行微服务调用时，都会出现以上日志标记内容，开发人员可以根据这些日志标记实现跟踪匹配，每组日志标记一共由 4 个部分所组成。

- 微服务应用名称：在 application.yml 中定义的 spring.application.name 配置项内容。
- TraceID：由 Sleuth 生成的一个唯一 ID，用来标识一条请求链路，一条请求链路包含多个 SpanID。
- SpanID：基本工作单元（例如部门微服务单元、网关微服务单元等）。
- Zipkin 接收状态：如果当前微服务使用 Zipkin 接收日志信息则返回 true，否则返回 false。

10.1.1 搭建 Zipkin 服务

搭建 Zipkin 服务

视频名称 1002_【了解】搭建 Zipkin 服务

视频简介 Zipkin 是由 Twitter 公司公布的开源项目，同时在 GitHub 也提供了项目源代码。本视频通过具体的操作实例为读者讲解 Zipkin 的项目打包与服务部署操作。

虽然 Spring Cloud Sleuth 提供了链路跟踪的实现支持，但是所有的链路跟踪都是基于日志信息实现匹配的，这样开发者实际上也很难清晰地观察出微服务调用的顺序，所以最佳的做法是引入一个链路数据的采集服务，该服务在实现链路数据存储的同时也可以提供可视化的解决方案。

在 Spring Cloud Sleuth 中较为常见的数据采集工具就是 Zipkin，该组件是由 Twitter 公司提供的开源项目，可以实现数据采集、持久化存储、跟踪链查找以及图形化展示等功能，如图 10-2 所示。

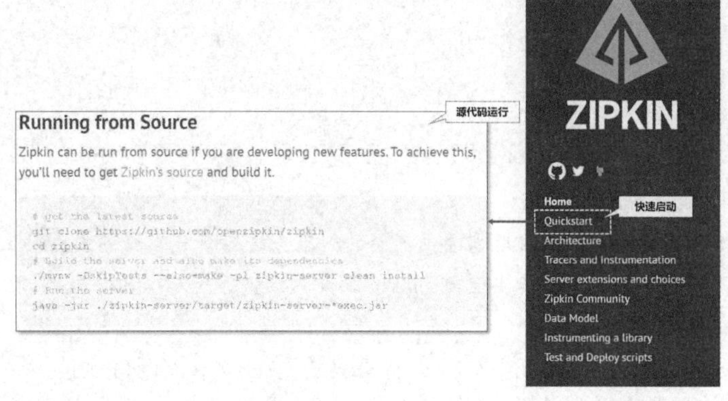

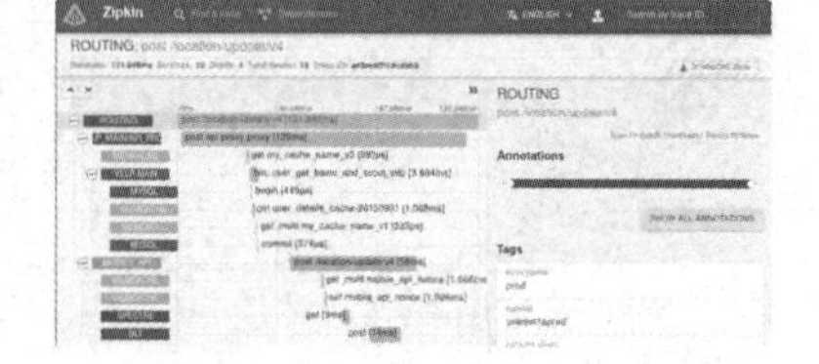

图 10-2 Zipkin 首页

Zipkin 官方文档中给出了 Docker 和源代码运行两种方式。为了便于读者理解，本次演示将通过 Zipkin 源代码手动打包的方式来运行 Zipkin 服务，GitHub 平台上的 Zipkin 源代码如图 10-3 所示。下面通过具体的操作步骤为读者演示 Zipkin 组件的打包与服务部署操作。

（1）【Git 客户端】Zipkin 是一个开源项目，开发者可以通过 Git 工具进行代码下载。

```
git clone git@github.com:openzipkin/zipkin.git
```

（2）【Maven 编译】Zipkin 源代码是基于 Maven 开发的，这样就可以使用 Maven 工具编译 Zipkin 源代码。

```
mvn clean package -DskipTests
```

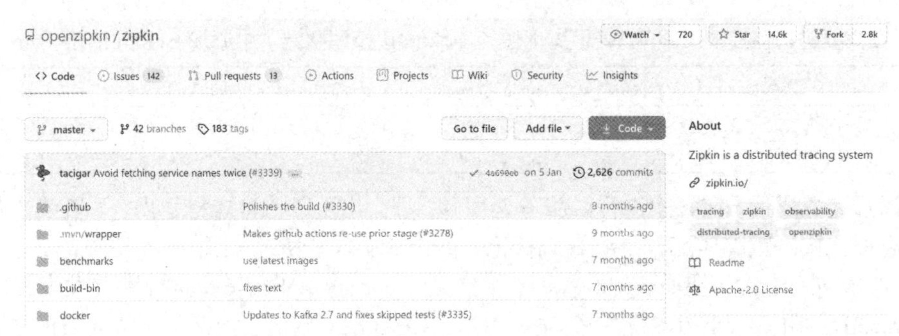

图 10-3 Zipkin 源代码

Zipkin 源代码编译完成后会在“${Zipkin 源代码目录}/zipkin-server/target/”目录中生成一个 Zipkin 服务端的可执行程序文件（zipkin-server-2.23.3-SNAPSHOT-exec.jar）。本次将基于此文件来运行服务

（3）【Java 命令】通过 java 命令直接运行 zipkin-server-2.23.3-SNAPSHOT-exec.jar。

```
java -jar zipkin-server-2.23.3-SNAPSHOT-exec.jar
```

（4）【本地系统】修改 hosts 配置文件，增加主机名称配置。

```
127.0.0.1   zipkin-server
```

（5）【浏览器】Zipkin 启动后会自动占用 9411 端口。通过浏览器访问 zipkin-server:9411 服务，即可进入图 10-4 所示的界面。

图 10-4 Zipkin 首页

10.1.2 微服务日志采集

微服务日志采集

视频名称 1003_【了解】微服务日志采集

视频简介 Zipkin 作为数据采集服务，开启后需要由不同的微服务进行日志数据的发送。本视频介绍对已有的微服务进行修改，使其实现与 Zipkin 服务对接操作。

此时的 Zipkin 仅仅是一个独立的应用，而在现实的开发中还需要对所有的微服务进行改造（添加依赖与相关配置），使其可以将每次的跟踪链路数据发送到 zipkin-server 服务之中，如图 10-5 所示。下面通过具体的实现步骤进行操作演示。

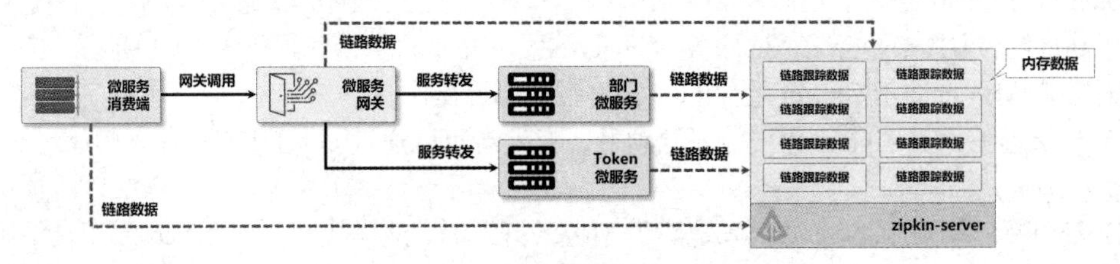

图 10-5 Zipkin 接入

（1）【microcloud 项目】修改 build.gradle 配置文件，在部门微服务、网关微服务、Token 微服

务以及消费微服务中添加 Zipkin 依赖库，以便于实现数据采集配置。

```
implementation('org.springframework.cloud:spring-cloud-starter-zipkin')
```

（2）【采集子模块】修改部门微服务、网关微服务、Token 微服务以及消费微服务中的 application.yml 配置文件，添加 Zipkin 数据采集配置。

```
spring:                                          # Spring配置
  zipkin:                                        # Zipkin配置
    base-url: http://zipkin-server:9411          # 服务地址
  sleuth:                                        # Sleuth配置
    sampler:                                     # 数据抽样配置
      percentage: 1.0                            # 定义抽样比率，默认为0.1
```

抽样比率如果采用默认的 0.1，则表示每 10 次访问进行一次抽样。本次为了便于读者观察将每一次的访问都进行了记录。

（3）【Zipkin 控制台】依次启动每一个微服务，随后通过 zipkin-server:9411 控制台查看监控数据，如图 10-6 所示。

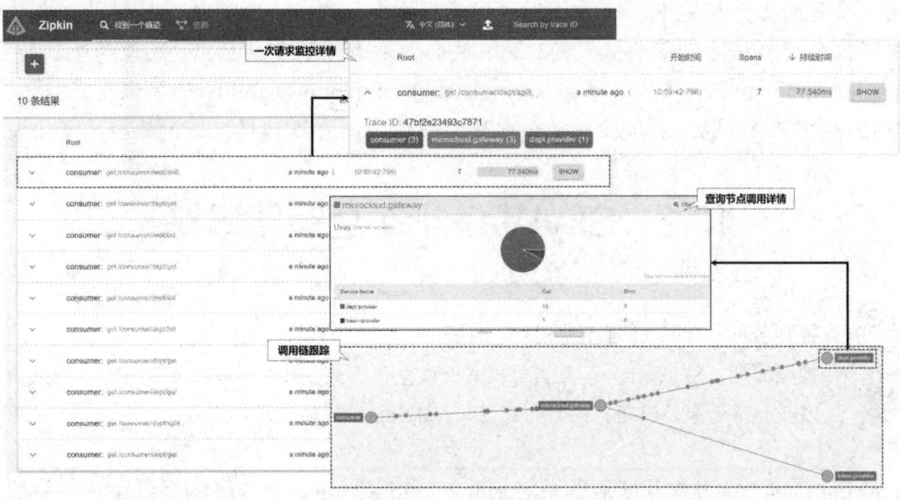

图 10-6　Zipkin 跟踪

10.1.3　Zipkin 数据持久化

视频名称　1004_【了解】Zipkin 数据持久化

视频简介　Zipkin 为了便于持久化存储提供了 MySQL 的解决方案。本视频介绍对已有的 Zipkin 采集服务进行修改，使其可以实现基于 MySQL 链路数据存储支持。

在默认情况下，所有发送到 Zipkin 中的日志数据全部保存在 Zipkin 应用的内存之中。这样一旦发生 Zipkin 服务死机问题，所收集的链路数据就会丢失。Zipkin 内部提供了数据库的存储支持，可以将链路数据保存在 MySQL 之中，如图 10-7 所示。下面通过具体的操作步骤来介绍如何实现这一功能配置。

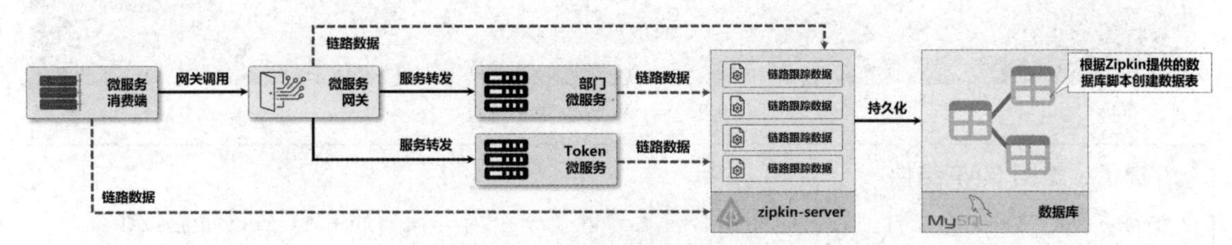

图 10-7　Zipkin 数据持久化存储

（1）【MySQL 数据库】创建 Zipkin 数据库。

```
DROP DATABASE IF EXISTS zipkin;
CREATE DATABASE zipkin CHARACTER SET UTF8;
USE zipkin;
```

（2）【MySQL 数据库】在“${Zipkin 源代码目录}/zipkin-storage/mysql-v1/src/main/resources”目录下可以找到数据库创建脚本文件（mysql.sql），通过此文件创建本次操作所要使用的数据表。

（3）【Java 命令】重新启动 Zipkin 服务，启动时设置 MySQL 连接信息。

```
java -jar zipkin-server-2.23.3-SNAPSHOT-exec.jar --STORAGE_TYPE=mysql \
     --MYSQL_HOST=localhost --MYSQL_TCP_PORT=3306 --MYSQL_USER=root \
     --MYSQL_PASS=mysqladmin --MYSQL_DB=zipkin
```

启动完成后，所有微服务发送到 Zipkin 上的链路数据都会自动存储在 MySQL 数据库之中，这样在服务出现问题后就可以自动恢复所记录的链路数据。

提示：防止日志数据过大。

在实际项目的生产环境下，随着并发访问量的增加，Zipkin 所接收的数据量也会非常庞大。这样一来不仅会给数据存储带来极大的压力，也会带来大量的重复数据。最佳的做法是修改微服务中的 application.yml 配置文件，通过 spring.sleuth.sampler.probability 选项配置采样率。如配置为 1.0，则采样率为 100%，采集服务的全部追踪数据。而在默认情况下的采样率为 0.1（10%）。

10.2　SkyWalking 全链路跟踪

SkyWalking 简介

视频名称　1005_【理解】SkyWalking 简介

视频简介　SkyWalking 是当前项目开发中使用最多的全链路跟踪组件。本视频为读者介绍 Zipkin 所带来的问题，同时介绍 SkyWalking 组件的使用特点。

虽然 Spring Cloud 官方推荐使用的是 Spring Cloud Sleuth + Zipkin 的全链路跟踪模式，但是由于相关技术出现较早，因此所提供的功能较为单一。考虑到代码的侵入性以及简洁性的问题，现代的 Spring Cloud 开发会更多地使用 Apache 所提供的 SkyWalking 组件来实现全链路跟踪，如图 10-8 所示。

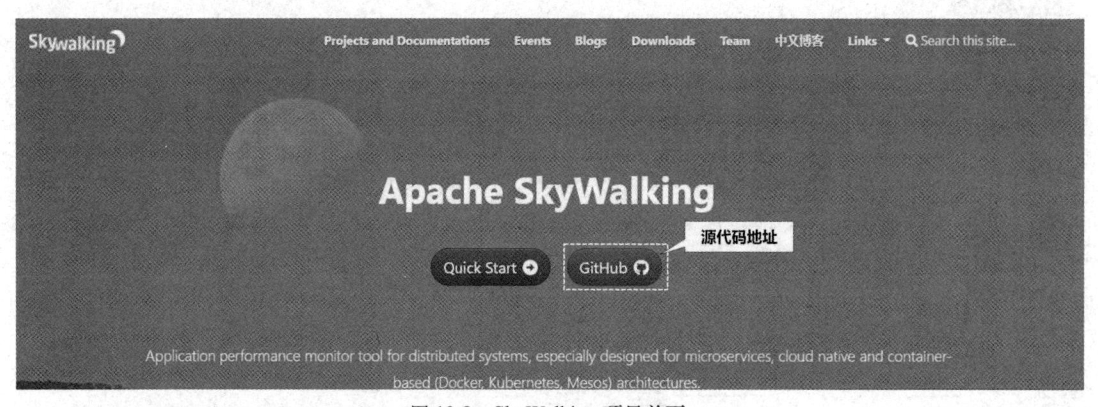

图 10-8　SkyWalking 项目首页

提示：美团 CAT 组件。

除了 SkyWalking 之外，国内的美团公司也开源了一个内部使用的 CAT 全链路监控组件，但是该组件对代码的侵入较大，不如 SkyWalking 的使用机制简单，所以没有被笔者采用。

Apache SkyWalking是一个开源、免费的服务监控平台，可以实现服务数据的收集、分析、聚合以及可视化管理，同时SkyWalking提供了一种简单的方式实现微服务的整合应用，其实现架构如图10-9所示。该组件一共提供了4个逻辑部分：探针、平台、存储及界面。

- **探针**：与终端服务捆绑在一起使用，将所有采集到的数据转化为SkyWalking适用的数据格式以实现数据传输（基于HTTP或gRPC）。在Java中可以直接基于字节码的方式进行监控植入，从而实现无侵入的代码收集。
- **平台**：提供数据采集的支持，可以实现数据的聚合、数据分析以及可视化界面的生成。在该平台中可以方便地实现存储终端的整合。
- **存储**：采用开放式的存储设计，支持主流的存储设备（MySQL、TIDB、PostgreSQL、ElasticSearch），或者根据自己的需要进行自定义存储的实现。
- **界面**：SkyWalking提供了丰富的可视化界面，UI功能非常“酷炫”。

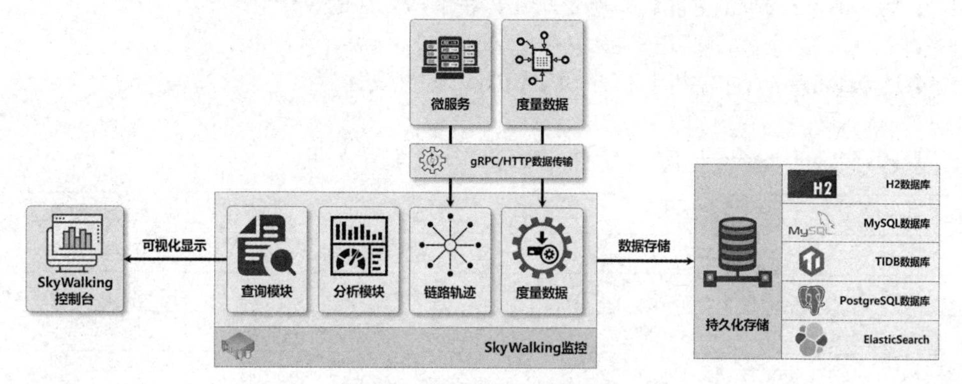

图10-9　SkyWalking实现架构

10.2.1　SkyWalking服务安装与配置

SkyWalking服务
安装与配置

视频名称　1006_【理解】SkyWalking服务安装与配置

视频简介　SkyWalking官网提供了完整的工具组件包。本视频为读者讲解组件的获取，并讲解SkyWalking与ElasticSearch存储的整合配置。

SkyWalking官方提供了项目发布的GitHub地址，开发者可以根据需要下载源代码，也可以直接通过官方地址下载打包完成的开发组件。本次为了简化处理，直接使用了支持ElasticSearch 7的发布版，如图10-10所示。

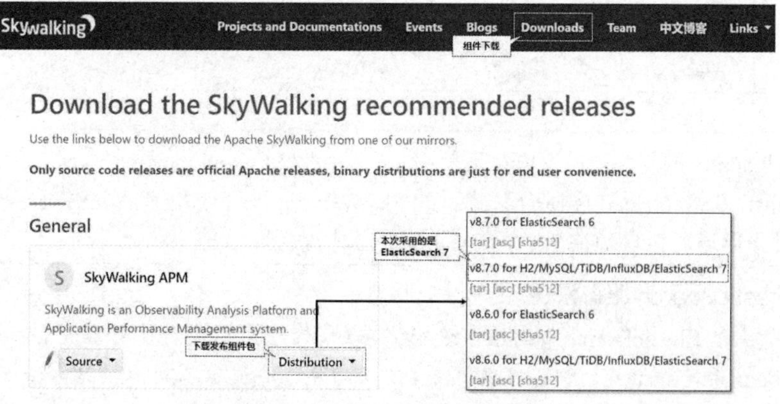

图10-10　SkyWalking组件下载

通过图 10-10 的下载提示可以发现，SkyWalking 支持多种不同的数据存储终端。考虑到实际的应用、扩展性等因素，本次将演示基于 ElasticSearch 7 的服务配置。下面通过具体的操作步骤介绍如何实现。

（1）【本地系统、skywalking-server 主机】本次将进行一个新的虚拟机配置。修改本地的 hosts 文件，增加相关的主机配置。

```
192.168.190.161elk-server
192.168.190.162skywalking-server
```

（2）【本地系统】将下载得到的 apache-skywalking-apm-es7-8.7.0.tar.gz 开发包上传到 skywalking-server 主机之中，组件保存的父目录为/var/ftp。

（3）【skywalking-server 主机】将 SkyWalking 组件解压缩到/usr/local 目录之中。

```
tar xzvf /var/ftp/apache-skywalking-apm-es7-8.7.0.tar.gz -C /usr/local/
```

（4）【skywalking-server 主机】为便于配置，将解压缩后的目录更名。

```
mv /usr/local/apache-skywalking-apm-bin-es7/ /usr/local/skywalking
```

（5）【skywalking-server 主机】打开 SkyWalking 配置文件。

```
vi /usr/local/skywalking/config/application.yml
```

（6）【skywalking-server 主机】修改 application.yml 配置文件。

`storage:`	数据存储配置
`  selector: ${SW_STORAGE:elasticsearch7}`	配置存储终端为 ES
`  elasticsearch7:`	ES 的相关接入配置
`    clusterNodes: ${SW_STORAGE_ES_CLUSTER_NODES:elk-server:9200}`	ES 服务地址
`    user: ${SW_ES_USER:"elastic"}`	ES 用户名
`    password: ${SW_ES_PASSWORD:"elastic"}`	ES 密码

（7）【skywalking-server 主机】启动 SkyWalking 服务。

```
/usr/local/skywalking/bin/startup.sh
```

程序执行结果：

```
SkyWalking OAP started successfully!
SkyWalking Web Application started successfully!
```

SkyWalking 服务启动后会占用 8080（HTTP 管理端口）、11800（微服务接入的探针端口，用于 gRPC 通信）、12800（访问 Collector 端口）3 个端口，可以通过“netstat -nptl”查看这几个端口是否已被占用。如果它们已经被相关的服务进程占用，则表示 SkyWalking 服务启动成功。

（8）【skywalking-server 主机】SkyWalking 服务启动后会占用 8080、11800、12800 端口，修改防火墙规则配置其访问。

添加访问端口：

```
firewall-cmd --zone=public --add-port=8080/tcp --permanent
firewall-cmd --zone=public --add-port=11800/tcp --permanent
firewall-cmd --zone=public --add-port=12800/tcp -permanent
```

重新加载配置：

```
firewall-cmd -reload
```

（9）【SkyWalking 控制台】服务启动后通过浏览器访问 SkyWalking 控制台首页，如图 10-11 所示。

（10）【ElasticSearch Head 控制台】此时的 SkyWalking 已经使用 ElasticSearch（简称 ES）作为存储终端，服务启动后会自动在 ES 中创建相应的数据索引，如图 10-12 所示。

提示：ElasticSearch 服务搭建。

本项目是基于 ElasticSearch 7 的组件进行搭建的，管理界面使用了 ElasticSearch Head 前端应用，包括 Kibana 和 Logstash，都已经配置成功。对 ES 服务搭建不熟悉的读者，可以通过本套丛书中的相关图书来进行学习，本书不再对此内容进行过多阐述。

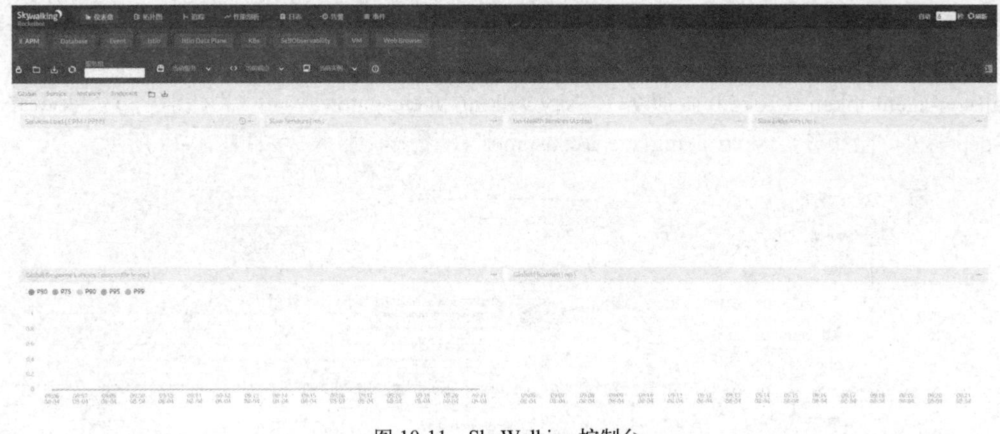

图 10-11　SkyWalking 控制台

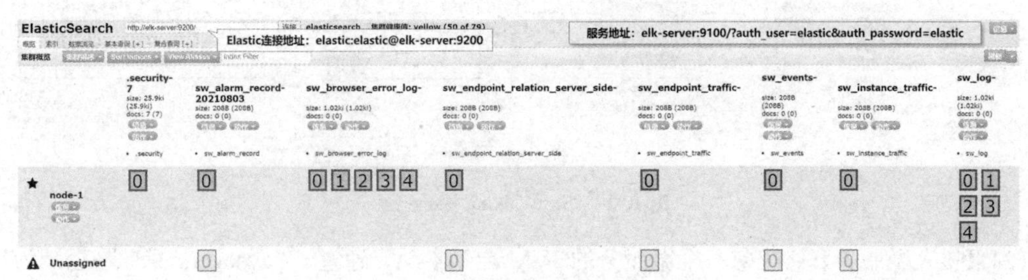

图 10-12　ElasticSearch Head 控制台

10.2.2 微服务接入

微服务接入

视频名称　1007_【理解】微服务接入

视频简介　SkyWalking 为了实现无侵入的接入模式，提供了探针的接入策略。本视频在已有的 SkyWalking 应用的基础之上介绍链路数据的发送以及可视化显示处理。

SkyWalking 组件内部提供了探针支持，利用探针的机制可以与要跟踪的代码进行整合，从而实现最终的服务接入处理，如图 10-13 所示。而用户所获取的 SkyWalking 组件的内部已经提供了探针的应用 skywalking-agent.jar 文件（复制其 agent 父目录），开发者直接基于此程序包即可实现将微服务的监控数据发送到 SkyWalking 应用之中。下面将通过具体的操作对已有的服务运行进行改造。

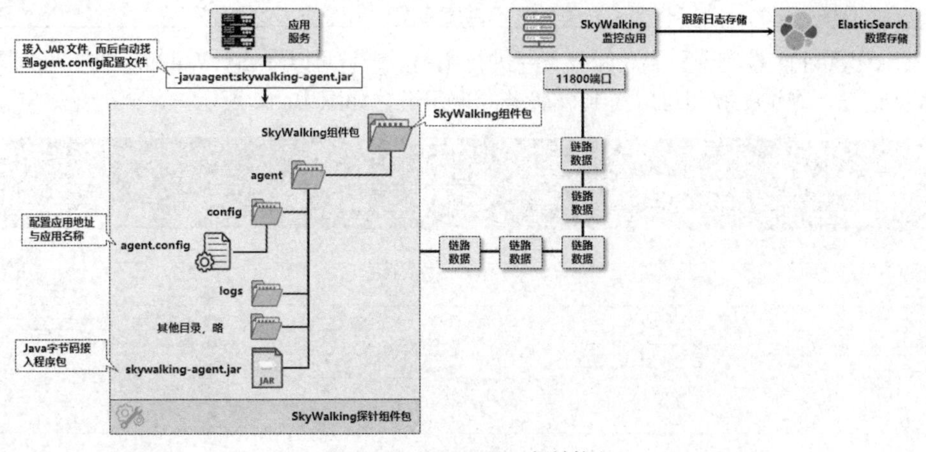

图 10-13　SkyWalking 探针接入

（1）【本地系统】在一套完整的微服务中，存在众多的微服务处理节点，每一个节点都需要提供一个探针工作目录。将 agent 目录复制 5 份，名称分别为 skywalking-agent-admin（管理微服务）、skywalking-agent-token（Token 微服务）、skywalking-agent-gateway（网关微服务）、skywalking-agent-dept（部门微服务）、skywalking-agent-consumer（消费端微服务），如图 10-14 所示。

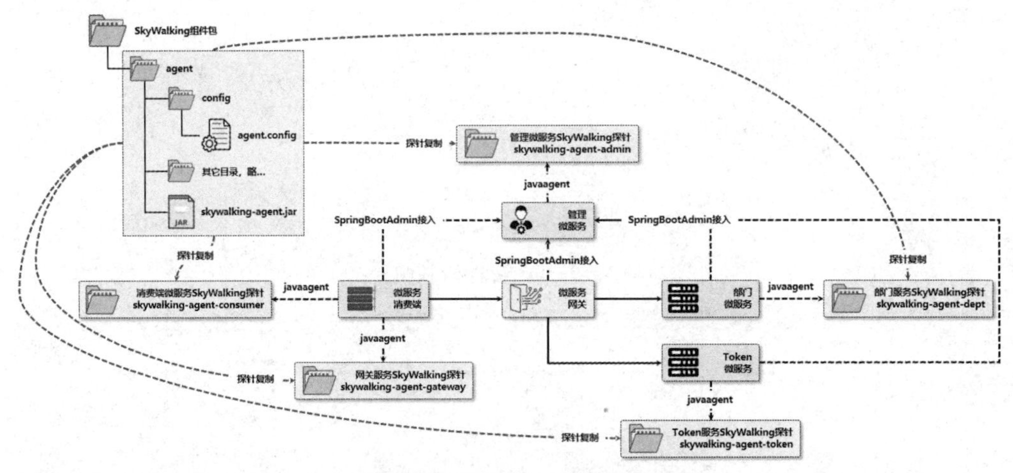

图 10-14　SkyWalking 探针配置

（2）【接入微服务】修改每一个微服务的 JVM 启动参数，通过“-javaagent:探针路径”为每个微服务设置匹配的探针路径、微服务名称、SkyWalking 服务地址等信息。

微服务	启动参数

管理微服务	`-javaagent:H:/workspace/skywalking-agent-admin/skywalking-agent.jar`
	`-Dskywalking.agent.service_name=gateway`
	`-Dskywalking.collector.backend_service=skywalking-server:11800`

网关微服务	`-javaagent:H:/workspace/skywalking-agent-gateway/skywalking-agent.jar`
	`-Dskywalking.agent.service_name=gateway`
	`-Dskywalking.collector.backend_service=skywalking-server:11800`

Token 微服务	`-javaagent:H:/workspace/skywalking-token/skywalking-agent.jar`
	`-Dskywalking.agent.service_name=token.provider`
	`-Dskywalking.collector.backend_service=skywalking-server:11800`

部门微服务	`-javaagent:H:/workspace/skywalking-dept/skywalking-agent.jar`
	`-Dskywalking.agent.service_name=dept.provider`
	`-Dskywalking.collector.backend_service=skywalking-server:11800`

消费端微服务	`-javaagent:H:/workspace/skywalking-consumer/skywalking-agent.jar`
	`-Dskywalking.agent.service_name=consumer`
	`-Dskywalking.collector.backend_service=skywalking-server:11800`

（3）【SkyWalking 控制台】启动微服务并进行调用后，就可以在 SkyWalking 首页见到图 10-15 所示的数据信息。查询拓扑图可以得到微服务的调用链，如图 10-16 所示。

图 10-15　微服务链路监控数据

图 10-15 所给出的监控数据中实际上有 SkyWalking 的 3 个核心数据项，管理者可以根据这些数据项来观测微服务的调用信息。这些配置项作用如下。

- 服务（Service）：表示对请求提供相同行为的一系列服务应用。每一组服务应用在启动时可以设置唯一的服务名称，当未设置时会由 SkyWalking 进行定义。
- 服务实例（Instance）：每一组监控都由若干个服务实例所组成。每个实例代表一个应用服务，本质上每一个应用都属于一个系统中的服务进程。
- 端点（Endpoint）：对于特定服务所接收的请求路径，例如，HTTP 请求中的 URI 路径，或者 gRPC 通信中的服务名称。

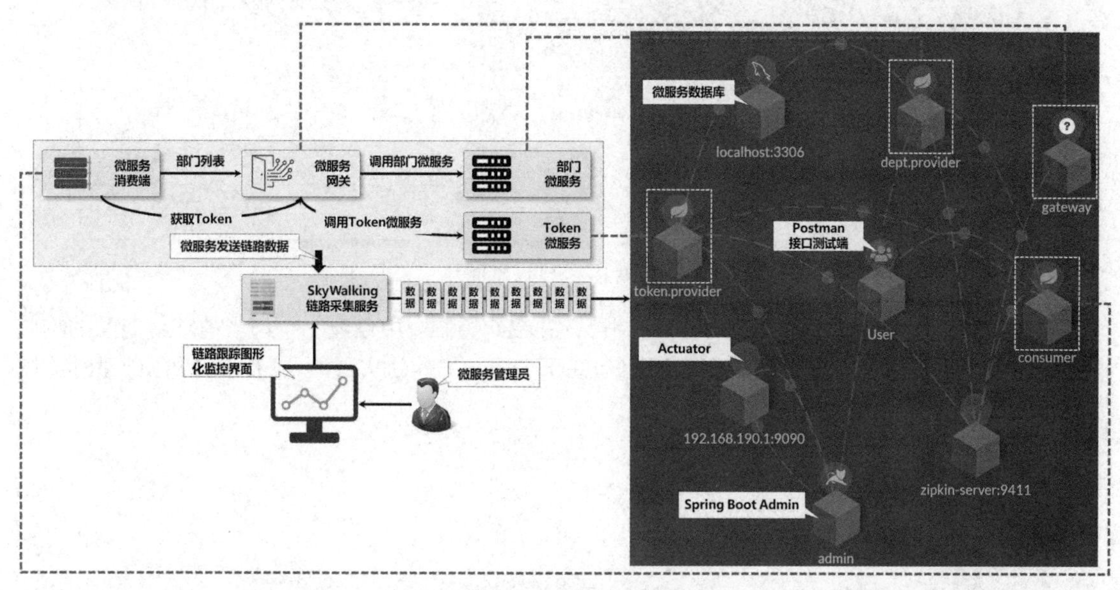

图 10-16 微服务拓扑

10.3 Sentry

视频名称 1008_【了解】Sentry 服务简介

视频简介 为了进行及时的错误排查，需要对微服务中的异常进行记录。在项目开发中开发者可以基于 Sentry 实现异常记录。本视频为读者讲解 Sentry 的主要作用。

在微服务的设计与开发中，很难保证所编写的代码不出现任何异常，而微服务庞大的调用链形式为异常的排查与修复带来了极大的麻烦。为了解决这样的问题，在开发中可以通过 Sentry 这一开源项目来实现异常的收集与预警处理，如图 10-17 所示。

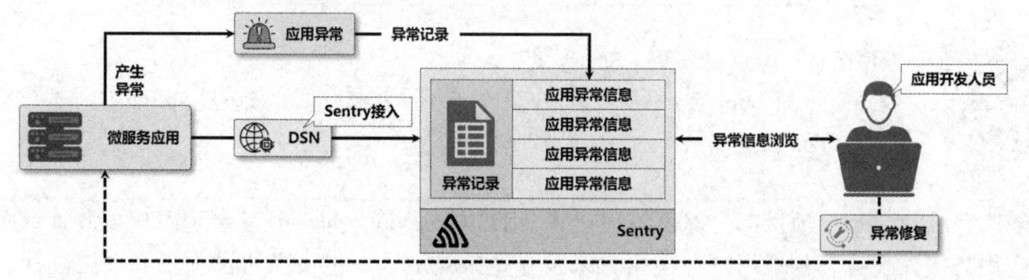

图 10-17 Sentry

Sentry 是一个免费的 SaaS 平台应用程序，开发者只需要登录 Sentry 首页，如图 10-18 所示，注册自己的账户后就可以直接在项目中进行应用。

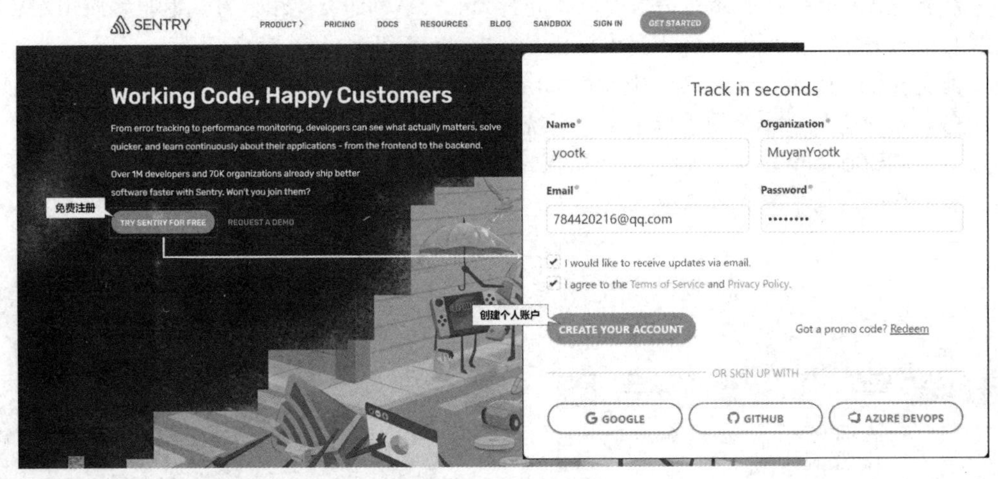

图 10-18　Sentry 首页

用户注册成功后可以进入 Sentry 的控制台，并在此控制台中创建一个与当前环境匹配的响应来实现异常信息的采集。例如，本次的应用是基于 Spring Cloud 实现的，所以可以在创建时选择“Spring Boot”项目类型，如图 10-19 所示。

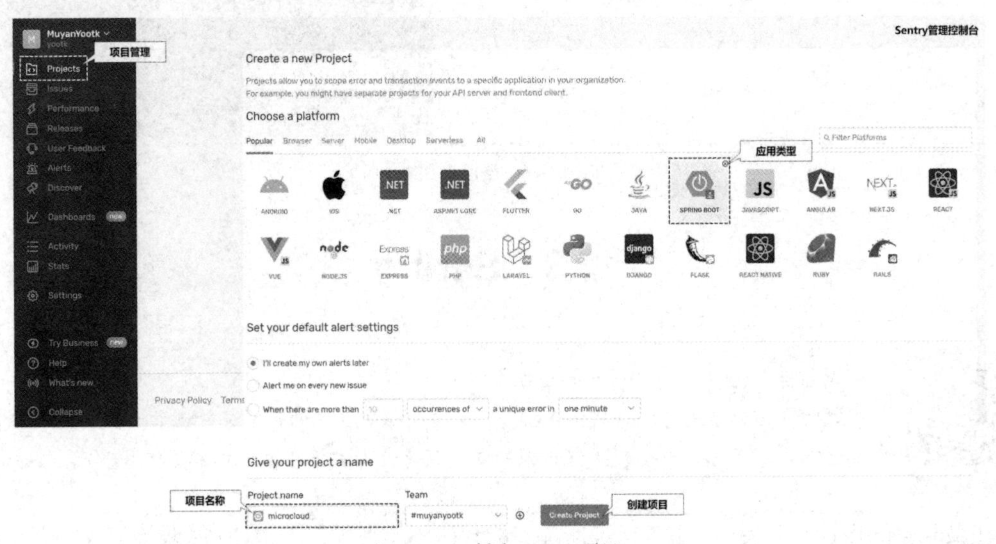

图 10-19　创建 Sentry 项目

10.3.1　Sentry 服务接入

Sentry 服务接入

视频名称　1009_【了解】Sentry 服务接入

视频简介　Sentry 的数据主要来自项目应用。本视频介绍对前面的 Spring Cloud 项目应用进行改造，使其可以将异常信息发送到 Sentry 应用之中。

创建完成的 Sentry 项目都会提供给用户一个 DSN 的访问地址，开发者可以根据此地址实现微服务应用的接入配置。如果用户在使用时未及时记录此地址，也可以通过项目的设置查找，如图 10-20 所示。

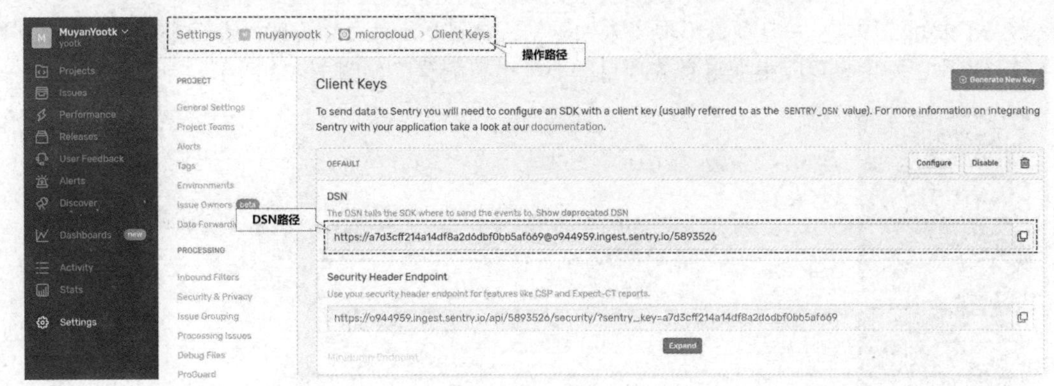

图 10-20　客户端接入地址

Sentry 采用无侵入的方式实现了客户端的接入处理，又提供了更加方便的“sentry-spring-boot-starter”依赖库，这样开发者通过 application.yml 的配置项定义即可实现服务的对接。下面通过具体的步骤介绍如何实现。

（1）【microcloud】修改 build.gradle 配置文件，添加 Sentry 依赖库。

```
implementation('io.sentry:sentry-spring-boot-starter:5.0.1')
```

（2）【微服务】修改需要接入 Sentry 的微服务配置文件 application.yml。

```
sentry:
  dsn: https://a7d3cff214a14df8a2d6dbf0bb5af669@o944959.ingest.sentry.io/5893526
```

（3）【Sentry 控制台】微服务出现异常之后，这些异常信息会自动发送到 Sentry 中进行记录，如图 10-21 所示。

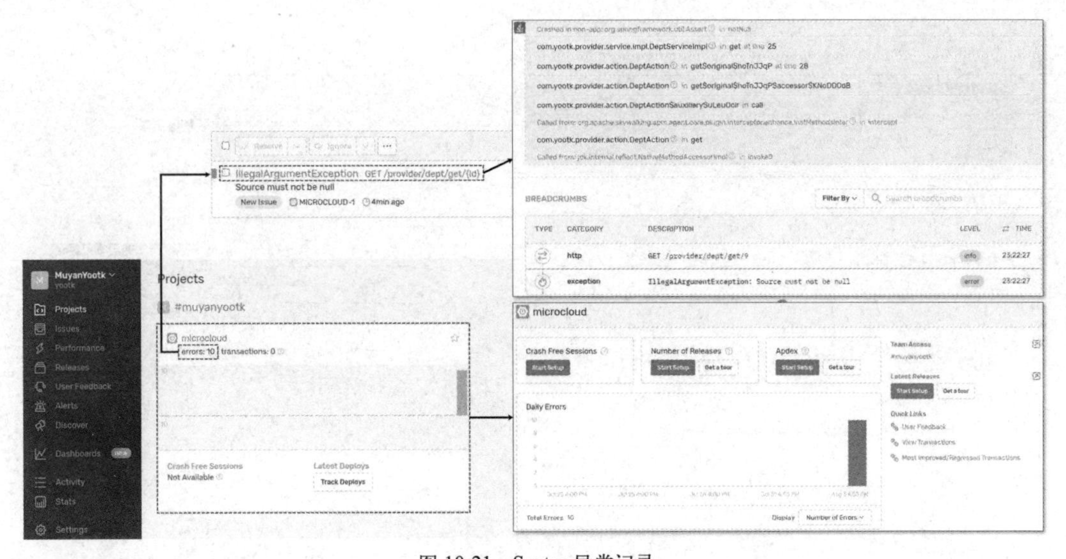

图 10-21　Sentry 异常记录

10.3.2　Sentry 异常警报

Sentry 异常警报

视频名称　1010_【了解】Sentry 异常警报

视频简介　为了可以及时地进行异常处理，Sentry 还提供了预警机制的配置。本视频为读者演示 Sentry 预警机制的配置与触发操作。

项目中的异常一般会有两类，一类是已规划异常（会出错，但是不会影响到程序执行结果），另一类属于未知的异常。一旦项目出现了未知异常，就需要及时通知管理员进行排查。Sentry 支持异常预警操作，开发者可以根据自身的项目需要来进行配置，如图 10-22 所示。

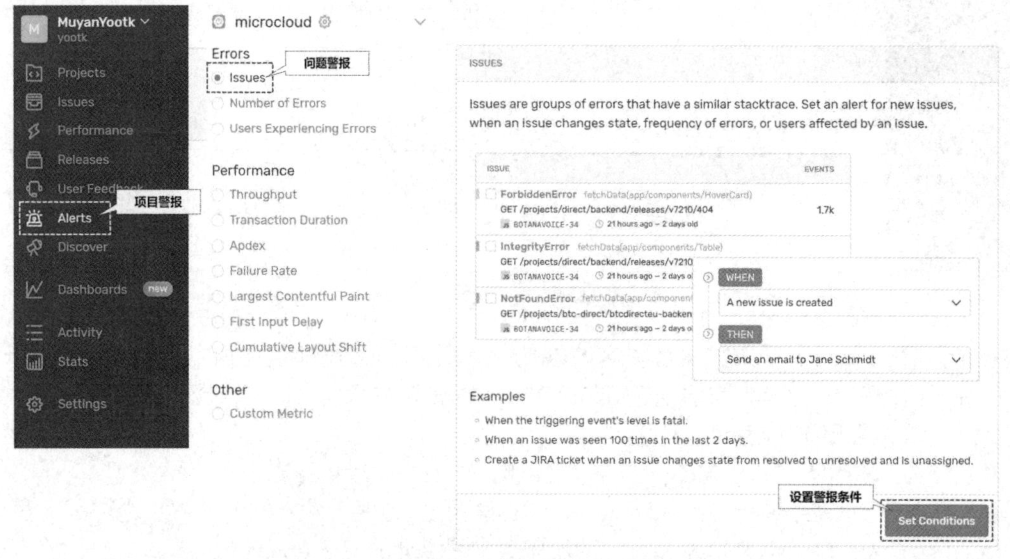

图 10-22　项目警报设置

在进行警报设置时需要设置警报的名称、警报的触发条件以及警报间隔等信息。本次将以新异常出现为警报触发的条件（可以先删除先前测试时产生的异常项），具体设置如图 10-23 所示。

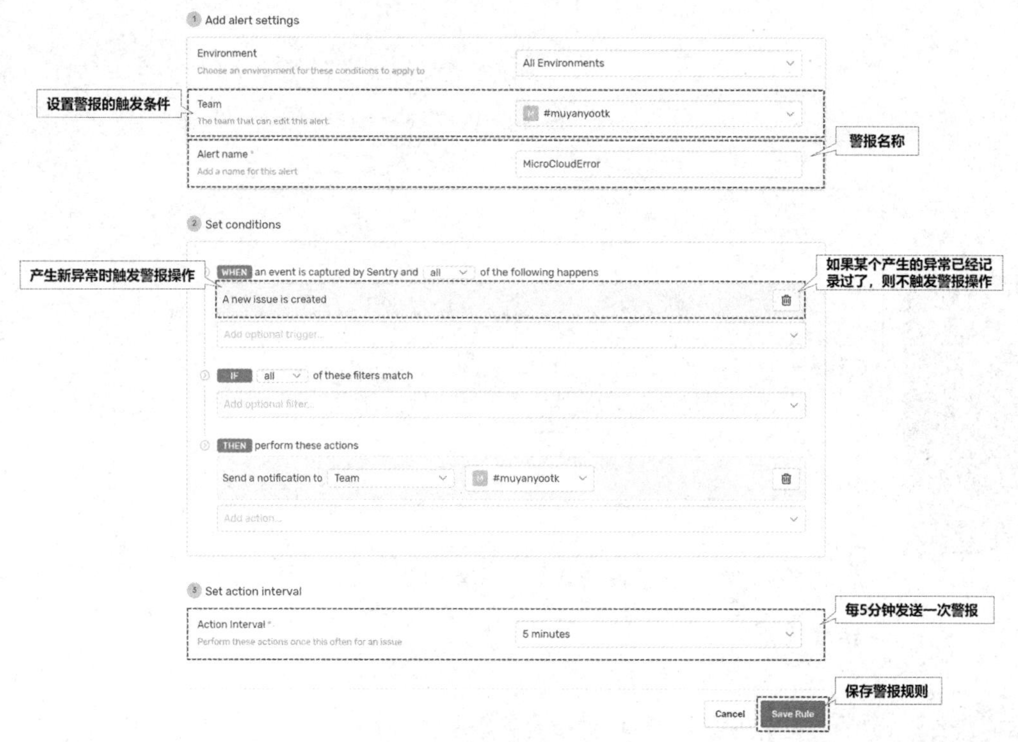

图 10-23　设置警报触发条件

警报环境设置完成后，一旦产生了新的异常，Sentry 则会自动向用户注册邮箱发送警报信息，如图 10-24 所示，这样管理员就可以及时收到异常信息，而后进行排查。

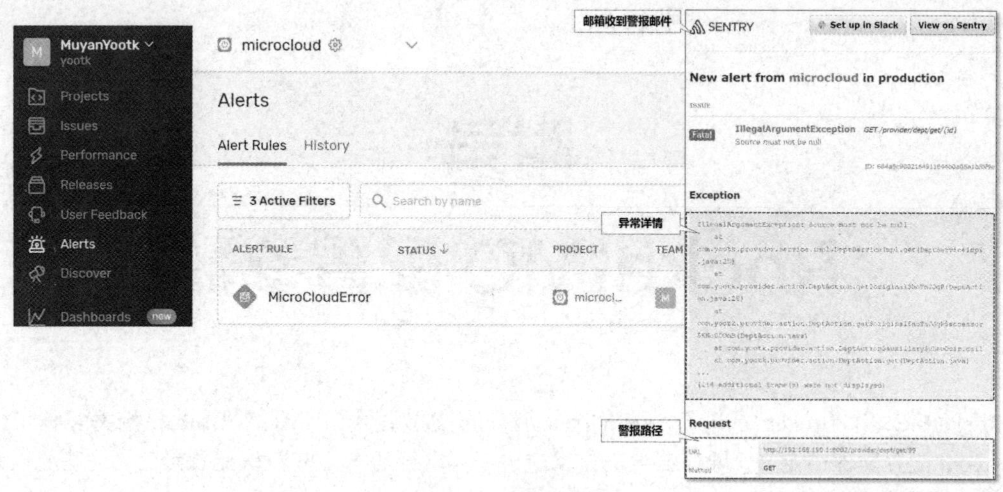

图 10-24 收到警报邮件

10.4 本章概览

1．Spring Cloud Sleuth 是 Spring Cloud 内置的全链路监控跟踪工具，结合 Zipkin 可以实现可视化的链路监控。

2．SkyWalking 是一款全链路的跟踪工具，其功能比 Zipkin 的更加强大，代码也可以采用无侵入的方式进行接入。

3．Sentry 是一款异常收集与分析工具，便于开发人员进行异常的调试以及异常报警操作。

附录

RocketMQ 配置参数

RocketMQ 中 Broker 是进行消息生产、消费的核心组件，所有的 Broker 相关配置可以在 broker.conf 配置文件中定义。除了必要的定义之外，其内部还有一些默认配置参数，这些配置参数的相关信息如附录-1 所示。

附录-1 Broker 配置参数

序号	参数名	类型	描述	默认值
001	aclEnable	boolean	是否开启 ACL 安全机制	false
002	accessMessageInMemoryMaxRatio	int	访问消息在内存中的比率	40
003	adminBrokerThreadPoolNums	int	处理控制台管理命令线程池数量	16
004	autoCreateSubscriptionGroup	boolean	是否自动创建消费组	true
005	autoCreateTopicEnable	boolean	是否自动创建主题	true
006	bitMapLengthConsumeQueueExt	int	ConsumeQueue 扩展过滤 bitmap 大小	112
007	brokerClusterName	String	Broker 集群名称	Default Cluster
008	brokerFastFailureEnable	boolean	是否支持 Broker 快速失败。设置为 true 会立即清除发送消息线程池、消息拉取线程池中的等待任务，直接返回系统错误	true
009	brokerId	int	Broker 服务 ID 标记，0 表示主节点，大于 0 表示从节点	0
010	brokerIP1	String	Broker 服务地址	
011	brokerIP2	String	BrokerHAIP 地址，供 Slave 同步消息的地址	
012	brokerName	String	Broker 服务器名称	broker-a
013	brokerPermission	int	Broker 权限，6 表示可读可写	6
014	brokerRole	enum	Broker 角色，包括 SLAVE、ASYNC_MASTER、SYNC_MASTER	ASYNC_MASTER
015	brokerTopicEnable	boolean	Broker 名称是否可以用作主题	true
016	channelNotActiveInterval	long	通道未使用间隔	60000
017	checkCRCOnRecover	boolean	文件恢复时是否校验 CRC	true
018	cleanFileForciblyEnable	boolean	是否支持强行删除过期文件	true
019	cleanResourceInterval	int	清除过期文件线程调度频率	10000
020	clientAsyncSemaphoreValue	int	客户端同步信号量	65535
021	clientCallbackExecutorThreads	int	客户端回调线程池大小	8
022	clientChannelMaxIdleTimeSeconds	int	客户端通道最大空闲时间	120
023	clientCloseSocketIfTimeout	boolean	超时后是否关闭连接	false
024	clientManagerThreadPoolQueueCapacity	int	客户端管理线程池任务队列初始大小	1000000
025	clientManageThreadPoolNums	int	服务端处理客户端线程池容量配置	32

续表

序号	参数名	类型	描述	默认值
026	ClientOnewaySemaphoreValue	int	客户端单向信号量	65535
027	clientPooledByteBufAllocatorEnable	boolean	客户端缓冲分配	false
028	clientSocketRcvBufSize	long	客户端 Socket 接收缓冲区大小	131072
029	clientSocketSndBufSize	long	客户端 Socket 发送缓冲区大小	131072
030	clientWorkerThreads	int	客户端工作线程数量	4
031	clusterTopicEnable	boolean	集群名称是否可在主题使用	true
032	commercialBaseCount	int	单次消息传输量基数	1
033	commercialBigCount	int	单次消息传输最大量	1
034	commercialEnable	boolean	单次传输启用	true
035	commercialTimerCount	int	单次传输时间量	1
036	commercialTransCount	int	单次传输计数	1
037	commitCommitLogLeastPages	int	一次提交至少需要的脏页数量，默认 4 页，针对 Commitlog 文件	4
038	commitCommitLogThoroughInterval	int	Commitlog 两次提交的最大间隔，如果超过该间隔，将忽略 commitCommitLogLeast Pages 直接提交	200
039	commitIntervalCommitLog	int	commitlog 提交频率	200
040	compressedRegister	boolean	存储压缩配置	false
041	connectTimeoutMillis	long	连接超时时间	3000
042	consumerFallbehindThreshold	long	消息消费堆积阈值，在 disableConsumeif ConsumeIfConsumerReadSlowly 为 true 时生效	17179869184
043	consumerManagerThreadPoolQueue Capacity	int	消费端管理线程池任务队列大小	1000000
044	consumerManageThreadPoolNums	int	消费端管理线程池大小	32
045	debugLockEnable	boolean	是否支持 PutMessage 锁输出信息	false
046	defaultQueryMaxNum	int	查询消息默认返回条数	32
047	defaultTopicQueueNums	int	主题在 Broker 上创建队列数量	8
048	deleteCommitLogFilesInterval	int	删除 commitlog 文件的时间间隔	100（单位：ms）
049	deleteConsumeQueueFilesInterval	int	删除消费端队列文件的时间间隔	100（单位：ms）
050	deleteWhen	String	每天执行删除过期文件的时间	04
051	destroyMapedFileIntervalForcibly	int	销毁 MappedFile 被拒绝的最大存活时间	120000
052	disableConsumeIfConsumerReadSlowly	boolean	当消费组消息消费堆积时是否禁用该消费组继续消费消息	false
053	diskFallRecorded	boolean	是否统计磁盘的使用情况	true
054	diskMaxUsedSpaceRatio	int	commitlog 目录所在分区的最大使用比率，如果 commitlog 目录所在的分区使用比率大于该值，则触发过期文件删除	75
055	duplicationEnable	boolean	是否允许重复复制	false
056	enableCalcFilterBitMap	boolean	是否开启位映射	false
057	enableConsumeQueueExt	boolean	是否启用 ConsumeQueue 扩展属性	false
058	enablePropertyFilter	boolean	是否支持根据属性过滤，true 表示使用基于标准的 SQL92 模式过滤消息	false
059	endTransactionPoolQueueCapacity	int	处理提交和回滚消息线程池线程队列大小	100000
060	endTransactionThreadPoolNums	int	处理提交和回滚消息线程池	24
061	expectConsumerNumUseFilter	boolean	布隆过滤器参数	32

续表

序号	参数名	类型	描述	默认值
062	fastFailIfNoBufferInStorePool	boolean	从 transientStorePool 中获取 ByteBuffer 是否支持快速失败	false
063	fetchNamesrvAddrByAddressServer	boolean	是否支持从服务器抓取 NameServer	false
064	fileReservedTime	String	文件保留时间	120（单位：ms）
065	filterDataCleanTimeSpan	long	清除过滤数据的时间间隔	86400000（单位：ms）
066	filterServerNums	int	Broker 过滤服务器数量	0
067	filterSupportRetry	boolean	消息过滤是否支持重试	false
068	flushCommitLogLeastPages	int	一次刷盘至少需要的脏页数量	4
069	flushCommitLogThoroughInterval	int	两次刷盘的最大间隔，如果超过该间隔，则执行刷盘操作	10000（单位：ms）
070	flushCommitLogTimed	boolean	使用 await()方法等待脏页刷新完成，如果为 true 表示使用 Thread.sleep()方法等待	false
071	flushConsumeQueueLeastPages	int	一次刷盘队列至少需要的脏页数量	2
072	flushConsumeQueueThoroughInterval	int	消费端两次刷盘的最大间隔，如果超过该间隔，则忽略	60000
073	flushConsumerOffsetHistoryInterval	int	刷盘时偏移量历史间隔	60000
074	flushConsumerOffsetInterval	int	刷盘时偏移量间隔	5000
075	flushDelayOffsetInterval	long	延迟队列拉取刷盘间隔	10000
076	flushDiskType	enum	刷盘方式，默认为 ASYNC_FLUSH（异步刷盘），可选择 SYNC_FLUSH（同步刷盘）	ASYNC_FLUSH
077	flushIntervalCommitLog	int	commitlog 刷盘频率	500
078	flushIntervalConsumeQueue	int	consumuQueue 文件刷盘频率	1000
079	flushLeastPagesWhenWarmMapedFile	int	用字节 0 填充整个文件的，达到配置页数则使用异步模式刷盘一次	4096
080	forceRegister	boolean	是否强制注册	true
081	haHousekeepingInterval	int	Master 与 Slave 长连接空闲时间，超过该时间将关闭连接	20000（单位：ms）
082	haListenPort	int	Master 监听端口	10912
083	haMasterAddress	String	Master 服务器 IP 地址与端口号	
084	haSendHeartbeatInterval	int	Master 与 Slave 心跳包发送间隔	5000
085	haSlaveFallbehindMax	int	允许从服务器落后的最大偏移字节数，超过该值则表示该 Slave 不可用	268435456
086	haTransferBatchSize	int	一次主从同步传输的最大字节长度	32768
087	heartbeatThreadPoolNums	int	心跳线程池线程数	8
088	heartbeatThreadPoolQueueCapacity	int	心跳线程队列数量	50000
089	highSpeedMode	boolean	高速模式	false
090	listenPort	int	服务端监听端口	10911
091	longPollingEnable	boolean	是否开启长轮询	true
092	mapedFileSizeCommitLog	int	单个 conmmitlog 文件	1073741824
093	mapedFileSizeConsumeQueue	int	单个 consumequeue 文件大小，默认 30 万×20，表示单个 Consumequeue 文件中存储 30 万个 ConsumeQueue 条目	6000000
094	mappedFileSizeConsumeQueueExt	int	ConsumeQueue 扩展文件大小	50331648（单位：Byte）
095	maxDelayTime	int	最大延迟时间	40
096	maxErrorRateOfBloomFilter	int	布隆过滤器参数	20

续表

序号	参数名	类型	描述	默认值
097	MaxHashSlotNum	int	单个索引文件哈希槽的个数	5000000
098	maxIndexNum	int	单个索引文件索引条目的个数	20000000
099	maxMessageSize	int	默认允许的最大消息体	4194304
100	maxMsgsNumBatch	int	一次查询消息最大返回消息个数	64
101	maxTransferBytesOnMessageInDisk	int	一次服务端消息拉取，消息在磁盘中传输允许的最大字节数	65536
102	maxTransferBytesOnMessageInMemory	int	一次服务端消息拉取，消息在内存中传输允许的最大传输字节数	262144
103	maxTransferCountOnMessageInDisk	int	一次服务端消息拉取，消息在磁盘中传输允许的最大条数	8
104	maxTransferCountOnMessageInMemory	int	一次服务端消息拉取，消息在内存中传输允许的最大条数	32
105	messageDelayLevel	String	延迟队列等级（s=秒，m=分，h=小时）	1s 5s 10s 30s 1m 2m 3m 4m 5m 6m 7m 8m 9m 10m 20m 30m 1h 2h
106	messageIndexEnable	boolean	是否支持消息索引文件	true
107	messageIndexSafe	boolean	消息索引是否安全，默认为 false，文件恢复时选择文件检测点最小值与文件最后更新对比。如果为 true，文件恢复时选择文件检测点保存的索引更新时间来对比	false
108	messageStorePlugIn	String	消息存储插件地址默认为空字符串	
109	namesrvAddr	String	NameServer 地址	
110	notifyConsumerIdsChangedEnable	boolean	消费者数量变化后是否立即通知 RebalenceService 线程，以便马上进行负载均衡配置	true
111	offsetCheckInSlave	boolean	Slave 端是否使用 Offset 检测	false
112	osPageCacheBusyTimeOutMills	long	putMessage 锁占用超过该时间，表示 PageCache 忙	1000
113	pullMessageThreadPoolNums	int	服务端处理消息拉取线程池线程数量，默认为 16 加上当前操作系统 CPU 核数的两倍	16 + Core × 2
114	pullThreadPoolQueueCapacity	int	消息拉取线程池任务队列初始大小	100000
115	putMsgIndexHightWater	int	设置消息索引存放的最大值	600000
116	queryMessageThreadPoolNums	int	服务端处理查询消息线程池数量，默认为 8 加上当前操作系统 CPU 核数的两倍	8 + Core × 2
117	queryThreadPoolQueueCapacity	int	查询消息线程池任务队列初始大小	20000
118	redeleteHangedFileInterval	int	重试删除文件时间间隔	120000（单位：ms）
119	regionId	String	消息区域	Default Region
120	registerBrokerTimeoutMills	int	注册 Broker 超时时间	6000（单位：ms）
121	registerNameServerPeriod	int	Broker 注册频率	30000
122	rejectTransactionMessage	boolean	是否拒绝事务消息	false
123	rocketmqHome	String	RocketMQ 主目录	
124	sendMessageThreadPoolNums	int	服务端处理消息发送线程池数量	1
125	sendThreadPoolQueueCapacity	int	消息发送线程池任务队列初始大小	10000
126	serverAsyncSemaphoreValue	int	异步消息发送最大并发度	64
127	serverCallbackExecutorThreads	int	Netty 业务线程池个数	0
128	serverChannelMaxIdleTimeSeconds	int	网络连接最大空闲时间。如果连接空闲时间超过此参数设置的值，连接将被关闭	120

续表

序号	参数名	类型	描述	默认值
129	serverOnewaySemaphoreValue	int	发送单向消息请求并发量	256
130	serverPooledByteBufAllocatorEnable	boolean	ByteBuffer 是否开启缓存	true
131	serverSelectorThreads	int	I/O 线程池线程个数	3
132	serverSocketRcvBufSize	int	Netty 网络 Socket 接收缓存区大小	131072
133	serverSocketSndBufSize	int	Netty 网络 Socket 发送缓存区大小	131072
134	serverWorkerThreads	int	Netty 业务线程池个数	8
135	shortPollingTimeMills	long	短轮询等待时间	1000（单位：ms）
136	slaveReadEnable	boolean	从节点是否可读	false
137	startAcceptSendRequestTimeStamp	int	开始发送请求的时间戳	0
138	storePathCommitLog	String	CommitLog 存储目录	用户目录
139	storePathRootDir	String	Broker 存储目录	用户目录
140	storePathConsumeQueue	String	ConsumerQueue 存储目录	
141	storePathIndex	String	Index 存储目录	
142	storeCheckpoint	String	Checkpoint 存储目录	
143	abortFile	String	Abort 存储目录	
144	syncFlushTimeout	long	同步刷盘超时时间	5000（单位：ms）
145	traceOn	boolean	是否开启跟踪	true
146	transactionCheckInterval	long	事务检查周期	60000（单位：ms）
147	transactionCheckMax	int	事务检查次数	15
148	transactionTimeOut	long	事务检查超时时间	6000（单位：ms）
149	transferMsgByHeap	boolean	消息传输是否使用堆内存	true
150	transientStorePoolEnable	boolean	Commitlog 是否开启事务存储池（transientStorePool）机制	false
151	transientStorePoolSize	int	缓存的 ByteBuffer 个数	5
152	useEpollNativeSelector	boolean	是否启用 EPoll I/O 模型，在 Linux 环境建议开启	false
153	useReentrantLockWhenPutMessage	boolean	消息存储到 commitlog 文件时获取锁类型。如果为 true，使用 ReentrantLock，默认使用自旋锁	false
154	useTLS	boolean	是否使用安全传输层协议	false
155	waitTimeMillsInHeartbeatQueue	long	清理 Broker 心跳线程等待时间	31000（单位：ms）
156	waitTimeMillsInPullQueue	long	清除消息拉取线程池任务队列等待时间	5000
157	waitTimeMillsInSendQueue	long	清除消息发送线程池任务队列等待时间	200（单位：ms）
158	waitTimeMillsInTransactionQueue	long	清理提交和回滚消息线程队列等待时间	3000（单位：ms）
159	warmMapedFileEnable	boolean	是否温和地使用 MappedFile。如果为 true，将不强制将内存映射文件锁定在内存中	false
160	connectWhichBroker	String	FilterServer 连接的 Broker 地址	
161	filterServerIP	String	FilterServer IP 地址	主机 IP 地址
162	compressMsgBodyOverHowmuch	int	如果消息 Body 超过该值则启用压缩	4096
163	zipCompresslevel	int	Zip 压缩方式	5
164	clientUploadFilterClassEnable	boolean	是否支持客户端上传代码	true
165	filterClassRepertoryUrl	String	FilterClass 服务地址	
166	fsServerAsyncSemaphorevalue	int	FilterServer 异步请求并发量	2048
167	fsServerCallbackExecutorThreads	int	处理回调任务的线程池数量	64
168	fsServerWorkerThreads	int	远程服务调用线程池数量	64